THE COSMIC VOYAGE

THE
COSMIC
VOYAGE
Through Time
and Space

1992 Edition

WILLIAM K. HARTMANN

Wadsworth Publishing Company
Belmont, California
A Division of Wadsworth, Inc.

Astronomy Editor: Anne Scanlan-Rohrer
Development Editor: Mary Arbogast
Editorial Assistant: Leslie With
Production Editors: Deborah Cogan, Gary Mcdonald
Designer: MaryEllen Podgorski
Managing Designer: Cynthia Schultz
Print Buyer: Barbara Britton
Art Editor: Nancy Spellman
Copy Editors: Charles Hibbard, Jennifer Gordon
Technical Illustrators: Catherine Brandel, Victor Royer,
Darwen Hennings, Vally Hennings, Lisa Sliter,
Salinda Tyson
Cover Designer: Cynthia Schultz
Compositor: Graphic Typesetting Service, Los Angeles
Color Separator: Rainbow Graphic Arts Co., Ltd.,
Hong Kong
Printing: Arcata Graphics/Hawkins

Frontispiece: The inner core of the Orion Nebula, photographed in wavelengths of infrared light (1.2 to 2.2 μm wavelength). As explained in Chapter 5, this is called a false color image because the colors are not those that the eye can see. However, the image gives a realistic picture of the nebula as it might be seen by someone sensitive to infrared light: The bluer stars and clouds are the warmer ones, and the redder stars and clouds are cooler. The image reveals some 500 stars in an area about 0.6×0.6 parsecs, in the center of the nebula. For more information on the Orion Nebula, see Chapter 20. (Photo from NASA Infrared Telescope, Hawaii; courtesy Mark McCaughrean, NASA Goddard Space Flight Center.)

Cover: Total eclipse of the Sun, as photographed by the author from the island of Hawaii, July 11, 1991. The bright ring of light is the corona, which is the glowing outer atmosphere of the Sun, visible to the eye only during a total solar eclipse. Its leftward portion is dimmed by thin clouds, which blocked the eclipse entirely for many viewers.

Back Cover: A large crater formed by an asteroid impact on the surface of the planet Venus, imaged by the U.S. Magellan spacecraft using radar equipment that "sees" through Venus's global cloud layer. The crater is 50 km (30 miles) across, and is tentatively named Barton, after Clara Barton, founder of the Red Cross. The beautiful double ring structure is typical of impact craters in certain size ranges, due to mechanical processes during the impact. (NASA, Magellan Project, Jet Propulsion Laboratory.)

2 3 4 5 6 7 8 9 10——96 95 94 93 92

**Library of Congress
Cataloging-in-Publication Data**

Hartmann, William K.
    The cosmic voyage: through time and space /
    William K. Hartmann.—1992 ed.
       p.   cm.
    Includes bibliographical references and index.
    ISBN 0-534-17376-4
    1. Astronomy.   I. Title.
QB43.2.H36 1992
520—dc20   91-29140

For Teachers and Students

With human footprints on the Moon, radio telescopes listening for messages from alien creatures (who may or may not exist), technicians looking for celestial and planetary sources of energy to support our civilization, orbiting telescopes' data hinting at planetary systems around other stars, and political groups trying to figure out how to save humanity from nuclear warfare that would damage life and climate on a planet-wide scale, an astronomy book published today enters a world different from the one that greeted books a generation ago. Astronomy has broadened to involve our basic circumstances and our enigmatic future in the universe. With eclipses and space missions broadcast live, with the USSR occupying a permanent space station, and with American and Soviet leaders discussing cooperative missions to Mars, astronomy offers adventure for all people—an outward exploratory thrust that may one day be seen as an alternative to mindless consumerism, ideological bickering, and wars to control dwindling resources on a closed, finite Earth.

Today's astronomy students not only seek an up-to-date summary of astronomical facts; they ask, as people have asked for ages, about our basic relationships to the rest of the universe. They may study astronomy partly to seek points of contact between science and other human endeavors: philosophy, history, politics, environmental action, even the arts and religion.

Science fiction writers and the special effects artists on recent films help today's students realize that the unseen worlds of space are real places—not abstract concepts. Today's students are citizens of a more real, more vast cosmos than conceptualized by students of a decade ago.

In designing this book, the Wadsworth editors and I have tried to respond to these developments. Rather than jumping at the start into the murky waters of cosmology, I have begun with the viewpoint of ancient people on Earth and worked outward across the universe. This method of organization automatically (if loosely) reflects the order of humanity's discoveries about astronomy and provides a unifying theme of increasing distance and scale.

This arrangement aims to give an unfolding, ever-expanding panorama of our cosmic environment. We hope it unfolds like a story in which each chapter provides not only a new facet but also a growing understanding of the relationships among the elements of the whole.

The subtitle refers to three separate cosmic voyages that we undertake simultaneously. First, we travel through historical time, where we see how humans slowly and sometimes painfully evolved our present picture of the universe. Second, we journey through space, where we see how our expanding frontiers have revealed the geography of the universe. Beginning with an Earth-centered view, we study the Earth–Moon system, the surrounding system of planets, the more distant surrounding stars, our own vast galaxy, and the encompassing universe of other galaxies. Finally, we travel back through cosmic time. Familiar features of the Earth are typically only a few hundred million years old. The solar system is about 4.6 billion years old. Our galaxy is roughly 13 or 14 billion years old. The universe itself began (or began to reach its present form) an estimated 14 billion years ago.

Because astronomy touches many areas of life and philosophy, I have allowed the text to encompass a wide range of relevant topics, including space exploration, financing of science, cosmic sources of energy, the checkout-counter's barrage of astrology and other pseudoscience, and the possibility of life on other worlds, as well as the conventional "hard science" of astronomy. This variety of topics shows how basic scientific research touches all areas of life—I hope in a way that lets readers ponder the relation between science and priorities in our society.

Using This Book

The arrangement of text material into nine parts and twenty-eight chapters, plus a section of optional enrichment essays, should give instructors some flexibility in tailoring a course according to their interest. For example, those who are not much interested in historical development could use Part A only as assigned outside reading.

Each part gives some historical background, describes recent discoveries and theories, and then discusses advances that might occur if society continues to support research. This more or less chronological approach has several purposes. Since there is often a certain logic to the order in which discoveries were made, historical emphasis may help readers remember the facts. Second, historical discussion allows us to introduce basic concepts in a more interesting way than by reciting definitions. It makes life richer to realize that some seemingly modern concepts descend from knowledge of ancient millennia and have thousand-year-old names. Third, there is a widespread fallacy that the only progress worth mentioning is that of the last few decades. Astronomy, of all subjects, shows clearly that, to paraphrase Newton, we see as far as we do because we stand on the shoulders of past generations. Exploration of the universe is a continuing human enterprise. As we try to maintain and improve our civilization, that is an important lesson for a science course to teach.

Another principle we have followed is to treat astronomical objects in an *evolutionary* way, to show the sequence of development of matter in the universe. Stars, pulsars, black holes, and other celestial bodies are linked in evolutionary discussion, rather than listed as different types of objects detected by different observational techniques. I have also not hesitated to mention nonscientific approaches to cosmology and evolution, such as "creationist" concepts recently encouraged in two states' school systems by their state legislatures but later thrown out after courtroom battles.

In this new 1992 edition, various updates and improvements have been added, including news about Venus from the ongoing Magellan mission, and some cosmological updates on galaxy distribution and evolution.

Illustrations and Text

Because color imagery plays an increasingly important role in astronomy, we use color figures throughout the book. Emphasis is on true-color imagery to give students the best conception of astronomical objects' appearance, though false-color image enhancement techniques are also described.

More specifically, teaching aids incorporated in the book include the following:

1. In addition to the classic large telescope photos and recent NASA photos, I have included three other categories of illustrations:

a. Photos from recently published research papers, kindly provided by various authors and institutions.

b. Photos by amateurs with small and intermediate instruments, often used to show sky locations of well-known objects in the large-telescope photos. These can help readers to visualize and locate these objects in the sky, a difficult task if based on classic large-telescope photos alone. Photographic data provided with many of these pictures may be used in setting up student projects in sky photography.

c. Scientifically realistic paintings show how various objects might look firsthand to observers in space. Discussion of features shown in the paintings illustrates a synthesis of scientific data from various sources.

2. Key concepts are shown in **boldface** type. These are repeated in *Concepts* lists at the ends of chapters and defined in a *Glossary* at the end of the book.

3. Limited numbers of references to technical and nontechnical sources appear in the text. They are there partly to help students and teachers find more material for projects and partly to help instructors emphasize that statements should be verifiable. These sources are included in the *References* section.

End-of-Chapter Materials

1. *Chapter Summaries* review basic ideas of the chapter and sometimes synthesize material from several preceding chapters.

2. *Concepts* lists include the important concepts appearing in **boldface** in the text. Reviewing the *Concepts* is a good way for the student to review the content of each chapter.

3. *Problems* are aimed at students with nonmathematical backgrounds.

4. *Projects* are intended for class use where modest observatory or planetarium facilities are available. The intent is to get students to do astronomical observing or experimenting.

Supplementary Material

1. *Enrichment Essays* can be used or not as instructors wish. These include essays on *Pseudoscience and Nonscience* and *Astronomical Coordinates and Timekeeping Systems.*

2. *Appendixes* are included on *Powers of 10, Units of Measurement,* and *Supplemental Aids in Studying Astronomy.*

3. The *Glossary* defines all terms included in the *Concepts* lists, as well as other key terms.

4. The *References* section includes all sources mentioned in the text. Nontechnical references useful for student papers are starred (*); widely available journals and magazines are emphasized in this group, including most astronomy articles appearing in *Scientific American* in recent years.

5. The *Index* includes names and terms.

6. *Star Maps* for the seasons are found after the index. Since more detailed, larger maps are usually available in classrooms or laboratories, these star maps have been simplified, emphasizing the plane of the solar system and the plane of the galaxy and indicating major constellations mentioned or illustrated in the text.

Ancillaries

1. An *Instructor's Manual* contains course outlines, suggestions for planetarium programs, main concepts, outlines, and test questions for each chapter, and sources of supplies, films, books, and charts.

2. *Testing software* for both the IBM-PC and Macintosh environments is available.

3. *Wadsworth Astronomy Transparencies* consist of one-, two-, and four-color acetates showing illustrations from Wadsworth astronomy texts.

4. *Voyages through Space and Time* by Jon Wooley, Eastern Michigan University, contains projects for use with VOYAGER, the Interactive Desktop Planetarium, and is for sale to students.

5. *Introductory Astronomy Exercises* by Dale Ferguson, Baldwin-Wallace College, contains a wide variety of laboratory exercises and is for sale to students.

6. *Color slides* are available in two different sets: "Portrait of the Solar System" and "A Color Tour of the Milky Way."

In addition, the following are available through an adoption incentive program:

1. *Software* for classroom demonstrations or laboratory instruction is available in two formats: PC Planetarium, for IBM-PCs and compatibles, and VOYAGER, the Interactive Desktop Planetarium for Macintosh.

2. The *videotape series* "The Astronomers" is available as individual tapes or as part of a series.

For more information about any of the above, contact your Wadsworth/Brooks-Cole sales representative.

Acknowledgments

My thanks go to many people who helped produce this book. I have tried to incorporate as many of their suggested corrections and improvements as possible; final responsibility for weaknesses remains mine.

The Cosmic Voyage is a short version of *Astronomy: The Cosmic Journey,* 4th Edition. The names of those reviewers who criticized parts of various editions for research comprehensiveness and teaching potential are listed on the following page. I also thank a wide circle of colleagues, especially in Tucson and at the University of Hawaii, for helpful discussions.

In locating and providing photographs, Dale Cruikshank, Walter Feibelman, Steven Larson, Alfred McEwen, Alan Stockton, Don Strittmatter, Alar Toomre, and Dave Webb were helpful. Kelly Aggen and Jurrie Van der Woude (Jet Propulsion Laboratory), David Moore (National Optical Astronomy Observatory), Margaret Weems (National Radio Astronomy Observatory), and David Malin (Anglo-Australian Observatory) gave kind assistance with their institutional files. I especially thank Chesley Bonestell, Dennis Davidson, Don Dixon, David A. Hardy, James Hervat, Paul Hudson, Pamela Lee, Ron Miller, Jim Nichols, Mark Paternostro, Kim Poor, Adolf Schaller, Andrei Sokolov, Richard Sternbach, and Brian Sullivan for stimulating discussion about paintings of astronomical subjects. Floyd Herbert and Mike Morrow helped me in making several wide-angle sky photos with clock-driven mounts.

I thank Elaine Owens and Catherine Date for assistance in coordinating this and other research projects during preparation of the manuscript.

The staff at Wadsworth Publishing Company has worked hard to create a useful and beautiful book, and made the job a pleasure at the same time. I thank Anne Scanlan-Rohrer, Mary Arbogast, Gary Mcdonald, Deborah Cogan, MaryEllen Podgorski, Cindy Schultz, and Nancy Spellman for their friendship and professional help in shepherding *The Cosmic Voyage* through various editorial stages. Thanks to all of them as well for exceptional work on production and design, cheerful dispositions, hospitality, and care for the subject, not to mention the harried author.

William K. Hartmann
Tucson

Reviewers

I appreciate the help and advice of the following reviewers and advisors who have made comments or suggestions on various portions of this or previous editions.

Helmut A. Abt

Bill T. Adams, Jr.

Richard A. Bartels

Henry E. Bass

Lee Bonneau

David R. Brown

Warren Campbell

Robert J. Chambers

Kwan-Yu Chen

Clark G. Christensen

Leo Connoly

Dale P. Cruikshank

Donald R. Davis

Robert J. Doyle

Robert J. Dukes, Jr.

John Fix

Benjamin C. Friedrich

F. Trevor Gamble

Owen Gingerich

Edward Ginsberg

Jeff Goldstein

Richard Greenberg

Hubert E. Harber

Alan Harris

Thomas G. Harrison

Gayle Hartmann

Floyd Herbert

George H. Herbig

Ulrich O. Hermann

Lon Clay Hill, Jr.

J. R. Houck

Jay S. Huebner

William W. Hunt

Stuart J. Inglis

Donald B. Krall

Nathan Krumm

Francis Lestingi

Leonard Muldawer

Gerald H. Newsom

Charles J. Peterson

Terry P. Roark

Randy Russell

John L. Safko

John M. Samaras

John Schopp

Richard Schwartz

Robert F. Sears, Jr.

Myron A. Smith

Richard C. Smith

Ronald Smith

Michael L. Stewart

Kjersti Swanson

Donald J. Taylor

James E. Thomas

Scott D. Tremaine

Peter D. Usher

Stuart J. Weidenschilling

Walter G. Wesley

Ray J. Weymann

Richard M. Williamon

R. E. Williams

Warren Young

Contents
in Brief

Detailed Contents

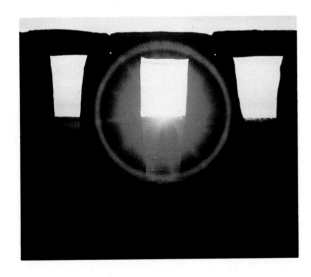

P A R T   B

**TWO METHODS FOR EXPLORING SPACE:
UNDERSTANDING GRAVITY AND
UNDERSTANDING LIGHT   51**

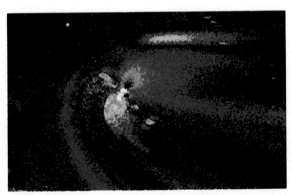

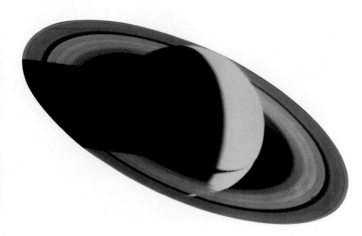

PART E

STARS AND THEIR EVOLUTION 225

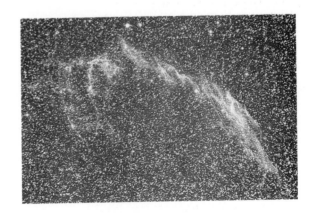

THE COSMIC VOYAGE

Invitation to the Cosmic Voyage

If you awoke one day to discover that you had been put on a strange island, your first project after getting food and water would be to try to find out where you were. This is just what has happened to all of us. We have all been born on an island in space: Earth. We are all passengers on a cosmic voyage. Astronomy is the process of finding out where we are—our place in space and our position in the unfolding history of the universe.

This definition of astronomy would not always have satisfied people. At one time we thought we knew where we were: at the center of the universe, with the Sun, stars, planets, and satellites all revolving around us. Modern astronomy contradicts that old theory. The definition is not fully satisfactory now, either, because it doesn't really explain what astronomers do.

OUR DEFINITION OF ASTRONOMY

Oddly enough, as science progresses, astronomy becomes harder to define. When astronomy was restricted to observations by earthbound viewers, it was easily defined as "the science of objects in the sky." But now that we have actually begun to explore space, other disciplines have become involved. Is the astronaut who chips a rock sample of the Moon practicing astronomy? What about the researcher who studies the sample once it reaches Earth? What of the nuclear physicist who measures properties of nuclear reactions going on at the center of stars? What about the biologist interested in life on Mars? Their work should be included in our definition of astronomy.

There is a popular misconception that as various scientists pursue new research, knowledge proliferates into an endless complex of specialized disciplines. Indeed, students are usually taught this way, with advanced courses probing into narrower and narrower specialties.

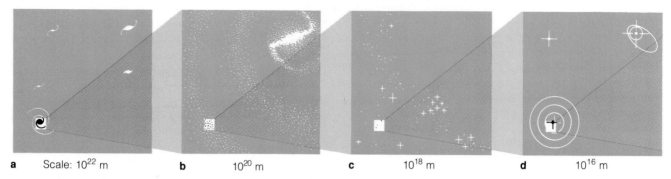

a   Scale: 10^{22} m   b 10^{20} m   c 10^{18} m   d 10^{16} m

Figure I-1 A journey through the universe from large-scale structure of galaxies to the human scale. The change in scale from one picture to another ranges from a factor of 100 to a factor of 100,000. In cartoon form, **a** shows galaxies; **b** and **c,** parts of our Galaxy; **d,** our solar system and the nearest star; **e,** part of our solar system; **f,** the Earth–Moon system; and **g,** human observers. Scale in meters is expressed as powers of 10.

According to this idea, research is like climbing a tree, starting near the trunk and proceeding to ever more specialized branches of knowledge.

But science can also be compared to working our way *down* the tree of knowledge. At first we see a bewildering variety of seemingly unrelated phenomena—twigs of knowledge. After analyzing these twigs, we see that they meet in a branch. As workers map details on the branch, someone with vision may discover that it is joined to another branch. As we work our way toward the trunk, more and more branches join. By grasping the relations between the major branches close to the trunk, we can better understand the higher branches and twigs. This is why different specialties are connecting with astronomical research; astronomy is becoming more general.

A historical example illustrates this process. Around 1600, Galileo and others conducted many experiments to understand seemingly unrelated types of motion, such as the motion of falling objects and the motion of the Moon. The wide variety of motions and accelerations seemed impossible to explain in a simple way. However, by about 1680, Isaac Newton realized that they were connected by simple relationships among mass, acceleration, and gravitational attraction. As a result of Newton's insight, today's college students can learn to understand motions better than the Galileos of the past!

In the same way, students can grasp the relationship between Earth and the cosmos better than the explorers who mapped astronomical frontiers in the past. Better than any generation in human history, we can begin to sense where we are, at least in relation to other physical objects in the universe.

This book treats astronomy not as a narrow set of academic observational results or abstruse physical laws, but as a voyage of exploration with practical effects on humanity. A broad view of the major discoveries of the last years, the last decades, and even the last centuries is at least as important as the latest factual detail, because the broad view helps us understand how scientific research will continue to affect us and perhaps offer us new options for living. In an era of energy crises, food shortages, and environmental threats, when badly applied technology has created such horrors as biological weapons and hydrogen bombs, educated people are called on to make judgments about their own actions—the kinds of jobs they have, the materials they consume, and the actions of organizations they work for. In short, we are called upon to judge what kind of civilization should continue. Thus we all need to know how our world is affected by its cosmic surroundings and to understand the scientific and social procedures for obtaining and extending that knowledge.

Therefore, this book is organized according to the following definition of **astronomy:**

Astronomy is the study of all matter and energy in the universe, emphasizing the concentration of this matter and energy in evolving bodies such as planets, stars, and galaxies, and fully recognizing that we observers— humanity—are part of the universe, and that our home, the Earth, is only one of the many places in the universe, but also the special point from which our voyage of exploration has started.

The reason for emphasizing humanity is that astronomy influences how we think about ourselves and our

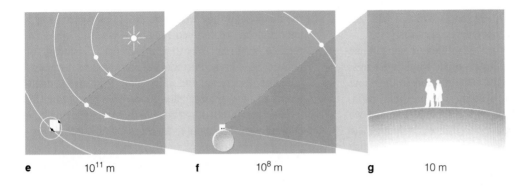

e 10^{11} m   f 10^8 m   g   10 m

role in the universe. The growing perception of environmental problems in the 1970s and 1980s has shown that we *must* look at Earth as a single astronomical body in order to survive. As the poet Archibald MacLeish said after one of the Apollo flights, we have now seen the Earth as "small and blue and beautiful" and ourselves as "riders on the Earth together." Thus this book regards astronomy as including space flights and studies of rocks from our neighbor planets, just as much as telescopic observations of remote galaxies.

A SURVEY OF THE UNIVERSE

The ensuing chapters start with subjects on Earth and then move outward through space. Before beginning, however, we give a brief preview of the universe, starting at the largest scale.

The **universe** is everything that exists. To the best of our knowledge, the universe consists of untold thousands of clusters of galaxies (Figure I-1). **Galaxies** are swarms of billions of stars. Galaxies differ in form. Some are football shaped, some irregular. Our galaxy, like many others, is a disk with about a hundred billion stars arrayed in curving spiral arms. Most galaxies, including ours, are surrounded by a halo of **globular star clusters,** each a spheroidal mass of hundreds of thousands of stars. If we made a model of our galaxy big enough to cover North America, Earth would not be as big as a basketball, or even a BB. It would be about the size of a large molecule.

Scattered inside our galaxy, mostly in the spiral arms, are groupings of stars called **open star clusters** and

clouds of dust and gas called **nebulae.** Individual stars are scattered at random. Each **star** is an enormous ball of gas, mostly hydrogen. In the center of a typical star, atoms are jammed together so closely that their nuclei interact to create the heat and light of the star.

Three hundred million billion kilometers (200,000, 000,000,000,000 mi) from the center of our galaxy and 40 thousand billion kilometers from the nearest star is the **Sun.** A hydrogen sphere more than a million kilometers in diameter, the Sun is the center of the **solar system**—the small system made up of the Sun and its family of orbiting rocky and icy **planets.** Circling around most planets are smaller bodies called **satellites.** Between planets are various other small bodies called **asteroids** and **comets.**

About 150 million kilometers from the Sun is the **Earth.** The **Moon,** Earth's satellite, is only 384,000 km away. Earth itself is nearly 13,000 km across, a tiny speck among myriads of larger and smaller specks in the universe.

A WORD ABOUT MATHEMATICS

Our preceding survey of the universe shows how hard it is to express astronomical quantities using ordinary units of measurement. For this reason the system of expressing numbers as **powers of 10** is useful. For example, instead of being written as the unwieldy number 300,000,000,000,000,000 km, the distance from the Sun to the galactic center becomes simply 3×10^{17} km; the distance from Earth to the Sun is 1.5×10^8 km. A reader uncertain about this system should consult

Appendix 1 at the end of the book. We will use this system occasionally, as convenience dictates.[1]

When discussing physical quantities, it helps to use a consistent set of units. For example, the English system of inches, ounces, etc., must not mix in the same equation with the metric system. The metric system, long used by scientists, is becoming universal and we will use it here, often giving English equivalents. The metric system is *much* easier to use than the English system because its units come in multiples of 10. One centimeter = 10 millimeters, 1 meter = 100 centimeters, and so on. Gone are complicated conversions like 1 mile = 5280 feet. Specifically, in this book we will give preference to the **SI metric system**—the units adopted in the Système International (International System), such as meters, kilograms, and so on.[2] In some cases, astronomical units such as parsecs will be explained and adopted. These units of measurement are summarized in Appendix 2.

SCIENCE, PSEUDOSCIENCE, AND NAMES OF PEOPLE

An important characteristic of the scientific method is that if a scientist makes a claim, it must be verifiable by other people. If other scientists can't observe the same thing, using telescopes, atom smashers, or whatever equipment applies to their field, the claim is rejected. In today's society, we are bombarded by pseudoscientific claims, from tabloid stories to come-ons of TV hucksters. How are we to distinguish good science from pseudoscience? One way is that scientific assertions are backed by evidence that can be checked. When the claim rests on earlier data, this is done by listing the other publications that can be checked for verification. Usually the name of the author and the date of the publication are listed, for example, "(Smith, 1989)," and then the publication is listed at the end of the book or article so that an interested reader can find it and read more about the subject. This is the style used in this book.

With so many distorted claims being publicized by the media as if they were scientific, this referencing is important in helping you recognize whether you have reliable material. Whether in science, business, or politics, readers and listeners should demand documentation of assertions, and, if in doubt, look up the documents to judge their quality. Aside from promoting this intellectual tradition, we believe that the extensive bibliography will be useful to teachers and to students preparing term papers.

Students are sometimes distressed to find these names, thinking they are supposed to learn names and dates. The names are only to document assertions, not to be memorized. The emphasis in this book is on the exciting nature of the universe that astronomy is revealing, not on the astronomers themselves. Many of the Greek and Renaissance astronomers mentioned in Chapters 2 and 3 are major figures in our Western intellectual heritage, however, and students should learn about them.

A HINT ON USING THIS BOOK

When reading a text, many students underline everything that seems important. My publishers and I have taken care of part of that job by putting terms that represent important concepts in boldface and then listing them in order at the end of the chapter. A useful study procedure is to read a chapter through and then go over the concept list. Try to define each term and use it in a sentence. If you have trouble with a term, go back and check it in the chapter or look it up in the glossary. If you are comfortable with all the terms in each concept list, you will have gone a long way toward mastering the material and will be able to enjoy many articles about astronomy in newspapers and magazines.

FACE TO FACE WITH THE UNIVERSE

The universe is immense; we are perhaps audacious even to attempt studying it. Yet something noble in the human spirit urges us to seek knowledge of the universe and our place in it.

Throughout history, scientific discoveries in the heavens have sometimes illuminated and sometimes conflicted with religious, metaphysical, and philosophical conceptions. Persecution, legal action, even wars

---

[1] We also use the term *billion* here to mean 1,000,000,000, or 10^9, which is the usual American meaning. The British use *billion* to mean 1,000,000,000,000, or 10^{12}.

[2] Besides the familiar abbreviations like *km, m, cm, mi, ft, in., min,* and *mo,* we also use *y* for year, *h* for hour, and *s* for second.

have resulted from the challenge to old beliefs by new knowledge. In this book we present scientific theories and the *evidence* for them with the understanding that our knowledge is limited and slowly improving. New evidence is what usually distinguishes new knowledge from old superstition.

Astronomy gives us a new cosmic perspective from which many conflicts of the past now seem inconsequential. For example, we have given up the medieval arguments over how many angels can dance on the head of a pin. Astronomy unifies humanity by showing us that we are all in this together, facing the unknown—a word that aptly describes much of the universe, its past, and its future.

The universe is no abstract concept; it consists of real, physical places. This book offers descriptions of those places, often with photos or scientifically realistic paintings to show how we think they look.

Suppose that we were able to explore these places and that in all these vast reaches, among all the stars and galaxies, we found no one else—no living creatures anywhere except ourselves. Or suppose that we discovered nonhuman intelligent life—alien civilizations, perhaps advanced and incomprehensible or perhaps very simple. Our contacts with them could alter our history far more profoundly than the first contacts between Native Americans and Europeans in 1492.

These are two wildly different possibilities. One is likely to be true. Either boggles the mind. We are just barely entering the century in which we may be able, if research continues, to decide which scenario is more realistic.

CONCEPTS

astronomy	planet
universe	satellite
galaxy	asteroid
globular star cluster	comet
open star cluster	Earth
nebula	Moon
star	powers of 10
Sun	SI metric system
solar system	

PROBLEM

1. Scientific discoveries are sometimes described as conflicting with religious or philosophical beliefs. Before continuing this book, examine your present beliefs or expectations and comment on which of the following thoughts best match your own as you begin this course:

a. There are concepts in astronomy that conflict with my own religious or philosophical views.

b. Astronomical concepts are consistent with my religious or philosophical views; there is no problem.

c. If there is a conflict, scientific hypotheses will evolve and eventually become consistent with my religious or philosophical views.

d. Science measures the reality of the physical world only; there is no need for this to be consistent with an inner psychological or spiritual reality.

e. Science tries to reflect how all nature behaves; thus science and religious philosophy of daily life should be consistent with each other if they are to contribute to a well-integrated personality.

Prehistoric "standing stones" erected between 2800 and 2200 B.C. at Stonehenge on the plain of Salisbury in southern England are aligned on the Sun's position at sunrise on the date of the summer solstice and symbolize early humanity's attempts to interact with the cosmos. (Photo by author.)

CHAPTER 1

Prehistoric Astronomy: Origins of Science and Superstition

In 1906 the American astronomer Percival Lowell wrote:

Smoke from multiplying factories . . . has joined with electric lighting to help put out the stars. . . .
today few city-bred children have any conception of the glories of the heavens which made of the Chaldean shepherds astronomers in spite of themselves.

This observation is even more apt today. As a result, it is hard for us to realize how important the sky once was in daily life.

We have several reasons to begin our cosmic voyage with the most ancient discoveries. First, they are among the most basic discoveries about Earth's relation to the universe. Second, there is a certain historical logic. For instance, early recognition that Earth moves around the Sun had to come long before our modern space journeys and discoveries from space telescopes. Third, certain conceptions and misconceptions of everyday life, such as the 360° circle or the superstitions about astrological signs, have evolved from ancient traditions.

THE EARLIEST ASTRONOMY: MOTIVES AND ARTIFACTS (c. 30,000 B.C.)

Archaeology and anthropological studies of the few remaining primitive tribes have shed light on the earliest astronomical practices. Imagine yourself an intelligent hunter of 30,000 y (years) ago. Agriculture is not yet practiced. You have to depend on hunting and gathering for food. Celestial phenomena are not your most immediate concern, but you know that certain celestial cycles are important. You want to know when berries on the mountain will ripen and when migratory birds will arrive at a nearby lake. Such knowledge requires cumulative day counts and knowledge of the seasons, best derived from sky phenomena. When you travel far, you want to be able to find your way home. Celestial objects are your only reliable guides in unfamiliar landscapes.

If you are a woman, you also want to know in what season the child you may carry will be born and when your next menstrual period will occur. Living mostly outdoors, ancient women undoubtedly used the $29\frac{1}{2}$-d (days) cycle of the Moon's phases (for example, from one full moon to the next) to keep track of their 27- to 30-d menstrual cycles. Some experiments indicate that menstrual cycles of women sleeping in the light of an artificial moon become synchronized to its phases, and some scientists have even speculated that the human menstrual cycle evolved to match the period of the Moon's phases (Luce, 1975).

For such reasons, early people began to keep records of events in the sky (Figure 1-1). This is supported by modern aborigines' calendar sticks that record dates, or counts of days, by carved notches. For example, a lunar calendar stick carved by Indian Ocean islanders shows a series of grooves almost identical to grooves carved on prehistoric artifacts of 30,000 y ago. And North American Indian calendar sticks similarly record

Figure 1-2 Portion of a calendar stick carved by Tohono O'odham Indians of Arizona. The stick, begun in 1841 and passed on to the carver's son, carries a record of social and natural events, including an earthquake, for the years 1841 to 1939. (Calendar stick from collection of Arizona State Museum, University of Arizona. Photo by author.)

historical events and other natural events by sequences of carvings. The calendar stick in Figure 1-2, for example, records an earthquake and other events. Because the Indians who made it had no written language, it can be "read" only by the carver. This calendar stick is dramatic evidence of how records of natural events may have first accumulated, thus allowing the discovery of astronomical cycles.

CALENDAR REFINEMENTS (10,000–3000 B.C.)

Around 10,000 B.C., an agricultural revolution brought two new incentives for understanding the sky. First, to know when to plant, people had to be able to predict seasonal changes days in advance. Second, in the stable villages made possible by agriculture, people kept records; the mere existence of records gave thinkers an incentive and opportunity to discover seasonal and annual cycles of celestial events.

People began to get a better sense of the yearly cycle by counting days or Moon cycles (from one full moon to another) between certain dramatic annual events. In Egypt, for instance, the Nile flooded at a certain time each year. At high northern latitudes people looked eagerly for the Sun to return to the northern sky after the cold, short days of winter. In some early societies, day counts of this cycle gave estimates of about 360 d, later refined to 365 and then $365\frac{1}{4}$ d.

Figure 1-1 Paleolithic evidence suggests that ancient people tabulated the phases of the Moon. The configuration shown here, called "the old moon in the new moon's arms," is seen in the evening sky a few days after a new moon, when the portion not lit by the Sun is dimly lit by light reflected off the Earth. (Photo by author.)

Accumulated observational experience revealed the correlation between this annual cycle of seasons and the apparent movements of the Sun. The Sun does not follow the same path across the sky every day. In the summer at northern latitudes (for example, Europe and the United States), the Sun is high at midday; in the winter, it is low in the southern sky at midday. Thus careful observations of the Sun's motion relative to other features of the sky would have helped determine the exact number of days in the year and the exact dates of such observations.

Organized observations of this kind were made throughout the world thousands of years ago. Mesopotamian calendars were being designed by 4000 to 3000 B.C. In recent decades, anthropologists have found modern pretechnological people using similar means to create sophisticated systems of time measurement.

OTHER EARLY DISCOVERIES

Before describing ways in which these principles were utilized and extended, we pause to describe other aspects of the sky discovered or defined by early people. These aspects are important aids in describing many astronomical phenomena.

North Celestial Pole   Camp out for a night and you will discover that the stars slowly wheel around a single point in the sky. In the Northern Hemisphere that single point is called the north celestial pole; in the Southern Hemisphere, the south celestial pole. The wheeling of the stars is due to Earth's rotation on its axis, and therefore one complete circuit takes about 24 h (hours). The two **celestial poles** can be thought of as the projections of the Earth's axis onto the sky, or the spots targeted by vertical searchlights at the North and South Poles. You can visualize the sky as a giant dome overhead with the north celestial pole always in the northern direction, as seen in Figure 1-3. Recognizing the celestial pole helped in navigation, because it indicates the north or south direction and can be used to measure an observer's latitude (a point to which we will return shortly).

North Star   A bright star, **Polaris** (shown in Figure 1-4) happens to lie near the north celestial pole. Thus it stands almost still all night above the northern horizon as other stars wheel around it. Its position makes it an excellent reference beacon.[1] Because of a dynamical

---

[1]Note that a magnetic compass points only approximately north, because Earth's magnetic field is not aligned with true north. Because the field varies, the compass needle also varies in direction by small amounts from year to year. In any case, the compass was not available to the ancients. It was apparently first invented by the Olmecs in Mexico around 1400–1000 B.C. Compasses were also invented in China by 300 B.C. to A.D. 100 (Carlson, 1975).

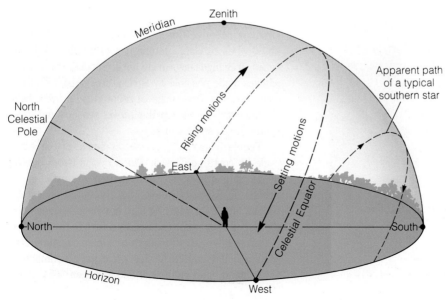

Figure 1-3 General properties of the sky as defined by the Earth's daily rotation. The sky is shown as seen by observers at midnorthern latitudes, such as the United States.

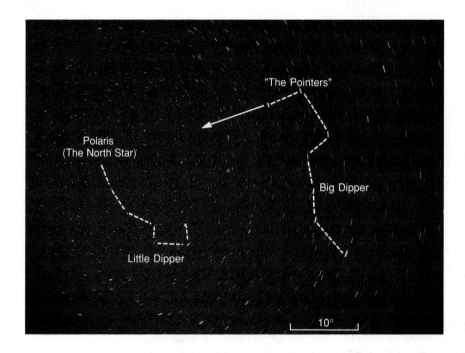

"The Pointers"

Polaris
(The North Star)

Big Dipper

Little Dipper

10°

Figure 1-4 Two notable constellations of the northern sky, once labeled the greater and lesser bears, are more commonly known today as the Big Dipper and the Little Dipper. They are important to outdoor people because the star at the end of the Little Dipper's handle is Polaris, the North Star. It always hangs directly above the north point on the horizon, as seen from the Northern Hemisphere. It can be found by using "the pointers"—the two end stars in the Big Dipper's bowl. Note subtle color differences among stars. These are due to stars' temperature differences and will be discussed in Chapter 16. (Photo by author, 35-mm camera with wide-angle 24-mm lens, f2.8, on commercially available 3M ASA 1000 film.)

phenomenon called **precession**, Earth's polar axis points in somewhat different directions in different historical eras. For nearly 1000 y, Polaris has been located within a few degrees of the north celestial pole, but in other times other stars have been nearer the pole. If a bright star is located near the north celestial pole, it is called the **North Star.**

Zenith and Meridian Wherever an observer stands, the point directly overhead is called the **zenith.** The **meridian** is an imaginary north–south arc from horizon to horizon through the celestial pole and the zenith. The zenith and meridian are shown in Figure 1-3. The meridian marks the highest point that each star reaches above the horizon on any night; it is the dividing line between the rising stars and setting stars. For this reason it is especially useful in timekeeping. In the daytime, the moment when the Sun crosses the meridian is called noon; in very early times, therefore, people used this astronomical concept to divide the day.

Celestial Equator As shown in Figure 1-3, the imaginary circle lying 90° from both the north and the south celestial poles is the **celestial equator.** It is the projection of the Earth's equator onto the sky and is parallel to the daily east-to-west paths of the stars as they wheel around the celestial poles. The celestial equator divides the sky into two equal halves.

The 360° Circle Our division of the circle into 360 units, or degrees, follows a Babylonian practice of about 3000 y ago. It probably comes from the ancient estimate of 360 d for one year, and from the fact that 360 is easily divisible into 2, 3, 4, 5, or 6 equal parts. The decimal (10-based) system of counting and writing numbers later replaced the sexagesimal (60-based) system in most uses *except* the expression of angles and time.

The Ecliptic By observing the relative positions of Sun and stars at dawn or dusk, one can establish that the Sun appears to shift nearly a degree to the east each day relative to the stars. (This can be confirmed by noting that the Sun "moves" 360° in 365 d.) The cause of this shift is Earth's motion around the Sun. Ancient observers found that the Sun traces the *same path* through the stars year after year. This yearly path of the Sun among the stars is a circle extending all the way around the sky and is called the **ecliptic.** We now know that the ecliptic is the same as the plane of Earth's orbit around the Sun.

The Planets Ancient observers—priest-astronomers and wakeful shepherds alike—found that five of the brighter starlike objects were not fixed like the rest. They came to be known in the Western world as **planets,** from the Greek word for "wanderers." Not until after the telescope was invented, in about 1610, was there any proof that these planets were globes like our Earth or that three faint ones had gone undiscovered. For this reason the ancients spoke of five planets, whereas we speak of nine (counting our own).

The Zodiac Ancient observers tracking the paths of the planets from night to night among the stars discovered that the planets never stray out of a narrow band, about 18° wide, that goes all the way around the sky and follows along the ecliptic. Many star patterns, or constellations, along this zone were said to resemble animals, so the zone came to be called the **zodiac** (after Greek for "animals"—the same root as *zoo*). The constellations within the 18° zone came to be known as the signs of the zodiac. They include such familiar examples as Scorpius (the scorpion) and Leo (the lion)—see the star maps on the last pages of the book.

ORIGIN OF THE CONSTELLATIONS

As memory aids for learning and locating stars, early observers named groups of stars for their resemblance to familiar animals, objects, and mythical characters— such as Scorpius (which really resembles a scorpion),

Cygnus the swan (see Figure 1-5) or Orion (the hunter), and many others. These groups are **constellations.** People in different cultures saw different patterns, usually derived from their local mythologies. An interesting exception is the star group called Ursa Major, which was seen as a bear in Europe, Asia, North America, and even ancient Egypt, where there are no bears. For this reason, historian Owen Gingerich (1984) suggests the bear identification may have originated in ice-age Europe or Asia and spread from there.

Many of our familiar constellations are described in historic records from as early as 420 B.C., but the conceptualizations of these constellation patterns must be much older (Gingerich, 1984). According to one astronomical scholar and sleuth (Ovenden, 1966), the constellations' characteristics give clues to their origin. For instance, they have rough symmetry around the ancient north star of 2600 B.C., and they do not extend far into the Southern Hemisphere. The constellations' designers must have been Northern Hemisphere observers because, as shown in Figure 1-6, parts of the far southern sky cannot be seen from the Northern Hemisphere. Such clues suggest that most of our familiar constellations were designed around 2600 ± 800 B.C. by seafaring navigators who lived at around latitude $36\frac{1}{2}°$. They

Figure 1-5 Color view of the summer and autumn constellation of Cygnus, the swan, popularly known as the Northern Cross. In pre-Christian times, the constellation was viewed as a swan flying downward in this view, with the crossbar forming the wings and a long neck extending toward the head, which coincides with the foot of the cross. The photo was made with an ordinary 35-mm camera fixed on a tripod with a 1-min exposure; rotation of the Earth during 1 min causes the stars to leave trails $\frac{1}{4}°$ long. Notice that the stars have different colors, a fact that will be explained in Chapter 16. (Photo by author on commercially available 3M 1000 film, 55-mm lens, f2.8.)

Figure 1-6 Due to Earth's rotation, a stationary camera making a time exposure at night of the northern or southern sky will reveal the circular tracks of the stars across the sky. **a** Northern sky. Polaris is the bright star near the center of the circular star traces. Because this photo is made at latitude 20°N, stars within 20° of the north celestial pole never set. **b** Southern sky. In this photo from the same location, stars within 20° of the south celestial pole never rise. The stars in the south make rainbowlike arcs across the sky as the Earth turns during the night. (Both photos by author from Mauna Kea Observatory, illuminated by moonlight. Stationary 35-mm camera with 15-mm wide-angle fisheye lens; approx. 20-min exposure on commercially available 3M ASA 1000 film.)

a

b

may have been the Minoan sailors from the island of Crete. These ancient constellations were probably then handed down to the Greeks, and then to us. We should not assume that "it all started with the Greeks." We have received knowledge and traditions from even earlier societies!

As shown in the four star maps at the end of this book, different constellations are visible in the evening sky during different seasons. Today they no longer have mythical significance, but it is pleasant to greet them each year as old friends who signal the arrival of each new season.

THE SEASONS: SOLSTICES, EQUINOXES, AND THEIR APPLICATIONS

Most people think the Sun rises in the east and sets in the west. This is roughly true, but as the seasons progress, the Sun rises and sets at different points on the horizon. Consider an observer in the United States or elsewhere in the Northern Hemisphere. In summer, the Sun rises in the *north*east, passes high overhead at noon, and sets in the *north*west. In winter, the Sun rises in the *south*east, passes low in the southern sky at noon,

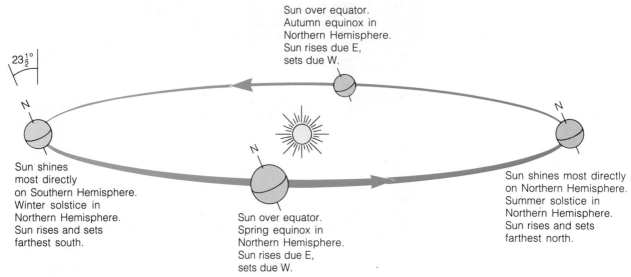

Sun over equator.
Autumn equinox in
Northern Hemisphere.
Sun rises due E,
sets due W.

Sun shines
most directly
on Southern Hemisphere.
Winter solstice in
Northern Hemisphere.
Sun rises and sets
farthest south.

Sun over equator.
Spring equinox in
Northern Hemisphere.
Sun rises due E,
sets due W.

Sun shines most directly
on Northern Hemisphere.
Summer solstice in
Northern Hemisphere.
Sun rises and sets
farthest north.

Figure 1-7 The cause of the seasons. Earth's axis is tipped at $23\frac{1}{2}°$ to the plane of its orbit in a constant direction as it moves around the Sun. Therefore the Sun shines more on the Northern Hemisphere during one season (northern summer) and more on the Southern Hemisphere 6 mo (months) later (northern winter). See text for further discussion.

and sets in the *south*west. On only two dates, the beginning of spring and the beginning of autumn, does the Sun rise due east and set due west (see Figure 1-3, where such a path is shown).

The reasons for this have to do with the **cause of the seasons:** Earth's equator is tipped by $23\frac{1}{2}°$ to the plane of the ecliptic, as seen in Figure 1-7. As Earth orbits the Sun during a given year, its axis always stays pointed in the same direction—toward the North Star (or, to be more precise, toward the north celestial pole). Thus, as seen in the figure, the Northern Hemisphere is tipped toward the Sun for part of the year and away from the Sun during the other part of the year. When the Northern Hemisphere is tipped toward the Sun, the Sun rises and sets further north along the horizon; the Sun also rises higher in the sky and shines more directly down on us; the weather grows warmer and we call the seasons spring and summer.

Thus there are four special days each year, called **equinoxes** and **solstices:**

Spring Equinox (First Day of Spring, About March 21) Sun rises due east, sets due west. An important day to ancients in the Northern Hemisphere because it marked the return of the Sun to the northern

half of the sky, bringing warmth to the Northern Hemisphere. Also called the vernal equinox.

Summer Solstice (First Day of Summer, About June 22) Sun rises and sets farthest north. Most hours of daylight in Northern Hemisphere, ensuring warm weather. Most important day of year in many ancient northern calendars.

Autumn Equinox (First Day of Autumn, About September 23) Sun again rises due east, sets due west. It is moving into the southern half of the sky. Cold weather on the way in the Northern Hemisphere.

Winter Solstice (First Day of Winter, About December 22) Sun rises and sets farthest south. Fewest hours of daylight in Northern Hemisphere.

Part of this sequence can be seen in Figure 1-8, a series of photos looking west at sunset and taken one month apart. In the first photos, around winter solstice, the Sun sets in the southwest (left), but as we move toward summer, the sunset position slides to the north (right). We will call this cyclical shift in the sunset and sunrise positions from summer to winter solstice the **solstice principle.**

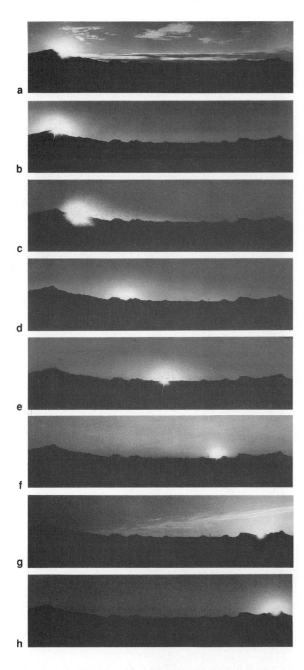

Figure 1-8 Movement of the sunset position along the horizon during an 8-mo period. Moving south in **a** the Sun reaches winter solstice in **b.** Here the setting position reverses direction and proceeds north again in **c**, reaching the vernal (spring) equinox in **e**. Summer solstice is reached in **h.** The author's former home, from which these pictures were made, happened to occupy the focus of a "natural Stonehenge," where the two solstice positions were marked by prominent mountain peaks. Such situations may have triggered the prehistoric idea of building solstice markers to reset calendars twice a year. (Photos by author.)

Ancient Applications of the Solstice Principle

Observations of solstices and equinoxes were very important to ancient civilizations as a way to keep an accurate calendar. With no dated newspaper arriving on the doorstep each day, how could early people know the date? Without a date, how would they know the best day for spring planting or important ceremonies? The answer is that in many societies, the year was considered to begin at one of the solstices. Priest-astronomers observed the day of solstice by observing the Sun's rising or setting position on the horizon, as in Figure 1-8. Then they counted the number of days from that date as the year progressed. For example, Spanish records show that the Peruvian Incas used this method and sent runners through the Inca empire to announce the beginning of the new year when the solstice was observed.

One of the earliest archaeological records of this practice comes from the famous English monument **Stonehenge.** Around 2500 B.C., prehistoric builders constructed a large circular embankment with a broad avenue leading outward for about $\frac{1}{2}$ km ($\frac{1}{4}$ mi) directly toward the horizon position of the summer solstice sunrise. Figure 1-9 shows a map of this early design. Subsequently, as late as 2100 to 1500 B.C., workers brought huge stones from as far as 380 km away (240 mi) and erected them to form the famous structure in the center of the ring (Figure 1-10). A large stone (called the "heel stone") was placed at a distance down the avenue, so that an observer at the center of Stonehenge could see the Sun rise over the distant stone *on the day of summer solstice,* as shown in Figure 1-11 (p. 118). Most archaeologists believe that Stonehenge was once used as a ceremonial center for calibrating the calendar each year. Additional astronomical functions, including lunar observations, have been suggested for Stonehenge but these are more controversial. There is some indication that Stonehenge's purpose gradually shifted from precise observation to ceremony; by the Christian era, its original function had been forgotten.

In other parts of the world, astronomically aligned temples and monuments from ancient times have been discovered that may also have been used to help calibrate the calendar. Examples include the Temple of Amon-Re in Egypt (c. 1400 B.C.), and temples in Cambodia, Bolivia, and Mexico.

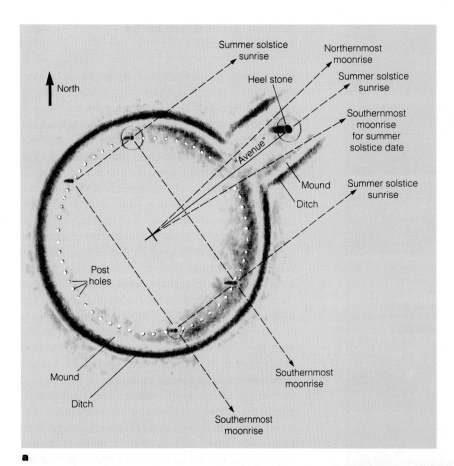

North

Summer solstice sunrise

Northernmost moonrise

Heel stone

Summer solstice sunrise

Southernmost moonrise for summer solstice date

"Avenue"

Mound

Ditch

Summer solstice sunrise

Post holes

Mound

Ditch

Southernmost moonrise

Southernmost moonrise

a

Figure 1-9 a Aerial plan of the original Stonehenge construction around 2500 B.C. An outer ditch, mound, and ring of posts or stones were cut by an avenue leading toward summer solstice sunrise. The outer ditch ring could barely contain a football field. As seen from the center, the summer solstice Sun rose over the heel stone, partway down the avenue. Selected moonrises could also be observed and timed. **b** Aerial photo shows modern Stonehenge. Central ring of giant stones was added around 2000 B.C. White path is a twentieth-century addition. The avenue leads out of the right of the picture, and the heel stone is near the road. (Georg Gerster.)

b

Figure 1-10 Stonehenge (rear) and its "heel stone" (foreground), which marks the position of sunrise on the date of summer solstice. (Photo by author.)

New Year's Day and the Solstice Principle in Ancient Europe

Why does our New Year start on January 1? Is there anything special about that date in terms of Earth's motion around the Sun? No. The date seems associated with ancient European celebrations of winter solstice, about December 22. Probably the early Christian church tried to co-opt these pagan celebrations with Christmas and New Year's.

Remnants of the older tradition persisted into historic times. In Europe, some older communities started the year on March 25 (nearly spring equinox). The early Christian church set a feast day on this date, commemorating the conception of Jesus—possibly another attempt to co-opt the pagan calendar. In *Tess of the D'Urbervilles,* novelist Thomas Hardy records that farm laborers' contracts in nineteenth-century England were still calculated from that date. Such a yearly calendar, starting with the spring equinox and celebrating winter and summer solstice, is probably a holdover from Europe's early prehistoric roots in the days of Stonehenge!

Even Christian cathedrals inherited some of the ancient solstice/equinox traditions. The old (330–1506) and new (1506–) cathedrals of St. Peter in Rome were reportedly oriented so accurately to the east that on the morning of the spring equinox "the great doors were thrown open and as the Sun rose, its rays passed through the outer door, then through the inner door, and pen-etrating straight through the nave, illuminated the High Altar" (Lockyer, 1894). Certain other Christian cathedrals and missions were oriented toward sunrise on the day of the saint to whom they were dedicated. This sounds suspiciously like temple orientation traditions in ancient Egypt and elsewhere. Probably these ancient traditions were absorbed by the early Christians from their cultures. Ghostly fingers of ancient practices reach into the present to touch us!

Modern Applications of the Solstice Principle

A totally different application of the solstice principle has appeared in modern times. As fossil energy supplies dwindle, we are more concerned about utilizing solar energy in our buildings. The cheapest way is by clever building orientation and roof overhang to let sunlight in during winter and keep it out during summer. In a sunny climate, most of the heat input through windows and walls comes in afternoon and morning when the Sun is low and strikes these surfaces full on. (Midday heat input through roofs must be minimized by thick insulation.) As shown in Figure 1-12, house design can utilize the solstice principle. Southeast- and southwest-facing windows, for example, can be unshaded to provide "free" warmth around winter solstice, and northwest-facing windows can be covered with external shields in summer to prevent late afternoon sunlight from pouring unwanted heat into the house.

ASTROLOGY: ANCIENT ORIGINS OF A SUPERSTITION

If you lived in a society where the calendar was determined by observations of the stars, where priest-astronomers counted days from the summer solstice in order to announce when to plant crops, and where the death of a king was remembered as happening in "the year when Mars passed through the constellation of Gemini," you could see how easy it might be to make a serious philosophical error. Instead of realizing that the stars offer a practical way to date and coordinate human affairs, you might come to believe that the stars *control* or at least influence human affairs. This superstition is called **astrology.** We call it a superstition for the same reason that we say it is superstitious to believe in invisible fairies in your garden: No experimental evidence

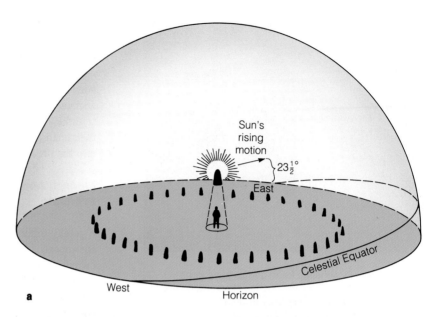

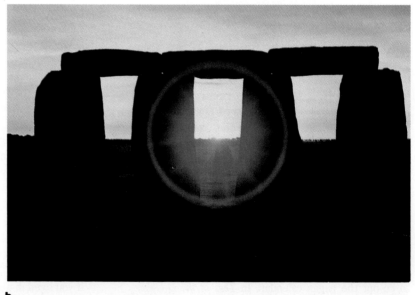

Figure 1-11 The Stonehenge principle. **a** The Sun's position at solstice is 23½° from the celestial equator. The rising (or setting) Sun on this date is thus as far as possible from due east or west, respectively. This position on the horizon is marked by a pillar or other prominent marker. Additional markers in a ring could allow observations of other selected risings or settings during the year. **b** Sunrise on the morning of summer solstice from the center of Stonehenge, showing the Sun's position over the top of the distant heel stone. (Georg Gerster.)

supports this belief, and predictions based on it have no better than random accuracy.

An occupational hazard of being an astronomer is to be introduced at parties as an astrologer. This mistake is annoying because astrology is not part of modern astronomy but a pseudoscience associated with astronomy as it was practiced 3000 y ago.

The pseudoscientific nature of astrology can be understood best by exploring astrology's roots. Astrol-

ogy can be traced back at least 3000 y, when it flourished with other ancient magical beliefs. A common form of ancient magic was to associate patterns in nature with patterns of human events. Most people today would scoff at having their futures read from flight patterns of migrating birds or designs in the bloody intestines of freshly sacrificed animals. Yet these were popular forms of divination in Babylon and Rome when astrology was flourishing, and they have the *same basic logic*. In astrol-

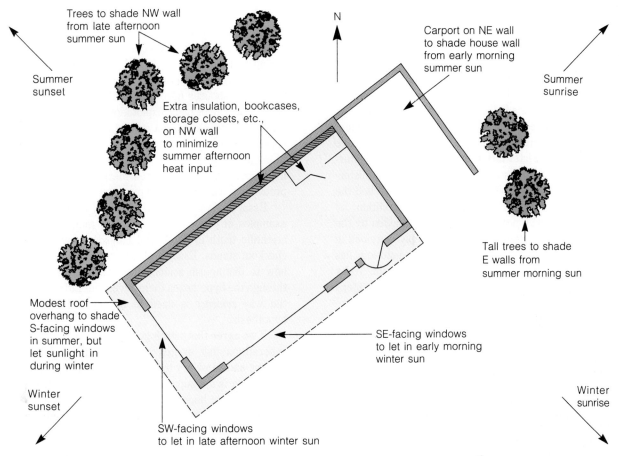

Figure 1-12 The solstice principle as applied to environmental architecture and landscaping. Northeast and northwest walls are shaded to reduce heat from the summer morning and afternoon sun. Southern windows allow sunshine to enter and heat the house in winter.

ogy, the pattern in nature is made by the planets as they move through the imaginary figures of the constellations.[2]

When priest-astronomers were observing signs in the sky to calibrate the calendar or determine when to plant crops, it must have been easy for the "man in the street" to conclude mistakenly that celestial signs forecast *human* events.

Human minds are quick to assume that if one event follows another, the second is caused by the first. Sometimes this is true and sometimes not. Centuries ago, logicians identified and named this error with the Latin phrase *"post hoc, ergo propter hoc"* ("following this, therefore *because* of this"). Yet this type of error is still common today. Historical evidence suggests that astrology grew out of this error. For example, if a king died a few days after an eclipse, some early observers reasoned that an eclipse causes or foretells the death of a king.

In some societies, early astronomical knowledge later degenerated into myth and pseudoscience. For exam-

---

[2]An interesting example of this all-too-easy mental connection—between changing patterns in the sky and mystic patterns of human relationships—was given by French diarist Anaïs Nin during a 1935 visit to a planetarium. A practicing psychoanalyst, she had been thinking of the "human tangles" of her patients. Her diary records: "The planetarium . . . restored to me a sense of space and I could detach myself from the haunting patients. . . . I saw, instead of stars, relationships moving like constellations, moving away and towards each other . . . turning, moving, according to an invisible design, according to influences we have not yet been able to measure, analyze, contain."

ple, the best astronomical alignments at Stonehenge were probably those constructed first, around 2500 B.C. Later construction, around 1800 B.C., muddled the design. In the nineteenth and twentieth centuries we see the monument used for mystical rites by costumed Druids and self-proclaimed witches. Here is an example of the loss of knowledge of original astronomical purpose.

In the same way, records of astronomical events, originally gathered by accurate observation, may have degenerated as later interpreters added mystical interpretations of astronomical events to accounts of historical incidents, describing them as if one controlled the other, and thus leading to astrological superstition.

In medieval times, astrology was forbidden by the Church as magic, yet it persisted, deeply ingrained in people's thinking. Since the Renaissance, astrology has continued to flourish alongside scientific thought. Today the ancient magic of 3000 y ago arrives on our doorsteps in the astrology columns of our daily papers!

Does Astrology Work?

A scientific test of astrology is to examine the success of its predictions. It has failed in many spectacular cases. For example, a grouping of all known planets in Libra in 1186 led astrologers to predict disastrous storms, because Libra was associated astrologically with the wind. People dug storm cellars in Germany; the archbishop of Canterbury ordered fasting; the palace in Byzantium was walled up; people fled to caves in the Near East. But the conjunction of planets passed without incident. Similarly, in 1524, all known planets clustered in Aquarius, the Water Bearer. This time astrologers predicted a second Deluge, but the month of the conjunction passed without disaster and reportedly was drier than usual (Ashbrook, 1973).

Similarly, pseudoscientists and astrologers predicted cataclysmic earthquakes and other disasters due to the so-called Jupiter effect—increased gravitational stresses caused by a rough alignment of Jupiter, Earth, and other planets on the same side of the Sun in March 1982. The month and year came and went without the "predicted" cataclysms.

Some researchers have made statistical studies of astrological predictions to see if they are more successful than random predictions. Jerome (1975) reviewed several of these studies and concluded that "legitimate statistical studies of astrology have found absolutely no correlation between the positions and motions of the celestial bodies and the lives of men."

We might just smile at astrology as another amusing human foible if it weren't for the nagging suspicion that it threatens healthy civilizations. For example, a 1975 Gallup poll indicated that about 12% of all Americans take astrology quite seriously. Prominent politicians and politicians' spouses have reportedly consulted astrologers. In an age of space exploration and serious nuclear threats, do we really want ancient magic to be an influence in human affairs? It is sad to realize that millions of people seeking advice about real problems in their lives spend their (sometimes meager) money on the products of astrologers and pious con artists.

Essay A, in the back of this book, describes other examples of pseudoscience that are merchandised as scientific truth in books and tabloids at supermarket checkout stands. Essay A also discusses in more detail how to distinguish science from pseudoscience—not through the hypotheses themselves, but rather through the way *evidence* is used to confirm or refute the hypothesis.

If we agree that astrology is little more than ancient superstition that wastes people's money and mental energy, should we try to suppress it? In a free country the answer is not suppression. In a free country with a free press, we hope that educated, pragmatic citizens will pick and enjoy the best from among competing published materials and look for evidence of accuracy before committing themselves to a system of belief.

ECLIPSES: OCCASIONS FOR AWE

So far we have treated the Moon and the Sun as independent bodies that can be tracked for practical purposes. Cycles of the Moon divide the year into about 12 equal months; solstices and equinoxes, defined by sunrise positions, divide the year into four seasons. Ancient people who discovered these facts may have gained a sense of well-being: At least *some* things in their environment could be counted on from year to year. Calendars could be made and agriculture regulated. The predictable, friendly Sun could be worshiped as a deity, always providing light and warmth.

What, then, could be made of **eclipses,** the sporadic occasions when the Sun or more commonly the Moon disappears while above the horizon? In the few minutes of a total solar eclipse the sky turns dark, stars can be seen in daytime, and an uncanny chill and gloom settle over the land, as indicated in Figure 1-13. During

Figure 1-13 The March 1970 total solar eclipse as seen from the village of Atatlan, Mexico. The sky has darkened dramatically, giving a general appearance of dusk. (Photo by author.)

the hours of a total lunar eclipse, the Moon turns blood-red. Small wonder that the Greek root of eclipse, *ekleipsis,* means abandonment or that the ancient Chinese pictured a solar eclipse as a dragon trying to devour the Sun. Eclipses must have seemed an unpredictable menace to the scheme of things.

We know eclipses awed early people. In Greece, the poet Archilochus, observing the solar eclipse of April 6, 648 B.C., was moved to write: "Nothing can be sworn impossible . . . since Zeus, father of the Olympians, made night from midday, hiding the light of the shining Sun, and sore fear came upon men." Herodotus reports that a war between the Lydians and the Medes stopped when an eclipse surprised the competing armies during a battle in the 580s B.C. A hasty peace was cemented by a double marriage of couples from the opposing camps.

Cause and Prediction of Eclipses

Whatever their emotions and motives, some early people learned how to predict eclipses—and this was an early step toward recognizing that we live on a world among worlds orbiting in space.

To understand how eclipses can be predicted, we should first understand the **causes of eclipses.**

Two important things to remember about eclipses are: (1) They happen when the shadow of one celestial body falls on another, and (2) what you see during an eclipse depends on your position with respect to the shadow. If you see an eclipse at all, you are either in the shadow (a celestial body has come between you and the source of light) or you are looking at the shadow from a distance as it falls on some surface.

Earthbound observers see two types of eclipses: solar and lunar. **Solar eclipses** occur when the Moon, on its $29\frac{1}{2}$-d journey around the Earth, happens to pass between Earth and the Sun as illustrated in Figure 1-14. A **total solar eclipse** occurs if the Moon completely covers the Sun as seen by an earthbound observer. A **partial solar eclipse** occurs if the Moon is "off center" and covers only part of the Sun. By coincidence, the Moon happens to have the same angular size as the Sun—about $\frac{1}{2}°$, or a little less than the angular size of a little fingernail at arm's length. Therefore, the Moon usually just covers the Sun during a solar eclipse. But if the Moon happens to be at the farthest point in its orbit, it has a smaller angular size than usual and doesn't quite cover the Sun. This causes an **annular solar eclipse,** in which the observer sees a ring (Latin *annulus*) of light, which is the rim of the Sun, surrounding the Moon.

Lunar eclipses occur when the Moon, on its journey around Earth, passes through a point exactly on the opposite side of Earth from the Sun. This point lies in the shadow cast by Earth (Figure 1-15). It takes the Moon a few hours to pass through Earth's shadow, during which time the Moon usually turns an astonishing copper-red because the only sunlight reaching the Moon

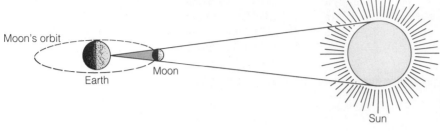

Figure 1-14 Geometry of an eclipse of the Sun (not to scale). The shadow of the Moon falls on Earth.

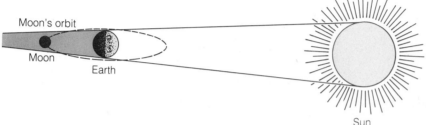

Figure 1-15 Geometry of an eclipse of the Moon (not to scale). The Moon passes through Earth's shadow.

passes through Earth's atmosphere, as seen in Figure 1-16. In effect the Moon is illuminated only by the colored light of a sunset.

Eclipses can happen in other parts of the solar system, as other moons and planets cast shadows on each other.

Umbral and Penumbral Shadows

In solar and lunar eclipses, the shadows of Earth and the Moon are not sharply defined. Because the Sun as seen from Earth has an angular size of $\frac{1}{2}°$ instead of being a tiny point source, sunlight is slightly diffuse; its rays come from slightly different directions.

Thus shadows—including eclipse shadows—cast by sunlit objects near Earth are not sharply defined. The inner, "core" area of a shadow—the part that receives no light at all—is the **umbra,** named from the Latin word for shadow. The outer, fuzzy boundary is the **penumbra.** An observer in the umbra sees the Sun entirely obscured; an observer in the penumbra sees the Sun only partly obscured. Figure 1-17 shows some varieties of solar eclipses, illustrating these principles. If you hold your hand at eye level above smooth ground in sunlight, its shadow will have a central, dark umbra and a penumbra about a centimeter wide. If you hold your hand high and spread your fingers, their shadows will be indistinct. An ant in the umbra would see a total solar eclipse; an ant in the penumbra would see a partial solar eclipse.

The relative sizes of umbra and penumbra depend on the distance between the shadow-casting body and

Figure 1-16 A nearly total lunar eclipse. Sunset-colored reddish light refracted through Earth's atmosphere colors the dim, umbral part of the shadow (right). The silvery crescent on the left is in the penumbral part of the shadow, partly lit by the Sun. This photo gives a good impression of the visual appearance of the lunar eclipse. (Photo by Stephen M. Larson.)

the surface on which the shadow appears. During a lunar eclipse, Earth's shadow on the Moon has an umbra several times the Moon's diameter and a considerably larger penumbra. Total lunar eclipses, with the Moon in the umbra, can last up to $1\frac{3}{4}$ h.

On the other hand, the Moon's umbral shadow on Earth is no larger than 267 km (166 mi) in diameter. Due to the motion of this shadow across Earth, total solar eclipses cannot last more than $7\frac{1}{2}$ min. In an annular eclipse, the Moon is too far away to produce any umbral shadow, and the eclipse has no true total phase, as seen in Figure 1-17 (case C).

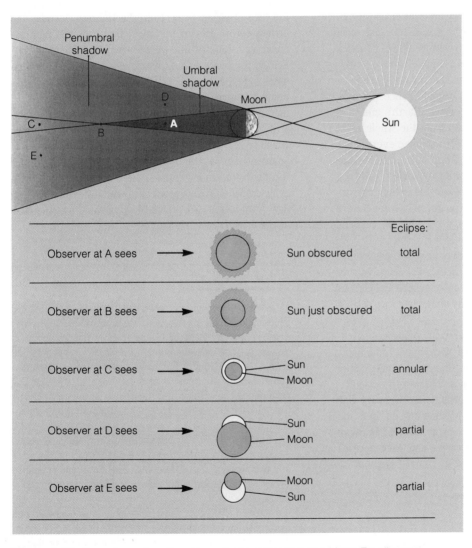

Figure 1-17 Geometry and configurations of solar eclipses by the Moon. Top diagram shows umbral and penumbral shadows of the Moon. Observers at different points see different kinds of eclipses, as shown here. By chance, the tip of the umbra (B) lies very close to Earth's surface.

Frequency of Eclipses

By keeping records of eclipses over many generations, ancient astronomers realized that they occurred at regular time intervals. Discovery of these repeating intervals several thousand years ago marked the beginning of human ability to predict seemingly mysterious celestial events.

One such interval is called the **saros cycle** (a Greek word from an earlier Assyrian-Babylonian word). It is 18 y and 11 d. Early astronomers discovered that if, in a given year, a particular sequence of eclipses occurred, a similar sequence would probably occur after one saros. (Shorter periodicities, such as 41 and 47 mo, also exist but are less accurate.) These cycles are due to complex interplays of Earth's and Moon's orbits.

Because the Moon's umbral shadow on Earth is so small, a fixed observer has only a small chance of seeing any given solar eclipse. But because Earth's umbra is large and the Moon can be seen from the whole night hemisphere of Earth, half of Earth will see every lunar eclipse (barring cloud cover). Total lunar eclipses are therefore relatively common for any observer, and they were thus easier than solar eclipses for ancient astronomers to predict. Modern astronomers use computers and orbital theory, not cycles, to predict all eclipses accurately.

On the average a given location may witness a lunar eclipse nearly every year and a partial solar eclipse nearly every other year, but a total solar eclipse only about once every four centuries. Many of the dates are related by cycles.

Let us now return to the viewpoint of ancient earthbound peoples. Once records were made and kept, astronomers could benefit from their community's past observations of eclipses. If they discovered periodicities, they could make rough predictions. We know this happened by 700 B.C. in the Mediterranean area and probably by about A.D. 500 in Central America, because records found in Assyrian libraries and Mayan cities discuss the prediction of eclipses.

Two Examples of Eclipses in History

Thales (c. 580 B.C.) Historical records indicate that one of the first known astronomers, Thales of Miletus, predicted the eclipse that stopped the battle between warring Greek factions in the 580s B.C. Thales may have made this prediction by knowing of the saros cycle from Mesopotamian records, or he may have discovered a useful 3-saros periodicity of 669 lunar months, which would have been prominent in the eclipse records of his region for the preceding 125 y.

The Crucifixion An interesting example of the use of eclipses to date historical events is given by Humphreys and Waddington (1983) and Schaefer (1989), who attempt to find the date of the crucifixion of Jesus. References in the gospels of the New Testament suggest the date of Nisan 14 or 15 in the Hebrew calendar, on or just before Passover. Calculating back through calendar revisions to find years when this date fell on a Friday, the two scholars restrict the number of possible dates to four Fridays between A.D. 27 and A.D. 34. Additional historical references suggest restricting the date further to April 7, 30, or April 3, 33. At this point the scholars invoke a speech of Peter, reported in Acts, which refers to a blood-red moon, possibly referring to an event at about the time of the crucifixion. Calculating the dates of lunar eclipses, they find that a partially eclipsed Moon rose over Jerusalem on the night of Friday, April 3, A.D. 33, which they conclude was the date of the crucifixion.

SUMMARY

Unaided by telescopes, unknown geniuses of prehistoric times made many of the most basic astronomical discoveries. They (1) recognized and tracked five planets; (2) discovered the celestial poles; (3) learned to use solstitial risings and settings to formulate calendars; (4) designed constellations as memory aids for learning the sky; (5) recognized the ecliptic and the zodiac as the paths of the Sun and planets, respectively; and (6) discovered eclipse-related cycles.

Stable societies encouraged astronomy, and vice versa. Astronomy contributed to civilization by creating calendars and encouraging record keeping. The discovery of eclipse cycles in particular required record keeping over many years.

Around 2600 B.C. there may have been a golden age in the Mediterranean world, when the modern constellations were invented on or near Crete and the Stonehenge solstitial observatory was built in England. By around 1400 B.C. temples with astronomical alignments were being built in Egypt.

Although the recognition of solstices, eclipse cycles, and so on did have applications in ancient societies, it did not lead to visualization of Earth's movement through space. As we will see in the next chapter, the first glimmers of a Sun-centered system came in Greek times, around 200 B.C. These concepts were not fully clarified until the most recent "moments" in the long story of human civilization. Of the 150 generations that have lived since astronomy emerged around 3000 B.C., only the last dozen generations (since about A.D. 1600) have well understood the concept of Earth as a ball moving around the Sun.

CONCEPTS

celestial poles
Polaris
precession
North Star
zenith
meridian
celestial equator
ecliptic
planet
zodiac
constellation
cause of the seasons
equinox
solstice

solstice principle
Stonehenge
astrology
eclipse
causes of eclipses
solar eclipse
total solar eclipse
partial solar eclipse
annular solar eclipse
lunar eclipse
umbra
penumbra
saros cycle

PROBLEMS

1. Suppose you are standing facing north at night. Describe, as a consequence of the Earth's rotation, the apparent direction of motion of each of the following:
   a. A star just above the north celestial pole
   b. A star just below the north celestial pole
   c. A star to the left of the north celestial pole
   d. A star on the northern horizon

2. Answer the parts of Problem 1 but for an observer in the Southern Hemisphere facing south and looking near the south celestial pole.

3. What is your latitude? What is the approximate elevation of Polaris above your horizon?

4. How many degrees from the south celestial pole is the southernmost constellation that you can see from your latitude?

5. Solar eclipses are slightly more common than lunar eclipses, but many more people have observed lunar eclipses than solar eclipses. Why?

6. Does the Moon cast an umbral shadow on Earth during an annular solar eclipse? Why or why not?

7. How would a lunar eclipse look if Earth had no atmosphere? Compare the Earth's appearance from the Moon during such an imaginary eclipse with its actual appearance from the Moon.

8. Why would lunar and solar eclipses each occur once per month if the Moon's orbit lay exactly in the ecliptic plane?

9. Compare ancient European and American cultures of around A.D. 200 to 1000.
   a. How much did each know about eclipses?
   b. How many years apart did they achieve a comparable ability to predict eclipses?
   c. When, if at all, did they begin to understand the causes of eclipses?
   d. During this period, were the two cultures' advances in technology similar to their astronomical advances?
   e. Do you think that either of these two types of advances (or the two combined) offer a valid measure of cultural achievement?

PROJECTS

1. Starting on the date of the new moon (shown on many calendars), observe the sky at dusk and record whether the Moon is visible. Repeat each evening for several weeks and record the Moon's appearance. Repeat at the next new moon. How many days is the Moon visible between the new moon and the first quarter? Between the first quarter and the full moon? How might these counts relate to clusters of grooves, such as 3, 6, 4, 8, . . . , found on prehistoric tools? Can you prove or only speculate that the prehistoric records are lunar calendars?

2. From the preceding project, determine how much later the Moon sets or rises each night.

3. If a planetarium is available, arrange for a demonstration showing the following:
   a. The position of the north celestial pole
   b. The position of the celestial equator
   c. The daily motion of the stars
   d. The prominent constellations
   e. The daily motion of the Sun with respect to the stars
   f. The position of the ecliptic
   g. Planetary motions

4. Measure the angle of elevation λ of the North Star above the horizon. A protractor can be used as shown to make the measurement. Compare the result with your latitude.

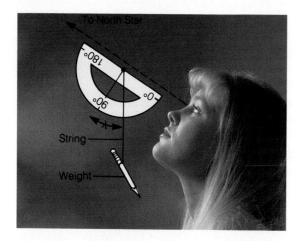

5. Identify a bright planet and draw its position among nearby stars each night for several weeks. (A sketch covering about 10° × 10° should suffice.) Does the planet move with respect to the stars? How many degrees per day? (The latter result will differ from one planet to another and from one week to another.)

6. From a viewing area with a clear western horizon, chart sunset positions with respect to distant hills, trees, or buildings for several days and demonstrate the motion of the sunset point from day to day. Do the same for a few days around winter or summer solstice and demonstrate the reversed drift of the sunset position. How accurately can you measure the solstice date in this way? Does the Sun approach the horizon vertically or at an angle?

7. Using a light bulb across the room for the Sun, a small ball for the Moon, and yourself as a terrestrial observer, simulate total, partial, and annular solar eclipses.

8. Using the same props, demonstrate a total eclipse of the Moon. Show why a lunar eclipse occurs only during a full moon.

9. If an eclipse of the Moon occurs while you are taking this course, observe it. Compare the visibility of surface features on the Moon before the eclipse, in the penumbra, and in the umbra. Confirm that the curved shadow of Earth defines a disk bigger than the Moon. What color is the Moon and why? Would the Moon be lighter or darker during eclipse if there were an unusually large number of storm clouds around the "rim" of the Earth, as seen from the Moon?

10. Make a star trail photo like that of Figure 1-6 by setting your camera at night on some interesting foreground, opening the lens wide open, focusing on infinity, and exposing on a "fast" film, like Tri-X, for an hour or more.

CHAPTER 2

Historic Advances: Worlds in the Sky

Most of pre-Greek astronomy comes from unknown ancient geniuses, whose work was passed down in the form of tradition and myth. But the Greeks made startling advances with this early astronomy, and their work is known from historical records. Can you figure out how to measure the diameter of the Earth or the relative distances of the Moon and Sun without telescopes or electronic devices? The Greeks did. Passed on by the Romans, Greek astronomy was one of the main foundations of modern Western culture. This chapter will describe those pioneering, fundamental discoveries and then finish with some notes on astronomy in non-Western cultures.

EARLY COSMOLOGIES AND ABSTRACT THINKING (2500–100 B.C.)

Neither Greek scholars nor their systematic observations burst upon the Mediterranean scene from nowhere. The *idea* of thinking about abstract physical concepts can be traced back to early **cosmologies,** or theories about the origin and nature of the universe. For example, certain Egyptian cosmologies assigned different roles to different godlike personages who interacted with the real world. The Memphite theology (c. 2500 B.C.) spoke of an intelligence that organized the "divine order" of the universe (see Figure 2-1). This kind of thinking represented a step toward astronomical science because it assigned different gods, intelligences, or forces to different kinds of natural events and sought relations between them. Early naturalists expressed these relations in myths; later naturalists expressed them in "laws," or generalizations derived from repeated observations.

It is interesting to trace how cosmological thought evolved. Many ideas seem to have carried over directly from the world of 3000 y ago to us today.

Figure 2-1 An ancient Egyptian conception of the universe. Stars are distributed in different shells around, the universal center. The shells are separated by boundaries depicted as goddesses. This conception may have carried over to Greek and medieval times, when the planets and stars were also pictured as distributed in concentric shells. (Giraudon/Art Resource.)

Out of Egypt, around 1300 B.C., escaped a tribe of nomads whose book of monotheistic religious thought is the core of Western religious principles today. Some scholars believe their Psalm 104 may be a direct transcription of a hymn to the Sun written by Pharaoh Akhenaten himself and still preserved (Pritchard, 1955). The pharaoh praised the Sun in almost biblical terms: "How manifold it is, what you have made! . . . You created the world according to your desire. . . ."

The Hebrews recorded a similar cosmology: "In the beginning God created the heavens and the earth. . . ." This cosmological theory asserted that Earth was created in six stages, or days: (1) light, day, night; (2) sky; (3) dry land, ocean, plants; (4) Sun, Moon, stars; (5) sea creatures, birds; and (6) humans. This cosmology dominated much Western art and science.

Cosmological theories of this type, which were especially common in the Middle East, stimulated new questions about relations between phenomena in the universe. Out of this thinking came an important new idea. Regardless of arguments over different gods, facts could be learned about nature by systematic observations and experiments from which repeatable results could be obtained. A naturalist in Alexandria could get the same results as a naturalist in Athens. **Science** (from Latin "to know") is simply the process of learning about nature by applying this technique: Questions are formulated that can be answered by observations or experiments, which are then carried out.

This system of scientific observation and recording of nature was developed to the greatest extent in Greece. In addition to richly provocative Egyptian and Hebrew philosophies, the Greek world received a legacy of astronomical concepts, such as the ecliptic, the zodiac, solstices, and the saros cycle, as well as generations of astronomical observations of eclipses and planet motions from Mesopotamian (and perhaps European) sources. This inheritance, plus the Greeks' inclination to philosophize about natural phenomena, led to a Greek renaissance.

THE SYSTEM OF ANGULAR MEASUREMENT

One of the most important legacies that the Greeks received from the earlier world was the sexagesimal (60-based) system of measuring angles and time, which was mentioned in Chapter 1. Just as there are 60 min of time in one hour and 60 s (seconds) in a minute, the system of angular measurement uses the following definitions and symbols:

$$1 \text{ degree } = 1° = \tfrac{1}{360}\text{th of a circle}$$
$$60 \text{ minutes of arc } = 60' = 1°$$
$$60 \text{ seconds of arc } = 60'' = 1'$$

To give a better idea of 1 second of arc: It is the angular size of a tennis ball seen at a distance of about 8 mi.

Applying this numerical system for measuring angles, the Greeks developed not only rules of geometry (such as Euclid's geometry and Pythagoras' famous theorem about the hypotenuse of a right triangle), but also ways of measuring phenomena in the sky. With sighting devices, Greeks and other early observers measured the positions of planets relative to the fixed pattern of background stars and the number of degrees that the Sun rose above the southern horizon as it crossed the meridian at noon in different seasons. These measures first revealed the *detailed* movements of celestial bodies.

An important concept is the difference between linear measure and angular measure. **Linear measure** gives the actual length of something in linear units such as inches, meters, or miles. **Angular measure** gives the angle covered by an object (or, alternatively, the apparent separation between two objects) at a given distance from the observer, in angular units such as degrees. The verb **subtend** refers to the angle covered by such an object: For example, we might say a distant object subtends 1°. A useful rule of thumb is that your thumbnail at arm's length subtends about 1°. The disks of the Sun and Moon always subtend $\frac{1}{2}°$. The pointers in the Big Dipper (see Figure 1-4) subtend about 5°. As the Greeks knew, when you see a distant object (such as a ship at sea or the Moon), you cannot directly measure either its linear size or its distance, but only its angular size. We unconsciously *infer* linear distance of many objects by recognizing the object (such as a ship) and knowing roughly how big it is; we similarly *infer* the linear size by estimating an object's distance and judging it must be as big as a house, a dog, and so on. The treachery of such inferences shows up, for example, when people report unfamiliar aerial objects such as a bright meteor, called a fireball. People commonly report that a fireball "looked as big as a dinner plate," but this statement is literally meaningless; they really perceive only angular size, not linear size. To specify angular size correctly, they would have to say, "It looked as big as a dinner plate at a distance of 50 feet" or "It looked twice the angular size of the Moon." Similarly, the common report that a fireball "must have landed just over the hill" is almost always wrong. The speaker misjudges the distance because he *assumes* a certain linear size after he *observes* only the angular size. Fireballs are typically in the upper atmosphere 60 mi from observers!

As we will see in a moment, Greek thinkers carefully separated angular measures from linear measures, and used angular measures together with clever logic to estimate linear sizes and distances of the Sun, Moon, and other objects.

The word **resolution** refers to the smallest angular sizes that can be discriminated with optical systems. The human eye, for example, can resolve an angle of about 2′; thus, we can see details covering about $\frac{1}{15}$ of the lunar disk. The largest planetary disks subtend only about 1′ and are thus too small for the eye to resolve.

These principles help us analyze photographs and other images. For example, as shown in Figure 2-2, a typical snapshot subtends an angle of about 40°, about the portion of a scene that the eye concentrates on. A 35-mm camera is so named because it uses film 35 mm wide and makes a negative about 35 mm in width. The typical 40°-wide snapshot can resolve 2′ details and thus presents a view comparable to what the eye sees. However, 40° views made with smaller cameras (such as instamatic and disk cameras) are limited by the inherent graininess of film and often do not resolve as much as the eye can see from the same spot. TV images suffer a similar limitation, resolving less detail than the eye can see in the typical 40° view. Conversely, modern motion pictures photographed and projected with 70-mm film have a dramatically realistic presence because, like life, they can present the eye with a 40° view containing more detail than the eye can resolve.

The ancient system of angular measure, using degrees, minutes of arc, and seconds of arc, is still used today by surveyors, engineers, navigators, astronomers, and others. Modern large telescopes can often resolve details as small as 0.5″, but rarely can do better than this because of the shimmering quality of heat waves in the atmosphere. The orbiting Space Telescope, planned for operation in the 1990s, should routinely be able to resolve 0.05″ to 0.1″. Later in this book we will encounter more applications of angular measurement.

EARLY GREEK ASTRONOMY (c. 600 B.C. to A.D. 150)

Around 600 B.C. the Greeks began vigorously applying logic and observation to learn about the universe. They talked more of tangible physical "elements" and less of metaphysical relations. They used geometric principles, including applications of angular measurement, to measure cosmic distances as well as farmyards.

One of the first known Greek thinkers was Thales of Miletus (a Greek-dominated town in present-day

a

b

c

d

Figure 2-2 These views toward the northeast from the center of Stonehenge (compare map, Figure 1-9) illustrate angular measurements applied to photography with different lenses. **a** This view using an ultra-wide-angle fisheye lens subtends a horizontal angle of 120°. **b** This view with a standard wide-angle lens subtends 65°. **c** This view with a normal lens subtends 40°; this is about the field of view of most snapshots, postcard views, paintings, and so on. **d** This telephoto view subtends only 15°. (Photos by author with a 35-mm camera. Lens focal lengths of *a–d*: 15 mm, 24 mm, 50 mm, and 135 mm, respectively. Curvature of pillars at edges of *a* is distortion common with ultra-wide-angle lenses.)

Turkey). Living about 636–546 B.C., Thales was a noted statesman, geometer, and astronomer. He is best known for predicting the peacemaking solar eclipse mentioned in Chapter 1.

Thales' school produced several notable thinkers. For example, Anaximander (611–547 B.C.) made astronomical and geographical maps; speculated on the relative distances of the Sun, Moon, and planets from our Earth; and argued that the matter from which things are made is an eternal substance.

The Pythagoreans: A Spherical, Moving Earth (c. 500 B.C.)

Pythagoras (flourished 540–510 B.C.), famous for his theorem on right triangles, was also one of the first experimental scientists. Pythagoras proposed the unusual idea that Earth is spherical. He may have gotten this idea by studying the phases of the Moon. The **terminator,** the line separating the lit side of the Moon (or any planetary body) from the unlit side, changes its curvature as the Moon's phases progress, thus revealing that the Moon is spherical rather than flat, as shown in Figure 2-3. By analogy, then, Earth and other astronomical bodies would also be spherical.

In southern Italy, Pythagoras founded a school that had wide influence around 450 B.C. It is unclear, however, which thinkers should be credited with which ideas in this school. Pythagoras himself put the Earth at the center of the universe, but later Pythagoreans proposed that it moves, together with the Moon, the Sun, the five planets, and the stars, around a distant central "fire."

a   Gibbous   Full   Gibbous   Quarter   Crescent   b

Figure 2-3 Evidence of the Moon's true nature. **a** Its phases correctly suggested to some Greeks that the Moon is not a disk but a sphere illuminated by the Sun. **b** The Earth's curved shadow on the Moon during every lunar eclipse suggested that Earth too is spherical.

This system predates by more than 2000 y the correct model of the planets, moving around the Sun (see Chapter 3). The idea of a spherical Earth persisted among some Greeks, though it was not universally accepted, and it was eventually lost.

Anaxagoras (500?–428 B.C.) is credited with deducing the true cause of eclipses. Thereafter, the observed roundness of the Earth's shadow on the Moon (Figure 2-4) undoubtedly helped to establish the theory that Earth itself is a spherical body. After residing in Athens for 30 y, Anaxagoras was charged with impiety and banished for saying that the Sun was an incandescent "stone" even larger than Greece.

Aristotle: The Earth at the Center Again (c. 350 B.C.)

The most influential Greek scientist-philosopher was Aristotle (384–322 B.C.). His views were built on earlier knowledge but were biased in favor of absolute symmetry, simplicity, and an abstract idea of perfection. Aristotle believed the universe was spherical and finite, with the Earth at the center. Planets and other bodies moved in a multitude of spherical shells centered on the Earth. The shells were supposed to turn with varying rates, which explained the observed changeable motions of the planets.

Aristotle is credited with founding modern scientific investigation. His school at Lyceum (a grove near Athens) contained a library, a zoo, and lavish physical and biological research equipment paid for by his onetime pupil Alexander the Great, then ruler of Greece. In the Middle Ages, when research lapsed, Aristotle came to be regarded as the final authority, and his placement of Earth at the center of the solar system turned out to delay progress in astronomy.

Figure 2-4 Contrary to popular conception, the Moon is visible during the day for part of the month, as well as at night. Thus its phase can be studied in relation to the Sun, showing that the phases match that of a sphere illuminated by the Sun. This helped early observers realize that the Moon is a spherical world. (Photo by author.)

Aristotle was right about several important astronomical ideas, however.

1. He thought the Moon is spherical.

2. He argued that the Sun is farther from Earth than the Moon because:

   a. The Moon's crescent phase shows that it passes between Earth and the Sun.

   b. The Sun appears to move more slowly in the sky than the Moon. (This second argument is not rigorous, but the first is.)

3. He thought Earth is spherical because:

a. The curvature of the Moon's terminator rules out its being a disk, and Earth is probably like the Moon in this respect.

b. As a traveler goes north, more of the northern sky is exposed while the southern stars sink below the horizon—a circumstance that would not arise on a flat Earth.

The apparent motions of the Sun, the Moon, and the stars around Earth could be explained, said Aristotle, either by their actually moving around us or by Earth moving. But Aristotle concluded that Earth is stationary and gave a very powerful argument. If Earth were moving, we ought to be able to see changes in the relative configurations of the various stars, just as, if you walk down a path, you see changes in the relative positions of nearby and distant trees. If you line up a tree in the middle distance with a very distant tree and then step to one side, the nearby tree will seem to shift to the side of the distant one. Such a shift in position due to motion is called **parallax,** or a *parallactic shift.* If Earth were moving in a straight line, we would see a continuous parallactic shift of the nearer stars with respect to more distant stars; and if Earth were moving around some distant center, we would see a periodic parallactic shift back and forth among the stars. But a visual survey of the stars and the constellations over time showed no evidence of such a shift. So, reasoned Aristotle, Earth must not move.

Aristotle's reasoning was sound, but the stars are too far away to produce noticeable parallactic shifts for the unaided eye during a human lifetime. In the same way, distant mountains show little parallactic shift from a car speeding down an interstate highway, even though nearby trees whiz by in seconds. Stellar parallaxes were sought for years and not discovered until 1838.

Aristotle died shortly after being forced to leave Athens for allegedly teaching that prayer and sacrifices to the Greek gods were useless.

Aristarchus: Relative Distances and Sizes of the Moon and the Sun (250 B.C.)

Aristarchus (310?–230? B.C.) of Samos (an island off present-day Turkey) cleverly extended the Greek methods of seeking quantitative data. His only surviving work is "On the Sizes and Distances of the Sun and Moon," although his other astronomical works are quoted by other Greek authors.

He devised a way to measure the relative distances of the Sun and Moon from Earth, based on the geometry of the Moon's orbit and phases. From this he correctly inferred that the Sun is much farther away than the Moon. Aristarchus also formulated a way to measure the relative sizes of Earth and Moon. He concluded that the Moon is one-third as big as Earth and that the Sun is about seven times as big as Earth. The correct figures are closer to one-fourth and 100, but Aristarchus was on the right track.

Aristarchus made still another contribution. Because he thought the Sun is much bigger than Earth, he guessed (without many supporting observations) that the Sun, not Earth, must be the central body in the system. For this an outraged critic declared he should be indicted for impiety.

Although Aristarchus made some quantitative errors, he was nonetheless far ahead of later scholars who thought Earth is flat. Aristarchus correctly visualized the Moon in orbit around a spherical Earth and Earth in orbit around the Sun, and he developed a method of measuring interplanetary distances. These ideas were not confirmed for another 2000 y!

Eratosthenes: Earth's Size (200 B.C.)

As Greece declined and Rome prospered, Greek scholars became resident intellectuals in many parts of the Mediterranean world. Eratosthenes (276?–192? B.C.) was a researcher and librarian at the great Alexandrian library in Egypt. He reportedly completed a catalog of the 675 brightest stars and measured the $23\frac{1}{2}°$ inclination of Earth's polar axis to the ecliptic pole,[1] as shown in Figure 2-5. As described in Figure 1-7, this tilt causes our seasons.

Eratosthenes is most famous for using angular geometric relations to measure Earth's size. Told that at summer solstice the Sun shone directly down a well near Aswan, he noted that the Sun's direction was off vertical by $\frac{1}{50}$ of a circle on the same date at Alexandria (Figure 2-6). He realized this difference had to be due

---

[1]As can be seen in Figure 2-5, the angle between Earth's polar axis and the ecliptic pole is the same as the angle between Earth's equatorial plane and the plane of the ecliptic. The ecliptic pole is defined by any perpendicular to the ecliptic plane.

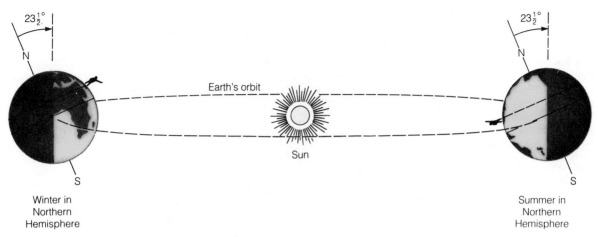

Figure 2-5 The seasons are caused by the $23\frac{1}{2}°$ angle of tilt between Earth's North Pole and the north pole of the plane of Earth's orbit around the Sun. As shown by the human figure, an observer at a fixed northern latitude finds the noontime Sun more nearly overhead in summer than in winter. This difference, which would not arise if the tilt were zero, makes summer days hotter and longer. Eratosthenes measured the difference between the summer and winter noontime Sun's elevation and was able to use it to deduce the $23\frac{1}{2}°$ tilt angle. (The angular difference between the summer and winter noontime Sun elevations is twice the tilt angle—a statement the student may try to verify.)

to the curvature of the Earth and concluded that Earth's circumference was 50 times the distance from Alexandria to the site of the well. Measuring that distance, he multiplied by 50 and got an estimate of the Earth's circumference that was probably within 20% of the right answer. This Greek master clearly understood the shape and approximate size of Earth 1700 y before Columbus!

Hipparchus: Star Maps and Precession (c. 130 B.C.)

From his observatory on the island of Rhodes, Hipparchus (160?–125? B.C.) observed the positions of astronomical bodies as accurately as possible and compiled a catalog of some 850 stars. His exhaustive observations—all done, of course, without a telescope—along with material he inherited from Babylon, enabled him to predict with reasonable accuracy the position of the Sun and Moon for any date. Hipparchus has been called antiquity's greatest astronomer.

The most important discovery attributed to Hipparchus is precession. Comparing his own measurements of star positions with materials handed down to him from centuries before, Hipparchus found that, with respect to the background stars, there had been curious shifts in the positions of the north celestial pole, the

vernal and autumnal equinoxes, and other coordinates. The whole celestial equator was oriented somewhat differently with respect to the stars! Could the old maps be wrong? Hipparchus concluded instead that the whole coordinate system of the celestial equator and the poles was drifting slowly with respect to the distant stars. This drift came to be known as **precession** or *precession of the equinoxes.*

In modern terms, *precession is the result of a wobble of the spinning Earth due to forces produced by the Sun and the Moon.* Just as a spinning top describes a conical wobble when it is pulled downward by the force of the Earth's gravity, the spinning Earth's polar axes describe a conical wobble with respect to the fixed stars. Thus, as mentioned in the last chapter, in different millennia different stars become the North Star; a complete cycle takes about 26,000 y.

Ptolemy: Planetary Motions (A.D. 150)

Claudius Ptolemy (flourished c. A.D. 140) was another scholar associated with the Alexandrian library. His fame as an astronomer is based on a 13-volume work, *The Mathematical Collection.* Passed on to the Arabs after the destruction of the library, the work became known

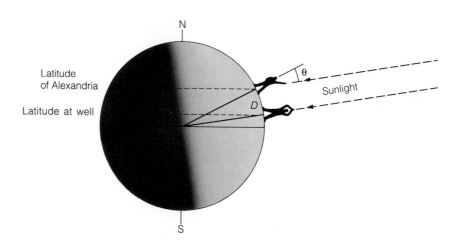

Figure 2-6 The geometry of Eratosthenes' measurement of the size of Earth. When the Sun was directly over a certain well, Eratosthenes measured the Sun's angle θ from the zenith at Alexandria, a known distance D away. He found D was $\frac{1}{50}$ of the way around Earth. He could thus find Earth's circumference and diameter.

as *al-Megiste* (The Greatest). European translations were called the *Almagest,* and the book was famous for more than a thousand years.

Ptolemy extended Hipparchus' star catalog to 1022 entries, correcting older reported star positions to compensate for precession.[2] But his best-known contribution was a method for predicting the positions of the Sun, Moon, and planets, called the **epicycle theory,** or *Ptolemaic theory.* Following Hipparchus, Ptolemy incorrectly assumed that Earth is near the center of the planetary system. In order outward from Earth, he placed in *circular* orbits the Moon, Mercury, Venus, the Sun, Mars, Jupiter, and Saturn.

In spite of its usefulness in ancient times for predicting planet positions, Ptolemy's theory was incorrect. Today we know that planets move in elliptical orbits around the Sun, not in circles around the Earth. Ptolemy has been criticized for abandoning Aristarchus' Sun-centered system and for biasing his theory toward a supposed "perfection" of the circle, thus delaying the introduction of the true elliptical orbits. But the system fitted the observations available in Ptolemy's time. Aristotle's old argument still seemed true: Earth could not be moving around a distant center because no one had observed parallax. Certainly no one had yet realized they were observing elliptical motions. And Ptolemy's system worked fairly well. The trouble with Ptolemy's choice of Aristotle over Aristarchus lies not in its being

a mistake, but in its historical effects: The *Almagest* became the Bible of ancient astronomy, and the erroneous Earth-centered system held sway for 1400 y.

The Loss of Greek Thought (c. A.D. 500)

Alexandria, where Cleopatra first fascinated Julius Caesar and Mark Antony around 47 B.C., was the world's intellectual center by A.D. 250. With Rome's fall and the world in disorder in 410, maintaining the great library of ancient discoveries became more and more difficult.

Among the last guardians of the old knowledge in Alexandria was one of the first known woman astronomers, Hypatia (c. A.D. 375–415). Widely admired for her learning and eloquence, she wrote a commentary on Ptolemy's work and invented astronomical navigation devices, but was murdered by a mob during one of the riots that plagued Alexandria during its decline. In A.D. 640, after a 14-mo siege by the Arabs, Alexandria fell. The library buildings were burned and the best collection of Greek knowledge was lost. Because there was no printing, there were few other reference books. Table 2–1 summarizes the dramatic Greek advances.

ANCIENT ASTRONOMY BEYOND THE MEDITERRANEAN

With the fall of Alexandria in A.D. 640, in the West the rate of new discoveries declined and Europe slipped into the Dark Ages. But intellectual progress occurred in other cultures.

---

[2]Correction for precession continues today. For precise setting of a large, modern telescope, a star's coordinates published for "epoch 1950" or any other year must be corrected for the current date, or the star could be missed. The correction is generally done by a computer operating as part of the telescope.

TABLE 2-1

Astronomical Discoveries of the Greeks

Observation	Inference	Observer Commonly Quoted
Curved lunar terminator	Moon round	Pythagoreans
Round shadows during lunar eclipses	Earth round	Pythagoreans
Crescent phases of Moon	Moon between Earth and Sun	Aristotle
Different stars at zenith at different latitudes	Earth round	Aristotle
No evident stellar parallax observed by naked eye	Distance Earth moves is small compared with distances to stars	Aristotle
Relative sizes and angles of Moon and Earth's shadow	Moon smaller than Earth; Sun bigger than Earth	Aristarchus
Angle from first quarter to last quarter Moon slightly less than 180°	Sun tens of times farther away from Earth than Moon	Aristarchus
Relation between angular shift of zenith distance of Sun and linear distance traveled on Earth	Calculable circumference of Earth	Eratosthenes
North celestial pole's shift with respect to constellations	Precession	Hipparchus

Islamic Astronomy

Much of the Alexandrian knowledge passed into the hands of the Arabs. About a century after Muhammad, around A.D. 760, Islamic leaders in the new capital of Baghdad began to sponsor translations of old Greek texts. The next known measurement of the Earth's circumference, made near Baghdad in 820, was only 4% too large. Similarly, Arab astronomer Muhammad al-Battani (c. 850–929; known later in Europe as Albategnius) made only a 4% error in his measurement of the eccentricity, or noncircularity, of the Earth's orbit. (He would have called it the Sun's orbit around the Earth.) By A.D. 1000 the Islamic empire had spread to Spain, and astronomical tables were published with the 0° reference longitude in Córdoba (rather than in Greenwich, England, as in the modern longitude system, introduced when Britannia ruled the waves).

Astronomy in India: A Hidden Influence

Astronomical practices in India date back to about 1500 B.C. The first known astronomy text, describing planetary motions and eclipses and dividing the ecliptic into 27 or 28 sections, appeared around 600 B.C. By this time India had contact with the Mesopotamian and Greek worlds, and influences probably traveled both ways.

Texts dating from around A.D. 450 use Greek computational methods and refer to the longitudes of both Alexandria and Benares, a major Indian astronomical center. Arabs who later visited India wrote of Brahmagupta (588–660?) as one of the greatest Indian astronomers, who reportedly helped introduce Greco-Indian astronomy to the Arabs.

Unfortunately, most records of this fertile early period of Indian astronomy were destroyed during invasions in the 1100s. The great center at Benares was destroyed in 1194, and various university libraries of Buddhist and other ancient literature were burned in religious wars. A massive observatory—one of the world's five major observatories by the 1700s—was reestablished at Benares in later centuries (and damaged again by invading religious fanatics).

Astronomy in China: An Independent Worldview

In legend, Chinese astronomers were predicting eclipses before 2000 B.C.; scholars estimate the time as being closer to 1000 B.C. Thus Chinese astronomy flourished about the same time as Greek astronomy. Both were probably influenced by the same Middle Eastern cultures. Ancient Chinese observations include records of Halley's comet and the world's best lists of the mysterious "guest stars" (exploding stars that we will encounter in later chapters) from 100 B.C. to the present, still consulted by modern astronomers.

A Chinese statement of this period is apt (quoted by Needham, 1959): "The Earth is constantly in motion, never stopping, but men do not know it; they are like people sitting in a huge boat with the windows closed; the boat moves but those inside feel nothing." Contrast this with Aristotle's view of the same era, which, for most Westerners, put a stationary Earth at the universe's center. Unfortunately, these advanced ideas had little influence on Western astronomy until after the Renaissance.

Native American Astronomy: Indian Science Cut Short

Many people underestimate the sophistication of American Indian civilizations. The highest levels were reached in Central America, where the Mayans, around A.D.

Figure 2-7 A prehistoric American Indian observatory, Casa Grande, Arizona, dating from c. A.D. 1350. Windows, originally on an upper floor in this four-story adobe structure, were cut in different shapes, apparently to facilitate astronomical observations. The view through the window at sunset on the summer solstice shows it was built so that the sunset position on the horizon was revealed by a diagonal view through the cylindrical shape. This orientation allowed determination of the day of the solstice each year. Similar methods were used by Pueblo Indians in historic times to calibrate their calendars. (Photo by author.)

400, developed a written language, used a complex calendar, recorded positions of planets, and predicted eclipses. Mayan astronomy was well organized and state-supported. One inscription records a conference of astronomer-priests at Copán, Honduras (probably on May 12, 485) to discuss their calendar system. The Mayans passed on much of their protoscience to later Central American cultures, but most of it remains unknown because the Spanish conquerors were determined to obliterate "heathen" culture and burned most of the Mayan manuscripts in 1562. One of three priceless Mayan manuscripts that still survives is a record of observed and predicted solar eclipses, motions of Venus, and other astronomical data.

Astronomical knowledge spread from Central America into North America. Even within historic times, priests in southwestern pueblos were charged with studying sunrise and sunset positions to predict solstice dates for religious ceremonies (Zeilik, 1985). Specially cut windows were used for such observations, allowing sight-lines to solstice or other special sunrise and sunset positions on the horizon (Figure 2-7).

The Mayans created a strange calendar that was still in use when the Europeans arrived. They celebrated the beginning of the new year on July 26! What could have led them to this choice? Once again, it was astro-

a   b

Figure 2-8 The Mayan site of Edzna lies almost exactly at the latitude where the Sun passes overhead on July 26, the Mayan New Year's Day. In the main temple courtyard **a** the Mayans erected a stone device to measure this event. At the precise moment when the Sun shines straight down from the zenith, the shadow of the top of the stone covers the entire shaft. At other dates and times (as in photo **b**), part of the shaft is in sunlight. (Photos courtesy V. H. Malmstrom, Dartmouth College.)

nomical observation, but in a non-European tradition. In the tropics, but not in Europe, the Sun can pass directly overhead at noon. Perhaps because their horizons were obscured by dense jungle, the Mayans paid as much attention to *zenith* observations as to observations of the Sun's position on the horizon. Near the latitude of a major Mayan site, Edzna, in Yucatan, the Sun passes through the zenith, on its way to more southern latitudes, at noon on July 26. Recent archaeological studies show that Edzna was a major city of some 20,000 people in the first centuries A.D. In the courtyard in front of the main, five-story pyramid, a cleverly designed stone pedestal (Figure 2-8) allowed priests to measure the important "New Year's Day" when the Sun passed through the zenith (Thomsen, 1984; Malmstrom, 1987, private communication). Probably it was in or near this prehistoric city that early Mayan astronomers first codified, and then ceremonialized, July 26 as their New Year's Day.

The Mayans' "alien" astronomy is perhaps the most fascinating example of incipient Native American science. Although it survived until historic times and produced written records of complex planetary observations, astronomical conferences, eclipse predictions, and calendars, all of this was stamped out by the zealous Europeans who were destroying non-Christian practices.

SUMMARY

We have seen how various cultures, preliterate and technological alike, moved toward certain concepts about Earth as a world among other worlds in space. Some of these concepts were purely practical; some, abstract. These movements came in fits and starts and were scattered throughout the world. Progress toward knowledge has not been continuous. Cultures have advanced and regressed, depending on their stability and vigor.

Many of the discoveries reviewed here dealt with the relation of Earth, the Moon, and the planets. Table 2-1 furnishes a good review of key observations by the Greeks. Their advances, among all those of antiquity, were most important in influencing Western scientific thought. Although models of the universe differed among cultures, all moved toward discovering astronomical relationships.

CONCEPTS

cosmology	subtend
science	resolution
degree	terminator
minute of arc	parallax
second of arc	precession
linear measure	epicycle theory
angular measure	

PROBLEMS

1. How critical can we be of early theorists who believed the Earth is at the center of the universe? Explain, considering the following questions:

a. Did they have any basis for not putting the Earth at the center?

b. Did either possibility fit the available observations?

c. Did any of the Greeks *prove* that any celestial bodies do or do not revolve around a central Earth?

2. As the Moon goes through its phases:

a. Why is its terminator usually curved?

b. At what lunar phase is the terminator straight?

c. At what lunar phase is the terminator not seen?

3. Do you think Aristotle's faith in symmetry and perfection helped or hindered his investigations of the universe?

4. How does Hipparchus' discovery of precession prove the existence of earlier, careful astronomical records of stars' positions?

5. Do you think destructive events such as the pillaging of the Alexandrian library or the sacking of the observatory and library at Benares are significant or insignificant in world history? (The answer, of course, requires that you define *significant*.)

6. Does the Sun pass through the zenith every day on the equator? If not, on what dates does it do so?

PROJECTS

1. Using a distant, strong light source such as the Sun or a light bulb, a small ball to represent the Moon, and your eye to represent a terrestrial observer, show that crescent phases of the Moon prove that it passes between the Earth and the Sun.

2. If you travel far enough during vacation to change your latitude significantly, compare measurements of the elevation of the North Star (or the Sun during daytime, taking care not to stare directly at it) made from various points in your trip. A sighting device like that described for the projects in Chapter 1 can be used. (Point the device at the Sun by watching its shadow; don't look directly at the Sun.) How accurately can latitude be determined in this way? Measure the number of kilometers or miles corresponding to your change of latitude, and estimate the circumference of the Earth. (This method is similar to Eratosthenes'.) The project can be done as a class effort, with different people's reports of elevation angles plotted against their latitude to give a curve showing how elevation angle changes with latitude. Coordinate with your instructor.

3. With a camera and fairly fast black-and-white film, such as Tri-X, make time exposures of the night sky. Try different exposures, such as 1 min, 5 min, and 1 h. Make one series including the North Star, one toward the eastern or western horizon, and one toward the southern horizon. Explain the patterns made by the trails.

4. During a camping trip or late-evening outing, pick an equatorial constellation in the sky and follow its motion. Make a series of sketches at different hours, showing its position with respect to the horizon. Do the same for a circumpolar constellation and contrast the results.

5. Using star maps, trace out the position of the celestial equator in the sky. Compare this with the position of the Sun's path, the ecliptic.

CHAPTER 3

Discovering the Layout of the Solar System

Until 1500, the planets were just orbs of light moving among the stars—gods in the sky. A revolutionary change in this view came from Europe in the 1500s and 1600s. As a result of that change, which is called the Copernican revolution, we now conceive of the **solar system** as a system of worlds with the Sun in the center and all the planets orbiting around it.

The change in our conception of the planets came from an explosion of knowledge during the Renaissance—the two or three centuries of intense intellectual exploration culminating in the 1500s.

INFERIOR AND SUPERIOR PLANETS

How did scholars begin to map the layout of the solar system? Their first clue came from the fact that Mercury and Venus always lie within an angle of some tens of degrees of the Sun and never appear on the opposite side of the sky from the Sun. This indicates that they lie generally at less distance from the Sun than Earth does. For this reason, Mercury and Venus came to be called **inferior planets,** whereas all the other planets, more distant from the Sun than Earth is, came to be called **superior planets.**

This explains why Mercury and Venus are prominent in the sky only at dawn and dusk, as seen in Figure 3-1. The terms *"morning star"* or *"evening star"* apply to them at these times, respectively. These terms are in quotes because they are misnomers. They usually refer to the brightest *planet* in the evening sky, not to a *star.* Mercury is rarely very prominent, but Venus is often by far the brightest "starlike" object in the sky, as shown in Figure 3-1.

Proof of the fact that Mercury and Venus can pass between the Sun and the Earth is provided by **transits,** or passages of an inferior planet directly between Earth and Sun so that the planet is silhouetted against the

Figure 3-1 Venus as the "morning star." Venus often appears brighter than any other planets or stars, as seen from Earth. Because its orbit lies between Earth and the Sun, Venus never appears very far from the Sun in our sky and is thus often prominent at dawn or dusk. In this pre-dawn view, the Sun is somewhat below the horizon. (25-s exposure with a fixed 35-mm camera, 24-mm wide-angle lens at f2.8 on 3M ASA 1000 film, using a "star filter" that artificially creates radiating crosshairlike lines around light sources. (Photo by author.)

Sun's disk. Mercury is too small to be seen at this time without a telescope, but a sharp observer looking through fog or smoked glass can see Venus as a tiny black spot moving across the Sun. Such transits may have helped early astronomers to estimate relative positions of the planets. The next transit of Mercury is November 14, 1999, and the next transit of Venus, June 8, 2004.

PROBLEMS WITH THE PTOLEMAIC MODEL OF THE SOLAR SYSTEM

From the time of Ptolemy to the 1500s and 1600s, most astronomers accepted Ptolemy's model of the solar system, which placed Earth in the center with the Sun and planets moving around it. As shown in Figure 3-2, Ptolemy took into account observations of Mercury and Venus by placing their orbits between Earth and the supposed orbit of the Sun. As shown in the figure, the **Ptolemaic model** assumed that each planet moved in a small circular path, or **epicycle,** whose center moved in turn on a circular orbit around Earth. The main purpose of Ptolemy's theory was to allow prediction of the positions of planets in the sky for the use of navigators, astrologers, and others. For this purpose, the theory worked well enough for more than a thousand years. By adjusting the sizes and rates of motion of the epicycles and circular orbits, Ptolemaic scientists could compute future positions of planets quite accurately.

As observations got better, however, scientists found that they had to make increasingly complicated adjustments to the theory in order to get correct answers. They even had to add tiny epicycles onto the larger epicycles. In 1252, Spain's King Alfonso X funded a special almanac of predicted planetary positions. Remarking on the complexity of the calculations formulated by his astronomers, he reportedly commented that had he been present at the creation, he could have suggested a simpler arrangement.

In 1340, the English scholar William of Occam proposed what became a famous principle, with the peculiar name of **Occam's razor.** It was called a "razor" because it helped scientists in any field cut through the thickets of competing theories. In brief, it said

> **Among competing theories, the best theory is the simplest theory—that is, the one with the fewest assumptions.**

By the 1500s the best scientists had noted that older predictions of planetary positions were now in error by a degree or so, and they began to wonder if there might not be a simpler theory that could give better results than the patchwork Ptolemaic assembly of epicycles. One observer sought a new model that he said might be more "pleasing to the mind." This observer was Nicolaus Copernicus, father of the Copernican revolution.

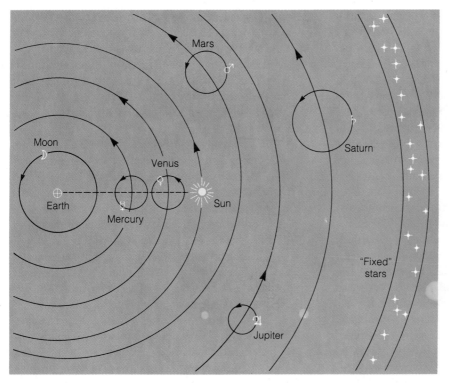

Figure 3-2 The solar system as it might have been conceived by a Ptolemaic astronomer between A.D. 100 and 1500. The symbols on the chart are ancient astronomical (and astrological) symbols for the planets. Ptolemaic astronomers thought that the Sun and other planets moved around a stationary, central Earth, in circular orbits. Superimposed on the larger orbits were smaller circular motions called epicycles.

THE COPERNICAN REVOLUTION

The **Copernican revolution** was an intellectual revolution that abolished the old theory of an Earth-centered universe with the discovery that the Sun is at the center of the solar system, with Earth moving around it. The Copernican revolution took about a century and a half, from roughly 1540 to 1690. It involved five very famous scientists: Copernicus, Tycho, Kepler, Galileo, and Newton.[1]

Copernicus' Theory

Nicolaus Copernicus (Figure 3-3) was born February 19, 1473, the son of a Polish merchant. During his university education in Italy, he became excited by the surging scientific thought of that country. He associated with several astronomers and mathematicians and made his first astronomical observations at age 24. A few years

---

[1]For historical reasons, Tycho Brahe and Galileo Galilei are commonly referred to by their first names.

Figure 3-3 Nicolaus Copernicus (1473–1543). Copernicus is holding a model with a central Sun circled by Earth, and Earth by the Moon. (The Bettmann Archive.)

later, a cathedral post gave him the economic security to continue his observations. At age 31 he observed a rare conjunction, or passage of planets close to each other as seen in the sky, that brought all five known planets as well as the Moon into the constellation of Cancer. He found that their positions departed by several degrees from an earlier set of Ptolemaic predictions.

Familiar with several classical alternatives to Ptolemy's system, Copernicus soon realized that the solar system would be simpler and the prediction of planetary positions easier if the Sun were placed at the center and Earth placed as one of the Sun's orbiting planets. In 1512 he circulated a short comment (*Commentariolus*) containing the essence of his new thesis: The Sun is the center of the solar system, the planets move around it, and the stars are immeasurably more distant. This comment was not widely distributed, however, and few of Copernicus' acquaintances realized that his work would scandalize and revolutionize the medieval world.

He continued his studies but, fearing controversy, delayed publication for many years. Finally, encouraged by visiting colleagues, including some in the clergy, he allowed the *Commentariolus* to be more widely circulated. News of Copernicus' work spread rapidly. Late in life, Copernicus prepared a synthesis of all his work, *De Revolutionibus (On Revolutions,* 1543). In this book he laid out and explained the evidence about the solar system's arrangement:

Venus and Mercury revolve around the Sun and cannot go farther away from it than the circles of their orbits permit. . . .

According to this theory, then, Mercury's orbit should be included inside the orbit of Venus. . . .

If, acting upon this supposition, we connect Saturn, Jupiter, and Mars with the same center, keeping in mind the greater extent of their orbits . . . we cannot fail to see the explanation of the regular order of their motions.

This proves sufficiently that their center belongs to the Sun.

Now the stage was set for turmoil. Church officials and most intellectuals held that Earth was at the center. The printer of *De Revolutionibus,* a Lutheran minister, had tried to defuse the situation by extending its title to *On the Revolutions of Celestial Orbs,* as if to imply that the Earth was not necessarily included. He had also inserted a preface stating that the new theory need not be accepted as physical reality but could be seen merely as a convenient model for calculating planetary positions. From a philosophical viewpoint, we might accept this, but it did not deter medieval critics. Already Copernicus had come under fire from Protestant fundamentalists: In 1539 Martin Luther had called him "that fool [who would] reverse the entire art of astronomy. . . . Joshua bade *the Sun* and not the Earth to stand still."

In a world of strong dogmas, tampering with established ideas is dangerous. The Reformation era of ideological clashes was no exception. In the 1530s Michael Servetus had been criticized for his writings on astrology and astronomy; in 1553 he was burned at the stake as a heretic for professing a mysterious theology that offended both Protestants and Catholics.[2]

Copernicus himself missed the height of the violent debate. The first copies of his book were reportedly delivered to him on the day of his death, in 1543, at age 70. But the Copernican revolution was under way.

In 1575, a correspondent wrote to the astronomer Tycho Brahe: "No attack on Christianity is more dangerous than the infinite size and depth of the heavens." Was the Earth to be taken as merely a minor province of the universe? Where was heaven? In 1616, the Catholic church banned reading of *De Revolutionibus* "until corrected." It was corrected in 1620 by removing nine sentences asserting that it contained actual fact, not just theory.

These sad incidents illustrate the continuing problem of reconciling differences. As Will Durant wrote:

The heliocentric astronomy compelled men to reconceive God in less provincial, less anthropomorphic terms; it gave theology the strongest challenge in the history of religion. Hence the Copernican revolution was far profounder than the Reformation; it made the differences between Catholic and Protestant dogmas seem trivial.

Tycho Brahe's Sky Castle

Tycho Brahe (Figure 3-4) was a flamboyant naturalist who wore a silver nose to cover a dueling mutilation.

---

[2]Both Protestants and Catholics were involved in outrageous suppression. John Calvin masterminded Servetus' execution, although, in a fit of moderation, he recommended beheading instead of burning. Servetus, a man of wide learning and varied interests, had improved geographic data on the Holy Land and discovered blood circulation in the lungs.

Figure 3-4   Tycho Brahe (1546–1601), as shown in an old print. Silver plate on his nose covers a dueling scar.

Figure 3-5   Johannes Kepler (1571–1630). (The Bettmann Archive.)

With funds from the king of Denmark, he built the first modern European observatory, named Uraniborg (Sky Castle), at his island home near Copenhagen. From his observations, all made with the naked eye (the telescope had not yet been invented), he made catalogs of star and planet positions. By demonstrating that stars and other bodies show no angular shift in position as our position shifts with the rotation of Earth, Tycho proved that stars and planets were many times farther away than the Moon, for which he *could* detect a shift (called *parallax*).

At age 16 Tycho noticed errors in predicted planetary positions in the same Alfonsine Tables that Copernicus had used. At age 25, in 1572, Tycho saw a temporarily bright exploding star (further discussion in Chapter 19). By demonstrating that it had no parallactic shift, he disproved the popular belief that it was an object in the Earth's atmosphere or near the Earth–Moon system. In 1577 he observed a bright comet and showed that it too was a remote object, far beyond the Moon.

These discoveries were critical in overturning pre-Copernican theories. They meant that new objects could appear in the supposedly unchangeable heavens and that planets could not be attached to crystalline spheres, because such spheres would be smashed by the comets.

These observations inspired Tycho to catalog the precise positions of the stars and planets, which he did between 1576 and 1596. Unable to convince himself that the Earth could move, Tycho invented a compromise solar system in which the Earth was central and stationary, but other planets were placed in the correct sequence from the Sun.

His pension withdrawn by the king of Denmark, Tycho moved to Prague in 1599, where he was joined in 1600 by a 30-year-old assistant named Johannes Kepler. When Tycho died in 1601, Kepler inherited the great compendium of Tycho's data and its potential for fruitful analysis.

Kepler's Laws

Devoutly religious, Johannes Kepler (Figure 3-5) believed that the universe must be governed by simple mathematical harmony. Many of the...

first went to work on the orbit of Mars, whose movement had plagued astronomical theorists since Ptolemy. He found something astonishing: After all the centuries of debate over the arrangement of circular orbits, the orbit that fitted Mars' motion best was not a circle at all, but an ellipse. **Ellipses** are roughly egg-shaped figures that can range from nearly circular to highly flattened, elongated loops. Each ellipse is symmetric around two inner points called **foci** (singular **focus**). Kepler found that Mars' orbit is an ellipse and that the Sun lies exactly at one focus. Eventually, this was found to be true of every planet's orbit. Although the planets' orbits are elliptical, they are only slightly so. That is, they look nearly like circles, and that is why the Ptolemaic theory worked as well as it did.

Kepler went on to discover two other related principles, and these *three laws of planetary motion* were published in two books, *New Astronomy* (1609) and *The Harmony of the Worlds* (1619). **Kepler's laws** describe how the planets move (without attributing this motion to any more general physical laws), show that the Sun is the central body, and allow accurate prediction of planetary positions:

> **1.** **The shape of each planet's orbit is an ellipse with the Sun at one focus.**
>
> **2.** **The line between the Sun and the planet sweeps over equal areas in equal time intervals, as the planet moves around the Sun.**
>
> **3.** **The ratio of the cube of the semimajor axis to the square of the period (of revolution) is the same for each planet.**[3]

In the case of the Earth, the semimajor axis, or average distance from the Sun, is defined as 1 **astronomical unit** (AU). Other planets' distances are measured in multiples of this unit.

Kepler's laws not only described the solar system more accurately and simply than Ptolemy's theory, but they also allowed more accurate *prediction* of the positions of planets in the sky from year to year. It is interesting to note that Kepler's first law demolished the old Greek idea that planets could move only in circles. The

second law demolished their idea that velocity had to be uniform, because this law requires that planets move faster when they are closer to the Sun and slower when they are farther away.

Galileo's Observations

Kepler's laws might not have gone so far in establishing the Copernican model of the solar system had it not been for the contemporary invention of the telescope and for extensive observations by an Italian scientist, **Galileo Galilei** (Figure 3-6). Unlike Kepler, Galileo had a superbly practical turn of mind. For example, after reportedly watching the regular swing of a lamp in the Pisa cathedral, he applied the periodic motion of the pendulum to regulate clocks. As early as 1597, Galileo wrote to Kepler: "Like you, I accepted the Copernican position several years ago. . . . I have not dared until now to bring [my writings on the subject] into the open."

Galileo perfected the telescope and began astronomical observations with it in late 1609. By 1610 he had made some of the most important observations until that time of the solar system. For example, he found four satellites orbiting Jupiter—proving at last that some bodies do not revolve around the Earth.

Moreover, Galileo's telescope showed that the planet Venus undergoes a variety of phases, from crescent to

of an ellipse is its longest diameter; the semimajor
use most planets in the solar system have
emimajor axes of their orbits are essen-

Figure 3-6 Galileo Galilei (1564–1642). (The Bettmann Archive.)

nearly full. To have a full phase, Venus would have to be on the far side of the Sun. In the Ptolemaic theory, Venus' epicycle was wholly between the Sun and Earth, however, implying that only crescent phases could exist. *Here, then, was proof that the Ptolemaic model of Venus' orbit was wrong.* The Copernican model, however, fit the observation. As a final example, Galileo discovered mountains on the Moon and emphasized that the Moon was a *world,* with geological features, like Earth. These discoveries electrified European intellectuals.

Because Galileo wrote in Italian rather than Latin, he built a popular following outside the universities. Academics and churchmen saw in him a threat, and Galileo soon found himself being attacked from local pulpits. His invitations to critics to look through his telescope and see for themselves led nowhere. Some looked and said they saw nothing; some refused to look; some said that if the telescope had been worth anything, the Greeks would have invented it.

From 1613 to 1633 Galileo was in frequent contact with church authorities, even in Rome. In 1616 a cardinal ordered Galileo not to "hold or defend" Copernican theory, though he could discuss it as a "mathematical supposition." In 1632, Galileo's great book *Dialogue of the Two Chief World Systems* appeared. It featured a fictionalized debate between Copernican and Ptolemaic advocates. The next year, 69-year-old Galileo was ordered to Rome to stand trial before the Inquisition for teaching Copernican theory.

The Inquisition jurors were inclined to be lenient only if Galileo repudiated his work. The elderly Galileo saw no point in getting himself killed; his book was already published and he had faith that intelligent people could see plain truth through telescopes or in print. So he recited a prepared recantation and was sentenced to prison, a sentence commuted by the pope to house arrest on Galileo's own estate, where he died in 1642.

Newton's Synthesis

In spite of the Inquisition, evidence for a Sun-centered solar system accumulated so rapidly that the Copernican revolution was almost complete. The main element still lacking was an overall theoretical scheme that would draw together Kepler's empirical laws of planetary motion into a concise physical explanation of the behavior of the solar system. To be intellectually satisfying, this theory needed to start with a few universal principles and show that Kepler's elliptical orbits and the Galilean satellite motions *had* to exist as a consequence of these princi-

Figure 3-7 Isaac Newton (1642–1727). (The Granger Collection, New York.)

ples. The man who achieved this synthesis—the man usually considered the greatest physicist who ever lived—was **Isaac Newton** (Figure 3-7).

Isaac Newton was a father of physics and astronomy. Between the ages of 23 and 25, while attending Cambridge, he almost single-handedly developed calculus, discovered the principle of gravitational attraction and certain properties of light, and improved the design of the telescope. Newton once said that he made his discoveries "by always thinking about them," a trait that no doubt contributed to his reputation for absentmindedness.

At age 41, Newton began writing his famous *Principia,* a revolutionary compendium of physics, and published it three years later in 1687. He became president of the Royal Society at 60, died at 84 in 1727, and was buried in Westminster Abbey. Of Newton, Alexander Pope wrote:

Nature and Nature's laws lay hid in night:
God said, "Let Newton be!" and all was light.

The Moon must be attracted to Earth by some force, Newton thought, because it does not travel in a straight line, as it would if no force were pulling on it. Reasoning in this way, Newton was able to deduce one of the most important discoveries in the history of science. It is called **Newton's law of universal gravitation:**

> **Every particle in the universe attracts every other particle with a force proportional to the product of their masses and inversely proportional to the square of the distance between them.**

This statement means that if you could double the mass of the Sun, the Sun's gravitational attraction to Earth would double; but if you doubled the distance between them, the force on Earth would *decrease* by a factor of four (the square of two). Similarly, if you tripled the Sun's mass, the force would triple, but if you tripled the distance, the force would decrease by a factor of nine.

We explore some of the ramifications of this law in the next chapter. Here we will simply note that the attraction of every particle for every other particle gave, at last, a quantitative explanation of why the planets follow orbits around the Sun instead of flying off into interstellar space in a straight line: The massive Sun, in the center of our solar system, *attracts* the planets. If the Sun suddenly vanished, the planets would indeed fly away!

Once Newton discovered that masses attract each other gravitationally, he concluded that gravity is the *only* force that keeps the planets moving around the Sun. But this conclusion led to another riddle for natural philosophers of the day: How could the Sun influence the planets in their orbits if it always stayed at such a great distance from them?

Newton answered these questions of "action at a distance" with three simple laws of motion and his law of gravity. These laws are the basis of most modern physics except for the corrections made necessary by work on relativity during the present century. These laws were enunciated in Newton's book *Principia* in 1687. *They are quite unlike Kepler's three laws.* They are not merely empirical rules based on observation, but *fundamental postulates* from which Kepler's laws and many other phenomena can be predicted. **Newton's laws of motion** are:

> **1. A body at rest stays at rest and a body in motion moves at constant speed in a straight line unless a net force acts on it.**

> **2. For every force acting on a body, there is a corresponding acceleration proportional to and in the direction of the force and inversely proportional to the mass of the body. In other words, force = mass × acceleration.**

> **3. For every force (sometimes called action) on one body, there is an equal and opposite force (called reaction) acting on another body.**

The properties of elliptical orbits follow from Newton's laws. An important exercise in advanced astronomy courses is to derive all three of Kepler's laws from Newton's laws. This exercise shows that if Newton's laws are true, the Copernican theory and Kepler's laws also have to be true. Newton's laws thus tidied up the miscellaneous observations of preceding centuries and completed the Copernican revolution.

By the time of Newton's death, at age 84 in 1727, the solar system was conceived essentially as we see it today, lacking only the discovery of the three outer planets, Uranus, Neptune, and Pluto. Subsequent astronomical observations have shown that Newton's laws also apply in all other parts of the universe that we can see. They correctly predict properties of certain pairs of stars that orbit around each other and properties of stars orbiting around galaxies.

BODE'S RULE

A curious relationship discovered by the German astronomer Johann Titius and popularized by his colleague Johann Bode, in 1772, is helpful in memorizing the distances of the planets from the Sun. **Bode's rule** is: Write down a row of 4s, one for each planet, and add the sequence, 0, 3, 6, 12, 24, and so on, doubling each time, as shown in Table 3-1. Then divide the sums by 10 to get the number of astronomical units between each planet and the Sun. Because Bode's rule, unlike Kepler's laws, does not necessarily follow from Newton's laws, it is considered more descriptive than explanatory, and not a law of physics. It is merely a handy way to remember planetary positions.

In 1781 Bode's rule was strengthened with the discovery of Uranus at its predicted position, about twice as far from the Sun as Saturn. Astronomers then noted that the rule predicted a planet between Mars and Jupiter. German observers, nicknamed "celestial police,"

TABLE 3-1

Bode's Rule: Distances of Planets from the Sun

	Mercury	Venus	Earth	Mars	Asteroids	Jupiter	Saturn	Uranus	Neptune	Pluto
	4	4	4	4	4	4	4	4	4	4
	0	3	6	12	24	48	96	192	—	384
Predicted distance	0.4	0.7	1.0	1.6	2.8	5.2	10.0	19.6	—	38.8
Actual distance	0.4	0.7	1.0	1.5	2.8	5.2	9.5	19.2	30.0	39.4

Note: All distances are expressed in astronomical units (1 AU = average distance of Earth from the Sun).

set out to find the missing planet, but an Italian observer beat them to it. The Italian discovered the first and largest asteroid, Ceres, at just the right distance! Ceres might have become known as the smallest planet, except that within a few years the "celestial police" found three more asteroids at about the same distance. Today we know of thousands of asteroids between Mars and Jupiter. The asteroids more or less confirmed Bode's rule.

In 1846, the discovery of Neptune somewhat reduced the credibility of Bode's rule by putting a planet closer than predicted, though the 1930 discovery of Pluto did put a small planet at roughly the next predicted position.

THE SOLAR SYSTEM AS WE KNOW IT TODAY

A fantastic philosophical and scientific advance was wrought by Copernicus, Tycho, Kepler, Galileo, and Newton, as summarized in Table 3-2. To see its effect, compare the Ptolemaic system shown in Figure 3-2 with Figure 3-8, the solar system as perceived today. Earth is no longer at the center but relegated to an orbit like any other planet's. None of the planets follow epicycles. Jupiter has moons of its own and Saturn has rings.

The solar system as we know it today is more complex and interesting than the simple system of Sun, Earth, and five other planets known to the ancients. First, we see the orbits of eight major planets (solid curves) spaced evenly and more or less obeying Bode's rule. Four small planets, including Earth, are close to the Sun; four much larger "giant" planets orbit further away from the Sun. We will study the properties of all these planets in later chapters.

Second, we note a number of interplanetary bodies. Many of these are asteroids—rocky bodies only a fraction the size of the smallest planets and located mostly in a belt between Mars and Jupiter. Others are comets. A typical comet's orbit is shown at top left in Figure 3-8. Note its high ellipticity.

TABLE 3-2

Five Key Figures in the Copernican Revolution

Nicolaus Copernicus	1473–1543	Proposed circular motions of planets around Sun
Tycho Brahe	1546–1601	Recorded planets' positions
Johannes Kepler	1571–1630	Analyzed Tycho's records; deduced elliptical orbits and laws of planetary motion
Galileo Galilei	1564–1642	Made telescopic discoveries supporting Copernican model
Isaac Newton	1642–1727	Formulated laws of gravity and motion and used them to explain elliptical planetary orbits

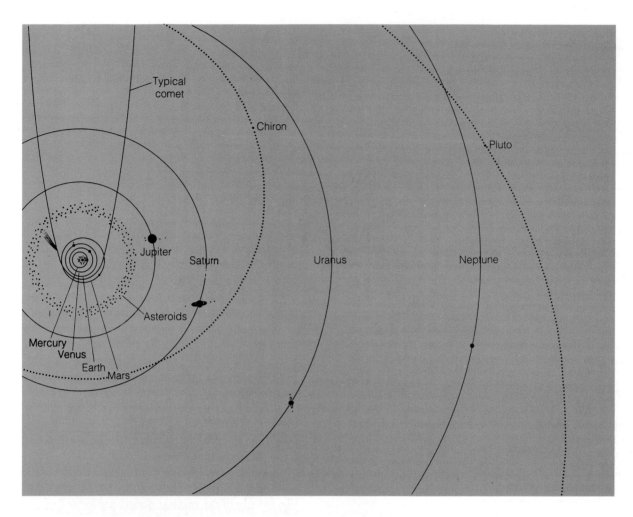

Figure 3-8 The arrangement of the solar system as it is now known, shown approximately to scale. Orbits of the nine main planets and a typical comet are shown, plus positions of typical asteroids. The orbits of Pluto and the unusual asteroidlike object Chiron are also shown as dotted. Note eccentricities of the orbits of Pluto, Mercury, and Mars.

Third, we note that Pluto has a considerably more elliptical orbit than other planets. Moreover, Pluto crosses inside Neptune's orbit and is the only planet that crosses another's orbit. Pluto is also much smaller than the other planets—even smaller than our Moon and only about 60% as big as Mercury. Its orbit is also unusually inclined, running well "above" the plane of the solar system, where the other planets' orbits are concentrated. Pluto is thus unusual among the planets. In 1977 astronomers discovered another object (now called Chiron) one-tenth as big as Pluto on an eccentric orbit between Uranus and Saturn. Many astronomers

believe that more Pluto-sized bodies may be found in the outermost solar system.

For these reasons, astronomers are beginning to suspect that they should classify Pluto not as a true planet but only as the largest known interplanetary body. After all, Pluto is only about three times as big as the largest known asteroid, a 1000-km-diameter object named Ceres. Thus, although Pluto has been called the ninth planet since its discovery in 1930, it may be "demoted" in the near future, depending on new discoveries. Let us remember that we have not yet learned everything there is to know about the solar system!

SUMMARY

The planets are tiny specks circling the Sun. If you backed off far enough to see the system as a whole, the outer giants would hardly be noticeable and the inner planets would be lost in the glare of the Sun. This conception of the solar system was accepted only after one of the major intellectual upheavals in human history took place about four centuries ago. The key to this Copernican revolution was the work of five scientists listed in Table 3-2, which you should review. Of special importance were Kepler's three laws, which described how planets moved, and Newton's laws of motion and gravity, which revealed the underlying forces that explain Kepler's laws.

CONCEPTS

solar system	Johannes Kepler
inferior planet	ellipse
superior planet	focus
transit	Kepler's laws
Ptolemaic model	astronomical unit
epicycle	Galileo Galilei
Occam's razor	Isaac Newton
Copernican revolution	Newton's law of universal
Nicolaus Copernicus	gravitation
Tycho Brahe	Newton's laws of motion
	Bode's rule

PROBLEMS

1. To an observer north of the plane of the solar system, do the planets appear to revolve around the Sun clockwise or counterclockwise? Which way to an observer south of the plane? (*Hint:* All planets revolve in the same direction as Earth rotates, from west to east.)

2. Can the full moon ever occult Venus (pass between Earth and Venus)? Draw a sketch to show why or why not.

3. One of Galileo's discoveries was that Venus, like the Moon, goes through a complete cycle of phases, from narrow crescent to full. How did this disprove the Ptolemaic model, which restricts Venus to positions between the Sun and Earth?

4. Which planet can come closer to Jupiter: Earth or Uranus? (*Hint:* Use Bode's rule.)

5. If Venus, Earth, Mars, and Jupiter are in a straight line on the same side of the Sun, what phenomenon does an observer on Earth see? What would an observer on Mars see? An observer on (or near) Jupiter?

6. Do you think Copernicus and other major figures in the Copernican revolution would have viewed themselves as social or political revolutionaries? Why? Contrast the causes of scientific revolutions and political revolutions. What roles do factual discoveries, strong personalities, controversy, and publicity play in each? Which are more dangerous to human life? Which have the more lasting effects?

PROJECTS

1. On a piece of typewriter paper try to make a scale drawing of the orbits of the planets, based on orbital radii listed in Table 8-1. Which orbits are hard to show clearly? What size dots could represent Earth and Jupiter at this scale?

2. Make a large wall chart showing the orbits of the planets out to Saturn in scale. Mark the motion of each planet in one-day or one-week intervals, as appropriate. Using the *Astronomical Almanac*, an astronomy magazine, or a similar source, locate each planet's relative position for the current date. Update the chart during the semester and watch for alignments, days when planets seem to pass near each other as seen from Earth. Confirm these in the night sky.

3. If a planetarium is nearby, arrange for a demonstration of planetary motions from night to night.

4. With a telescope of at least 2-in. aperture, examine Jupiter and its four prominent satellites discovered by Galileo. (On any given night, one or more satellites may be obscured by Jupiter or its shadow.) By following Jupiter from night to night, confirm that the satellites move around Jupiter. This proves Copernicus' statement that not all celestial objects move in Earth-centered orbits.

Two Methods for Exploring Space: Understanding Gravity and Understanding Light

NASA's Infrared Telescope Facility is only one of numerous state-of-the-art observatories at 14,000-ft elevation on the summit of the extinct volcano Mauna Kea, in Hawaii. Sun and wind carve grotesque shapes in winter ice deposits. (Photo by author.)

CHAPTER 4

Gravity and the Conquest of Space

In Part A of this book we saw how humans arrived at a conception of Earth moving around the Sun among the planets. But how can we proceed from there to learn more about the physical nature of the astronomical bodies around us? To explore our space environment, we need to use a variety of tools: both figurative tools, like physical theories of gravity and light waves, and literal tools, like telescopes and spaceships. Note that by *space environment* I mean not empty space but the rich diversity of planets, moons, gas, dust, energy fields, and stars that stretch in all directions outward from Earth.

There are two ways to explore our space environment. We can actually go there, or we can interpret light signals coming from there. In this chapter we will concentrate on the first method: how humans learned to understand, and then overcome, the bonds of gravity, so that we can send instruments and people to distant worlds in space. But some celestial bodies are too far away to visit, and the next chapter will show how we use light waves—nature's messages from space— to gain information about them.

DREAMS OF ESCAPING EARTH

As soon as humans began thinking about the arrangement of worlds in space, they began to dream of being able to leave Earth and soar through the heavens. Amazingly, fictional flights to the Moon appear in literature as far back as Greco-Roman times and recur throughout European history. The Greek satirist Lucian (c. A.D. 190) had one of his characters put on vulture and eagle wings, take off from Mt. Olympus, and fly to the Moon to learn how the stars came to be "scattered up and down the heavens carelessly."

Many fictional accounts of voyages to the Moon and planets can be found throughout subsequent centuries.

Figure 4-1 The rocket was still not recognized as the best mechanism for space travel by the late 1800s, when French artist Gustave Doré made this illustration of a lunar voyage.

As shown by Figure 4-1 from the 1800s, flight to the Moon was not just an idea conjured up by twentieth-century engineers upon hearing the knock of technological opportunity!

NEWTON'S LAW OF GRAVITATIONAL FORCE

One of the greatest accomplishments of Isaac Newton was to recognize some simple principles that describe how gravitational attraction works. This allowed him to calculate the force that one body exerts on another. This breakthrough allows all humans to master their environment in many ways: The astronomer can calculate the orbital motion of a moon around a planet; the rocket scientist can calculate the power needed to lift a cargo into orbit; the civil engineer can calculate the stresses in a bridge spanning a river.

Newton realized that the material in Earth exerts an attraction on any material nearby, whether an apple, a stone, or a person. When he thought about the Moon,

he realized that it, too, must be attracted toward Earth. Because the Moon moves in a curve around Earth, Newton reasoned, a force must be acting on it, and that force must be coming from Earth to keep the Moon circling around Earth. Newton realized that he could calculate from the Moon's known orbital motion how fast it "fell away" from a hypothetical straight line. He then compared this acceleration rate to that of a falling body at the Earth's surface. After gathering accurate data, Newton showed that whereas the Moon is 60 times farther from Earth's center than Earth's surface is, its gravitational acceleration is about 1/3600, or $1/60^2$, of the acceleration experienced at Earth's surface. As mentioned in the last chapter, the force diminishes as the *inverse square of the distance*. Gravitational force is thus said to follow an **inverse square law.**

Many phenomena in nature follow an inverse square law. This is not surprising. Let us look at inverse square laws a little more closely to see why. If either a force or a substance spreads out from a point in straight lines in all directions, it must become less concentrated as it gets farther from that point. Light, radio waves, and water spray from a lawn sprinkler are examples. Newton, in a leap of imagination, concluded that gravity works the same way. Earth (or any individual atom of it) acts like a source of gravity, but the farther away you go from the source, the weaker the force.

The inverse square relation is illustrated by Figure 4-2. Imagine light from a candle shining into a pyramid-like section of space. If the pyramid is cut at the point where its base has an area of 1 cm^2, then all the candlelight entering the tip of the pyramid passes through this square centimeter. Twice as far from the light, the base of the pyramid is twice as wide and so has four times the area, but it receives the same amount of light. The original radiation is now dispersed over 4 cm^2. Thus the base receives one-fourth as much light per unit area at twice any given distance. Similarly, three times as far from the light, the light is dispersed over nine times the area, and the base receives one-ninth as much light per unit area. Hence light radiation follows the inverse square law.

Newton had discovered how distance affects gravity, but he did not know what else affected it. He eventually showed that the gravitational attraction between two bodies is proportional to the amount of material in each body—that is, to its **mass**. The more mass, the more attraction. Thus if the mass of either body doubles, the force between them doubles; but if the dis-

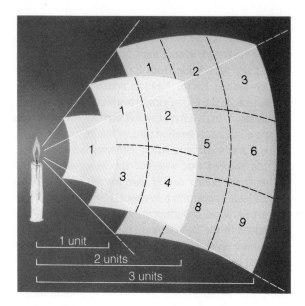

Figure 4-2 The inverse square law. At twice the distance from the source, the light is spread over four times the area. At three times the distance, the light is spread over nine times the area.

tance between their centers doubles, the force drops by a factor of 4.

Mass should not be confused with *weight.* The weight of an object is merely the gravitational force with which the Earth pulls the object against Earth's surface. *Weight* is merely a convenient name for gravitational force. The weight of an object would be different on different planets because the gravity of different planets differs. The mass, or amount of material, of a body remains the same regardless of the body's location.

Circular Velocity: How to Launch a Satellite

Newton realized that an object could be launched into orbit around Earth. His *Principia* contains a diagram of an Earth projectile fired from a cannon on a mountaintop, with the barrel pointed parallel to the ground (Figure 4-3). In keeping with his first law of motion, the projectile would keep moving forward in a straight line except that the force of gravity pulls it toward the ground. If the launch speed is too slow, the cannonball falls to the ground near the cannon (curve *A* in the figure). At a higher speed it travels farther (curve *B*). At a high enough speed, it curves toward the ground, but the surface of Earth, being round, curves away at the same

rate. Thus the projectile never reaches the ground, but travels all the way around Earth and returns to the mountaintop, as shown by cases *C, D,* and *E,* explained in the figure. Those are all elliptical orbits, in accord with Kepler's laws. Case *D* is the special class of ellipse that is a circle; the particular speed at which an object must move parallel to the surface of a body in order to stay in circular orbit around it is called the **circular velocity.**

The farther from Earth or other central body, the less the force of gravity that must be overcome, and therefore the lower the circular velocity. At Earth's surface, 6378 km from Earth's center, the circular velocity is 8 km/s, or nearly 18,000 mph. The Moon, 384,000 km from that same center, moves at only about 1 km/s in its circular orbit.

In short, launching a satellite is a seventeenth-century idea! All we had to do is get the satellite above the atmosphere and fire it parallel to the ground at 8 km/s (Newton's mountaintop would not have been high enough because air resistance would retard the satellite, mak-

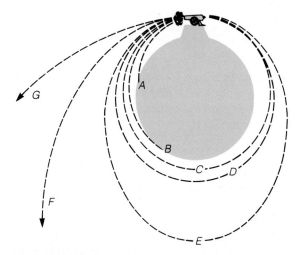

Figure 4-3 Newton realized that a projectile launched parallel to the surface of Earth from a mountaintop could travel varying distances depending on its launch velocity. At low speed, it would drop along curve *A*, hitting nearby. At a higher speed, it could travel halfway around Earth on curve *B*. A slightly higher speed would put it in an elliptical orbit with a low point (or perigee) at *C*. A still higher speed, called circular velocity, would put it in a special type of elliptical orbit, the circular orbit *D*. A slightly higher speed would create an elliptical orbit with the farthest point (or apogee) at *E*. Escape velocity would create a parabolic orbit *F* that never returns to the Earth. A still higher speed creates a hyperbolic orbit *G*, which also never returns.

ing it fall back to the ground; and no shell can be accelerated to the necessary speed in a cannon barrel).

Escape Velocity

Study the orbits in Figure 4-3. Each complete orbit B to E is an ellipse. The point closest to Earth is called the **perigee,** and the farthest point is called the **apogee** (the suffix *-gee* comes from the Greek root for "Earth," as in *ge*ology; *peri-* and *apo-* come from the Greek roots for "near" and "far," respectively). Now, if the projectile is launched at higher and higher speeds, as in the sequence B to E, the apogee gets higher and higher. A high enough launch speed would send the projectile to an apogee far beyond the Moon. A projectile launched at high enough speed would follow an open curve in the shape called a **parabola** (curve F), and would never come back. This unique speed, which allows the object to escape Earth forever, is called the **escape velocity**, or *parabolic velocity*. At a point near Earth's surface, the escape velocity is about 11 km/s; it is less at more distant points. Launched at a still higher speed, a body travels a similar curve called a **hyperbola** (such as curve G) and does not return. Thus a launch speed exceeding escape velocity is called a *hyperbolic velocity*.

Every body in the universe, each planet, moon, or star, has its own gravitational field. Hence a unique escape velocity applies at the surface of each body.

ROCKETS AND SPACESHIPS

How can a body be propelled to circular velocity or escape velocity? Jules Verne, in *From the Earth to the Moon*, imagined using a 900-ft-long cannon and 400,000 lb of explosive. But as just noted, a cannon is impractical.

A more realistic technology has since been devised. A spacecraft must carry its own means of propulsion and operate in the vacuum of space. The second point rules out propeller or jet aircraft. Around the turn of the century, several experimenters and visionaries realized that rockets were ideal. Their use was first recorded in China and Europe in the 1200s.

Rockets work essentially by Newton's third law of motion, that for every action, there is an equal and opposite reaction (see Chapter 3).

For example, if you sit on a wagon and throw a large mass (like a cinder block) out the back, the wagon coasts forward; the force needed to expel the mass causes an opposite force on the vehicle. In the same way, the force used to expel high-velocity gases out the back of a rocket nozzle pushes the rocket forward with equal force. This force is called **thrust**.

The Russian experimenter Konstantin Tsiolkovsky, beginning in 1898, and the American Robert Goddard, in the 1920s, studied and fired rockets. Both were mavericks, however, and their work was almost ignored by their contemporaries. In the 1920s in Germany, Hermann Oberth, who remarked that Verne's book was an inspiration, published several books on rocket-powered space travel. Oberth's work attracted a group of enthusiasts, including Wernher von Braun, whose astronautical experiments were converted into the V-2 guided missile program under the Nazis. At the end of World War II, about 125 German rocket experts, including von Braun, moved to the United States and continued working on rocket technology.

The First Satellites

After several secret postwar studies of satellites had been made, President Eisenhower announced in 1955 that the United States would use a nonmilitary rocket called Vanguard to launch a satellite during the International Geophysical Year (1957–1958). Within days, Soviet scientists announced their plan to launch satellites larger than the American one. This plan was not taken seriously in the West: Americans viewed the Soviets as the Soviets portrayed themselves in their poster art—as unsophisticated, shirt-sleeved tractor drivers.

On October 4, 1957, the Soviet Union astonished the world by launching the first artificial satellite, the 83-kg (184-lb) instrumented sphere. It was named Sputnik I (Russian for "satellite"). In November the half-ton Sputnik II went up, carrying a dog as a biological test. Because it was easily visible to the naked eye when it passed overhead at dusk, Sputnik II electrified the Western citizenry. In December, under hasty orders, American technicians tried to launch a small satellite in one of the Vanguard test rockets, but as millions watched on live television the rocket blew up on the launch pad.

These three months produced a crisis in Western confidence and soul searching in American education. The first American satellites went up in early 1958 and the space age was underway. The Soviets generally launched much larger spacecraft, but the U.S. satellites used more sophisticated instruments, which allowed more data-gathering with less weight.

Figure 4-4 The launch of Vostok I and Yuri Gagarin, the first human to circle the Earth, in April 1961.

The first satellites were designed primarily to probe the nearby environment of space. Among their discoveries were the **Van Allen belts** (doughnut-shaped zones of energetic atomic particles surrounding Earth) and Earth's slight bulge in the Southern Hemisphere, which gives our planet a slightly pear shape.[1]

The First Space Flights of Astronauts

American engineers concentrated on miniaturizing precision instruments, whereas Soviet engineers, lacking the technology to miniaturize, concentrated on rocket power. The Soviets' large, powerful rockets gave them an edge during early space missions. After putting the first probe on the Moon and photographing the Moon's far side for the first time in 1959, the Russians began to test the biological possibilities of space flight with dogs, some of which they recovered from orbit. On April 12, 1961, in a 5-ton craft, a 27-year-old Russian, Yuri Gagarin, became the first person to orbit Earth, which he did in 108 min (see Figure 4-4).

The first American rocket flights with humans aboard followed within months. One scientific result of these flights was to allay fears that the Van Allen radiation belts or meteoroids might prevent human space flight.

THE DECISION TO EXPLORE THE MOON: SCIENCE AND NATIONAL POLICY

Between 1957 and 1961, American planners had roughed out the technical requirements and timetables for a lunar voyage. Though President Kennedy, who took office in January 1961, sought national goals that would spur creative effort, he was at first skeptical about a lunar program, particularly because of the existing Russian lead.

---

[1]Ironically, it is now known that Sputnik II first detected the Van Allen radiation belts. But because the Russians did not have a global tracking network and did not tell other countries how to decode the radio signals, the Russians themselves did not get enough tracking data from Sputnik II to recognize the belts. NASA researcher A. J. Dessler (1984) has noted: "Because of their perceived need for secrecy, the Russians missed making one of the most dramatic discoveries in space science." Instead of being named the Van Allen belts (after an Iowa scientist who built the detectors), they would have been called the Vernov radiation belts (for the corresponding Russian scientist)!

Figure 4-5 Apollo 14 departs, carrying three astronauts on their way to the Moon in 1971. (Photo by author.)

However, he became convinced that only by directly challenging the Soviet space achievement could the United States rise to its own best creative efforts. On May 25, 1961, the goal was set in Kennedy's extraordinary speech before Congress:

The dramatic achievements in space which occurred in recent weeks should have made clear to us all, as did the Sputnik in 1957, the impact of this adventure on the minds of men everywhere. . . .

I believe that this nation should commit itself to achieving the goal, before this decade is out, of landing a man on the Moon and returning him safely to Earth.

After Kennedy's assassination in November 1963, the NASA program, especially the Apollo Moon-landing program (Figure 4-5), became almost a memorial to the president. This prevented funding cutbacks until the program was completed in 1972, after six successful lunar landings. A few proposed additional landings were then canceled and funding for planetary exploration began a long-term decline.

AFTER APOLLO

The last remaining Apollo spacecraft was used in a short-lived effort at global cooperation in space exploration. In 1975, this spacecraft, carrying three American astronauts, linked in orbit with a Russian Soyuz spacecraft carrying two cosmonauts. The explorers shared hand-shakes, conducted joint experiments, and televised pictures of each others' countries to audiences below. The Apollo–Soyuz project was remarkable; the challenges included an initial political agreement between Presidents Nixon and Kosygin in 1972, design of a mating tunnel to dock the two vehicles, travel of engineers and astronauts between the two countries, and learning rudiments of each other's languages by astronauts and technicians. The program proved that if political leaders are willing to set challenging, cooperative goals, technical communities in the involved countries can respond with enthusiasm—a more hopeful situation than having these communities engage in bomb-building!

While the Soviets developed a permanent space station, American engineers developed a fleet of four Space Shuttles designed to fulfill most of the nation's launch needs for scientific, commercial, and military payloads (Figure 4-6). Its first 24 flights, from 1981 through 1985, were highly successful. Milestones included flights of the first American woman and black astronauts; launch of satellites for Indonesia, West Germany, and other countries; flight of the European Space Agency's Spacelab, crewed by a European-American team; and demonstration of an efficient drug-manufacturing technique utilizing weightless conditions. During the twenty-fifth shuttle launch, however, on January 28, 1986, an explosion destroyed one of the four shuttles, killing the racially mixed crew of two women and five men. Nearly all of America's spaceflight eggs had been put in the shuttle basket, to save money, and few rocket boosters were

Figure 4-6 The Space Shuttle gave humans new capability to launch, construct, and repair payloads in space. This vertical view, looking down onto Earth, shows the shuttle Challenger against a background of clouds during a 1983 flight. The U.S. space program was disastrously delayed when this spacecraft was demolished by an explosion during launch in 1986. (NASA.)

being manufactured. Suddenly there was no way to launch America's commercial and scientific payloads into space. A crisis temporarily paralyzed the American space program, which did not gear up again in earnest until the 1988 launches of the redesigned shuttle. In 1988, the Soviets successfully launched their own space shuttle, apparently intended to support their large *Mir* ("peace") space station.

SPACE EXPLORATION: COSTS AND RESULTS

The results of space exploration will be clearer to our grandchildren than to us, but we can at least compare some costs and benefits.

Costs

While we geared up for a space program from 1959 to 1969 (the year of the first lunar landing), the NASA budget (including traditional aircraft research as well as spacecraft development) was about 2.5% of the total U.S. budget. During the American military buildup of the 1980s the NASA budget dropped to only about 1% of the total. In the 1970s and 1980s funding for all scientific research (civilian and military) averaged around 6% of the total U.S. budget. In contrast, two departments—Defense, and Health, Education, and Welfare—*each* spent about one-third of the total budget.

Some critics have asked whether society might be improved by canceling space exploration and spending the money on programs to alleviate such social problems as poverty, illness, malnutrition, and the energy crisis. But as the preceding figures show, the science budget is too small to have much impact even if it were diverted to these areas. Most analyses show that the few percent of the budget devoted to research more than pays for itself—and is indeed the cutting edge that leads us into the technological future that is being simultaneously shaped in Japan, the Soviet Union, and elsewhere.

Practical Results

Seven centuries elapsed between the first use of rockets (probably in a Mongol battle in 1232) and their first scientific application (in an atmospheric research flight conducted by Goddard in 1929). Constructive benefits have come only in recent years.

Research for spacecraft design, as well as for ground-based astronomical instruments, has been an important driving force in developing industries, products, and inventions. These include a host of miniaturized electronics, new imaging devices, and contributions to medical technologies such as CAT scan equipment.

The first weather satellite, launched in 1960, led to nearly continuous monitoring of weather conditions. TV weather reports now routinely display satellite photographs. Thousands of lives and billions of dollars have been saved through the use of these photos to predict storms on land and sea.

Satellite photography has other uses as well. Amazingly enough, in some remote areas of the world—such as the Amazon jungles of Brazil and certain parts of Ethiopia—satellite photos are better than existing maps.

Ultraviolet and infrared satellite sensors are being used in agriculture, forestry, and prospecting. They can spot certain crop blights by subtle color changes before the infestations are detected on the ground. Moreover, they can detect and track pollutants in rivers, coastal waters, and the atmosphere, as seen in Figure 4-7, and locate mineral deposits by "reading" geologic features and subtle soil colorations.

Another important practical consequence of space flight is the communications satellite. Communications satellites, such as the Indonesian satellite launched by the shuttle, have provided the first effective communication between central governments and outlying villages in some Third World countries. Events ranging from Chinese ballet and Olympic games to outbreaks of war are now broadcast between continents. In 1983, U.S. and Russian scientists collaborated on live TV hookups between Washington and Moscow to hold an international conference on nuclear disarmament. Such TV conferences are now more common. Buckminster Fuller and other writers have predicted that this improved communication will strengthen our sense of human community on this planet, producing a "global village"— just as the bickering American colonies eventually came to accept a common identity.

Intercontinental telephone communications are made possible by a group of satellites orbiting far above the equator about 42,000 km (26,000 mi) from Earth's center. With an orbital period of 24 h, such satellites stay fixed in relation to a transmitting station. This is why backyard satellite-TV dish antennas are pointed at a fixed spot in the sky: the position of the distant satellite. A broadcaster's TV signal beamed at the satellite is thus retransmitted to other parts of Earth.[2]

The weightless, clean, and airless environments in space vehicles also offer opportunities to test new man-

Figure 4-7 View of the Texas and Louisiana Gulf Coast shows hazy air over land, contrasting with clear air over ocean (lower right foreground). Plumes of agricultural smoke drift out to sea (right center) and polluted water empties out of Galveston Bay (lower left corner). Researchers are concerned about CO_2 buildup in atmosphere from widespread deforestation by burning, revealed by satellite photos such as this. (NASA photo from 1966 Gemini crewed spacecraft.)

ufacturing techniques in space. These applications are likely to expand in the next decade. The drug processing equipment tested on the shuttle is being developed commercially, and many observers believe it will become the first self-supporting space industry.

Intangible Results

Space exploration has had two important intangible results. First, it creates a cosmic perspective. By no coincidence, the ecological movement emerged just at the time of the first lunar flights. Astronauts, on Christmas Eve 1968, radioed from lunar orbit that the "vast loneliness . . . of the Moon . . . makes you realize just what you have back there on Earth." One of them noted that Earth was "the only color in the universe—very fragile . . . it reminded me of a Christmas tree ornament." The cosmic perspective also opens the possibility of human survival even in the event of a natural or human-caused catastrophe on Earth. Rocket pioneer

---

[2]This system was first proposed in the late 1940s by science fiction writer Arthur C. Clarke. In practice today, it causes curious pauses in transcontinental phone calls. If your call goes by normal ground links from the East Coast to the West Coast, the gap between your question and your friend's answer is about 0.2 s, the normal response time for the brain to frame an answer. However, if your call is beamed up to a satellite about 36,000 km above the surface, then these radio waves must complete two round trips (144,000 km) at the speed of light (300,000 km/s) between question and answer. This takes roughly 0.5 s. Added to the 0.2 s mentioned above, it gives a characteristic awkward pause of 0.7 s between verbal exchanges.

Figure 4-8 Astronaut working outside the U.S. Skylab space station in 1973 gave an early demonstration of our ability to work in space. (NASA.)

Figure 4-9 Astronauts working in the cargo bay of the Space Shuttle in 1983 symbolize our growing ability to operate in the space environment. A cloudscape on Earth's surface fills the background. (NASA.)

this planet will there be unoccupied land, cultural isolation, freedom from bureaucracy, freedom for people to get lost and be on their own. Never again on this planet. But how about somewhere else?

LOOKING TO THE FUTURE

Every planet in the solar system as far out as Neptune, most satellites, and Halley's comet have been studied at close range by spacecraft. New missions are under way. Perhaps the most important to general astronomy is the Hubble Space Telescope, launched by the U.S. in 1991. The telescope was flawed by an optical manufacturing problem—a gross embarassment to U.S. engineering and industry. However, its position above the shimmer of the atmosphere, together with special image-processing techniques, allows it to return data that exceed capabilities of ground-based telescopes. Astronauts may visit it in a few years to install improvements.

In other U.S. program developments, the Magellan robotic probe is orbiting Venus and making superb new maps of its surface structures. The Galileo robotic mission is flying to Jupiter and will drop a probe into Jupiter's atmosphere. Problems with Galileo's antenna may limit the data it can transmit back to Earth. The Mars Observer mission in 1992 will orbit Mars and provide

Wernher von Braun commented shortly after the first lunar landing (Lewis, 1969): "The ability for man to walk and actually live on other worlds has virtually assured mankind of immortality."

Second, space exploration provides a frontier and a sense of adventure that is important to human well-being (Figures 4-8 and 4-9). After several decades during which frontiers seemed to be closing all around us on the Earth, we now see a new frontier opening above us. Princeton physicist Freeman Dyson (1969) noted:

We are historically attuned to living in small exclusive groups, and we carry in us a stubborn disinclination to treat all men as brothers. On the other hand, we live on a shrinking and vulnerable planet which our lack of foresight is rapidly turning into a slum. Never again on

new data on Mars' atmosphere, surface, and composition. A joint U.S.–European mission, called Cassini, in the late 1990s will study Saturn and parachute a probe onto its enigmatic giant moon, Titan.

In the U.S. astronaut program, President Reagan in 1984 called for a permanent U.S. space station, larger than the Soviet Mir station. It has been downsized from initial plans, but may be built in the 1990s (Figure 4-10). In 1989, on the 20th anniversary of the first manned lunar landing, President Bush called for a scientific base on the Moon, leading to human exploration of Mars. Many U.S. space efforts are now geared toward this Moon–Mars initiative.

Other nations are rapidly expanding their space exploration efforts. In the 1980s, the U.S. seemed to lose some of its leadership, with the Challenger disaster and resulting long delays in launch schedules. Japan, the USSR, and the joint European Space Agency (ESA) mounted brilliantly successful robotic probes to Halley's comet in 1986. In 1987, the Soviets tested a huge new rocket, Energia, largest in the world today. In 1988, they returned useful new data from Mars' satellite, Phobos, with a robotic probe that maneuvered close to Phobos before contact was lost due to a mechanical flaw. Also in 1988, they successfully flew a new space shuttle and broke records by having cosmonauts live in the Mir space station for a full year—long enough to fly to Mars. With Mikhail Gorbachev's policy of *glasnost* (openness), the Soviets abandoned their former secrecy and announced an ambitious future program, with heavy emphasis on Mars. They plan a major robotic Mars mission in 1994 or 1996, probably involving small landers and balloons in Mars' atmosphere. They also proposed other missions, including joint U.S.–Soviet robotic missions to bring back samples of Martian soil and rock. Finally, they have discussed cosmonaut flights to the Moon and Mars, beginning around 2000. The future of this ambitious program is in doubt, however, because of the Soviet economic problems.

Many observers believe that space exploration and astronomical projects in space will become increasingly international—partly because of their expense and partly because scientists, business leaders, and politicians are becoming increasingly involved in international projects as global relationships improve. For example, although a pact coordinating U.S. and Soviet cooperation was allowed to lapse in 1982, a new pact was signed in 1987, spelling out areas of future cooperation—extensive exchange of astronomical data, continued cooperation in studying Venus data, and coordination of Mars mis-

Figure 4-10 The United States plans to construct a space station in orbit around the Earth in the 1990s, to be serviced by the Space Shuttle. The Soviet Union already has a space station in orbit, which is being expanded by add-on modules. Such stations allow low-gravity and solar energy experiments, astronomical observations, and construction facilities for space probes. In this view the Sun is hidden behind one module of the station as astronauts construct a solar panel. (Painting by Ron Miller.)

sions, for example. Such pacts might lead to joint American–Russian human expeditions to the Moon and Mars after 2000. Soviet researcher V. Barsukov commented while presenting Soviet plans at a NASA meeting in 1989, "The time is past to argue over who is behind and who is ahead. With cooperation we will both be ahead."

Looking even further into the future, a U.S. presidential National Commission on Space presented a bold vision of the next 50 y—a vision calling for research colonies on the Moon and Mars with space stations and interplanetary vehicles to support them. Their report (National Commission on Space, 1986) compared this

Figure 4-11 A 1984 test of a free-flying astronaut "maneuvering unit" allowed astronauts to fly away from the Space Shuttle to pursue construction and repair activities. During such a flight, the astronaut is an independent satellite of the Earth controlled by small jets of compressed gas. (NASA photo.)

erating plants, though some analysts have expressed concern over the effects of the microwave beams that would carry the energy to the receiving stations. Although the cost of such a solar energy space program would be high (the Apollo program cost about $20 billion over 10 y), the United States now sends more than $50 billion each year to OPEC countries for oil. Once burned, the oil is gone for good, whereas each investment in space gives us new capabilities for the future (Arnold, 1980).

In a more visionary vein, we can imagine that industrial and manufacturing activity in space may reduce pollution on Earth. Because of the availability of iron from asteroids, hydrogen from lunar soil, energy from the Sun, and other space resources, space colonies may not only be feasible but may also pay for themselves (O'Neill, 1977). Someday the Earth may be appreciated as a Hawaii in a universe of Siberias, the only place where we can enjoy the natural environment unprotected. This view may encourage us to take many industrial activities into space, where waste products cannot harm planetary ecosystems. If organized with sufficient political freedom and flexibility, space colonies might even allow experiments in new systems of government, which might defuse political tensions in the closed societies of the Earth. They might also, in the long-term future, ensure human survival in the event of an environmental or military disaster on Earth.

enterprise to the opening of the American West by government survey expeditions and railroads in the 1800s.

These ideas promise the adventure of learning how to carry out activities in a new environment, as shown in Figure 4-11. But they go beyond adventure. Space operations may ultimately help solve energy, raw material, and pollution problems on Earth. For example, large satellites could collect pollution-free solar energy and beam it down to large (10-km) "antenna farms" that would replace power-generating plants on Earth. This proposal could provide energy with less environmental damage than that caused by coal-fired or nuclear gen-

SUMMARY

The idea of carrying people or instruments into space existed long before the technological possibility. Newton's theory of gravitational attraction made it possible to calculate how fast objects would have to go in order to orbit around Earth or escape from Earth. The development of rocket technology early in this century provided the means to reach these speeds. Political and social factors, including strong public support for space exploration, allowed the implementation of these means.

The technology of space travel has been used in four areas: (1) improving communication, weather prediction, and manufacturing processes; (2) helping us search for new nonrenewable resources, while helping us realize that Earth and its supply of resources are finite; (3) exploring other bodies in the solar system; and (4) improving telescopic observations of stars and galaxies far beyond the solar system. Space flight has provided a sense of adventure and exploration unmatched since the Renaissance voyages to the New World.

CONCEPTS

inverse square law	parabola
mass	escape velocity
circular velocity	hyperbola
perigee	thrust
apogee	Van Allen belts

PROBLEMS

1. If an astronaut flying in circular orbit just above the atmosphere wants to escape from Earth, how much additional velocity does he or she need?

2. Rocket engineers typically speak of the difference between one orbit and another in terms of velocity difference, for example, in meters per second. Why is this? (Consult Problem 1 if necessary.)

3. Because Earth rotates, the equatorial regions move at about 1600 km/h (1000 mph). Why is it easiest to launch a satellite in a west-to-east orbit over the equator?

4. You have just rowed a rowboat to a point at rest next to a dock. You step off the rowboat toward the dock.
    a. Why does the rowboat move away from the dock?
    b. In this instance, how are you analogous to the exhaust from a rocket?

5. Do you think a proposal to spend $20 billion to land a human being on the Moon for the first time would be endorsed by Congress or the public if it had happened this year instead of 1961? Explain your answer, accounting for similarities or differences in public attitudes during these two periods.

6. Do you believe that human flights to other planets, satellites, or asteroids will be common in the next century? Do you think this would have positive or negative effects on social progress, intellectual stimulation, the economy, the environment, the availability of energy and materials, and other characteristics of our civilization?

PROJECTS

1. Observe the Moon as it passes above or below a bright star. Remembering that the Moon is about 3480 km in diameter, observe how many of its own diameters it moves in 1 h, and calculate its orbital velocity.

2. Observe an artificial satellite. What illuminates it? Why are artificial satellites commonly seen at dusk or dawn and not at midnight? Observe whether the satellite passes into Earth's shadow and whether it reddens slightly as it does so.

CHAPTER 5

Light and the Spectrum: Messages from Space

If we send astronauts to the surface of the Moon or a robot probe past Neptune, we are reaching out to "touch" other parts of the universe. But most parts of the universe are too far away to reach in person or even with our space probes. To get information about these regions we have to rely on nature's messages from them reaching us in the form of light. In this chapter we will study light in all its forms. We will then apply this information in this and the next few chapters to see how information can be gained about planets. In later chapters we will expand on this information and apply it to stars and galaxies.

THE NATURE OF LIGHT: WAVES VS. PARTICLES

Suppose you stand by a quiet swimming pool where a cork is floating. You disturb the cork by jiggling it. You will notice that this disturbance causes a set of waves to move out across the water. The waves have a certain spacing from one crest to the next, called the **wavelength**. They move at only one fixed speed. As they move past a given point on the water's surface, that point moves up and down; the number of these up-and-down pulsations per second at any one spot is called the **frequency** of the wave.

These waves provide a useful analogy, but not a perfect description, of some **wavelike properties of light**. For instance, just as a water wave expands from its source, light spreads out in all directions from its source. **Visible light** has a tiny wavelength, around 400 to 700 nm (0.0000004 – 0.0000007 m—a nanometer, abbreviated nm, is a billionth of a meter; see Appendix 1). Radiation with still shorter wavelengths exists, but it is too deep-violet for our eyes to perceive; it is called **ultraviolet light**. Light with wavelengths longer

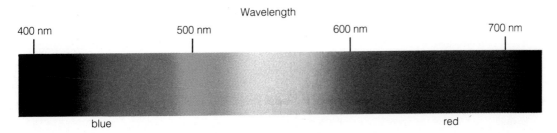

Wavelength

400 nm   500 nm   600 nm   700 nm

blue   red

Figure 5-1 The visible part of the spectrum—an array of colors from violet to red. The wavelength scale at the top is graduated in nanometers, or billionths of a meter.

than red light is too deep-red for the eye to perceive and is called **infrared light**. Extremely long wavelength infrared waves are called radio waves; radio waves are just another form of light that we cannot see. Just as the speed of water waves is constant, the **speed of light** through empty space is constant, about 300,000 km/s (186,000 mi/s). It has been measured, for example, by the time interval necessary to communicate with a distant spacecraft.

Now suppose you shoot a BB at something. The BB has a certain energy. Unlike a wave, which takes a while to pass a fixed object and then die out, the BB delivers all its energy to its target at the moment it hits. The BB is an analog for some **particlelike properties of light**. Neither a wave nor a particle alone is a perfect description of light. For instance, light energy is concentrated in individual units, instead of being spread out along the wave. The energy-containing unit of light is called the **photon**. It can be visualized roughly as a microscopic, BB-like particle that moves at the speed of light, yet has a certain wavelength associated with it. An important concept is that each color of light corresponds to a photon of different wavelength and energy. *The bluer the light, the shorter the wavelength and the more energetic the photon. Similarly, the redder the light, the longer the wavelength and the less energetic the photon.*

From the 1600s to the 1800s, scientists argued whether light is "really" a wave or a particle. For example, Newton argued for a particle theory of light, whereas the Dutch astronomer-optician Huygens (1625–1695) argued for a wave theory. Arguments of Huygens and others established that light definitely has many wavelike properties. For example, if your swimming pool has an inward-protruding corner or wall, you can observe that a water wave bends slightly as it goes past the corner, so that some energy (wave motion) reaches a target slightly behind the wall as seen from the wave's source. Light acts the same way, indicating a wavelike property. A BB would move in a straight line past the wall and not hit such a target. This phenomenon of light bending its path slightly as it passes an edge is called **diffraction**; it limits the sharpness that can be achieved in the image formed by a telescope.

On the other hand, Einstein described one of the important particlelike properties of light in 1905.[1] This is the so-called photoelectric effect—an effect in which light rays can knock electrons out of metal surfaces. The effect can be explained only if the light energy arrives in individual "packets"—that is, photons—rather than being "smeared out" along the wave.

In summary, light spreads out through space something like a wave in which energy is carried by tiny, particlelike photons. Which is light: a wave or a particle? Consider a platypus. It has some ducklike and some beaverlike properties, but it is neither. Similarly, light has some wavelike and some particlelike properties, but it is neither a pure wave nor a pure particle. In microscopic detail, light is a phenomenon not wholly familiar in terms of analogs in our everyday world.

THE SPECTRUM

An arrangement of all colors, in order of wavelength (or in order of photon energy), is called the **spectrum** (plural: spectra). Newton discovered he could see the spectrum of visible light by passing sunlight through a glass prism. The resulting band of colors, cast on a wall in a dark room, looks like Figure 5-1. Water droplets in a rainstorm act as little prisms and allow us to see the

---

[1]Interestingly, Einstein was awarded the Nobel Prize primarily for this work, not directly for his better-known work on relativity.

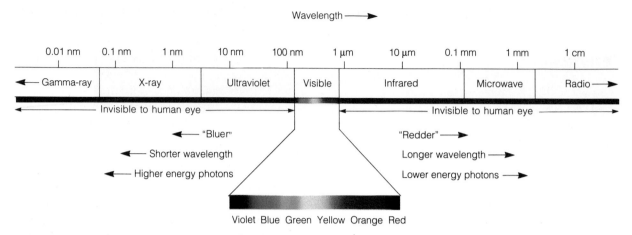

Figure 5-2 The spectrum. Top chart shows a wide range of wavelengths with names attached to different regions. Lower chart shows the order of colors in the narrow range that is visible to the eye.

spectrum—the same arrangement of colors from violet to red—in the rainbow. Newton's discovery proved that "white" light from the Sun is really made up of all the different colors. The spectrum extends to far shorter and longer wavelengths than we can see. Figure 5-2 shows the names attached to different parts of the spectrum. As this figure indicates, the radio waves we receive on our radios and TVs are simply long-wavelength versions of the light we see with our eyes.

Figure 5-3 shows an important way astronomers present this information. The band along the bottom (or in Figure 5-1) shows the kind of picture we would get by photographing a projected spectrum. But suppose we use a little photocell, like a photographer's exposure meter instead; we scan along the spectrum, left to right, to measure the amount of energy (that is, light intensity) at each wavelength. Then, if we plot the results as a graph (as in Figure 5-3), we can see the amount of light of each color. For example, Figure 5-3 shows that sunlight is most intense at greenish-yellow wavelengths. Astronomers often present the spectrum in this way.

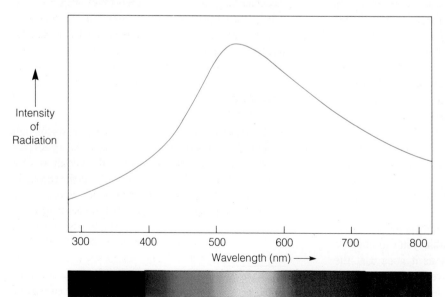

Figure 5-3 A representation of the spectrum of sunlight. The graph gives the intensity, or amount, of light at each wavelength and the corresponding colors along the bottom. The Sun's dominant radiation occurs at wavelengths we perceive as greenish-yellow; we perceive the blend of *all* the Sun's radiation at different wavelengths as yellowish-white.

Astronomers' study of the properties of the spectrum of different objects is called **spectroscopy**. *Spectroscopy is the most important method for learning about remote planets, stars, and galaxies.*

ORIGINS OF LIGHT: ELECTROMAGNETIC DISTURBANCES

Consider again the cork in the swimming pool. We had to disturb it to get a wave. If you put a second cork in the swimming pool and watched as you made a wave, you would see it bob up and down as the wave passed by. The cork would be a *detector* of the passing wave. In the same way, if you had a tiny enough electrically charged particle, and a tiny enough compass, you could detect light waves passing by. As the wave passed, the electric charge would vibrate with the frequency of the wave and the compass needle would oscillate rapidly with this same frequency. For this reason, physicists say that light is a disturbance of electric and magnetic **fields** in space. A field is an important concept in much of physics. A field is said to exist in a certain volume of space if some physical effect can be measured and assigned a numerical value *at every point* throughout that volume. For example, at any point near Earth, Earth attracts a given "test particle" with a measurable force; therefore a gravitational field is said to exist. Similarly, where a compass measures a certain response at any point, a magnetic field is said to exist. Also, when an electrically charged particle experiences a force at any point, an electric field exists.

Light involves pulsations in these electromagnetic fields (coexisting electric and magnetic fields). For this reason, we say that all light, of whatever wavelength— gamma-ray, X-ray, ultraviolet, visible, infrared, microwave, or radio—is **electromagnetic radiation**. The word *radiation* is used for light of any wavelength.[2]

If the second floating cork were hit by a BB or by

another drifting cork, this would disturb its motion and cause it to radiate a new wave. In the same way, a drifting electron can be disturbed in a way that releases electromagnetic radiation. Each electron has a certain amount of energy associated with it, by virtue of its motion, spin, and other properties. If an electron is disturbed in a way that reduces its total energy, it can give up the energy in the form of radiation—a photon that speeds away. The energy of this resulting photon is just equal to the energy the electron lost. In other words, the wavelength and color of the emitted radiation are exactly controlled by the amount of energy lost by the electron. As long as this electron is unattached to an atom, it can make a change in energy by any amount, and hence can radiate photons of any wavelength. Thus a mass of free electrons, if disturbed, could radiate an array of photons of all wavelengths. This produces a spectrum like Figure 5-1—a continuous band of colors called a **continuum**, or continuous spectrum.

Thermal Radiation

In fact, this type of radiation is emitted all the time, all around us. Electrons (and atoms and molecules) in all gases, liquids, and solid objects are in constant motion, jostling each other. **Temperature** is simply a measure of the average energy of these motions, which are thus called **thermal motions** (from the prefix *thermo-*, referring to heat). The higher the temperature, the faster the thermal motions and the more disturbed the atomic particles. Because the electrons of an object are constantly being disturbed by thermal motions, all objects continuously radiate a continuous spectrum of electromagnetic radiation. The hotter the temperature, the more radiation is given off. **Thermal radiation** is the type of radiation that depends on the heat of the radiating material.

Wien's Law

It is important to realize that the light you see from a rock or a planet in the night sky, or from this book, is *not* thermal radiation. It is **reflected radiation**—light from the Sun or some other source, bouncing off the object.

Why can't we see the thermal radiation given off by a rock, planet, or book? The basic reason is that they are too cold. Cold objects give off much less radiation than hot objects. More important, however, they give

---

[2] However, the word must be read with caution, because physicists and engineers sometimes speak of streams of atomic particles, such as protons or electrons, as radiation. For example, charged particles escaping from radioactive material are sometimes called "radiation," or better, "particle radiation." These atomic particles are not light or electromagnetic radiation. Other examples are "radiation belts" and "cosmic rays," both of which are particles, not electromagnetic radiation.

off the wrong color (or wavelength) of radiation for us to see. Let us explain with an important physical law discovered in 1898 by German physicist Wilhelm Wien (pronounced VEEN). It is called **Wien's law:**

> **The hotter an object, the bluer the radiation it emits.**

Remember that the bluer the radiation, the shorter the wavelength, so we could also state Wein's law by saying that the hotter an object, the shorter the wavelength of radiation it emits.

This effect is familiar in everyday life. Ordinarily, a nail emits no visible radiation. If you heat it over a stove, however, it begins to radiate a dull red light. Higher temperature leads to a bright orange-red glow. At a still higher temperature, the dominant color is "white hot," or yellowish-white. If the nail could be heated enough, its radiation would become distinctly bluish.

Despite appearances, the nail is radiating regardless of its temperature. When it is at room temperature, the light it radiates is so red (of such long wavelength) that we cannot see it because our eyes are not sensitive to infrared radiation. Similarly, a hot enough object radiates ultraviolet light, whose wavelength is too short for us to see. Though we cannot see infrared and ultraviolet radiation, we can build instruments to detect it; radios, for example, detect especially long wavelengths (usually called radio waves).

Measuring the Temperature of a Planet (or a Star)

Now we can see one of the ways that light carries messages from space. Instruments to detect infrared radiation of planets and stars were first built around World War II and are still being improved. With such infrared detectors, we can measure the temperature of a planet by measuring the color (wavelength) at which its thermal radiation peaks. The peak thermal radiation for all planets and moons, from Mercury to Pluto, lies in the infrared part of the spectrum. They are all too cold to emit appreciable visible light; we see them not by their *own* light but by the sunlight they reflect. Stars, on the other hand, are hot enough to emit visible light.

Nonthermal Radiation

The thermal radiation we have been discussing occurs when electrons change energy levels at random due to

disturbances arising from the *thermal* motions of atomic particles. For completeness, we should add that other types of disturbance can create a continuous spectrum of different shape than in Figure 5-3. These types of radiation, unrelated to thermal motions of atomic particles, are called **nonthermal radiation**. We will encounter some types of nonthermal radiation in later chapters.

EMISSION LINES AND BANDS

We've been describing ways to produce light as electrons are disturbed. This process, in which a particle emits a photon, is called **emission**. We started off picturing a free electron (like a cork floating in a pool), not attached to an atom. Now consider an electron orbiting the nucleus of an atom. As soon as we consider electrons attached to atoms, extraordinary modifications occur to the emission process, and these are critical to all of astronomy.

Atoms and Emission Lines

Figure 5-4 shows the schematic structure of an atom, with its central nucleus and some orbital paths available to a specific electron. This electron may be only one of many, depending on the element: hydrogen, helium, and so on. In some ways, the atom is like a tiny solar system—that is, a relatively massive central object and tiny orbiting particles. But there is an extraordinary difference. In the solar system we can put a rocket into any orbit we choose; each orbit would have its own velocity and hence energy. But in an atom, electrons can occupy only *certain* orbits. These are called **energy levels**, because each orbit has one specific energy. Physicists say the atom has quantized energy levels, because only certain quantities of orbital energy are possible. In Figure 5-4, the solid circle represents the orbit occupied by the electron, and the dashed circles are other allowed energy levels. At the top are a few of the infinite number of energy levels outside the atom, where energy levels are not discrete.

Now let us disturb the electron as before (hitting it with a photon or a neighboring atom) so that it drops to a lower energy level. (In fact, this process can also happen spontaneously without outside disturbance.) As it drops, it emits a photon whose energy equals the difference in energy between the two levels. But note

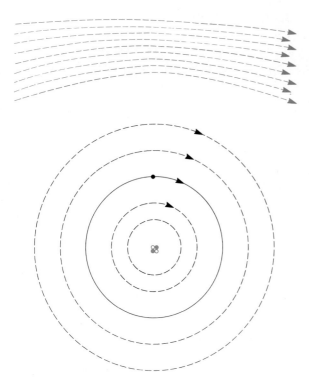

Figure 5-4 Schematic view of an atom's structure. If an electron (dot) finds itself in an orbit (solid line) around the nucleus of an atom, there are only certain other orbits, or "energy levels," that it can occupy in the atom (dashed circles). However, at large distances from the nucleus (top) virtually any orbits, or "energy levels," are possible.

that being inside an atom, the electron can make only certain, fixed changes in energy level, as shown in Figure 5-5. For instance, if it starts at level 3, it can drop only to levels 2 or 1. Thus it can emit only certain amounts of energy corresponding to these differences in energy level; it can emit only photons of certain wavelength or color. As a result, *each element can emit only certain wavelengths of light.* Figure 5-6 shows the emissions produced by a sample of hydrogen gas containing neutral (that is, uncharged) atoms, with their electrons at different energy levels. These emissions are called **emission lines** because they appear in a projected spectrum as *lines* or narrow bars of color.

Here, then, is an astounding and useful fact! If you see a certain set of emission lines, you can match it with a certain element and infer that atoms of *that* element are present and glowing in the distant object. You can tell something about the object's composition, even without having a sample!

If the electrons in an atom are at their lowest possible energy level, that atom cannot produce an emission line. This is because the electrons cannot drop to any lower energy level. An atom in which all electrons are at the lowest possible energy level is said to be in its **ground state.** An atom in which one or more electrons are at energy levels higher than the lowest available ones is said to be in an **excited state.**[3] Excited states

---

[3]Each energy level can be occupied only by a certain number of electrons. For example, the lowest level can take only two.

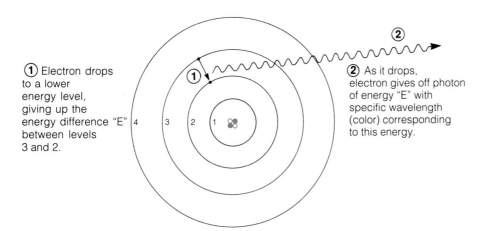

① Electron drops to a lower energy level, giving up the energy difference "E" between levels 3 and 2.

② As it drops, electron gives off photon of energy "E" with specific wavelength (color) corresponding to this energy.

Figure 5-5 Schematic view of an atom emitting a photon of light as an electron drops from a higher to a lower energy level.

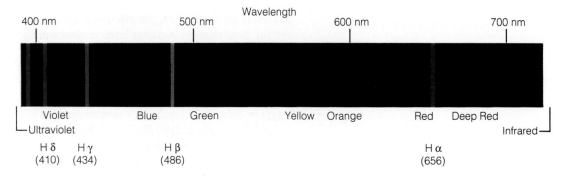

Figure 5-6 A visible-light spectrum consisting only of the emission lines produced by glowing, neutral hydrogen gas. Wavelength scale is given at the top. The red line, called the hydrogen alpha line, is especially prominent in many astronomical examples of glowing gas, causing many gaseous clouds in space to glow with reddish light.

usually last only a short time before the electrons "decay" to the ground state. Atoms generally need to be disturbed to produce and maintain excited states. For this reason, hot gases are more apt to produce emission lines than cold gases, because atoms in hot gases collide faster and more often.

Molecules, Crystals, and Emission Bands

The electron structure in a molecule is more complex than in an atom. The electron's path may take it around two or more nuclei. Thus the emission line structure from a molecule is not so simple as that from an atom. For example, in a gas containing water molecules (H_2O), we get more complex emission lines than in a gas containing single H and O atoms. The molecule has various ways of responding to a disturbance in addition to having

its electron change energy levels—for example, it may vibrate like two balls linked with a spring. As a result, the energy levels are vastly more numerous and the resulting emission lines blend together. Instead of sharp emission lines, we get blended clusters of barely separated emission lines over a range of wavelengths, as shown in Figure 5-7. The resulting broader emission feature from a molecule is called an **emission band**. The rest of the story is the same, though: A given molecule (such as H_2O) can produce only certain emission bands. Thus we can identify the molecules glowing in a given remote source just by detecting their emission bands.

For atoms and molecules to produce clear emission lines and bands, they must be detached from one another, as in gases. If the atoms were linked together, they would form molecules, and if the molecules were linked, they would form a solid or liquid. The electron structure of most solids and liquids is so complex that emissions

Figure 5-7 Schematic examples of emission lines and bands as they might appear in a spectrum. The molecular emission band shows a "fine structure" associated with the structure of the molecule.

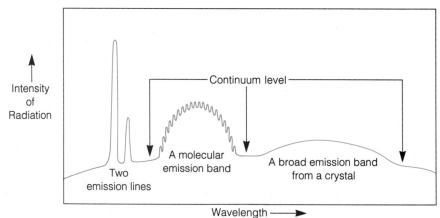

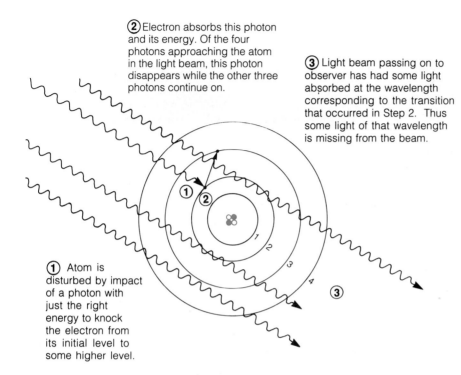

② Electron absorbs this photon and its energy. Of the four photons approaching the atom in the light beam, this photon disappears while the other three photons continue on.

③ Light beam passing on to observer has had some light absorbed at the wavelength corresponding to the transition that occurred in Step 2. Thus some light of that wavelength is missing from the beam.

Figure 5-8 Schematic view of absorption of a photon of light as an electron is knocked from a lower to a higher energy level in an atom.

① Atom is disturbed by impact of a photon with just the right energy to knock the electron from its initial level to some higher level.

are smeared out and are barely discernible. Most emission lines and bands arise from gases.

However, an important exception occurs among rock-forming minerals, which make up most planetary surfaces and interstellar dust grains. Most rock-forming minerals are crystals, which can be thought of as giant molecules. In a crystal, atoms are joined in a fixed pattern that simply repeats as the crystal grows bigger. Light can penetrate through the outer millimeter or so of a crystal, producing very broad, faint emission bands, as shown in Figure 5-7. Warm grains of silicate material near other stars have been identified from their emission bands, for example.

ABSORPTION LINES AND BANDS

Let us return, in Figure 5-8, to an atom with an orbiting electron. This time it gets disturbed by a passing light wave (photon) that bumps it up into a higher energy level. Energy was removed from the beam to do this. This process of energy removal from a light beam is called **absorption**.

Atoms and Absorption Lines

Only the specific energy corresponding to the specific transition (level 2 to 3 in the case of Figure 5-8) can be absorbed in an atom. Thus only a photon of specific energy or wavelength can be removed from the beam. A beam of light passing through a cloud of such atoms will have many of these photons removed, thus absorbing some light of that color. As a result, the light in a narrow interval of the spectrum is lost; this missing interval is called an **absorption line**. Various absorption lines can result from the various possible upward transitions—for example, level 2 to 3, 2 to 4, 2 to 5, 1 to 2, 1 to 3, and so on. The visible part of the spectrum, with absorption lines due to hydrogen, is shown in color in Figure 5-9.

Because the *intervals* between energy levels in a given atom are the same whether absorption or emission is occurring, the pattern of emission and absorption lines for a given element is the same. An element can be identified from either its emission lines or its absorption lines.

Molecules, Crystals, and Absorption Bands

Much of the preceding also applies to molecules and crystals. As light penetrates through a cloud of molecules (a gas) or through the upper millimeter of a crystal (in a rock surface), transitions of electrons among its numerous, closely spaced energy levels produce

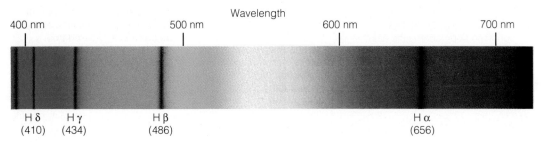

Figure 5-9 Visible portion of the spectrum showing absorption lines due to hydrogen. Such a spectrum would be seen if hydrogen gas lay between the observer and a light source with a continuous spectrum; the hydrogen absorbs only the specific "missing" colors. These absorption lines have the same positions as the emission lines in Figure 5-6.

absorption bands. These are analogous to the emission bands in Figure 5-7, except that they appear as dips (dark features) instead of bumps (bright features) in the spectrum. The molecules in the gas, or the composition of the crystal, can be identified if the absorption bands themselves can be detected and identified. The absorption bands of a substance have the same wavelength intervals as its emission bands.

ANALYZING SPECTRA

How to Measure the Atmospheric Composition of a Planet (or a Star) from a Distance

Now we see how light carries even more messages. Suppose we look at a planet with an atmosphere. Light passes into the atmosphere and back out, as shown in Figure 5-10. As the light passes through the gas, it may acquire absorption lines and bands that allow us to identify some of the atoms and molecules in the gas. Not all atoms and molecules have equally prominent lines and bands. Some, such as carbon dioxide (CO_2), are easy to detect, and others, such as nitrogen (N_2), are hard to detect by spectroscopy. Thus although the carbon dioxide of Venus and Mars was discovered by Earth-based spectroscopy decades ago, the abundant nitrogen of Saturn's moon, Titan, was not discovered until a Voyager spacecraft made a close pass by Titan.

How to Measure Properties of a Planet's Surface from a Distance

As mentioned earlier, the absorption bands caused by mineral crystals are broad, often shallow, and hard to

detect. For this reason, spectroscopy has only limited success in identifying planetary surface materials. There are two approaches: measuring colors and detecting actual absorption bands. Measuring the color of an object is essentially measuring the relative intensities of radiation in different parts of the spectrum. A red object looks red because it reflects more light at red wavelengths than at bluer wavelengths. A measurement of color often restricts the possible materials on a surface. For instance, most satellites of Saturn have a bluish-white color similar to ice, but inconsistent with the red rocks and minerals on Mars.

If we can detect a mineral absorption band, we can probably identify the mineral. For example, the satellites of Saturn just mentioned display absorption bands caused by crystals of frozen water, in the form of ice or frost, confirming the implication of the color studies that ice is a dominant material on their surfaces.

Furthermore, if we can measure the wavelength of the strongest thermal radiation, we can use Wein's law to get the temperature of the surface.

The Spectrum and Our Atmosphere: Seeing into Space from Earth's Surface

Figure 5-11 illustrates several important points about the relationship between astronomy and the Earth's atmosphere. First, clouds block much of our view of space. One reason for putting observatories on top of mountains is to get above the clouds. Another reason is to get above the dense, shimmery air that makes telescopic star images dance and twinkle when seen from ordinary, lower altitudes.

A still more important reason can also be seen in Figure 5-11. Different wavelengths of light from space

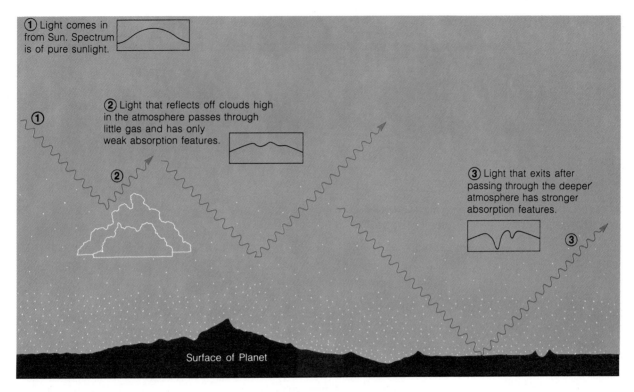

① Light comes in from Sun. Spectrum is of pure sunlight.

② Light that reflects off clouds high in the atmosphere passes through little gas and has only weak absorption features.

③ Light that exits after passing through the deeper atmosphere has stronger absorption features.

Surface of Planet

Figure 5-10 Production of absorption lines or bands in the spectrum of a planet as light passes through the planet's atmosphere. Identification of the lines or bands allows identification of at least some constituents in the atmosphere. The same process occurs as light radiates from a star outward through the star's surrounding gaseous atmosphere.

penetrate different distances downward through our atmosphere. Much infrared light, for example, gets absorbed a few kilometers above the ground by water vapor low in Earth's atmosphere. Observatories on mountain peaks above these low layers can thus detect wavelengths invisible from sea level. Telescopes in space, such as the Hubble Space Telescope, can do even better because they can detect *all* wavelengths arriving from distant objects.

THE THREE FUNCTIONS OF TELESCOPES

As described in Chapter 3, one of the great advances in the adventure of human exploration came with the invention of the telescope, around 1609. This seemingly magical instrument let us see distant objects in space as if from close range, for the first time.

A **telescope** is a device with three functions. First and most obvious, it magnifies the image of an object to a larger angular size than we perceive with our naked eye. The term **magnification** applies primarily to a telescope designed for visual observation; the magnification is the apparent angular size of a distant object seen through the eyepiece, relative to its apparent size seen by the naked eye. If a telescope makes something look 10 times larger, we say it has a magnification of $10\times$. The second function is **resolution**—the ability to discriminate fine detail. Whereas the eye can resolve angular details only a few minutes of arc across, a telescope might show detail a few seconds of arc across. The telescope's third function is **light-gathering power**—the ability to collect light and reveal fainter details than the naked eye can see. When light from a distant planet or star reaches Earth, a certain number of photons strike each square centimeter of Earth each second. The pupil of the eye has a diameter less than a centimeter and can receive only a limited number of photons per second. But a telescope collects *all* the photons striking a lens or mirror many centimeters across. This makes a much brighter image.

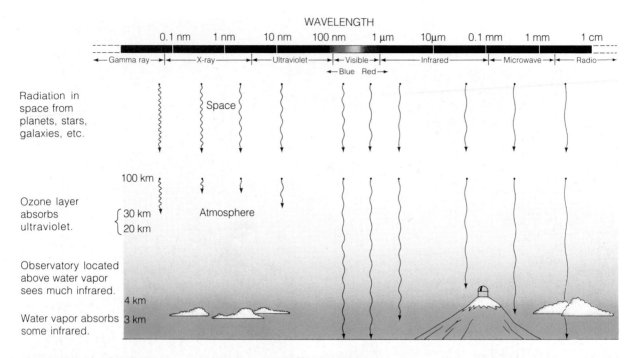

Figure 5-11 Electromagnetic spectrum is shown at top. All these types of radiation from various astronomical sources can be detected by a telescope in space (upper middle), but many of them are absorbed by various gases in our atmosphere (bottom).

Early telescopes were designed entirely for observers to look through. In modern professional telescopes, the human eye is replaced by instruments that can make more precise measurements. Thus modern astronomers rarely look through their giant telescopes! Nevertheless, the three functions still apply. A large telescope used with film to take pictures, for example, can be compared to an ordinary camera instead of to the eye: It obtains a more magnified image, a more clearly resolved image, and a brighter image than the camera.

Two Designs for Optical Telescopes

Two basic designs have been used for telescopes. The first to be built, the **refractor**, uses a lens to bend, or refract, light rays to a focus, as in Figure 5-12. Galileo first used this type astronomically in 1609. The second type, the **reflector**, uses a curved mirror to reflect light rays to a focus, as in Figure 5-13. Isaac Newton built the first reflector in 1668. Several reflector designs have since been constructed, but the simplest is Newton's, sometimes called the Newtonian reflector. In recent years new designs have combined lenses and mirrors. Often called **compound telescopes**, these designs are often very compact and portable.

Focal Length, Aperture, and the Telescope's Functions

In a telescope, the main lens or mirror is called the **objective**, as shown in Figures 5-12 and 5-13. The distance from the objective to the place where the image is focused is called the **focal length** of the objective (marked F in the figures). The diameter of the objective is called the **aperture**.

The three functions of a telescope are controlled by the focal length and the aperture. The longer the focal length, the greater the magnification and the bigger the image. (This principle also applies to camera lenses. Lenses with long focal length give big images and are called telephoto lenses.) The wider the aperture, the better the resolution. Note that magnification is not the same as resolution. Magnification alone only produces a big image, not necessarily a sharp image. By increasing focal length, we get a bigger image, but it may be blurry. If we increase the aperture (an expensive process, because we have to buy a bigger lens or mirror), we can get better resolution, that is, a sharper image. The third function, light-gathering power, is connected with aperture also: The larger the aperture, the more light collected.

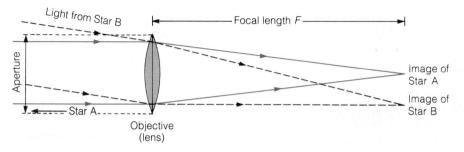

Figure 5-12 Cross section through a lens, showing image formation in a refractor (and most cameras). Light rays from two stars are focused into two images.

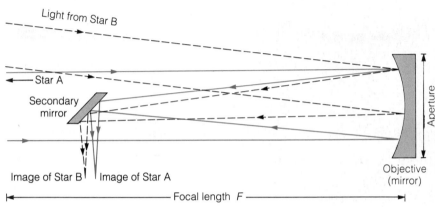

Figure 5-13 Cross section through a mirror system, showing image formation in a reflector. Light rays strike the curved mirror (right) and are reflected back toward the focus. The focus would normally lie in an inconvenient position in front of the mirror, but a secondary mirror is used to beam light rays to one side for easier access to the image. (The secondary mirror blocks a small fraction of the total light entering the telescope, making the image slightly fainter, but having no other effect.)

Radio Telescopes

Radio telescopes serve the same functions as optical telescopes. But because a 50-cm radio wave is a million times longer in wavelength than a visible light wave and carries less energy per photon, radio telescopes need larger surfaces to collect and focus enough energy to give strong signals. As shown in Figure 5-14, radio telescopes are reflectors, with a large curved surface (the "dish") of metal or wire mesh, that reflects the radio waves to a focus, where a smaller radio detector picks up the signal. Interesting techniques have been developed to link radio telescopes in tandem, combining their electronic signals. In effect, this allows several modest-sized radio telescopes in separate locations to act like one huge radio telescope.

USING VISUAL TELESCOPES

An optical telescope designed to be looked through, as opposed to a radio telescope or a camera, may be called a *visual telescope*. The optical systems of Figures 5-12 and 5-13 can be converted into visual telescopes simply by adding an eyepiece (and usually tubing to keep out

stray light), as shown in Figure 5-15. The eyepiece is simply a lens or system of lenses designed to magnify the image still more and allow the eye to examine it.

One of the first questions asked of backyard telescope enthusiasts is: "How far can you see with that thing?" This is not the right question to ask, because no telescope is limited by distance. Every optical system can see as far as there is an object large enough or bright enough to detect. Both the naked eye and the 5-m (200-in.) Palomar telescope can see the Andromeda galaxy 19 billion billion kilometers away—but the telescope shows more detail and fainter regions.

The three functions of a telescope—magnifying power, resolution, and light-gathering power—need to be considered when using a telescope. The magnifying power is controlled by the eyepiece chosen.[4] For instance, if the magnifying power is $100\times$, then the angular size of Jupiter, about 1 minute of arc across in the

---

[4]The magnifying power is given by the formula

$$\text{Power} = \frac{\text{focal length of objective}}{\text{focal length of eyepiece}}$$

For instance, if a 1-cm eyepiece is inserted in a telescope of objective focal length 100 cm, we get 100 power, often written $100\times$.

Figure 5-14 a One of the radio telescopes in the Very Large Array in central New Mexico. This view shows the underside of the "dish" (top), or curved reflector, that collects the radio waves. This telescope (along with its neighbors) can be repositioned in different arrays along the railroad tracks, giving different receiving qualities for different observing projects. (Photo by author.) **b** Schematic cross section of a radio telescope, showing similarity to the design of the reflector-type optical telescope.

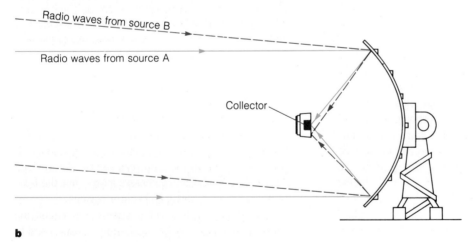

sky, looks about 100 minutes of arc across (nearly 2°) through the eyepiece. In theory, any telescope can be made to give any magnifying power, simply by inserting an appropriate eyepiece. In practice, though, magnifications greater than about 400× are rarely useful because of the shimmering of the atmosphere, which is magnified along with the image. This air quality, called **seeing,** varies from night to night and sometimes allows higher magnifying power.

The second function, resolution, or sharpness of the image, depends entirely on the aperture of the telescope. The bigger the aperture, the finer the details that can be seen. Due to fundamental properties of light waves, a telescope with a given aperture, A cm, cannot resolve details smaller than $\frac{A}{12}$ cm across. Thus a 12-cm (5-in.) telescope can resolve about 1 second of arc, but nothing smaller. Again, when the telescope gets much larger than about 50 cm (20 in.), the reso-

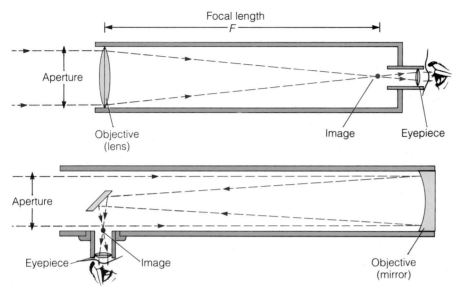

Figure 5-15 Cross sections showing how the systems of Figures 5-12 and 5-13 are converted to visual telescopes by the addition of an eyepiece. The eyepiece is a small magnifying lens (or several lenses mounted together) used to examine the image. Tubing from the objective (lens) to the eyepiece helps cut out stray light.

lution is limited by poor seeing—the shimmering of our atmosphere. Astronomers look forward to solving the problems of atmospheric seeing by orbiting large telescopes, like the Hubble Space Telescope, above the atmosphere.

The third function of the telescope, light-gathering power, also depends solely on the aperture: The larger the aperture, the brighter the image.

As can be seen from the previous discussion, the most exciting visual observing comes from using the telescope with the largest possible aperture. Unfortunately, costs of amateur and professional telescopes rise rapidly along with aperture. Generally, telescopes with aperture about 20 cm (8 in.) to 50 cm (20 in.) give inspiring views of planets, stars, and nebulae, and can be used successfully with magnifications of 200× or more.

Observing the Moon and Planets

Although any visual telescope can be used to observe the heavens, certain designs and techniques are best suited to certain purposes. To observe the Moon and planets, magnification is the most important function, because one wants to get a large enough image to see detail. Telescopes with rather long focal lengths, about 6 to 10 times the aperture, are desirable. Eyepieces giving magnifications of 100 to 200× are recommended. Some of the sights you can see with such equipment, in order of increasing difficulty, are the mountains and craters on the Moon, the phases of Venus,

the satellites of Jupiter, the satellites of Saturn, the rings of Saturn, the cloud bands on Jupiter, divisions in the rings of Saturn, the phases of Mercury, the polar ice caps of Mars, the dusky markings of Mars, the cloud bands on Saturn, and the dusky markings on Mercury. An aperture of 5 cm (2 in.) suffices to see the first five items on the list; an aperture of around 25 cm (10 in.) is recommended for all items and is usually necessary for the last one.

Observing Star Fields, Nebulae, and Galaxies

The most prominent star clusters, nebulae, and galaxies are faint but cover larger angular areas than planets; hence light-gathering ability is the most important function in observing them. Telescopes with shorter focal lengths of only about three to six times the aperture are useful, and magnifications of 20 to 100× are recommended. Some double stars and star clusters are interesting to study under higher magnifying power, but higher powers spread out the light too much to give good views of nebulae and galaxies, which are mostly very faint. As always, the largest possible aperture gives the best view.

Observing the Sun

A NORMAL TELESCOPE SHOULD NEVER, UNDER ANY CIRCUMSTANCES, BE POINTED AT THE SUN. An eye could be immediately burned or blinded, because the objective acts like a giant magnifying glass, concentrating solar light, heat, and ultraviolet radiation at a point near the eyepiece. Furthermore, the telescope is likely to be damaged, because the solar heat can crack the glass in eyepieces or secondary mirrors. Some telescopes can be modified for safe solar observing. One usually starts by covering the objective or open end of the telescope with an opaque card containing a hole, or a new aperture, as small as 1 or 2 cm. Then the image can be projected onto a white card so it can be viewed. Instructors should be consulted before attempting this procedure.

PHOTOGRAPHY WITH TELESCOPES

An image can be formed on a piece of photographic film instead of in an eyepiece. Development of properly exposed film then provides a permanent record of the observation. Many telescopes come equipped with photographic attachments for this purpose. In the case of a bright object like the Moon or a planet, the eye usually sees more than a photograph with the same telescope, because the eye can take advantage of moments of perfect seeing, whereas the photograph averages moments of poor seeing, which blurs the image. On the other hand, faint objects like nebulae and galaxies are usually better shown in photos, because light can be accumulated in long exposures whereas the eye cannot "store" light.

Many photos of star fields and nebulae in this book were taken with small telescopes or ordinary cameras. The reader should consult the picture captions, most of which describe the equipment and exposure used. Often, readers can duplicate or improve the results with their own equipment.

In large, modern observatories, the recording instruments are often electronic, instead of photographic or visual, as shown in Figure 5-16. Furthermore, computer-directed machines can point the telescope almost precisely at any known celestial object. Thus astronomers may spend nearly the whole night observing without looking through the telescope.

PHOTOMETRY

Pictures of astronomical objects are interesting, of course, but much astronomical work requires measuring an object's brightness—the amount of light coming

from it at all wavelengths or at selected ranges of wavelengths (such as blue to green). This is called **photometry**. By measuring precisely the amount of light at various wavelengths, photometry allows astronomers to determine temperature, composition, and other properties of a remote object. Until a few decades ago, this was done by measuring the size and density of the image on a photograph. Today it is done much more precisely with electronic devices. The basic device is a photomultiplier placed at the telescope's focus. Each photon of light collected by the telescope strikes the photomultiplier surface, which releases a shower of electrons with each photon impact. In some designs, each electron strikes a surface and releases more electrons, thus multiplying the effect into a weak electric current that can be measured. Measurement of the current tells the number of photons arriving from the astronomical object, thus giving a measurement of its light intensity.

In recent years, these techniques have been extended to video-imaging devices that can produce TV images of objects too faint to photograph. Most important of the modern detectors are *charge-coupled devices,* more commonly known as CCDs. These are extremely light-sensitive detectors, made possible by microelectronic technology. They are often used in arrays of, say, 500 $\times$ 500 light detectors, about the size of a postage stamp. An image can be created from the resulting 250,000 "dots"; each dot, called a picture element, or *pixel,* corresponds to one measurement of brightness, made by one detector. CCD imaging arrays can record in 5 min more faint detail than photographic film exposed for 100 min in the same telescope.

IMAGE PROCESSING

Extraordinary advances have occurred in the last decade in our ability to process images in order to gain maximum information from them. There are three main areas of interest: photographic images, digitized images, and false color images.

Photographic Images

Photographic images are composed of microscopic dots arranged in a random pattern; each dot represents a grain in the chemicals composing the film. The more light hitting a given grain, the larger and darker it appears in the final (negative) image. The negative image is then

a

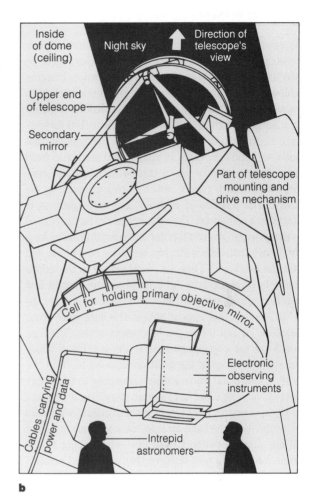

b

Figure 5-16 **a** A modern telescope, at Mauna Kea Observatory, Hawaii. In this telescope design, light is focused through a hole in the center of the main mirror, where the astronomer mounts electronic instruments to record data. Data are fed through cables to computer equipment, which performs initial analysis and stores data for further analysis. The dome (top) is open to night air in order to give the clearest observing conditions; temperatures inside the dome often drop below freezing during observing sessions. (Photo by author.) **b** Schematic diagram of *a*.

processed to make a positive print or transparency. Color films, such as Kodachrome, have several layers of light-sensitive chemicals with different color dyes to produce a color image from a single exposure. However, newer techniques allow us to combine separate black-and-white images made with different-colored filters—such as three images made with red, yellow, and blue filters in a spacecraft—to create a single color photo. Some of the color photos in this book were produced in this way. A major problem in classic photography has been that film

can reproduce only a limited range of contrast. If the bright head of a comet is 10,000 times brighter than its faint tail, an exposure long enough to record the faint details of the tail overexposes the head, giving the false impression of a bright blob rather than an intense star-like point surrounded by a diffuse glow. Recent photographic techniques, such as using variable-density "masks" that screen light from the brightest regions, allow improved images of high-contrast astronomical subjects.

Digitized Images

As described earlier under photometry, some modern sensors such as CCDs have a *grid* of dots, or pixels, in which the brightness is measured for each dot. Imagine a blowup of a newspaper photo. You can see each individual pixel as a dot. Now imagine that each dot is replaced by a number from zero (blackest tones) to 100 (brightest), representing the brightness of the light at that point in the image. Such an image need not come from a CCD; an ordinary photo can be scanned and converted into such pixels, which is called *digitizing the image*. The position of each dot and its brightness is stored in a computer. Then the image can be processed in many ways. We could order the computer to make a print such that zero is the blackest tone on a photo and 100 the brightest. Then we would preserve the entire range of brightness in the original picture, but at low contrast. Suppose we are interested in some subtle contrast features that appear only in the light gray areas between 70 and 90 on the brightness scale. Then we could ask the computer to make a new print in which the darkest tones of the print correspond to 70 in the original image and the brightest correspond to 90. Every part of the image that was fainter than 70 now appears as black; every part that was brighter than 90 now appears as white. We now have an extreme-contrast version of all the detail between 70 and 90—which might reveal some interesting new features that we missed before.

False Color Images

Another technique is to assign different colors to each brightness level. For instance, we could use violet for the dark tones of 1 to 10, blue for 10 to 20, green for 20 to 30, and so on. Or we could divide the brightness scale even finer, with yellow for 50 to 52, orange for 52 to 54, and so on. In this way, we could produce an image in which the colors have nothing to do with the original colors but are used merely as a code to allow us to separate features of slightly different brightness. An example is seen in Figure 5-17. Such images are called *false color images*. In recent years, they have become the darling of magazine art directors who are attracted by the flashy pictorial quality of the colors. Indeed, some false color images have been published without adequate explanation, and TV journalists have been heard to exclaim over the strange, "fried-egg" appearance of Halley's comet. Such pictures had nothing to do with the actual appearance of the comet. They were merely false color images with different rings of color representing different brightness.

For such reasons, "true color" images are used as much as possible in this book, though many texts and magazines are full of false color imagery. "True color" is in quotes because at some level of detail it is difficult to reproduce all the nuances of natural color, especially in faint objects. But the colors in most images in this book do give some impression of the actual colors of the objects.

SPECTROPHOTOMETRY

As we stressed in the first part of this chapter, it is important to measure not only the total amount of light coming from an astronomical object but also the intensity at each wavelength. In this process, called **spectrophotometry,** light is passed through a spectroscopic device that breaks it into different colors before it enters the photomultiplier. Then the photomultiplier can measure the amount of light in each color (that is, wavelength) range. Similarly, photos or TV images can be produced showing the object as seen in different colors (wavelengths).

Images at Selected Wavelengths

Photos or TV images can be produced showing an object as seen in different colors (wavelengths). One familiar example is when a photographer puts a red filter or blue filter over a lens, to show the scene as it appears in only red light or blue light. Similarly, detectors sensitive to infrared light, ultraviolet light, radio waves, or other wavelengths can be used to form an image, even though the eye is not sensitive to these wavelengths.

One example of this technique is shown in Figure 5-18. Image *a* is a photo of a giant canyon on Mars, made at ordinary visual wavelengths of about 0.6 μm. Image *b* is the same scene, but imaged at far infrared wavelengths of 11 μm. This is called a thermal infrared image, because this far infrared light is not visible reflected sunlight, but rather the invisible rays given off by the heat of the surface. The warmer the surface, the brighter it appears in this image. Typical noontime temperatures of the Martian soil measured in this image were about −6°C (21°F).

Notice what can be learned from comparing the two

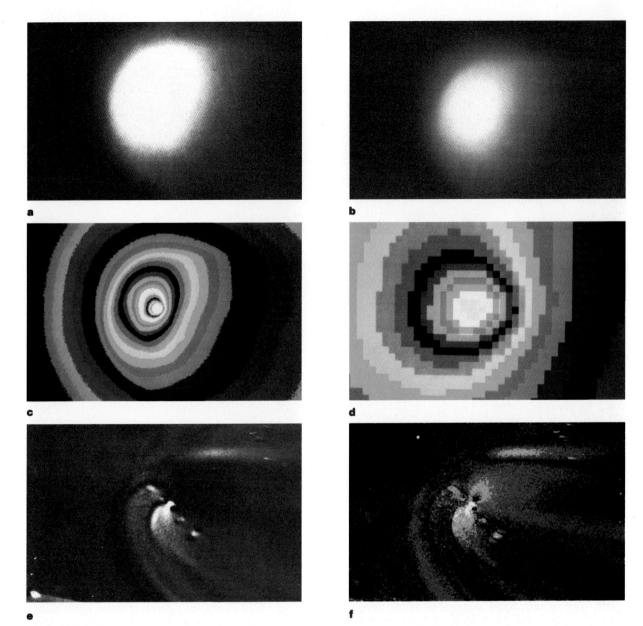

a

b

c

d

e

f

Figure 5-17 New image processing techniques are shown in these images of the head of Halley's comet. **a** A direct image of the comet's head taken by a CCD on a large telescope resembles an ordinary photo and shows an overexposed bright center surrounded by glowing gas. Image defects, such as the streak at right, are due to internal properties of the CCD. **b** Addition of several images and correction of flaws makes a cleaner image. The central region is still overexposed. **c** The same image in false colors, with different-colored bands used to represent different brightness levels. This image, though less realistic visually, reveals the smooth increase of brightness toward the center, even within the regions overexposed on print *a*. **d** An enlargement of the central part of *c* shows the individual pixels composing the image. **e** More sophisticated image enhancement has been used to exaggerate localized intensity boundaries within the seemingly uniform contours. This enhanced image reveals faint spiral jets, present in the earlier images but swamped by the comet's glow. Bright blobs are images of a star the comet passed during the several different exposures. The star, barely visible in *b* and *c,* is enhanced by the processing. **f** A false color version of *e* emphasizes the spiral jet structure and the increasing brightness toward the center, but it is even further removed from the appearance the comet would have to the naked eye, because the colors are arbitrary and the overall glow around the comet's head is suppressed to reveal the curved streaks. (All images by Stephen M. Larson, Lunar and Planetary Lab, University of Arizona, from observations made at Boyden Observatory, South Africa, on the day of Giotto probe encounter; see Giotto images in Chapter 13.)

a

b

Figure 5-18 Visible-light and thermal infrared images of a huge canyon on Mars, photographed in 1989 by the Soviet space probe Phobos-2. **a** Image in visible light (reflected sunlight). **b** Image made at far infrared wavelengths around 11 μm, using the infrared radiation deriving from heat of the surface. Cooler areas are dark; warmer areas are bright. Note that the infrared light penetrates better through the haze at the horizon (right). Comparison of the two images reveals surface properties (see text). North is at the top. (Courtesy A. Selivanov and M. Maraeva, Glavcosmos, USSR.)

images. Normally, a dark surface absorbs more sunlight and gets warmer than a bright surface, just as a black car gets hotter than a white car. So, other factors being equal, dark spots in image *a* should be bright (warm) spots in image *b*. This effect can be seen at the left end of the canyon, where the floor is dark in *a*, but bright in *b*. However it is not always so! The floor of the middle part of the canyon is dark in both pictures—dark and cool. One cause for such an effect is differences in surface materials. For example, sunlit rock surfaces warm up at a different rate than dusty soil, so that a light-toned soil may get warmer than a dark rock surface. Thus, careful study of the two pictures tells scientists not only the temperatures of different parts of the surface, but also the textures of the surface materials.

This book contains many examples of similar techniques of comparing visible and thermal infrared images, or images at other wavelengths, in various fields of astronomy (see photos of Orion Nebula, Figure 20-16, page 327).

LIGHT POLLUTION:
A THREAT TO ASTRONOMY

Figure 5-19 shows the pattern of urban lights across the United States as seen on a cloud-free night from space. Many observatories, especially in the West, are near rapidly sprawling cities. The glow from these cities lights up the sky, reducing the contrast between faint stars and the dark sky background. This is why it is so hard to get a good view of the sky from urban areas. Moreover, many types of fluorescent streetlights emit emission lines that interfere with astronomical work. These problems have rendered many famous observatory sites, such as Mt. Wilson and Palomar Mountain in southern California, nearly worthless for the most sensitive modern instruments. A classic solution to this problem has been for astronomers to move even further toward the frontier—as when Percival Lowell established his observatory in Flagstaff, Arizona, in the 1890s or when modern astronomers send telescopes into space.

Figure 5-19 A beautiful pattern that threatens modern astronomy. A montage of satellite photos shows North American urban lights as seen from space with no cloud cover. Skies over the Eastern megalopolis, southern California, and other regions are so strongly "polluted" with light that neither the eye nor the telescope can see the faintest stars that would otherwise be visible from these regions. (Photo by W. T. Sullivan III, courtesy National Optical Astronomy Observatories.)

Nonetheless, many cities have acted to protect the investment in nearby observatories by installing only streetlights that emit minimally disruptive wavelengths of light and hoods that block lights from shining up into the sky.

SUMMARY

Most of our knowledge of the universe comes from deciphering messages carried from extraterrestrial bodies by electromagnetic radiation. To do this we need to understand the nature of the radiation and how it is generated in atomic particles. Electromagnetic radiation is released in the form of photons, each of which has its own specific wavelength and energy. The light coming from an object, arranged in order of wavelength, is called the spectrum of the object. Each wavelength corresponds to a "color," ranging from too short (too blue) to see, through visible colors such as blue and red, to wavelengths too long (too red) to see. Photons of blue light have more energy than photons of red light.

Atoms and molecules each produce absorption and emission lines and bands in the spectrum as photons inter-act with them. These derive from the energy level structure of their electron orbits. Each element and compound has its own lines and bands. Materials in a distant object can be identified by obtaining the objects' spectrum and identifying emission or absorption lines or bands. Many gases in planetary atmospheres or stars can be identified in this way, but solid surfaces of planets can be only roughly characterized, because solid materials have only weak, poorly defined absorption bands.

Telescopes are devices that magnify the angular size of distant objects and, just as important, enable us to gather much more light than is gathered by the eye. Thus we can see objects much fainter than visible to the unaided eye. This chapter contains advice on viewing different astronomical objects with small telescopes. Astronomers bolt various kinds of instruments on large telescopes to analyze the light from planets, stars, and so on. One of the most important instruments is the spectrophotometer, which measures the amount of light at each wavelength over a range of wavelengths, allowing precise measurement of the spectrum of the object being observed. We are now entering an era when large telescopes are being placed in orbit above the atmosphere in order to give access to unblurred images throughout all wavelengths of the spectrum.

CONCEPTS

wavelength

frequency

wavelike properties of light

visible light

ultraviolet light

infrared light

speed of light

particlelike properties of light

photon

diffraction

spectrum

spectroscopy

fields

electromagnetic radiation

continuum

temperature

thermal motion

thermal radiation

reflected radiation

Wien's law

nonthermal radiation

emission

energy level

emission line

ground state

excited state

emission band

absorption

absorption line

absorption band

telescope

magnification

resolution

light-gathering power

refractor

reflector

compound telescope

objective

focal length

aperture

seeing

photometry

spectrophotometry

PROBLEMS

1. Photographic films are usually more efficient when measuring more energetic photons. Which would you expect to be easier to photograph: a faint blue, red, or infrared star?

2. The ground cools at night by radiating thermal infrared radiation into space. Explain why the air stays warmer on a moist, cloudy night than on a very dry, clear night.

3. A spacecraft passes close to a previously unknown planet. A camera system on the spacecraft photographs surface details quite clearly in all wavelengths except the absorption bands of carbon dioxide. At these wavelengths, the image is featureless. Interpret these results.

4. Why has there been more emphasis on getting ultraviolet and infrared telescopes into orbit than on getting radio telescopes into orbit?

5. You are looking at two pairs of binoculars. Both pairs have lenses with 50-mm aperture, but one has a magnifying power of $7\times$ and the other, $20\times$. Which pair:
   a. Gives the larger apparent image size?
   b. Gives a greater apparent brightness in the observed image?
   c. Is easier to hold still enough to observe? (Remember that the unsteadiness of your hands is magnified as much as the image.)

6. For your birthday, your parents offer you a choice of lenses for your 35-mm camera. One has a focal length of 24 mm and the other has a focal length of 90 mm. (Focal length numbers are marked on the inner rim of all camera lenses.)
   a. Which one would give a larger image but cover a narrower field of view in terms of degrees?
   b. Which one would give a smaller image but cover a wider angular field of view?
   c. Which one would be called a telephoto lens?

7. a. Why doesn't an atom in the ground state produce emission lines? **b.** Use this fact to explain why the colorful emission-line glows of certain nebulae occur near hot stars, but not in the cold gas of interstellar space.

PROJECTS

1. Compare views through binoculars of different sizes. (Binoculars carry a designation such as 7 × 35, where the first number is the magnifying power and the second is the aperture.) Confirm that at fixed power, larger apertures give brighter images.

2. Observe and sketch a rainbow. Verify that the colors appear in order of wavelength as seen in Figure 5-1.

3. Study a room light equipped with a dimmer switch. Turn the switch from bright to the faintest visible setting. Note the redder color of the light bulb at the dimmest position. Relate this effect to Wien's law.

4. Experiment with two lenses having different focal lengths, such as 20 cm and 5 cm. Measure their focal lengths by focusing sunlight on a surface. Then (repeating experiments that must have been done first by European lens-makers around 1600) hold or mount the two lenses to make a simple refracting telescope like that in Figure 5-15. Tubing is not essential. Estimate your telescope's magnification by comparing the magnified image with a naked-eye view, and then compare with the formula in footnote 4 on page 75. REMINDER: DO NOT LOOK DIRECTLY AT THE SUN WITH ANY TELESCOPE.

Exploring the Earth–Moon System

The only photo showing
Earth and its Moon from a great
distance was made by the Voyager 1
spacecraft from a distance of
7 million miles as it left on its flight
to Jupiter and the outer solar system.
Background stars are too faint
to be recorded in most space photos
unless they are special, long
exposures. (NASA.)

CHAPTER 6

Earth as a Planet

Some readers might wonder, "Why discuss Earth in an astronomy book? Why not relegate this material to a course in geology?" The answer is dramatized by Figure 6-1. The soil in your backyard is just as much a sample of planetary material as a piece of the Moon. People have taken a long time to realize this, and many people still don't recognize the implications of the astronomical discovery that we are spinning through space on a small, finite ball.

Earth is just one of a family of planets. Scientists in the 1980s have come to view each planet as a "natural laboratory" with a certain set of starting conditions. Each one gives an example of how a planet can evolve under one particular set of conditions (Figure 6-2). It is as if nature has given us several sets of planetary experiments. For example, we can ask why Earth came to have a nitrogen-oxygen atmosphere, whereas similar-sized Venus has a thick, carbon dioxide atmosphere. If we understand processes on Earth, it helps us understand processes we find on other planets. Conversely, as we study other planets, we come to understand processes of atmospheric evolution, geological evolution, and environmental change that affect Earth.

EARTH'S AGE

A variety of evidence shows us that Earth formed about 4.6 billion years ago (4,600,000,000 y) from particles orbiting around the newly formed Sun. This knowledge is only a few decades old. In the Middle Ages, for example, most scholars thought Earth was much younger, a supposition based partly on literal interpretations of biblical references. Around 1650 Archbishop James Ussher used this method to make a famous calculation that the cosmos was formed on Sunday, October 23,

Figure 6-1 The image of Earth hanging in space, as seen from partway to the Moon, provided humanity with a new perception of our planet as a fragile, finite globe. In this view, much of North America is under winter cloud cover, the deserts of the Southwest and Baja California are prominent. (NASA photo by Apollo 13 astronauts.)

Figure 6-2 Earth's mountain ranges are a unique feature of our planet. Crumpling of the crust creates uplifted areas, and subsequent erosion by water and wind creates types of sharp canyons and peaks not found on other worlds. Water, especially, carves valleys (right center) and deposits the eroded sediments at the valley mouth (light triangular deposit). (French Alps above Val d'Isère; photo by author.)

4004 B.C., and that humanity was created on Friday, October 28.

By the 1700s, however, the first geologists challenged these ideas; they measured sediments accumulated at the mouths of European rivers and noted that at the presently observed rate of accumulation, it would take much more than Ussher's 6000 years to accumulate the observed sediments.

Measuring Rock Ages by Radioactivity

Accurate measurements of ages of rock samples, leading to the 4.6 billion year age of Earth, come from the discovery of radioactivity. The technique is worth discussing in some detail because it allows us to date not only Earth, but also rock samples from cosmic sources such as the Moon and meteorites, thus shedding light on the age of our whole solar system and even the universe itself.

In 1896 French physicist Antoine Becquerel accidentally left some photographic plates in a drawer with some uranium-bearing minerals. Later he opened the drawer and found the plates fogged. Being a good physicist, he did not dismiss the event but investigated. He found that the uranium emitted "rays," which, like X-rays (discovered in 1895), could pass through cardboard. The rays turned out to be energetic particles emitted by unstable atoms. Radioactivity had been discovered.

A **radioactive atom** is an unstable atom that spontaneously changes into a stable form by emitting a particle from its nucleus. The original atom thus becomes either a new *element* (change in the number of protons in the nucleus) or a new *isotope* of the same element (no change in the number of protons but a change in the number of neutrons). The original atom is called the **parent isotope** and the new atom is called the **daughter isotope**.

The time required for half of the original parent isotopes to disintegrate into daughter isotopes is called the **half-life** of the radioactive element. If a billion atoms of a parent isotope were present in a certain mineral specimen, a half billion would be left after one half-life, a quarter billion after the second half-life, and so on. As examples, half the rubidium-87 atoms in any given sample change into strontium-87 in 50 billion years; uranium-238 changes (in a series of steps) into lead-206 with an effective half-life of 4.51 billion years; potassium-40 changes into argon-40 with a half-life of 1.30 billion years; and carbon-14 decays into nitrogen-14 with a half-life of only 5570 y.

Early in the twentieth century, physicists realized that here was a way to determine the ages of rocks—the date when a given rock formed. Suppose we could determine the *original* number of parent and daughter isotope atoms in a rock. In part, this can be done by measuring numbers of stable isotope atoms, which usually occur in certain proportions to the unstable isotope atoms in a given fresh mineral. Then, if we simply count the *present* numbers of parent and daughter isotope atoms in the rock, we can tell how many parent atoms have decayed into daughter atoms and hence tell how old the rock is. If half the parent atoms have decayed, the age of the rock equals one half-life of the radioactive parent element being studied. This technique of dating rocks is called **radioisotopic dating**, because radioactive atoms that decay are called radioisotopes.

Note that the quantity being determined in radioisotopic dating of rocks is the time since the rock began to retain the daughter element. In most cases being discussed here, this is the time since the rock solidified from an earlier molten material, such as lava. Once the rock solidifies, any daughter isotope atoms are trapped in the solid mineral structure.

Measuring Earth's Age

The oldest known Earth rocks, formed about 3.9 billion years ago,[1] were found in Greenland. Rocks that have survived so long are extremely rare. Thus, Earth must have formed more than 3.9 billion years ago. Without additional information, we would not be able to fix the age exactly. As we will see in the next few chapters, however, more information comes from outside Earth. Lunar rocks and meteorites show that *all the planets formed within a relatively short interval (about 50 to 90 million years) about 4.6 billion years ago* (Pepin, 1976). The **age of Earth** is therefore believed to be 4.6 billion years.

EARTH'S INTERNAL STRUCTURE

The world's deepest drill hole as of 1984 was a 12-km shaft in the Soviet Union (Kozlovsky, 1984). It has yielded interesting data on deep rocks, but no drill hole is deep enough to reveal Earth's deep internal structure. Our best clues about the interior come instead from waves that pass through the Earth's material. If you snap your finger on the surface of a calm swimming pool, waves radiate across the surface from the disturbance. Observers at other edges of the pool can gain information about the disturbance by observing the waves— for example, they can locate the disturbance by comparing the directions from which the waves come.

In the same way, earthquakes generate waves in Earth, and these waves carry information not only about the earthquake but also about the Earth's material. Waves traveling through the Earth (or other planets) are called **seismic waves**. Some travel along the surface and others penetrate through the Earth. The velocity and characteristics of the waves depend on the type of rock or molten material they traverse.

Studies of seismic waves have revealed two important types of layering in the Earth: chemical and physical. Compositional layering refers to layers of different composition. Physical layering refers to layers of different mechanical properties, such as rigid layers versus plastic or fluid layers.

Compositional Layering of Earth: Core, Mantle, Crust

Compositional layering was the first type of layering recognized. Seismic and other data indicate that Earth contains a central **core** of nickel-iron metal. The core's radius is about 3500 km, just over half Earth's radius (see Figure 6-3). The core is surrounded by a layer of dense rock, called the **mantle,** that extends most of the way from the core to the surface. Near the surface, the densities of the rocks are typically lower. The **crust** is a thin outer layer of lower-density rock about 5 km thick under the oceans and about 30 km thick under the continents.

The core–mantle–crust structure gives us important clues about the history of Earth and other planets. First, it shows the importance of **differentiation** processes—processes that separate materials of different composition from one another. Most geologists believe that the key differentiation process in the Earth was melting of much of the inner rock material after Earth formed. The source of the heat was radioactive minerals trapped in Earth as it formed. Gradually those minerals released heat as radioactive atoms decayed. The interior of Earth was so well insulated by overlying rock that the heat could not escape. The temperature rose until the rock melted. When the rock melted, heavy portions like metals flowed downward toward the center, while lighter, low-density minerals floated toward the surface, where they eventually solidified into a crust of low-density rock.

A simple example of differentiation that can produce such layering occurs during the smelting of metal ores. When the ore is melted, the metal sinks to the bottom of the vat (core) while the bulk of the rock fills the upper

---

[1]One billion years = 10^9 y = 1,000,000,000 (abbreviated "1 b.y."). (Note that the English use *billion* to mean 10^{12}! They usually write 10^9 as "one thousand million." Here we will use the American convention.)

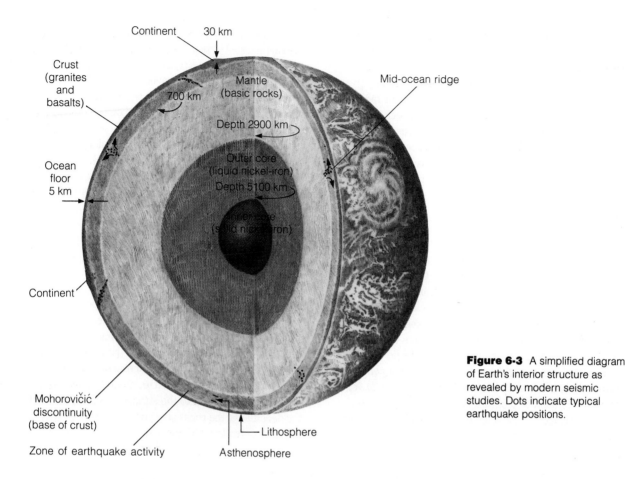

Continent   30 km

Crust
(granites
and
basalts)

700 km

Mantle
(basic rocks)

Mid-ocean ridge

Depth 2900 km

Ocean
floor
5 km

Outer core
(liquid nickel-iron)

Depth 5100 km

Inner core
(solid nickel-iron)

Continent

Mohorovičić
discontinuity
(base of crust)

Zone of earthquake activity

Asthenosphere

Lithosphere

Figure 6-3 A simplified diagram of Earth's interior structure as revealed by modern seismic studies. Dots indicate typical earthquake positions.

part of the vat (mantle). On the surface floats a thin scum of slag, or low-density rock (crust). Note that certain amounts of metals are chemically attracted to minerals in the crust and hence tended to stay with those minerals as they formed. Thus we can have iron and other dense ores in crustal rocks, even though much of Earth's iron is in the core.

As a more detailed example of this theory, we note that one of the lowest-density and most common minerals to form in cooling, molten rock is called *feldspar*. If the theory is right, feldspar should have formed and floated toward the surface. Indeed, we find that surface rocks of both Earth and the Moon are extremely rich in feldspar. One of the rock types formed from mixtures of minerals rich in feldspar is called **basalt**. Basalt is the common lava that erupts from volcanoes tapping the crust and upper mantle, and basalt is also common on the Moon's surface.

A still lower-density type of rock is **granite**, the light-colored quartz-rich rock commonly formed from molten materials in continents. In some ways, conti-

nents seem to be a low-density granitic "scum" floating on the denser rocks of the basaltic lower crust and upper mantle.

Support for these ideas also comes from meteorites, the stone and metal fragments that fall onto Earth's surface from space. As will be discussed further in Chapter 13, they are believed to be fragments of small interplanetary bodies that have broken apart during collisions among themselves. Interestingly, they show many of the rock types we have discussed: Nickel–iron meteorites are probably metal fragments from cores of these small worlds, while basaltlike meteorites are probably fragments of mantles and crusts. Thus we see that other worlds may have gone through a differentiation process similar to Earth's.

Physical Layering of Earth: Lithosphere and Asthenosphere

The second type of layering involves layers of different rigidity. This is thus a physical, rather than a chemical layering. It is extremely important in determining what types of features we see in the landscapes around us on Earth—and on other planets.

Layering by rigidity was created during the cooling of Earth. As you can imagine from the example of the smelter vat, the interior stays molten for a long time because it is hard for the heat to get out, but the surface cools fast because it is exposed to the atmosphere and surrounding space, and the heat can easily radiate away. Thus if a planet were melted and left to cool, a solid layer would form on the surface, while the center is still molten liquid. This picture is complicated by the fact that the layers at different depth have different compositions (because of differentiation) and are also at different pressures. They may thus have different solidifying temperatures, and they may actually form alternating layers of solid material, molten material, or partly molten slushy material. A partly molten layer can be visualized as a layer in which some low-melting-point minerals are melted, but high-melting-point minerals are not. (For example, if you heated a mixture of ice and sand, you would get a mixture of solid sand grains with water between them.)

The solid layer at the surface of Earth or any other such planet is called the **lithosphere** (from Greek roots meaning "rock shell"), and it is underlain by a partly melted layer that is much less rigid and less strong. As shown in Figure 6-3, the underlying partly melted layer is called the **asthenosphere** (from Greek roots for "weak shell").

Interplay between the lithosphere and asthenosphere determines the surface features of Earth, such as mountains, seafloors, and continents. To understand this, we need to review some principles of heat flow—to describe how heat gets out of the hot interior of Earth or any other planet, once the planet is heated by radioactivity. Heat always flows from hot to cooler regions, and it can flow by three methods. The first is radiation; an example is the radiant heat that reaches us through space from the Sun. The second is conduction; an example is the flow of heat through a metal cooking pot, whose handle might grow too hot to touch. The third is convection, which is heat flow by movement of the heat-carrying medium; an example is the buoyant ascent of warm air in a thundercloud or the rise of a hot-air balloon. If you put an inch-deep layer of cooking oil in a pan and heat it from below on a stove, you may be able to see patterns of currents, called convection cells, set up in the oil as it convects. See the Project, p. 100.

Scientists now believe Earth and other planets were fairly hot when they formed. Once each planet formed, its insides cooled by all three processes. Heat radiated from the surface into space; heat was conducted outward from the hot center through rock; and convection may have occurred in some of the more fluid layers where currents can flow, like the convection cells in the cooking oil. As Earth cooled, the surface layers solidified, forming a relatively rigid surface layer of rock—the lithosphere. In the Earth, the lithosphere is about 100 km deep, and the asthenosphere underlies it, at a depth of 100 to 350 km. Even though the asthenosphere is not totally molten, it seems to be plastic enough for sluggish convection currents to flow. These currents bring up hot mantle material, creating "hot spots" in the Earth's crust where volcanoes are likely. The currents also drag on the underside of the lithosphere, setting up stresses in it and tending to crack it into large pieces.

Just like arctic explorers crossing ice floes that are floating on the ocean, we are living on a rigid layer that "floats" on a more fluid base. Of course, we are not in quite so much danger of a cracking floor as the arctic explorer, but cracks do occur. Figure 6-4 shows a result. As the asthenosphere shifts, it can stretch the brittle lithosphere only so far before it cracks. This is what we perceive as an earthquake.

LITHOSPHERE AND PLATE TECTONICS: AN EXPLANATION OF PLANETARY LANDSCAPES

Tectonics is the study of movements in a planetary lithosphere, such as the movements that cause earthquakes, mountain building, and so on. For more than a century, geologists studied Earth's tectonics without recognizing the underlying nature of the forces that cause these movements, as just outlined. Only since the 1960s have geologists pieced together a real understanding of how these principles affect Earth and how they apply to other planets (Hurley, 1968). This new understanding is called the theory of **plate tectonics**.

As the asthenosphere drags on the more brittle lithosphere, it cracks the lithosphere into large, continent-scale pieces called *plates*. Further asthenosphere movements tend to drag and jostle the floating plates, sometimes pulling them apart from each other and sometimes pushing them into each other, as shown by the cross section in Figure 6-5. Cracks along the margins of plates are usually the sites of volcanoes and

Figure 6-4 Occasionally nature reminds us that we are living on a thin, brittle planetary lithosphere that is often broken by motions of underlying semifluid layers. This devastation was caused in 1906 in San Francisco by an earthquake and resultant fire. The quake involved movement on the San Andreas fault, which passes under the ocean just west of the downtown area. (Photo courtesy S. M. Larson.)

earthquakes. Volcanoes form because molten magma from below can squeeze up to the surface through the cracks. Plate collisions cause stresses—earthquakes—as plates rub together.

These stresses often cause violent fractures of the lithosphere rock layers. The fracturing events are better known as earthquakes, as California residents can attest. Fractures caused by earthquakes are called **faults**. Colliding plate regions are laced with faults like the famous San Andreas fault. Many earthquakes are caused by movements along these faults. As shown by Figure 6-4, the devastation can be extreme. Although earthquake prediction is only in its infancy, the U.S. Geological Survey has predicted a moderate to large earthquake on the central San Andreas fault by 1992 (near Parkfield, between Los Angeles and San Francisco) and a still larger earthquake in the Riverside–Palm Springs area of southern California by about 2020 (Silberner, 1987).

The theory of plate tectonics also explains why rocks older than 1 or 2 billion years are so rare on Earth. Most older surfaces have been crumpled beyond recognition or driven downward under other plates, to be remelted, mixed with mantle material, and perhaps reerupted as new lavas.

Now we can see a connection with the landscapes on other planets, which will be discussed further in later chapters. Smaller worlds, like Mars and the Moon, do not have well-developed crumpled mountain ranges or plate boundaries because they cooled faster than big worlds. Their lithospheres got thicker in the same amount of time. Thus their surfaces are more stable and more protected against asthenosphere currents far below. Lava does not so frequently gain access to the surface. Convection can't so easily drive plates apart or cause them to drift into each other. Therefore, the surfaces of Mars and the Moon have much more ancient structures (such as ancient impact craters) than Earth does.

OTHER IMPORTANT PROCESSES IN EARTH'S EVOLUTION

Volcanism is the eruption of molten materials from a planet's interior onto its surface. On Earth, the asthenosphere contains pockets of partly melted materials, as indicated by seismic wave analysis. This underground molten rock, called **magma**, is under pressure, often charged with gas such as steam, less dense than surrounding rock, and highly corrosive. Therefore it tends to work its way to the surface, especially in regions where fractures provide access. When it reaches the surface, it erupts and is then called **lava**. It may shoot in a foamlike form into the air (due to pressure from dissolved gas, like the spray from an agitated can of pop when it is opened) or ooze out and flow for many kilometers across the ground. If enough lava is erupted, it may accumulate into volcanic mountains. During intervals ranging from years to millions of years, volcanism thus creates new landforms ranging from flat lava flows to craters and volcanic peaks, as seen in Figures 6-6 and 6-7.

Space exploration has shown that volcanism is one of the most important processes forming landscapes on other planets. Some planets have huge lava flows and volcanoes. Study of Earth's volcanoes helps us understand these alien landscapes. Conversely, study of the other planets' volcanic features helps us understand relations we see among terrestrial volcanoes.

Among all known planets, Earth undergoes the most active processes of landform destruction. Largely, this activity is due to its thick atmosphere and flowing water, which other planets lack. **Erosion** includes all processes by which rock materials are broken down and transported across a planet's surface; such processes include water flow, chemical weathering, and windblown transport of dust. **Deposition** includes all processes by which the materials are deposited and accumulated; such

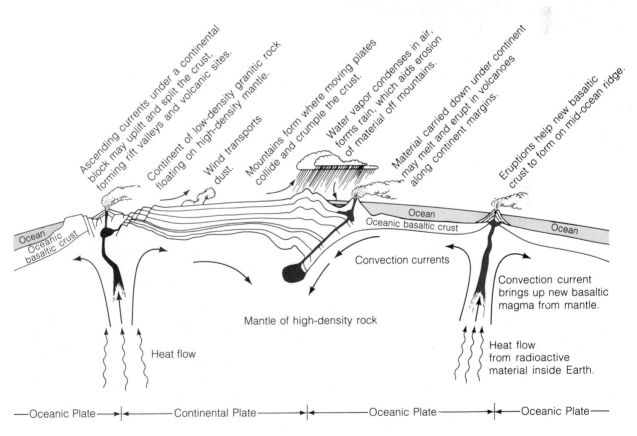

Figure 6-5 Schematic diagram of some of the processes making Earth more geologically active than most other worlds. Convective heat flow from the interior drives convection currents in the asthenosphere. These in turn crack the more rigid surface layers and move plates. Plate movements cause earthquakes, faulting, and the formation of rift valleys and crumpled mountain chains. In addition atmospheric processes and water flow cause rapid erosion. Vertical scale is exaggerated for clarity.

Figure 6-6 Volcanic activity on Earth builds varied landforms. This crater formed as the ground collapsed around an erupting vent; its light-colored floor is filled with windblown dust. It is on the flank of a shield volcano, a type found on several planets and named for its profile, as seen on the horizon. The rest of the landscape is covered by overlapping lavas and cinders and is dotted by volcanic cones. (Pinacate volcanic field, Mexico; photo by author.)

processes include deposition of sediments in lake and ocean bottoms and dropping of windblown dust in dune deposits. Most of those processes do not occur on most other planetary bodies.

EARTH'S MAGNETIC FIELD

If we move a compass throughout the region around Earth, we discover that there is a **magnetic field** defined by the fact that the compass is deflected in a direction roughly toward the North Pole. The compass does not point exactly at the true North Pole (defined by the Earth's spin) but at a point called the magnetic North Pole, which moves slightly from year to year and is presently located among arctic islands off northern Canada.

What causes Earth's magnetism? The interior of Earth cannot be magnetized like a giant bar magnet, because the interior regions are too hot to maintain magnetism. Rather, geophysicists believe that liquid iron in the outer core is slowly convecting, and that these

motions set up electric currents in the metal. A magnetic field will exist around any electric current, and Earth's field is thus attributed to currents in the liquid metal core. The presence or absence of a magnetic field thus becomes an important indicator of the presence of an iron core inside a planet.

EARTH'S ATMOSPHERE AND OCEANS

Just as Earth's interior and surface have dramatically evolved, so have the atmosphere and oceans—the linked layers of gas and fluid on the surface. The original atmosphere was probably rich in hydrogen (H) compounds, such as methane (CH_4), ammonia (NH_3), and water (H_2O), because hydrogen dominated the gas in the early solar system (see Chapter 14). This **primitive atmo-**

Figure 6-7 During Earth's history, eruptions of lava have created new land masses, pushing back the sea even as the sea eroded coastlines of old land masses. (Photo by author; 1988 Kalapana lava flow, Hawaii.)

Figure 6-8 Gases vented at volcanic fissures add water vapor, carbon dioxide, nitrogen, and other chemicals to the atmosphere. Volcanism was a major factor in altering Earth's original atmospheric composition. Yellow deposits are crystals of pure sulfur condensed from the hot, sulfur-rich gas on the cool surface rocks, just as water vapor from a teakettle may condense on a window. (Hawaii Volcanoes National Park, Hawaii; photo by author.)

sphere had less oxygen than today's atmosphere. Evidence for this is that sediments deposited more than about 2.5 billion years ago are less oxidized than modern sediments, indicating less oxygen was in the air. In addition, the earliest fossil life forms, dating from about 2.5 to 3.5 billion years ago, are types of algae found today only in oxygen-poor environments, such as salt marshes along seacoasts.

Indeed, most of Earth's present air is regarded as a **secondary atmosphere**—gas that was not originally present but was added by outgassing of Earth's interior from volcanic vents and fissures, as shown in Figure 6-8. Such volcanic gases are rich in water vapor

(H_2O), carbon dioxide (CO_2), and nitrogen (N_2). Today's atmosphere is 76% nitrogen and 23% oxygen by weight.

Scientists believe the oceans and atmosphere evolved in part from these volcanic gases. For example, the water vapor emitted by the earliest volcanism condensed into liquid water and eventually formed oceans. Because of Earth's unique temperature range between 0° and 100°C, Earth is the only planet with vast expanses of liquid water, as Figure 6-9 reminds us.

The carbon dioxide that was emitted at the same time mostly dissolved in the oceans, where it formed a weak carbonic acid solution. (Much of the carbon dioxide emitted by modern industrial processes does the same thing, with only a certain fraction of the total remaining in the air.) The carbonic acid reacts with rocks and sediments of the seafloor, eventually forming carbonate rocks. Through this direct chemical process, as well as by biological processes, most of the carbon dioxide has been tied up in carbonate rocks such as limestone. The nitrogen (N_2), being chemically inactive, remained in the air, forming the nitrogen observed today.

A small amount of water is always in the air as water vapor, and molecules of water are sometimes struck by powerful ultraviolet rays of sunlight and broken into

Figure 6-9 The most common "landscape" on Earth is unique in the solar system. An open ocean of liquid water covers about three-fourths of our planet. (Pacific Ocean; photo by author.)

hydrogen and oxygen atoms. The hydrogen, being the lightest element, tends to rise and eventually escape into space. The oxygen remains behind. At first, this was the only source of oxygen, and the atmospheric oxygen content increased quite slowly. Then about 2 or $2\frac{1}{2}$ billion years ago, abundant plant life evolved, consuming carbon dioxide and releasing oxygen. The atmospheric oxygen content thus increased more rapidly, eventually reaching the present level.

THE ASTRONOMICAL CONNECTION

Most people think of Earth as isolated from astronomical influences, having evolved independently as a result of its own internal forces. But recent research indicates a number of ways in which *external* astronomical forces influence Earth's evolution in dramatic ways.

Meteorite impact craters have been created by impacts of interplanetary bodies up to 10 km or more in diameter. The eroded craters, reaching at least 140 km (90 mi) across, have been recognized on Earth only in the last century. In the last two decades, dozens have been identified. Most are thousands or millions of years old and have been damaged by erosion, but Figure 6-10 shows a young, well-preserved example.

Impact craters formed at a much more rapid rate in the early days of Earth's history than they do today. Although the most ancient craters have been obliterated by erosion on the Earth, they are preserved in great numbers on the Moon and planets. The intense early bombardment of Earth may have affected Earth's early crustal structure, the early environment for forming

life, and the origin of the Moon (see next chapter). Layers of once-melted spherules, apparently droplets of melted rock sprayed out of impact explosions, have been found in 3.4-billion-year-old rock layers in Africa (Lowe and others, 1989), showing that Earth was bombarded by giant meteorites during its early history.

Until the 1980s, geologists considered such bombardment to be an unimportant effect, but recent evidence suggests that this astronomical connection was more important than earlier scientists ever dreamed. To understand this, note in Table 6-1 that geologists have divided Earth's history into periods, based on fossil life forms and on radioisotopic dates. This table is called the **geological time scale**. It chronicles shifts in Earth's biology and environment. Radical environmental changes occurred at certain times, marked by dramatic alterations in fossil life forms. The question is: What caused them? Some may have been caused by processes arising from within Earth itself, such as mountain building that changed sea levels or volcanism that destroyed large areas and changed climates with smoke fumes. But other changes may have had external, astronomical causes.

During the especially dramatic change 65 million years ago, about three-fourths of all species became extinct within a few million years—or perhaps much less! This extraordinary change marks the division between the Cretaceous period (see Table 6-1) and a

Figure 6-10 Meteor Crater, in Arizona, testifies to the effects of interplanetary debris hitting Earth. A meteorite impact formed this 1.2-km crater about 20,000 y ago. **a** Road and museum buildings (left) give scale in the aerial view. **b** View from rim at sunset shows strata distorted by the explosion and boulders thrown out onto the rim. (Photos by author.)

TABLE 6·1

Geological Time Scale

Era	Age (millions of years)	Period	Life Forms	Events
Cenozoic	0	Quaternary	Humanity	Technological environmental modification; extraterrestrial travel; ice ages
	3			
	65	Tertiary	Mammals	Building of Rocky Mountains
Mesozoic		Cretaceous	Extinction of many species of plants and animals, including dinosaurs	Large asteroid impact (65 million years ago); continents taking present shape
	130	Jurassic	Dinosaurs	
	180	Triassic	Reptiles	
	240			
Paleozoic		Permian	Conifers; extinction of many species of plants and animals	Building of Appalachian Mountains
	280	Pennsylvanian	Ferns	Pangaea breaking apart
	310	Mississippian		
	340	Devonian	Fishes	
	405	Silurian	Early land plants	
	450	Ordovician		
	500	Cambrian	Trilobites	Earliest abundant fossils (trilobites, etc.)
	570			
Proterozoic	640	Ediacarian	Small soft forms	
	1000		First macroscopic life forms; sexually reproducing life forms	Growth of protocontinents; scattered fossils
	2000			
	2600		Oxygen-producing microbes	Oxygen increasing in atmosphere
Archeozoic	3000			Crustal and atmospheric evolution
	3600		Microscopic life	Earliest fossils (algae)
				Oldest rocks; crustal formation?
				Heavy meteoritic cratering
	4500			
Formative	4600			Formation of Sun and planets (see Chap. 14)
Presolar	12,000			Formation of Milky Way galaxy (see Chap. 23)
	16,000			Origin of universe (see Chap. 27)

Source: Data in part from Schopf (1975); Morris (1987).

a

b

c

Figure 6-11 Impacts by interplanetary bodies as much as 10 km across and even larger may explain some sudden environmental changes found in the ancient geologic record. These views show the impact of a 10-km asteroid as seen from an altitude of 100 km (60 mi) in views 10 s before impact, 1 s before, and 60 s after. Massive amounts of dust being ejected in the final picture will settle into the atmosphere and be blown around the world by high-altitude winds. Chemical evidence suggests that such an impact 65 million years ago led to the extinction of dinosaurs and many other life forms, probably due to climate changes resulting from the dust pall in the atmosphere, which blocked most sunlight for many months. (See text; paintings by author.)

subperiod called the Tertiary. The radical extinctions at that time are called the **Cretaceous–Tertiary extinctions**. No one could explain the Cretaceous-Tertiary extinctions and environmental change until about 1980. Then, scientists discovered that a soil layer marking the transition contains unusually high amounts of elements common in many meteorites. Meteorites, which are stones that fall out of space onto Earth, are believed to be fragments of interplanetary bodies called asteroids. Within a few years the discovery was confirmed at sites around the world: Some unusual event had distributed the asteroidlike materials all over the Earth. From the data, scientists estimated that an asteroid 10 km across had crashed into Earth, as shown in Figure 6-11 (Alvarez and others, 1980; Ganapathy, 1980; Kerr, 1987a).

Would an asteroid's impact do the job of exterminating species and altering life on Earth? The soil layer was found to contain soot, indicating widespread fires (Wolbach and others, 1986). Calculations on the effects of a large impact indicate that dust from the impact and soot from the resulting wildfires would be thrown high into the atmosphere, blocking sunlight and making it too dark to see for a few months (Toon and others, 1982). This in turn would wipe out lower parts of the food chain, such as sunlight-dependent plankton and small plants, explaining eventual extinction of many higher species. After the large reptiles were wiped out, the hitherto obscure mammals emerged to inherit Earth! The effects of such giant impacts are discussed further in Chapter 28.

In an example of pure research dovetailing with

practical concerns, scientists have used the climate calculations to study effects of a cataclysm associated with nuclear war (Levi and Rothman, 1985; Schneider, 1987). They found that dust and soot raised by explosions and fires in all-out nuclear attacks would block the sunlight, causing temperatures to drop and damaging crops in both target countries and aggressor countries. This theory of **nuclear winter** has been widely publicized and has affected military strategists' thinking about nuclear war. The detailed climate calculations applied to the theory of an asteroid impact and the theory of nuclear winter symbolize our growing understanding of how our planetary climate could be altered by terrible, unforeseen events, both in the past and the future.

Still another example of an astronomical connection to Earth's history comes from studies of Earth's orbit. Many studies suggest that minor orbital changes are caused by gravitational forces of nearby planets, which cause changes in the Earth's tilt, distance from the Sun, and exposure to sunlight. These effects seem to explain most of the climatic changes associated with ice ages in the past million years or more (see review by Kerr, 1987).

As discussed in Chapter 15, certain disturbances on the surface of the Sun also appear to be correlated with disturbances in Earth's climate. Thus modern research suggests many links between astronomy and climate effects on Earth.

SUMMARY

Geological studies of strata and fossils have established a chronology of events in the history of the planet Earth, and radioisotopic dating of rocks has established the actual ages of these events, as summarized in Table 6-1. It is broadly divided into intervals called *eras*. Only the last 14% of Earth's history has yielded enough rock evidence, such as fossils and dates, to provide finer divisions, which geologists call *periods*.

Earth formed some $4\frac{1}{2}$ billion years ago. Giant meteorites struck much more often than today, scarring the primal landscape with great impact craters. The surface was lifeless.

Life probably originated within the first few hundred million years of Earth's history, because the earliest microscopic fossils date back about $3\frac{1}{2}$ billion years ago.

The heating of Earth's interior, probably due to radioactivity, led to a prompt internal melting within the first few hundred million years. This, in turn, caused differen-

tiation—a draining of metal toward the center to form a nickel–iron core and a floating of lighter minerals to form a low-density crust overlying a dense rock mantle.

As Earth cooled and solidified, the surface layer formed a rigid lithosphere overlying a more plastic asthenosphere. Currents in the asthenosphere broke the relatively thin lithosphere into *plates*. Continents repeatedly split and rejoined as rifts broke the lithosphere and plate motions caused crustal masses to collide.

Mountainous landscapes thus replaced the original, cratered surface, though vestiges of the old surfaces can still be found. With the coming of plants and animals, the landscape became familiar to modern eyes. Figure 6-9 reminds us that the dominant "landscape" of our planet is a vast expanse of water—a view unique in the solar system, as we will see in the next few chapters.

In the scale of planetary evolution, recent events such as the ice ages seem the merest of details. Larger climate changes have punctuated the past. They may have had various terrestrial and astronomical causes. For example, a large asteroid impact 65 million years ago probably was the main cause of the mass extinction of species at that time. It is important to understand such events. Even small shifts in climate can produce droughts, affecting world food production. Scientists studying the 65-million-year-old event recently concluded that dust and smoke from nuclear warfare would trigger a devastating climate change known as "nuclear winter"—yet another argument for stopping nuclear arms proliferation!

Let us conclude our summary of Earth's history by compressing events into a single day. Life evolved sometime in the early morning, but the fossil-producing trilobites that begin the traditional geological time scale in the Cambrian period did not appear until about 9:30 in the evening. By 10 P.M. there were fishes in the sea, and by 11 P.M., dinosaurs on the land. Mammals did not appear until about 11:40. Human beings, who have been here perhaps 2 million years (depending on your definition of *human*), made their appearance only 30 seconds before midnight. The last few thousand years—civilization—occurred in a tenth of a second—represented by the pop of a single flashbulb at midnight. The question is: What will be here a tenth of a second after midnight?

CONCEPTS

radioactive atom	age of earth
parent isotope	seismic waves
daughter isotope	core
half-life	mantle
radioisotopic dating	crust

differentiation
basalt
granite
lithosphere
asthenosphere
plate tectonics
fault
volcanism
magma
lava

erosion
deposition
magnetic field
primitive atmosphere
secondary atmosphere
meteorite impact crater
geological time scale
Cretaceous–Tertiary
  extinctions
nuclear winter

PROBLEMS

1. When were the last major earthquakes in your region? Is your region seismically active or inactive? How is it located with respect to the boundaries of tectonic plates?

2. If a rock sample can be shown to contain one-fourth of its original amount of radioactive potassium-40, how old is it?

3. In view of the preservation of ancient craters and the lack of folded mountain ranges on the Moon, would you predict the Moon to have more or less seismic activity than Earth? Discuss your reasoning.

4. Fossils of apelike predecessors of the genus *Homo* (such as *Australopithecus*), found in Africa, are believed to date back at least 2 to 3 million years. What percentage of the Earth's age is this? Can you accept, philosophically, that events happened during most of the history of the Earth before anyone was around to see, hear, or record them?

5. If the Earth is 4.6 billion years old, about how much more radioactive uranium-238 did it have when it formed? Would the heat-production rate from U-238's radioactivity when the Earth formed have been more, less, or the same as it is now?

6. The composition of Earth's atmosphere is probably much changed from what it first was.
  a. What happened to the abundant hydrogen atoms initially present or produced by breakup of molecules such as methane (CH_4)?
  b. Given that much ammonia (NH_3) was initially present and that ammonia molecules break apart when struck by solar particles in the atmosphere, account for one source of the Earth's now abundant nitrogen.
  c. What three gases were added abundantly by volcanoes, and where did these gases finally end up?

7. Describe what you would expect conditions on Earth to be if Earth were so close to the Sun that the mean surface temperature was above 373 K (100°C). What if Earth were far enough from the Sun that the mean surface temperature was below 273 K (0°C)? Explain why a view like Figure 6-9 is probably unique to Earth.

8. Suppose you could literally walk back through time, with 1 pace = 1000 y, and 1000 paces to a mile. How far would you have to walk to get back to:
  a. Cretaceous–Tertiary boundary
  b. Beginning of the Mesozoic
  c. Beginning of the Paleozoic
  d. Beginning of Earth

Locate some cities or other geographic features that approximate these distances from your college or university.

PROJECT

1. Place cooking oil an inch or two deep in a flat pan over low heat and under a single strong light. Because the oil does not readily boil, heat is soon transmitted to the surface in visible convection currents. Note how the currents divide the surface into cells, or regions of ascent, lateral flow, and descent. These cells are analogous to tectonic plates in Earth's surface layers. Sprinkle a slight skin of flour on the oil's surface to simulate floating continental rocks, and watch for examples of continental drift and plate collisions. Experiment with different depths of oil and different temperature gradients (by changing the heat setting), and record the results. Be careful not to turn up the heat too high (especially on a gas stove) to avoid having the cooking oil catch fire!

CHAPTER 7

The Moon

The same Moon that we see today shone down on the breakup of the continents, gleamed in the eyes of the last dinosaurs, and illuminated the antics of the first protohumans. The same Moon was seen by all the historical figures we have mentioned in preceding chapters: Stonehenge builders, Mayan eclipse observers, Aristarchus, al-Battani, Isaac Newton. They all asked what it is, where it came from. Today the Moon is a little different: It has footprints on it and we know some of the answers.

The Moon is the Earth's only natural **satellite**—a body that orbits around a larger body. (The term *moon* is also used generically to mean a satellite.) As a world, it is respectable: It has a diameter of 3476 km, about one-fourth the diameter of Earth. It is a rocky world splotched with dark gray flows of ancient lavas and dotted with craters formed by explosions when meteorites hit it in the ancient past. Ours is the first generation to have seen a few of the Moon's starkly beautiful landscapes. We know, not just from instruments but from the *experience* of our astronauts, that it has no air, no water, no weather, no blue sky, no clouds, no life.

In spite of the Moon's new familiarity, many people are still confused about even its simple phases. Can you recall which way the horns of the crescent point in the evening sky? Cartoonists often draw the horns pointing down toward the horizon. Not so. Because the fully illuminated edge of the crescent must face the Sun, which has just set below the horizon, the horns must point upward, away from the horizon. Scoff, too, at the novelist who describes the full moon rising at midnight! To be fully illuminated, the Moon must be opposite the Sun and hence the full moon must rise as the Sun sets.

In this chapter we will first consider the Moon's movements along its orbit, then examine the Moon's surface and the astronauts' discoveries, and finally explore the puzzling problem of the Moon's origin.

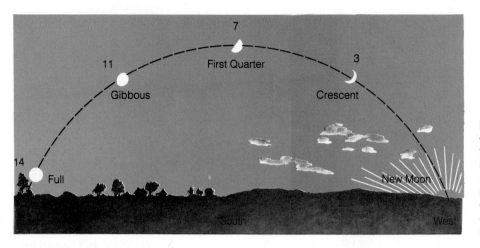

Figure 7-1 Wide-angle view of the sky, showing some of the phases of the Moon as seen by a Northern Hemisphere observer at sunset on the indicated days of the monthly lunar cycle, counting from the day of new moon.

THE MOON'S PHASES AND ROTATION

As can be seen by comparison of Figures 7-1 and 7-2, the Moon's **phases**, or shapes on different days, are directly caused by the Moon's motion in its monthly orbit around Earth. Let us say that on day 0 the Moon crosses the line between the Earth and the Sun. Here the Moon is nearly in front of the Sun, as seen from Earth, and is lost in the Sun's glare. On this day the Moon generally cannot be glimpsed. This is called the

date of **new moon**. A couple of days later, the Moon has moved far enough from the Sun to be glimpsed in the early evening sky; it is backlit by the Sun, giving it a crescent shape. On day 7, it is 90° from the Sun, a phase called **first quarter**, because it is a quarter of the way around its orbit. It looks half-illuminated at this time. For the next week it is more than half-illuminated, a phase called **gibbous** (hard *g*, as in *give*). On day 14 or 15, it is opposite the Sun and fully illuminated—a phase called **full moon**. This is the day on which the Moon rises at sunset and fills the night sky with its

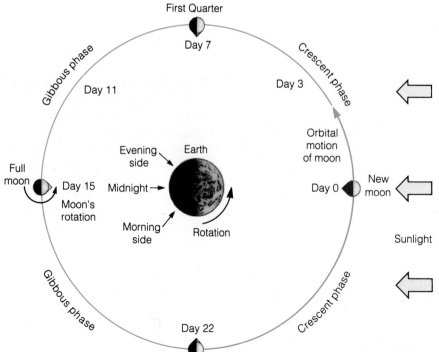

Figure 7-2 "Top view" of the Moon's motion around Earth. Synchronous rotation of the Moon is shown by a mountain (indicated by a triangle) that always faces Earth. The Moon completes one rotation during one complete revolution around Earth. Phases seen by an earthbound observer are indicated at different points in the orbit. First quarter is seen in the evening sky, third quarter in the early morning sky.

brightest possible light. Each evening it rises roughly an hour later than on the previous evening. For the next 2 weeks, the Moon rises after midnight and is visible primarily in the early morning sky. On day 22, the Moon is three-quarters of the way around its orbit and now in a half-lit phase called **third quarter**. On day 29, the Moon is back to the new moon phase.

The term *waxing moon* refers to the first 2-week period, when the Moon is growing more illuminated each day; *waning moon* refers to the second 2-week period, when the Moon is growing slimmer.

The Moon takes 29.5 d to complete one revolution around Earth relative to the Sun. Therefore the complete cycle of lunar phases takes 29.5 d.

Whenever the Moon is visible, no matter what its phase, we can always distinguish at least some of the dark lava plains that make up the features of the "man in the moon." This is because of a curious characteristic of the Moon's motion: As it revolves around Earth, it rotates in such a way that it always keeps the same side facing Earth, as shown by the stylized mountain in Figure 7-2. The rotation of any satellite in this way is called **synchronous rotation** because it is synchronized with the satellite's own revolution, as shown in Figure 7-2. How can the Moon rotate at all, you might ask, if it always keeps the same side toward Earth? The best answer is provided by an example. Put a chair in the middle of a room. The chair is the Earth; the walls, the distant stars. To represent lunar orbital motion, walk around the chair. If you walk around the chair always facing it, so that an Earthly observer in the chair never sees your back, you will find that you have faced all sides of the room during one circuit. In other words, you have rotated once on your axis and made one revolution around the chair. (If you put a strong light in one corner of the room and hold up a ball as you walk around the chair, the observer in the chair can see the cycle of phases on the ball.)

Even writers who should know better sometimes speak of the Moon's "eternally dark side." This mistake comes from a popular belief that the side hidden from us must always be dark. But the far side, just like the near side, has day and night—periods of sunlight and periods of darkness. This can be seen in Figure 7-2. Each period lasts about 2 weeks, because the Moon takes about 4 weeks to make a complete rotation. There is always a dark side, but it isn't necessarily the *far* side.

The Moon's synchronous rotation has another consequence. For an astronaut at any spot on the near side of the Moon, the Earth hangs forever in the same spot

a   b

Figure 7-3 Seen from the Moon, Earth goes through phases. **a** The crescent Earth rises, as seen from orbit over the cratered lunar highlands. **b** The gibbous Earth seen on another occasion from a similar position. (NASA photos by Apollo 17 and 8 astronauts, respectively, orbiting over the Moon.)

a

North

South

b

Figure 7-4 a At full moon, the rugged appearance of the terrain is minimized by the absence of shadows. Full lighting emphasizes different features, such as the bright rays emanating from the crater Tycho (bottom). **b** Several folklore figures, such as "the woman in the Moon," are formed by the pattern of dark lava plains. Squinting at *a* and *b* may help you see these features. (Photo from Lunar and Planetary Laboratory, University of Arizona.)

in the sky. (Imagine living on the lunar mountaintop in Figure 7-2; Earth would always be overhead.) Nonetheless, Earth goes through a complete cycle of phases every 4 weeks, as seen by a lunar astronaut. Figure 7-3 on page 103 shows Earth's crescent and gibbous phases as seen from the Moon.

Is there a reason why the Moon keeps one side toward Earth? When asking for a reason, a scientist normally means to ask: "Could the observation be explained by some more fundamental properties of nature, so that it becomes a special case of a more general phenomenon?" Here the answer is yes. In the 1780s and 1790s, Joseph Louis Lagrange and Pierre Simon de Laplace used Newton's law of gravity (Chapter 4) to show that if the Moon were slightly egg-shaped or football-shaped, gravitational forces would make the longest axis point toward Earth at all times. Confirming this, space vehicles in the last few decades have shown that one axis of the Moon *is* indeed about 2 or 3 km longer than the others and points steadily toward Earth.

SURFACE FEATURES OF THE MOON

Until the telescope was invented around 1609, no one knew much about the lunar surface features except for the gray patches that make up the face of "the man in the Moon" and "the woman in the Moon" (Figure 7-4). Some thought the Moon was a polished sphere. Thomas Harriot and Galileo Galilei, the first people known to have seen the Moon's features through a telescope, made their early observations in 1609 and 1610 (see page 44). They both recorded rugged regions with prominent

shadows cast by mountains and crater rims along the **terminator**, the line dividing lunar day from lunar night (see Figure 7-5). Details were less prominent under high lighting (full moon) or at the edge of the disk, called the **limb**. Galileo reported:

The Moon certainly does not possess a smooth surface, but one rough and uneven, and just like the face of the Earth itself, is everywhere full of vast protuberances, deep chasms, and sinuosities.

Figure 7-5 One of Galileo's first lunar telescopic drawings (left), made in 1610, compared to a photograph of the Moon at the same phase. Letters show corresponding features. Galileo detected mountains, plains, and craters (bottom). (Courtesy Ewen A. Whitaker, University of Arizona.)

These features were later recorded in better detail by telescopic photographs. Most of the roughness was caused by thousands of **impact craters**, or circular depressions caused by meteorite impacts (Figures 7-6 and 7-7), ranging up to 1200 km in diameter. Smaller ones (up to a few kilometers) are bowl-shaped, whereas larger ones have a more complex structure, such as central mountains or concentric rings of cliffs. Galileo found the dark gray patches that are visible to the naked eye and form "the man in the Moon" to be much smoother than the brighter, cratered areas, as seen in his sketch in Figure 7-5. He mistook these dark patches for seas. These "seas," as Galileo himself probably eventually realized, are actually vast plains, which we now know are covered with dark lava. **Dark lava plains** cover much of the front side of the Moon, but only 15% of the whole Moon. The bright regions are cratered, rugged **highland** areas. The highlands are older than the dark lava plains, and this explains why they have absorbed so many more impact craters. Figure 7-8 shows a view that encompasses cratered highlands and dark lava plains.

Galileo's work greatly strengthened the conception

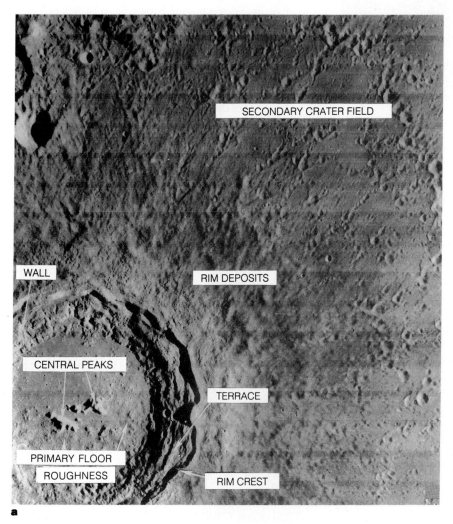

SECONDARY CRATER FIELD

WALL

RIM DEPOSITS

CENTRAL PEAKS

TERRACE

PRIMARY FLOOR

ROUGHNESS

RIM CREST

a

Figure 7-6 a The typical structures in a large lunar crater are seen in this Orbiter photograph of the crater Copernicus, which is about 90 km across. The main bowl was excavated by the impact of a meteorite a few kilometers in diameter. Rim deposits and secondary craters are caused by debris blasted out of the crater during its formation. (Courtesy James W. Head III, Brown University.) **b** An approximate cross section of such a crater.

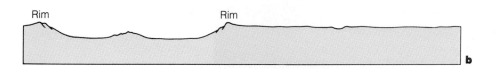

Rim   Rim

b

Figure 7-7 The sculpturing of the lunar highlands from countless impacts by meteorites of all sizes can be imagined from this nearly vertical view. This view, looking past the Apollo 16 command module down onto the Moon's far side, was taken from the landing module just after its undocking from the command module in order to proceed to the lunar surface. (NASA.)

that the Moon is not a magical disk in the sky but an Earth-like world, having familiar geological features such as mountains. He used this as a proof against those who argued that Earth should not be included among the planets.

The first reasonably accurate lunar map was produced by the German Johannes Hevelius in 1647. (A modern lunar map with names of certain prominent features is seen in Figure 7-9.) In 1651 an Italian priest, Riccioli, started the present practice of naming craters after well-known scientists and philosophers, such as Copernicus, Tycho, and Plato. The dark lava plains were given poetic and fanciful names, such as Sea of Tranquillity. Lunar mountains were named for prominent terrestrial ranges, such as the Alps. These mountains, however, are unlike terrestrial folded ranges; they are the rims of vast multiringed craters, called **basins**, which in turn contain the dark lava plains. Figure 7-10 illustrates one of these huge impact features and its probable subsurface structure. Bright streaks called **rays**, which radiate from various craters but show no relief, are fine debris blasted out of the craters. Most lava plains are merely flows of lava that have flooded ancient basins.

Astronomers of the 1700s and 1800s undertook a tantalizing endeavor, searching with ever larger tele-

Figure 7-8 The Apollo 12 lunar landing module sails over several terrain types on the Moon. Ancient cratered terrain dominates the right side. Somewhat younger, smooth lava flows, peppered with small craters and light-toned dust, cover the lower left corner. In the distance, more pristine, dark-colored lava lies near the horizon. (NASA.)

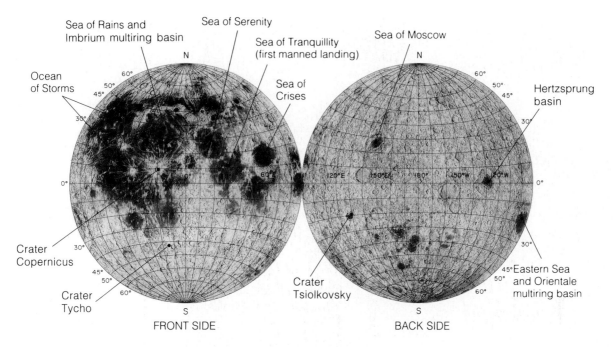

Sea of Rains and Imbrium multiring basin
Sea of Serenity
Sea of Moscow
Sea of Tranquillity (first manned landing)
Ocean of Storms
Sea of Crises
Hertzsprung basin
N
N
Crater Copernicus
Crater Tycho
Crater Tsiolkovsky
Eastern Sea and Orientale multiring basin
S
S
FRONT SIDE
BACK SIDE

Figure 7-9 Map of the front (left) and back (right) sides of the Moon, showing some prominent features. (U.S. Geological Survey, courtesy R. M. Batson.)

scopes with ever greater magnification for clues as to how the Moon formed.

Despite years of search, though, no substantial signs of change were found. The Moon is essentially dormant, preserving geological features from the early eras of solar system history.

Why is the Moon's geological character so much less active than Earth's? Why do craters dominate? Why are there no major mountain ranges? Where did the Moon come from? What would it be like to stand on the Moon? Such questions motivated actual journeys to the Moon.

FLIGHTS TO THE MOON

The first human device to reach the Moon was a Russian spacecraft that carried little scientific equipment and crashed into the Moon in 1959. The first close-up photos of the surface, from the American probe Ranger VII in 1964, showed that the surface was not craggy but covered by a gently rolling layer of powdery soil, scattered rocks, and shallow craters of various sizes. This type of soil cover (Figure 7-11 on page 110) is present nearly everywhere on the Moon and is called the **regolith** (rocky layer). The lunar regolith is typically 3 to 30

m (10 to 100 ft) deep and made primarily of rock fragments and dust blasted out of lunar craters as they were formed. Each well-preserved lunar crater is surrounded by a sheet of such pulverized debris, called an **ejecta blanket**. The regolith is therefore said to be composed of overlapping ejecta blankets. The powdery surface of the regolith is due to small meteorites (some microscopic), which are so abundant that they have "sandblasted" most of the upper few meters into fine dust.

Table 7-1 lists the six Apollo lunar landings and two earlier test flights. Twelve men walked on the Moon during the Apollo program from 1969 to 1972. Figures 7-11 through 7-15 on pp. 110–112 include scenes and samples from these missions. The first two landing missions were cautious tests that touched down on flat, smooth lava plains. The other four Apollo landings sampled a variety of complex and rugged sites. Astronauts collected many samples and placed various instruments in position to make measurements.

LUNAR ROCKS: IMPLICATIONS FOR THE MOON AND EARTH

As seen in Table 7-1, the collected rock samples proved that the lunar surface features are very old. Chemical

a

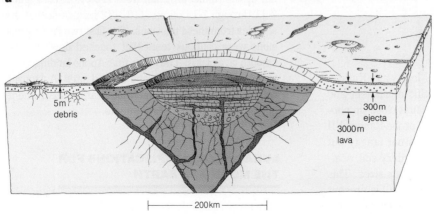

b

Figure 7-10 a The Orientale basin is the youngest and most dramatic multiring basin on the Moon. The outermost ring of cliffs is nearly 1000 km in diameter. Lava erupted to form dark "ponds" along fractures at the base of some cliffs. Map shows scale. (NASA photo from robot orbiter spacecraft.) **b** Hypothetical cross section.

evidence from the rocks showed that the Moon itself formed about $4\frac{1}{2}$ billion years ago, at the same time as Earth.

Lunar rocks older than about 4.2 billion years are rare, or heavily pulverized, because the oldest rocks were destroyed by impacts. Rocks younger than 3 billion years are also rare because by then the Moon had cooled enough that subsequent volcanism or rock-forming activity was infrequent. Because 3- to $4\frac{1}{2}$-billion-year-old rocks are rare on Earth, lunar rocks have given geologists a welcome insight into conditions in the Earth–Moon system during that formative era (Taylor, 1975).

Rocks are cosmically significant. They are solid materials that contain many clues to their long histories of crystallization, melting, recrystallization, and so on. Rocks thus tell the histories of their parent planets. Rocks can be analyzed in various ways:

1. By their elements and isotopes, which indicate the material from which the parent planet originally formed

2. By their minerals, which indicate the degree of differentiation that produced different chemical compounds inside the planet

3. By their structure, which indicates their environments through history

---

TABLE 7·1

Human Apollo Explorations of the Moon

Mission	Date	Landing Site	Results	Typical Rock Ages (billions of years)[b]
Orbital Missions				
Apollo 8	Dec. 24, 1968[a]	——	First lunar orbit. Orbital mapping. 115-km minimum altitude.	
Apollo 10	May 21, 1969[a]	——	Test of approach to approximately 17-km minimum altitude.	
Landing Missions				
Apollo 11	Jul. 20, 1969	Sea of Tranquillity	First landing. Samples of lava plains material.	3.5–3.7
Apollo 12	Nov. 18, 1969	Ocean of Storms	Samples of lava plains.	3.2–3.4
Apollo 14	Feb. 5, 1971	Fra Mauro (ejecta from Imbrium basin)	Samples of ejecta from Imbrium basin.	3.9
Apollo 15	Jul. 30, 1971	Edge of Sea of Rains at foot of Apennine Mts.	Samples of material from Apennine Mts., forming rim of Imbrium basin, and from lava plain. First use of roving vehicle.	up to 4.3 (upland) 3.3–3.4 (lava plain)
Apollo 16	Apr. 20, 1972	Lunar highlands near crater Descartes	First landing in highlands. Samples of highland materials.	3.8–4.2
Apollo 17	Dec. 11, 1972	Taurus Mts. at edge of Sea of Serenity	Samples from a region suspected of recent volcanism.	3.8 (lava plain) 4.2–4.4 (highland)

Note: In addition to samples mentioned above, lunar soil samples were returned to Earth by three Soviet probes without cosmonauts, Luna 16, 20, and 24. Ages were 3.3–3.4 billion years for basalt lavas and about 4.4 billion years for an anorthosite highland rock chip. Additional studies of isotopes in all the rock samples indicate that the Moon as a whole formed around 4.6 billion years ago.

[a]Date lunar orbit began. Dates for landing missions give the day of the landing.

[b]Date that crystalline rocks solidified, or breccias formed, is based primarily on rubidium–strontium results of Wasserburg, Papanasstasiou, Tera, and colleagues at the California Institute of Technology, and a summary by Taylor (1982).

Figure 7-11 Astronaut Harrison Schmitt collects lunar pebble samples with a "rake" during the Apollo 17 mission, final mission to the Moon. Schmitt stands here on dark lava plain soil; brighter hills of the lunar highlands can be seen in the distance. (NASA.)

Figure 7-12 A sample of basaltic lava from the Moon. This sample shows frothy texture caused when bubbles of gas formed inside the molten rock; similar textures are found in terrestrial lava. Other textures are also found on both the Moon and Earth. (NASA; Apollo 15 sample.)

4. By their radioisotopic ages, which indicate when the planet-forming, differentiating, and rock-altering processes occurred

The relative amounts of elements in lunar rocks are similar to those in Earth's mantle, except that the Moon is strongly depleted in **volatile elements and compounds**—the substances that are easily driven off by heating—such as water. This probably indicates that the lunar material was once strongly heated, driving off water and other volatile materials.

The lunar rocks' minerals are also similar to the rock minerals in Earth's mantle, except that the Moon's minerals are not as differentiated. For instance, there are no continental blocks of granitelike rock, as on Earth. These facts support the theory that the Moon has not been as geologically active as Earth. For example, it has never developed the plate tectonics that make Earth's geology so complex. The minerals indicate instead that the primordial lunar surface, $4\frac{1}{2}$ billion years ago, was a sea of molten rock, or magma. It is called a **magma ocean.** The magma ocean was slowly cooling. Low-density crystals accumulated into an anorthosite crust, and this was broken occasionally by lavas erupting from greater depths.

The structure of lunar rocks shows that many of them are composed of fragments of shattered rocks, welded together by heat and pressure. Such rocks are called **breccias** (BREH-chee-ahs). The lunar dust and the breccias testify to the intense cratering that pulverized much of the surface layers throughout lunar history.

The radioisotopic stages show that the magma ocean cooled and solidified, producing the anorthosite crust some $4\frac{1}{2}$ to 4 billion years ago. The lava plains formed when lavas broke through the crust and erupted, often in the floors of the giant impact basins, around 4 to $3\frac{1}{2}$ billion years ago. The ages also illustrate that there has been little geological activity in the last 3 billion years. It is extraordinary to realize that astronauts have walked and driven across landscapes whose craters and rocks have lain still and silent since long before trilobites crawled

Figure 7-13 Apollo 16 astronauts drove their battery-powered "rover" (rear) to the rim of this 40-m-diameter crater in 1972. Note that the crater is old enough to have been smoothed by the sandblasting effect of innumerable small meteorite impacts that pock it with smaller pits and blanket it in regolith. (NASA.)

Figure 7-14 Apollo 17 astronauts obtained samples from huge boulders, apparently dislodged in the past from bedrock further up the slopes of the highland hills near their landing site. (NASA.)

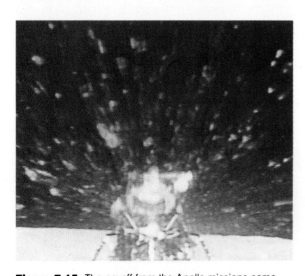

Figure 7-15 The payoff from the Apollo missions came with the returns to Earth. This blastoff from the Moon was televised to Earth by a remote camera set up by Apollo 16 astronauts. Debris flying away from spacecraft includes material blown off the descent stage (lower) by ignition of the ascent engine. (NASA.)

the seafloors of Earth, and whose mountains were in place even as life began evolving on Earth.

THE INTERIOR OF THE MOON

We can better compare the evolution of Earth and the Moon by acquiring information on the Moon's interior (see Figure 7-16). This comes from three sources.

First, the **mean density** of the Moon (total mass divided by total volume) is much less than that of Earth—3300 versus 5500 kg/m^3). This proves that the Moon is mostly rocky, like Earth's mantle, and that it lacks much of a dense, iron core.

Second, the Moon has virtually no magnetic field. This again indicates lack of a large, molten-iron core because scientists believe that planets' magnetic fields originate in currents in such cores. Properties of early lunar rocks indicate that the early Moon had a magnetic field roughly 4% as strong as Earth's present field. Thus, scientists conclude that the Moon probably has a very small iron core, which was molten in the past and is

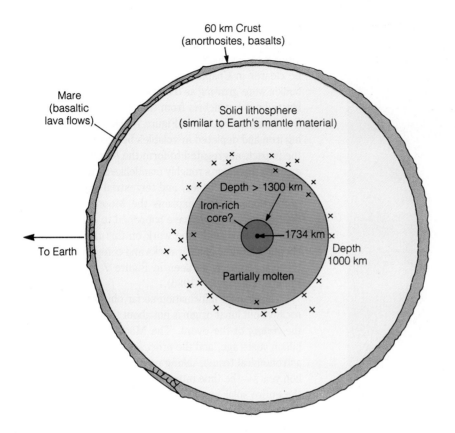

60 km Crust
(anorthosites, basalts)

Mare
(basaltic
lava flows)

Solid lithosphere
(similar to Earth's mantle material)

Depth > 1300 km

Iron-rich
core?

To Earth

1734 km

Depth
1000 km

Partially molten

Figure 7-16 Cross section of the Moon as revealed by Apollo and other sources. The crust is thinner on the front side than on the far side possibly due to big impacts on the front side. Fractures under large impact craters on the front side allowed lava to reach the surface and create more dark lava flows than on the far side. An earthquake zone *(x)* marks the bottom of the lithosphere. The existence of an iron-rich core is uncertain.

solid today. In turn, this indicates again that the Moon has differentiated somewhat, with iron moving toward the center, but not as much as Earth has.

Third, seismic data show that the Moon has fewer major quakes than Earth, again demonstrating that the Moon's interior is not convecting, or flowing, as actively as Earth's interior.

All these data show us that a small planetary body like the Moon cools off faster than a larger body, and thus does not have as much internal heat to drive geological activity such as plate tectonics, differentiation, or long-lived volcanism.

CRATERING OF THE MOON AND EARTH

As noted earlier, the Moon underwent an intense bombardment from about 4.6 to 4.0 billion years ago, with a cratering rate thousands of times higher than today's rate. All large impact basins, such as the one in Figure 7-10, as well as the heavily cratered highlands, formed

at that time. From about 4 to 3 billion years ago, the rate declined to the present level, which has been nearly constant since that time. Earth has had roughly the same cratering rate, and scientists assume that both Earth and the Moon experienced similar cratering histories. The Apollo data from the Moon thus help clarify the impact history of Earth, paving the way for recent theories that giant impacts may have altered climates and biological evolution (see Chapter 6).

The **early intense bombardment** of the Moon represents the final stages of the sweep-up of debris left over after planet formation (see Chapter 14). The lunar craters now visible may represent only the last interplanetary bodies to be swept up, with still earlier craters obliterated by the ones we now see.

WHERE DID THE MOON COME FROM?

The Moon's origin has long frustrated theorists. Prior to the Apollo missions, astronomers debated several

Figure 7-17 The formation of the Moon: a visualization of the events in Figure 7-18. This view shows the scene about 20 min after a giant impact blew out hot Earth-mantle debris that later formed the Moon. (Painting by author.)

theories, but none seemed to explain all the observations.

Apollo studies provided new information. Especially important was the finding that the Moon's bulk material is generally like the Earth's mantle, but with less volatiles. Another important result was that the proportions of different isotopes composing lunar oxygen exactly match those in terrestrial oxygen. Studies in the 1970s showed that oxygen in rocks from other parts of the solar system (such as meteorites) has different isotopic proportions. This establishes that lunar material formed at about Earth's distance from the Sun and rules out, for example, capture of a body from far away.

In 1984, an international conference produced a new leading theory called the **impact-trigger hypothe-**

sis.[1] It suggests that during Earth's formation, but after the iron core formed, Earth was hit by a very large interplanetary body—perhaps as large as Mars. (As will be clearer in Chapter 14, many sizable interplanetary bodies were growing as the planets formed.) The impact blasted out hot debris from the upper mantles of both Earth and the impactor (Figure 7-17). The debris, lacking iron and depleted in volatiles because of the heat of the impact, aggregated to form the Moon. This theory explains the Moon's roughly mantlelike composition, with little iron, few volatiles, and terrestrial oxygen-isotope proportions. It also explains the Moon's uniqueness: Such an impact might have happened to only one of the nine planets. Continuing work on this theory includes chemical studies of lunar rocks and computer models of the giant impact, as seen in Figure 7-18 (Hartmann, Phillips, and Taylor, 1986).

The firmest information so far obtained from lunar rocks about lunar origin is not about the mode but about the *timing* of the event. The Moon formed about 4.6 billion years ago, and the process was relatively fast, in astronomical terms, taking no more than about 20 million years—the time interval estimated for formation of the entire solar system.

We know also, from studies of the Moon's motion, that it was much closer to Earth when it formed and has been slowly moving away ever since. This motion is caused by a complex combination of gravitational forces between Earth and the Moon. These forces are called **tidal forces.**

RETURN TO THE MOON?

Ever since Apollo 17 blasted off the Moon in 1972, the Moon has been deserted. During the 1980s, a number of scientists and NASA planners began advocating a return to the Moon. A robotic satellite to orbit the Moon has been advocated but not funded in the United States; the Soviet Union and Japan are both reportedly planning one. Such a satellite would measure lunar rock com-

---

[1]This theory was first published in 1975 by D. R. Davis and myself and, essentially independently, in 1976 by A. Cameron and W. Ward, but little further work was done on it for several years. By the 1984 conference, new data were available and sophisticated computer models of impact were possible, and it was exciting to see our idea emerge from relative obscurity into the status of the leading theory!

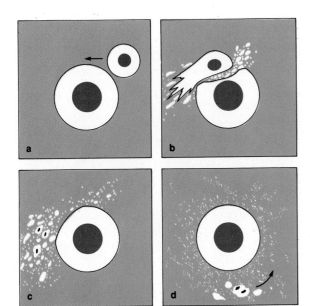

Figure 7-18 Schematic view of the impact-trigger hypothesis for lunar origin. A large interplanetary body approaches (*a*) and collides with the primordial Earth (*b*) following formation of iron cores (dark) in both bodies. Hot gas and condensing debris from the mantles of both bodies are thrown into near-Earth space (*c*). Part of the debris forms an orbiting cloud around Earth, in which the Moon begins to aggregate (*d*). (Adapted from computer models by A. Cameron, W. Benz, W. Slattery, M. Kipp, and J. Melosh.)

positions over the whole globe of the Moon, clarifying what resources might be available there.

Large-scale plans to establish research colonies on the Moon (and on Mars) have also been proposed by NASA and the USSR. Permanent lunar stations could produce oxygen from lunar rocks for breathing and for fuel (Thomsen, 1986). Extracting liquid oxygen fuel from lunar rocks for spaceships and space stations might ultimately be cheaper than hauling it up from Earth, because so much less energy is needed to launch material off the Moon than off Earth. Such lunar research stations would also clarify the Moon's and Earth's origin and evolution, and would conduct astronomical research more effectively than from low Earth orbit, where half the sky is blocked by the Earth.

Construction techniques for future lunar bases have already been tested on Earth. Although modules might be transported ready-made from Earth, 1986 tests on lunar soil showed that it is ideal for making concrete (Lin, 1986)!

SUMMARY

The Moon is an ancient planetary body, little disturbed since the formative days of the solar system. Much of the information in this chapter yields a chronological history of the Moon. Because the Earth shared much, if not all, of this history, we summarize the history of the Earth–Moon system in Table 7-2.

Origin Lunar rocks and meteorites reveal that the Moon and planets formed 4.6 billion years ago. The Moon's formation may have involved a giant impact. Earth and the Moon were close together shortly thereafter.

Duration of Formative Process Differences in ages among lunar and meteorite specimens indicate that the Moon and planets reached approximately their present sizes a few million to 90 million years after the formative process began.

Magma Ocean Analyses of lunar rocks indicate that the lunar surface was initially covered with a molten magma ocean several hundred kilometers deep. An initial magma ocean may also have formed on Earth, but evidence here has been destroyed.

Early Intense Bombardment Nearly all lunar rocks that formed before about 4 billion years ago have been pulverized by repeated bombardment. According to Apollo data, the cratering rate 4 billion years ago was 1000 times the present rate, and before that it was still higher. The Earth was presumably also bombarded at that time. Many large craters formed on Earth and the Moon during this period. The cratering rate declined to the current value by 3 billion years ago.

Tidal Movement of the Moon The Moon was once much closer to Earth than it is now. Analysis proves that from its moment of closest approach, it moved out quickly, reaching about half its present distance in only 100 million years, or 2% of the time since Earth's formation. It probably approached its present orbit between 4.5 and 4 billion years ago.

Heating and Differentiation Radioactive elements in Earth and the Moon heated their interiors, allowing heavy metal materials to sink to the center, probably within a billion years of the formation of these bodies. The Earth formed a large iron core, but the Moon had less iron and formed little or no core.

TABLE 7.2

History of the Earth–Moon System

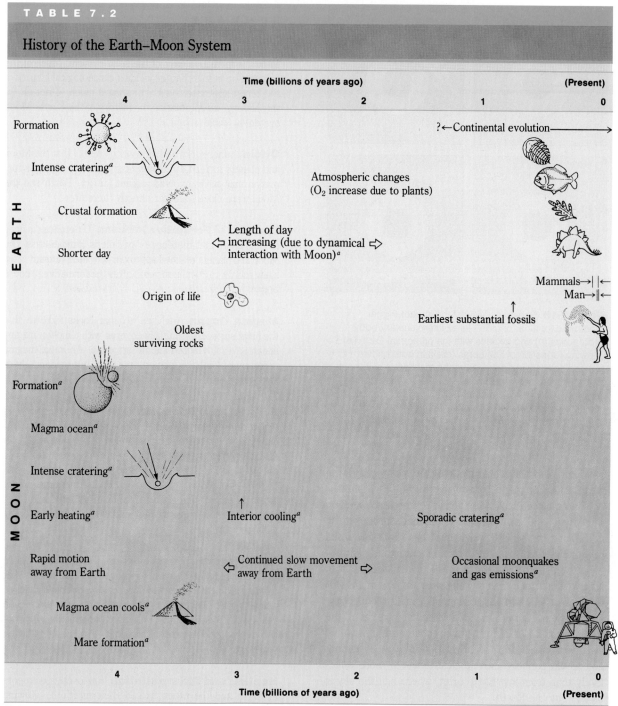

[a]Information discovered or improved through Apollo-related lunar research.

Lava Flows Basaltic lavas erupted in many places, especially where the crust was thinned by large impacts. Successive lava flows formed the dark plains 3.8 to 3.2 billion years ago. Because the Moon is smaller than Earth, it cooled more quickly and most volcanism died out around 3 billion years ago.

Lithosphere Evolution The Moon cooled off and formed a 1000-km-thick lithosphere that shielded its surface from internally caused changes. Convection currents continued in the Earth, breaking the 100-km-thick lithosphere into moving plates.

Sporadic Recent Cratering Occasional large impacts, involving meteorites a few kilometers across, excavated major craters, throwing bright rays of pulverized ejecta across the dark plains and old highlands (see Figure 7-4a). Smaller meteorite impacts churned the upper soil, producing the regolith. On Earth some of the large craters formed by meteorites during the last billion years have also been preserved (as described in Chapter 6).

Terrestrial Erosion The parts of terrestrial history studied by most geologists are only the tail end of Earth–Moon history. Because of plate tectonics and atmospheric erosion, most of Earth's surface rocks represent only the last 12% of Earth–Moon history, whereas most lunar surface structures date back through 84% of it. The Moon shows more clearly than Earth the early and middle parts of geological history and the combined results of internal geological processes (for example, partial melting) and external astronomical processes (impacts). The Moon has given us new understanding of processes that shaped our own world, and we have left our mark on it.

CONCEPTS

satellite	ray
phases	regolith
new moon	ejecta blanket
first quarter	magma ocean
gibbous	anorthosite
full moon	breccia
third quarter	volatile elements and compounds
synchronous rotation	
terminator	mean density
limb	early intense bombardment
impact crater	
dark lava plains	impact-trigger hypothesis
highland	tidal forces
basin	

PROBLEMS

1. At what time of day (or night) does:
   a. The first quarter moon rise?
   b. The full moon?
   c. The last quarter moon?
   d. The new moon?

2. When a terrestrial observer is recording a new moon, what phase would the Earth appear to have to an observer on the Moon?

3. Draw a diagram like Figure 7-2 and show approximate locations from where Figures 7-3a and 7-3b could have been taken.

4. Explain why the Moon keeps one side toward Earth.

5. Suppose an astronaut orbiting just above the atmosphere releases into an orbit of their own two ping pong balls just in contact with each other. Would you expect them to stay in contact with each other indefinitely? Why or why not?

6. Imagine you are selecting a lunar landing site.
   a. What type of feature might offer fresh bedrock where regolith layers have been stripped away?
   b. Would you expect the landscape inside a young 100-km-diameter crater to be rougher or smoother than inside an old crater of the same size?
   c. Where might astronauts land to seek evidence of recent volcanic activity?

7. Imagine you are an astronaut exploring the Moon.
   a. Would an ordinary compass work on the Moon? Why or why not?
   b. What celestial object could serve as a directional aid (like the North Star) for astronauts hiking on the Moon's front side?
   c. What property of this object's apparent motion would make it especially useful as a navigational aid during a lunar stay of several months?

8. If the ages of Earth and the Moon are nearly identical, as believed, why are most rocks found on the Moon so much older than Earth rocks?

9. By comparing pictures of lunar lava plains (3 to 4 billion years old) and highlands (4 to 4.5 billion years old), show that the meteoritic cratering rate was much higher during the first few hundred million years of lunar history than during the last 3 billion years.

PROJECTS

1. Observe the Moon at different phases with a telescope of at least 5-cm (2-in.) aperture. Locate and compare the texture of highland regions with dark lava plains. Compare visibility of detail near the terminator and away from the terminator, and explain the difference. Sketch examples of craters, ray systems, and mountains.

2. With a telescope, find a bright-ray crater, such as Tycho or Copernicus, and compare its appearance at full moon (high lighting) with its appearance near the terminator (low lighting). Why do the rays disappear under low lighting? Simulate this effect by scattering a thin dusting of white flour or powder with a raylike pattern onto a slightly darker, textured surface. Illuminate with a bright light bulb from above (full moon) and from the side at a low angle (low lighting), and compare the appearance.

3. Prepare a box with white flour several centimeters deep and a light dusting of darker surface powder (flush with the top edge of the box). Drop different-sized pebbles into the box to make craters. Illuminate with a bright light bulb from various angles and compare with the appearance of the lunar surface. Compare the number of craters needed to simulate a plain region and a highland region. Can the surface be saturated with craters if enough stones are dropped? What physical differences exist between this experiment and lunar reality? (Example: These stones hit at a few meters per second, whereas meteorites hit the Moon at several kilometers per second and cause violent explosions.)

The Solar System

Three worlds in one photo. Voyager 1 made this image showing giant, cloudy Jupiter in the background and two of its moons, Europa (white) and Io (orangish), passing in front of it. (NASA.)

CHAPTER 8

Introducing
the Planets—Mercury

If we could journey far beyond the Moon and look back, we would see the Sun and its family of planets—the **solar system**. We would discover that Earth is only the fifth largest of many worlds that orbit around the average-sized star we call the Sun. This is a far cry from the conception of 20 generations ago, when Earth was viewed as a kind of imperial capital of the universe—a unique stationary scene of human activities around which the Sun, Moon, planets, and stars moved. The exciting transition from this older idea to our modern conception began what astronomer Carl Sagan has called "the cosmic connection," the realization that we are only one part of a larger system of worlds—an idea still growing in our consciousness even today.

Chapter 3 described how the arrangement of the planets' orbits was discovered, and Figure 3-8 showed our modern knowledge of that arrangement. In this chapter we begin describing the other planets, their moons, and the interplanetary bodies. We will focus here not so much on their orbits as on their properties as worlds. The idea that we live in a system of worlds suggests a new conception of a large theater in which we may travel, investigate many new examples of geological, meteorological, and biological processes, and perhaps exploit new sources of energy and materials.

A SURVEY OF THE PLANETS

The solar system is defined as the Sun, its nine orbiting planets, their own satellites, and a host of small interplanetary bodies, such as asteroids and comets. Starting at the center of the solar system, the major bodies and their symbols are:

☉	**Sun**	⊕	**Earth**	♄	**Saturn**
☿	**Mercury**	♂	**Mars**	♅	**Uranus**
♀	**Venus**	♃	**Jupiter**	♆	**Neptune**
				♇	**Pluto**

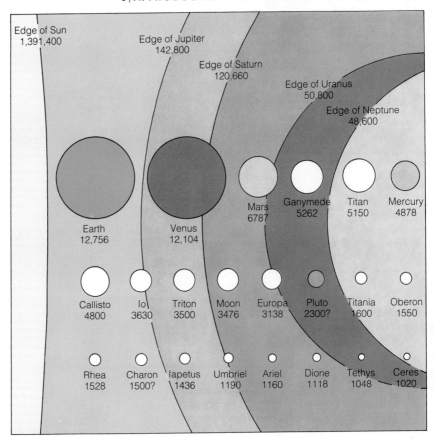

Figure 8-1 The 27 largest bodies in the solar system, shown to true relative scale. They include 1 star, 9 planets, 16 satellites, and 1 asteroid. Diameters are in kilometers.

The symbols, mostly derived from ancient astrology, are sometimes used as convenient abbreviations today. A traditional memory aid for this outward sequence is "*M*en *V*ery *E*arly *M*ade *J*ars *S*tand *U*pright *N*icely, *P*eriod." Surely today's students can do better![1] To avoid confusion about the positions of Saturn, Uranus, and Neptune, remember that the *SUN* is a member of the system, too.

Some simple facts about the bodies of the solar system are useful to remember. For example, the Sun is about 10 times the diameter of Jupiter, and Jupiter is about 10 times the diameter of Earth. Whereas the Sun is a **star**, composed of gas and emitting radiation by its own internal energy sources, **planets** are bodies at least partly solid, orbiting the Sun, and known to us primarily by reflected sunlight. **Satellites**, in turn, are solid bodies orbiting the planets.

The planets divide into two groups. **Terrestrial planets** are the four inner planets, Mercury through Mars. They most nearly resemble Earth in size and in rocky composition. The **giant planets** are the four large planets of the outer solar system, Jupiter through Neptune. Much larger than the terrestrial planets, they also have a different composition, being rich in icy or gaseous hydrogen compounds such as methane (CH_4), ammonia (NH_3), and water (H_2O). Pluto falls into neither category, being a special, mysterious case.

Table 8-1 presents a comprehensive list of data about objects in the solar system. Notice that some satellites are bigger than some planets! The diameters of the remotest satellites are uncertain because they are hard to observe, but the largest satellites in the solar system are Jupiter's moon Ganymede and Saturn's moon Titan, with diameters over 5000 km. They are both larger than the planets Mercury and Pluto. Planets and moons are not the only large bodies in the solar system. The large asteroid Ceres (diameter 1020 km), which orbits the Sun in a planetlike orbit between Mars and Jupiter, is larger than half the known satellites of the solar system. As Figure 8-1 shows, besides the Sun there are probably 26 worlds in the solar system larger than 1000 km across.

---

[1]A prizewinner in 1984 was created by a student at the University of Hawaii: *M*y *V*ery *E*rotic *M*ate *J*oyfully *S*atisfies *U*nusual *N*eeds *P*assionately.

TABLE 8·1

Objects in the Solar System
(Including the Sun, all planets, satellites known through 1989, the six largest asteroids, and Chiron)

Object	Equatorial Diameter (km)	Mass (kg)[a]	Rotation Period[b] (d)	Orbital Period (days unless marked)	Distance from Primary (10^3 km unless marked)	Orbit Inclination[b,c] (degrees)	Orbit Eccentricity	Escape Velocity (km/s unless marked)	Known or Probable Surface Material
Sun	1,391,400	1.99 (30)	25.4	—	0	—	—	617	Ionized gas
Mercury	4878	3.30 (23)	58.6	89	0.387 AU	7.0	0.206	4.2	Basaltic dust & rock
Venus	12,104	4.87 (24)	243R	225	0.723 AU	3.4	0.007	10.4	Basaltic & granitic rock
Earth	12,756	5.98 (24)	1.00	365	1.00 AU	0.0	0.017	11.2	Water, granitic soil
Moon	3476	7.35 (22)	S	27	384	18–29	0.055	2.4	Basaltic dust & rock
Mars	6787	6.44 (23)	1.02	687	1.52 AU	1.8	0.093	5.0	Basaltic dust & rock
Phobos	27 × 19	9.6 (15)	S	0.32	9.4	1.0	0.015	11 m/s	Carbonaceous soil
Deimos	15 × 11	?	S	1.26	23	2.8	0.001	6 m/s	Carbonaceous soil
Asteroids									
1 Ceres	1020	1.2 (21)?	0.38	4.6 y	2.77 AU	10.6	0.08	0.6	Carbonaceous soil
4 Vesta	549	2.4 (20)?	0.22	3.6 y	2.36 AU	7.1	0.09	0.3	Basaltic soil
2 Pallas	538	?	0.33	4.6 y	2.77 AU	34.8	0.24	0.3	Meteoritic soil
10 Hygiea	443	?	0.75	5.6 y	3.15 AU	3.8	0.10	0.2	Carbonaceous soil
511 Davida	341	?	0.21	5.7 y	3.19 AU	15.8	0.17	0.2	Carbonaceous soil
704 Interamnia	338	?	0.36	5.4 y	3.06 AU	17.3	0.15	0.2	Unidentified soil
Jupiter	142,800	1.90 (27)	0.41	11.9 y	5.20 AU	1.3	0.048	60	Liquid hydrogen?
J16 Metis	40	?	?	0.29	128	0.0	0.0	20 m/s	Rock?
J15 Adrastea	25 × 15	?	?	0.30	129	0.0	0.0	10 m/s	Rock?
J5 Amalthea	270 × 150	?	S	0.50	181	0.4	0.003	0.13	Sulfur-coated rock?
J14 Thebe	120? × 90	?	?	0.67	222	0.0	0.0	60 m/s	Rock?
J1 Io	3630	8.94 (22)	S	1.77	422	0.0	0.000	2.6	Sulfur compounds
J2 Europa	3138	4.80 (22)	S	3.55	671	0.5	0.000	2.0	H_2O ice
J3 Ganymede	5262	1.48 (23)	S	7.16	1070	0.2	0.001	3.6	H_2O ice, dust
J4 Callisto	4800	1.08 (23)	S	16.69	1883	0.2	0.008	2.4	Dust, H_2O ice
J13 Leda	8?	?	?	239	11,094	26.7	0.146	4 m/s?	Carbonaceous rock?
J6 Himalia	180	?	0.4	251	11,480	27.6	0.158	90 m/s?	Carbonaceous rock?
J10 Lysithea	40	?	?	259	11,720	29.0	0.130	20 m/s?	Carbonaceous rock?
J7 Elara	80	?	?	260	11,737	28.0	0.207	40 m/s?	Carbonaceous rock?
J12 Ananke	30	?	?	631	21,200	147R	0.17	16 m/s?	Carbonaceous rock?
J11 Carme	44	?	?	692	22,600	163R	0.21	20 m/s?	Carbonaceous rock?
J8 Pasiphae	70	?	?	735	23,500	148R	0.38	40 m/s?	Carbonaceous rock?
J9 Sinope	40	?	?	758	23,700	153R	0.28	20 m/s?	Carbonaceous rock?
Saturn	120,660	5.69 (26)	0.43	29.5 y	9.54 AU	2.49	0.056	36	Liquid hydrogen?
S15 Atlas	38 × 28	?	?	0.60	138	0.0	0.000	13 m/s?	Ice?
S16 Prometheus	140 × 74	?	?	0.61	139	0.0	0.002	50 m/s?	Ice?
S17 Pandora	110 × 66	?	?	0.63	142	0.0	0.004	35 m/s?	Ice?
S11 Epimetheus	140 × 100	?	S	0.69	151	0.3	0.009	50 m/s?	Ice?
S10 Janus	220 × 160	?	S	0.69	151	0.1	0.007	70 m/s?	Ice?

TABLE 8·1

Objects in the Solar System, *continued*
(Sources listed on following page)

Object	Equatorial Diameter (km)	Mass (kg)[a]	Rotation Period[b] (d)	Orbital Period (days unless marked)	Distance from Primary (10³ km unless marked)	Orbit Inclination[b,c] (degrees)	Orbit Eccentricity	Escape Velocity (km/s unless marked)	Known or Probable Surface Material
S1 Mimas	394	3.8 (19)	S	0.94	186	1.5	0.02	0.2	Mostly H_2O ice
S2 Enceladus	502	8.4 (19)	S	1.37	234	0.0	0.00	0.2	Mostly H_2O ice
S3 Tethys	1048	7.6 (20)	S	1.89	295	1.1	0.00	0.4	Mostly H_2O ice
S13 Telesto	≈25 × 11	?	?	1.89	295[d]	0	0	7 m/s?	?
S14 Calypso	30 × 16	?	?	1.89	295[d]	0	0	9 m/s?	?
S4 Dione	1118	1.0 (21)	S	2.74	377	0.0	0.00	0.5	Mostly H_2O ice
S12 Helene	36 × 20	?	?	2.74	377[d]	0.2	0.00	11 m/s?	Mostly H_2O ice
S5 Rhea	1528	2.5 (21)	S	4.52	527	0.4	0.00	0.7	Mostly H_2O ice
S6 Titan	5150	1.3 (23)	?	15.94	1222	0.3	0.03	2.7	Ices, liquid NH_3 & CH_4
S7 Hyperion	350 × 200	?	chaotic	21.28	1481	0.4	0.10	0.1	Ices?
S8 Iapetus	1436	1.9 (21)	S	79.33	3560	14.7	0.03	0.6	Ice and soil
S9 Phoebe	230 × 210	?	0.4	550.5	12,930	150R	0.16	0.1	Carbonaceous soil
Asteroid/comet 2060 Chiron[e]	350?	?	0.25	50.7 y	13.70 AU	7.0	0.38	0.1?	Carbonaceous soil (?) and volatile ices
Uranus	50,800	8.76 (25)	0.96?R	84.0 y	19.18 AU	0.8	0.05	21	?
U6 Cordelia	26	?	?	0.34	49.3	0	0	14 m/s?	Ice and soil
U7 Ophelia	32	?	?	0.38	53.3	0	0	17 m/s?	Ice and soil
U8 Bianca	44	?	?	0.44	59.1	0	0	23 m/s?	Ice and soil
U9 Cressida	66	?	?	0.46	61.75	0	0	35 m/s?	Ice and soil
U10 Desdemona	58	?	?	0.48	62.7	0	0	31 m/s?	Ice and soil
U11 Juliet	84	?	?	0.49	64.35	0	0	44 m/s?	Ice and soil
U12 Portia	110	?	?	0.52	66.09	0	0	58 m/s?	Ice and soil
U13 Rosalind	58	?	?	0.56	69.92	0	0	31 m/s?	Ice and soil
U14 Belinda	68	?	?	0.62	75.10	0	0	36 m/s?	Ice and soil
U15 Puck	160 x 150	?	?	0.76	85.89	0	0	126 m/s?	Ice and soil
U5 Miranda	484	7 (19)	S	1.41	130	3.4	0.02	0.4	H_2O ice, soil
U1 Ariel	1160	1.4 (21)	S	2.52	192	0	0.00	0.7	H_2O ice, soil
U2 Umbriel	1190	1.2 (21)	S	4.14	267	0	0.00	0.6	H_2O ice, soil
U3 Titania	1600	3.4 (21)?	S	8.71	438	0	0.00	1.1	H_2O ice, soil
U4 Oberon	1550	2.9 (21)	S	13.46	586	0	0.00	1.0	H_2O ice, soil
Neptune	48,600	1.03 (26)	0.92?	164.8 y	30.07 AU	1.8	0.01	24	?
1989 N6	54	?	?	0.30	48.2	4.5	?	26 m/s?	Ice and soil
1989 N5	80	?	?	0.31	50.0	<1	?	48 m/s?	Ice and soil
1989 N3	150	?	?	0.33	52.5	<1	?	74 m/s?	Ice and soil
1989 N4	180	?	?	0.40	62.0	<1	?	84 m/s?	Ice and soil
1989 N2	190	?	?	0.55	73.6	<1	?	106 m/s?	Ice and soil
1989 N1	400	?	?	1.12	117.6	<1	?	222 m/s?	Ice and soil
N1 Triton	2700	2.2 (22)	S	5.88	354	159R	0.00	1.5	NH_3 ice, CH_4 ice
N2 Nereid	340	?	?	360.2	5515	27.6	0.75	90 m/s	CH_4 ice, soil
Pluto	2300?	1 (22)?	6.4	247.7 y	39.44 AU	17.2	0.25	1.3	CH_4 ice
P1 Charon	1500?	1 (21)?	S	6.39	19	OR	0	0.8	CH_4 ice

Sources: See following page.

Until the 1970s, even the best telescopic views of Uranus, Neptune, Pluto, and all satellites beyond the Moon revealed only poorly perceived disks, like a pinhead held at arm's length. Virtually nothing was known about the features on these worlds. The satellites were especially anonymous—assumed to be cratered globes not much different from our own Moon. Perhaps the most astonishing discovery of solar system exploration in the last decade was the unforeseen variety among these worlds. As we will see in the next chapters, distant moons include a near-featureless iceball, a sulfur-orange world with dozens of active volcanoes, a world with one blackboard-black hemisphere and an opposing snowy-white hemisphere, a cloudy world where gasolinelike compounds may rain out of the sky, and a world that may have an ocean of liquid nitrogen! The solar system is not just nine planets and a sun; it has dozens of worlds, each with its own personality.

COMPARATIVE PLANETOLOGY: AN APPROACH TO STUDYING PLANETS

Planetology is the study of individual planets and systems of planets. In the early years of planetary studies through telescopes, each planet tended to be characterized as a world unto itself: Certain markings could be glimpsed on Mars; Jupiter had different markings; Saturn had rings; and so on. But as we have learned more about planets' surfaces, atmospheres, interiors, and evolution using spacecraft and sophisticated astronomical instruments, a new style of planetology has come into being. It is called **comparative planetology:** a systematic study of how planets compare with each other, why they are different, and why certain planets have certain similarities.

In comparative planetology, each planet and moon is regarded as an experiment that teaches us what type of environment evolves if you start with a certain mass, with a certain composition, at a certain distance from the Sun. A good example of this approach comes from comparing Earth and Venus. Here are two planets with nearly the same mass and size; but they have radically different atmospheres and climates, as seen in Figure 8-2. The comparative planetologist tries to understand why. Is it related to their different distances from the Sun? Or is there another factor? Such scientists then try to use this knowledge to clarify our understanding of Earth. Because the atmosphere of Venus has much more carbon dioxide than Earth's atmosphere, for example, we may learn something about the effects of the significant increase in CO_2 in Earth's atmosphere due to pollution.

More possibilities for comparative planetology may be glimpsed from Figures 8-1 and 8-2. Figure 8-2 gives a color comparison of several selected planetary bodies to scale. We see orange deserts on Mars contrasting with the gray lavas of the Moon. Jupiter's icy-gray moon Ganymede (lower left) contrasts amazingly with Saturn's same-sized moon, orange smog-covered Titan. Explaining such differences clarifies not only each world but also the whole system of worlds, including our own. Each world tells not only its own story but also a story of connections. In the next few chapters, we will explore these stories one at a time.

THE PLANET MERCURY

We will start our survey of the planets with the planet closest to the Sun—Mercury. Mercury is the second smallest planet in the solar system; it is only about 40% the diameter of Earth and about 40% bigger than the

TABLE 8·1   (Sources and Notes)

Sources: Data from tabulation by Chapman and others (*Proc. Lun. Plan. Sci. Conf.* 1979, *9*, inside covers); IAU announcements 3463–3476; Voyager team reports (*Science*, 1979, *204*, 964ff.; 1979, *206*, 934ff.; 1981, *212*, 159ff.); Reitsema, Smith, and Larson (*Icarus*, 1980, *43*, 16); Voyager press releases on Neptune satellites, 1986; Dunbar and Tedesco (on Pluto; *Astron. J.*, *92*, 1201); data tables in *Satellites*, ed. J. Burns, 1986; Thomas and others, 1989, on small Uranus satellites.

Notes: Numbers assigned to asteroids and outer planets' satellites indicate order of discovery, except for largest satellites. The tables in this book use a system of symbols to indicate data that is approximate (~), uncertain (?), and not available or not applicable (—).

[a]Numbers in parentheses are powers of 10.

[b]An R in this column indicates retrograde motion; S indicates that synchronous rotation has been confirmed.

[c]To ecliptic for planets; to planet equator for satellites.

[d]S12 in the leading Lagrangian point of Dione's orbit. S13 and S14 are in following and leading Lagrangian points of Tethys' orbit, respectively. Lagrangian points are stable points for small bodies in larger bodies' orbits, and are discussed further in Chapter 13.

[e]D. Tholen, W. Hartmann, and D. Cruikshank discovered anomalous brightening of this strange object in 1988, probably due to cometary activity. Karen Meech and M. Belton found a coma in 1989. Though it is catalogued as an asteroid, Chiron turns out to be the largest known comet nucleus!

▲ Venus   ▼ Earth

▼ The Moon   ▼ Mars

▼ Europa   ▼ Io

▲ Ganymede   ▲ Callisto

▲ Jupiter   ▼ Saturn

▲ Titan

Figure 8-2 The contrasting appearances of several selected worlds at the same scale. Colors are fairly realistic, except in the cases of Venus (where ultraviolet images are used to bring out the patterns of the relatively featureless yellowish-white clouds) and Saturn (where false color is used to emphasize differences among the whitish rings). (NASA photos.)

Moon. Because of its distance and smallness, its disk is very tiny when seen from Earth, and Earth-based telescopes reveal little detail. For this reason, little was known about Mercury prior to the era of space exploration.

In 1974 and 1975, the American spacecraft Mariner 10 sailed past Mercury on three different occasions. Due to Mariner 10's orbit, its cameras were able to record details on only about half the surface of Mercury. But this was sufficient to reveal that Mercury is a world much like the Moon, pocked with craters, marked with giant multiring basins and lava flows, and with virtually no atmosphere. From the standpoint of comparative planetology, therefore, Mercury gives a valuable example of a planet intermediate in size between the Moon and Earth. Before we go on to show how Mercury's size affects its surface features, we pause to consider some rather peculiar properties of Mercury's motions.

Rotation and Revolution of Mercury

There is a curious relationship between Mercury's rotation and its orbital revolution around the Sun. Due to complex orbital and gravitational effects, Mercury has gotten locked into a 59-d period of rotation, which is two-thirds of its 88-d orbital revolution period. The combination of these two rates means that the Sun moves very slowly across Mercury's sky, taking 176 "days" (two of Mercury's years) to go from noon until the next noon!

This can be seen better from Figure 8-3, in which we visualize Mercury at position 0, where the noontime sun is overhead at a certain mountain. By position 6, Mercury has gone once around the Sun in 88 d, but it is now midnight on the mountain, because the mountain faces directly away from the Sun. Another 88 d are necessary to bring the mountain back to noontime, as you can confirm by sketching in the next orbital trip, positions 7 through 12. Thus the Mercurian "day" (from sunrise to sunrise) is not the 59-d period of rotation but 176 d.

Mercury and Dr. Einstein

Mercury's orbital motion seems to contradict the laws of Kepler and Newton: The *perihelion*—the point nearest to the Sun—shifts in position slowly around the Sun from year to year. This movement is called **orbital precession**. Some precession had been predicted from Newton's laws, but observers found an excess shift of

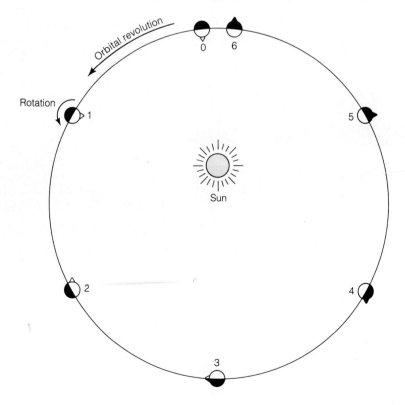

Figure 8-3 A view of Mercury's orbit showing the relationship between its rotation and revolution. The cartoons represent Mercury as a globe with one high mountain marking the rotation position. Shading shows the nighttime side. See the text for a discussion.

43 seconds of arc each century—a tiny amount, but enough to consternate orbital theorists!

The explanation came in 1915, when Albert Einstein showed that the great mass of the Sun disturbs the orbits of nearby planets in a way unpredicted by Newton's laws. Einstein's theory of relativity predicted almost exactly the excess precession observed—43.03 seconds of arc per century. Einstein predicted smaller excesses for Venus and Earth, and these too were confirmed by observation. Thus Einstein's contribution to solving the puzzle of Mercury's precession played a major role in the acceptance of his theory of relativity.

Surface Properties of Mercury

Because Mercury is so close to the Sun, its daytime surface is much hotter than Earth's. Measurements by infrared detectors of the thermal radiation from both the day and night side of Mercury show that temperatures in the upper few millimeters of soil range well above 500 K (441°F) in the "early afternoon" near perihelion to lows of about 100 K (−279°F) at night.

Mariner 10 produced the best available data about Mercury, including photographs of the surface (Figure 8-4) and magnetic measurements. The surface resembles the Moon's in its abundance of rugged craters and in its huge, multiringed craters known as basins. The most prominent basin, shown in Figure 8-5, is named the Caloris basin (in keeping with the high Mercurian temperatures). Its concentric ring structure, more than 1200 km in diameter, strongly resembles the Orientale ringed basin on the Moon (see Figure 7-10). Evidently the same impact and lava flow processes occurred on Mercury as on the Moon some $3\frac{1}{2}$ to $4\frac{1}{2}$ billion years ago. From this discovery, most scientists believe all planets suffered an intense bombardment by interplanetary debris at the close of the planet-forming process.

Mercury's landscape probably superficially resembles the lunar landscape: Rolling, dust-covered hills have been eroded by eons of bombardment by meteorites and covered by a regolith. Dusty lava plains and fault-cliffs remind us of violent ancient volcanism. Fresh impact craters might display rugged boulders and outcrops of craggy rocks. As shown in Figure 8-6, the sky is black, dominated during the nearly 88-d period of sunlight by a Sun looking about $2\frac{1}{2}$ times bigger in angular size than the Sun in Earth's sky. Besides the Sun, two jewellike planets, much brighter than any stars, occasionally dominate the sky. These are yellowish-white Venus and bluish Earth.

Figure 8-4 Mariner 10 spacecraft photo of Mercury shows a heavily cratered planet that resembles the Moon. (NASA.)

Figure 8-5 The Caloris basin on Mercury. The center of the basin lies in shadow out of the frame to the left. The left half of the frame is dominated by curved cliffs and fractures surrounding the impact site. The photo from top to bottom covers an area of about 1300 km; the cliffs are believed to be about 2 km (6000 ft) high. Compare with similar lunar features in Figure 7-10. (NASA.)

Figure 8-6 Landscapes on Mercury probably resemble those on the Moon. In this imaginary view, the Sun, which would appear $2\frac{1}{2}$ times larger than from Earth, is hidden behind a rock. The solar corona, or outer atmosphere of the Sun (see Chapter 15), dominates the sky. In the upper right, Venus and Earth (bluer) are seen in conjunction. Zodiacal light (see Chapter 13) stretches to the upper right. (Painting by the author.)

Mercury's Internal Structure and Tectonic Activity

Another comparison with the Moon relates to lithosphere structure. As we indicated in the last two chapters, the lithosphere's thickness is a key to a planet's surface evolution. If the planet is small, like the Moon, it cools rapidly and a thick, rigid lithosphere forms, preventing any internal activity from breaking through to disturb the surface. If the world is as big as Earth, however, it takes a long time to cool and only a thin lithosphere forms. Earth's lithosphere is so thin that it

is readily broken by faulting and volcanism associated with plate tectonics.

Scientists were thus interested to study the surface features of Mercury, because it is in between the size of the Moon and Earth. Using the philosophy of comparative planetology, we might predict that Mercury would have a thinner lithosphere and more signs of surface tectonic disturbance than the Moon. This turns out to be correct. Cliffs such as those in Figure 8-7 appear to mark huge faults, suggesting that the lithosphere was just thin enough to fracture when Mercury contracted as it cooled. As shown in Figure 8-7, the faults appear

a

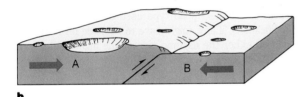

b

Figure 8-7 **a** A heavily cratered portion of Mercury resembles the lunar uplands. The cliff running from the upper left to the lower right, called Discovery Scarp, may be a fault caused by contraction of the planet following early heating and volcanism. The largest crater (left center) is about 125 km across. (NASA, Mariner 10.) **b** Schematic diagram showing probable geometry of Discovery Scarp.

to be compressional faults, formed because the surface was growing smaller as Mercury contracted. Aside from these faults, which are not found on the Moon, the rest of Mercury's surface is very lunarlike, with occasional patches of smooth lava filling the larger craters. There is some evidence that Mercury's cratered uplands contain more ancient lava flows than the Moon.

The magnetic field of Mercury is only about 1% that of Earth's, but it has the same general form, with a magnetic axis tilted about 7° to the planet's rotation axis. These results suggest that this field is generated internally, by weak currents in an iron core like Earth's

core. A large iron core also accounts for Mercury's high mean density (5500 kg/m³)—high compared to that of the Moon (3300 kg/m³), which lacks much of an iron core.

A LESSON IN COMPARATIVE PLANETOLOGY: SURFACE FEATURES VS. PLANET SIZE

We learned from studying the Moon that intense cratering occurred during the interval from 4.5 to about 4.0 billion years ago. Millions of large meteorites rained down onto the lunar surface. This **intense early bombardment** happened to all the planets—it was the great sweep-up of interplanetary debris left over after the planets formed. It allows us to explain some general features of planets' surfaces.

The basic idea here is a competition between the construction of *internally* derived features, such as volcanoes, and their destruction by the *external* bombardment. The intense early bombardment lasted for the first 500 million years of planetary history—from 4.5 to 4.0 billion years ago. Imagine all the planets starting with a magma ocean and hot interior, products of the formative process and heating by radioactive minerals buried in the planets. Volcanic mountains, lava flows, and faulted cliffs might form as long as a planet remains hot and active and its lithosphere is not too thick. If a planet or moon is tiny, it cools rapidly and forms a thick lithosphere, say, within 300 million years. Formation of volcanoes and tectonic features stops. The intense bombardment continues for another 200 million years, saturating the surface with craters and destroying traces of the constructional features—just as missiles pounding a city might reduce its buildings to a rubble of rocks and craters. Such a world would show only craters on its surface. On a slightly larger world, like the Moon, a few "last gasp" lava flows might escape from the interior through the lithosphere after 4 billion years ago, leaving dark plains on the surface. Because the intense early bombardment had already ended, these features would not be destroyed. On a still larger world, like Mercury, the cooling and lithosphere formation persists after 4 billion years ago. Cooling led to stresses in the lithosphere, and resulting fractures created the faulted cliffs. On Earth, of course, the interior remains hot and active to this day, continuing to break the lithosphere with faults, mountain chains, volcanic cones, and the like. Because of the small crater production rate in recent geological time, the construction processes, together with erosion from Earth's winds and rains, win out over the cratering. Relatively few meteorite craters can be found on Earth.

Thus we see a rule of thumb: The larger a world (or, more precisely, the more *massive* a world), the younger and more active its surface; the smaller a world, the older and more heavily cratered its surface. There are four worlds intermediate in size between Mercury and Earth: Mars, Venus, Jupiter's moon Ganymede, and Saturn's moon Titan. As we will see in the next few chapters, they tend to confirm our rule.

SUMMARY

According to present data, the solar system includes about 26 planetary worlds larger than 1000 km and a host of smaller bodies. These worlds show great variety, depending primarily on their mass but also on other conditions, such as distance from the Sun. Details of surface structure and composition are becoming clear for most of these worlds only since the advent of space exploration in the 1960s. A current approach to studying these worlds, called comparative planetology, is to examine them not as isolated cases but in comparison with each other and especially in comparison with Earth. In this way, we learn which forces determine such properties as surface geology.

Mercury is the closest planet to the Sun. It is the most Moonlike of the planets and is only 40% larger than the Moon. It has a heavily cratered surface that is broken in a few places by large, faulted cliffs produced by contraction. Mercury has virtually no atmosphere.

CONCEPTS

solar system	giant planets
star	planetology
planet	comparative planetology
satellite	orbital precession
terrestrial planets	intense early bombardment

PROBLEMS

1. Which planet can come closest to Earth? (*Hint:* See Table 8-1.)

2. How many times larger in diameter than Earth is the largest planet? How many times more massive than Earth is the most massive planet?

3. Which planet has the largest satellite?

4. Which planet has the largest satellite measured in terms of the diameter of its planet—that is, which satellite's diameter is the largest fraction of its planet's diameter?

5. Which planet can come closer to Uranus: Earth or Pluto?

6. Using Table 8-1, list some systematic differences in physical properties between terrestrial planets and giant planets. Consider size, temperature, density, number of satellites, and orbital properties.

7. List processes of planetary evolution that have occurred on both Mercury and the Moon. List the processes, if any, that are unique to each. Is there any indication of plate tectonic activity on Mercury? What might this indicate about heat flow and mantle convection inside Mercury?

8. Some of the best telescopic observations of Mercury are made during midday instead of after sunset or before sunrise. Why? Why are none made at midnight?

9. Suppose astronauts plan to land on Mercury wearing spacesuits of the Apollo type used on the Moon. What modifications, if any, might have to be made for using the suits on either the daytime side or the nighttime side of the planet?

PROJECTS

1. Try to see Mercury. This is likely to be harder than it sounds, because Mercury is prominent for only a few days during its orbit, when its orbital motion brings it to a maximum angular distance from the Sun, as seen from Earth. Even then, it is visible for only an hour or less each day just after sundown or before sunrise. Determine from an almanac or an astronomy magazine when Mercury comes to evening elongation and find a site with a very clear western horizon. It is best to start a few days before the elongation so you can become familiar with the background stars in the appropriate region of the sky.

2. If Mercury becomes visible during your studies, observe it with a telescope of at least 20 cm (8 in.) aperture. Try to detect the phase of the planet. Observe on several successive days, and see if you can detect a change in phase as Mercury circles the Sun.

3. If a planetarium is nearby, arrange a demonstration of the motions of Mercury as seen from space or from Earth.

CHAPTER 9

Venus

Venus is widely regarded as Earth's sister planet because its size is so similar to Earth's. Venus has about 95% of the Earth's diameter (see Figure 8-2) and about 82% of its mass. It has no moons.

From the viewpoint of comparative planetology, the similarities between Venus and Earth make us eager to compare the geological and atmospheric properties of the two planets. Because Venus approaches closer to Earth than any other planet, you might think that it would be one of the best-observed worlds. Its surface remained very mysterious until space probes could be sent there, however, because Venus is completely obscured by clouds. Indeed, the brilliant, yellowish-white, nearly blank cloud layer is the first feature to impress a telescopic observer. It prevented early astronomers from observing a single surface detail or even the rotational properties of the planet. Space probes without crews, both in orbit around the planet and on its surface, have revealed that the atmosphere is totally different from Earth's, although the surface has some similarities and some differences from the ground we walk on. Venus is an interesting challenge to comparative planetologists: Why should a planet close to Earth's orbit, and of nearly the same size, show so many differences? In this chapter we will search for answers.

THE SLOW RETROGRADE ROTATION OF VENUS

Radar signals, bounced off Venus from Earth, reveal that Venus has a rotation unlike that of Mercury, Earth, or Mars. Those planets all have **prograde rotation**—spin from west to east. Venus has **retrograde rotation**—spin from east to west (Figure 9-1). Venus' spin is also unusual in being very slow, taking 243 d to make a complete turn on its axis. These properties were discovered in 1962, when radar signals were first bounced

Figure 9-1 Rotation (motion around an internal axis) and revolution (motion around an external body). Prograde motion (top) is west to east. Most planets, including Earth, have prograde rotation and revolution. The rotation of Venus is retrograde, as diagrammed in the lower left figure.

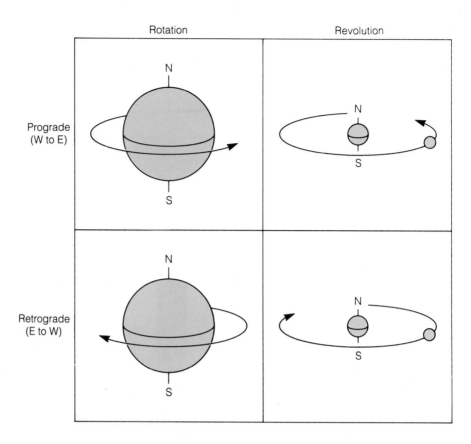

Rotation   Revolution

Prograde (W to E)

Retrograde (E to W)

Figure 9-2 A telescope view of Venus while passing nearly between Earth and the Sun. Backlighting by the Sun illuminates the atmosphere of Venus all the way around the disk. This effect, first recorded in 1761, not only proves the existence of an atmosphere but gives information on its structure. (New Mexico State University.)

off the planet. The cause of the unusual reverse spin may involve an ancient collision with a body larger than our Moon, of which there may have been many in the early solar system.

VENUS' INFERNAL ATMOSPHERE

The first proof that Venus has an atmosphere came as long ago as 1761, when the Russian scientist M. Lomonosov observed the backlit atmosphere extending around the disk, as shown in Figure 9-2. This phenomenon occurs when Venus is approximately between Earth and the Sun, so that sunlight backlights the upper atmosphere of Venus, revealing the haze layer ringing the planet.

The Venusian atmosphere intrigued scientists for two centuries after it was discovered.[1] What caused this

---

[1] Adjectives for planets are controversial. Some writers, claiming that "Venusian" is an ugly word, use "Cytherean" (from a name of Aphrodite), which is merely confusing. "Venereal" is already preempted by other areas of human endeavor. While "Venusian" (ve-NOO-sian) has the sanctity of science fiction tradition, many astronomers use "Venerian" or the noun "Venus" as an adjective.

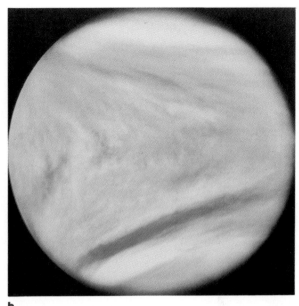

a

b

Figure 9-3 Circulation of Venus' atmosphere is shown in these two views taken 5 h apart. In **b**, the features have shifted to the west (left; note *V*-shaped dusky marking in left center). Although the planet itself takes 243 d to turn, 200-mph winds blow the clouds around the planet in about 4 d. This is a false-color image made in ultraviolet light. In visible light, the contrast is so low that the yellowish-white clouds look nearly featureless, although telescope observers have sometimes reported the bright polar cloud cap and adjacent dark band. (NASA, Pioneer Venus Orbiter.)

whitish shroud? What kind of planetary surface did it hide? In 1928 the American astronomer Frank Ross photographed dusky cloud patterns by using plates and film sensitive to ultraviolet. These patterns are shown better in spacecraft photos, such as Figure 9-3. They are formed by cloud layers differing in composition, particle size, or altitude.

In 1932 Mt. Wilson astronomers studied the spectrum of Venus and detected extraordinary amounts of carbon dioxide (CO_2, the same gas dissolved in carbonated soft drinks). Later data showed that Venus' atmosphere is about 96% CO_2, as seen in Table 9-1.

Still undiscovered was the composition of the opaque cloud layer and the air below it. In the 1940s and 1950s some writers imagined stormy clouds of water droplets, a surface swept by torrential rains, and vegetation like that of a Brazilian rain forest. Some supposed that the highly reflective clouds would shade the surface and moderate the climate in spite of Venus' closeness to the Sun. In the 1960s, however, when the planet's thermal radiation was measured at far-infrared and radio wavelengths, Wien's law revealed that the lower atmosphere

has a temperature of about 750 K (891°F)—hardly conducive to liquid water or life as we know it.

In 1970, **Venera 7**, a Soviet probe that was the first spacecraft to land successfully on another planet, transmitted data from the Venusian surface for 23 min. The data confirmed the high temperature and revealed an atmospheric pressure about 90 times as great as Earth's. Instead of our 101,000 N/m^2 (14.7 lb of force pressing on every square inch of surface) the pressure on Venus is about 9,000,000 N/m^2 (1320 lb/in.^{2}), equivalent to that endured by a diver nearly a kilometer (3000 ft) below the terrestrial ocean surface. Later Soviet spacecraft landings confirmed these hellish conditions.

Venus was soon revealed to be stranger yet. In 1972–1973 astronomers discovered that the clouds of Venus consist not of water droplets, like Earth's clouds, but rather of tiny droplets of sulfuric acid (H_2SO_4). In 1978 probes dropped by the American spacecraft Pioneer Venus showed that the clouds lie primarily in a high layer 48 to 58 km above the surface, unlike Earth's clouds, which are mostly less than 10 km high.

In 1985, two balloons (dropped by Russian probes

TABLE 9·1

Atmospheres of Venus and Earth

Venus		Earth	
Gas	**Percent Volume**	**Gas**	**Percent Volume**
CO_2	96.5	N_2	78.1
N_2	3.5	O_2	20.9
SO_2	0.015	H_2O	0.05 to 2 (variable)
H_2O	0.01	Ar	0.9
Ar	0.007	CO_2	0.03
CO	0.002	Ne	0.0018
He	0.001	He	0.0005
O_2	⩽0.002	CH_4	0.0002
Ne	0.0007	Kr	0.0001
H_2S	0.0003	H_2	0.00005
C_2H_6	0.0002	N_2O	0.00005
HCl	0.00004	Xe	0.000009

Sources: Oyama and others (1979); von Zahn and others (1983).

Note: Compositions are for near-surface conditions, with terrestrial data other than H_2O tabulated for dry conditions. CO_2 on Earth is probably increasing by 2% to 3% of the listed amount in each decade, because we are burning so much fossil fuel. This activity may be modifying Earth's climate.

on their way to Halley's comet) floated in the clouds for 46 h, measuring hurricanelike winds (150 mph) but relatively moderate conditions at this altitude ($T = 95°F$ and pressure like that on Earth's surface).

In spite of the differences, the clouds of Venus and Earth form in a similar way, in atmospheric layers where the temperature and pressure cause condensation of some relatively minor atmospheric constituent. On Earth, this constituent is H_2O, condensed either into droplets (lower clouds) or ice crystals (high cirrus clouds). On Venus, it is H_2SO_4 droplets, which begin to fall as they grow. If a droplet gets big enough, it falls out of the cloud deck, where it encounters much higher temperatures and evaporates. Thus Venus' weird sulphuric acid rain never reaches the ground. This explains why the clouds have a well-defined bottom surface, as detected by the Pioneer probes, and why the lower atmosphere and surface are clear, as found by the Russian Venera landers.

In 1978 the American and Russian Venus probes made yet another startling discovery—terrific blasts of lightning play among the clouds of Venus.

The Greenhouse Effect on Venus

Why is Venus so hot? Planets absorb sunlight and, following Wien's law, radiate infrared light. A planet's surface temperature is determined by the balance between the amount of visible sunlight it absorbs and the amount of infrared radiation it emits. If the solar energy absorbed each second is greater than the infrared energy radiated each second, the planet heats up. If the incoming amount is less than the outgoing amount, the planet cools. The mean, or equilibrium, temperature is reached when the two rates are equal.

The incoming rate is easily calculated as the total sunlight striking the planet minus the amount reflected. The total energy striking the planet per square meter in 1 s is called the **solar constant** for that planet. The outgoing radiation increases as the planet's surface temperature increases.

If the planet has no atmosphere, the situation is easy to predict. The surface rocks heat up until their temperature is so high that outgoing infrared radiation equals the incoming sunlight. But if the planet has an atmosphere that absorbs some incoming sunlight (as in Figures 5-10 and 5-11), the atmosphere and the surface both warm up and radiate infrared energy. Because of the nature of atmospheric gases, the outgoing infrared energy from the surface may not escape directly into space. In fact, CO_2 and H_2O gases absorb a great deal of the outgoing infrared, thus adding energy to the atmosphere and warming it even more. The warming continues until the amount of infrared escaping from the top of the atmosphere equals the amount of incoming sunlight. On Venus, the lower atmosphere has to reach about 750 K before this condition is met (Figure 9-4).

This heating is called the **greenhouse effect** because of its resemblance to the heating physics of a greenhouse. The glass panes of a greenhouse admit sunlight but block the escape of the infrared. (They also keep the warm air from escaping—an important function that, in the case of planets, is performed by gravity.) Hence the inside becomes warmer than the outside. The greenhouse effect explains why Venus is so hot: Its massive CO_2 atmosphere blocks outgoing infrared radiation. The greenhouse effect also explains why a cloudy night on Earth often stays warmer than a very clear night; the water vapor in the cloud layer blocks outgoing infrared radiation from the cooling Earth.

The greenhouse effect is of great concern to environmental scientists on Earth. Burning of fossil fuels has increased Earth's atmospheric CO_2 by perhaps 10% since 1860 (Walker, 1977, pp. 127–128). Of course, there is much less CO_2 here than on Venus, but greenhouse effects associated with this and other atmospheric changes could alter climates and agricultural productivity on Earth. Venus, with its CO_2–caused greenhouse effect, thus serves as a "natural lab" for understanding environmental change on Earth.

Why Venus Has a CO₂ Atmosphere

Why should Venus' atmosphere be mostly CO_2 instead of N_2 and O_2, like the Earth's? (Compare columns in Table 9-1.) The answer is clearer if we rephrase the question: Why does Earth *not* have a massive CO_2 atmosphere?

The groundwork for answering this question was laid in Chapter 6, where we saw that volcanic degassing of terrestrial planets produces secondary atmospheres.

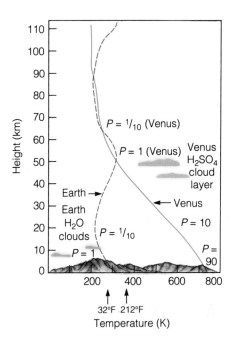

Figure 9-4 Temperature and pressure of Venus' atmosphere (solid curve) and Earth's atmosphere (dashed). The greenhouse effect greatly heats the lower atmosphere of Venus. Pressures (P) are measured in units of the Earth's surface pressure.

Volcanoes, especially those that release primordial lavas from the mantle, emit mostly H_2O and CO_2 gas, with some N_2. On Earth, the H_2O formed oceans, and the CO_2 dissolved in the oceans and ended up in carbonate rocks. This explains why Earth was left with a nitrogen-rich atmosphere.

If for some reason a terrestrial planet did not form oceans of water, however, the CO_2 would not dissolve in the oceans and would be left as a dominant atmospheric gas. Then if the H_2O molecules somehow disappeared, an atmosphere of nearly pure CO_2 would be left. This is what happened on Venus.

Many scientists assume that Venus also had primordial volcanoes that emitted much water vapor from its interior. (This view is somewhat controversial pending further surface chemical studies. Other scientists argue that Venus had noticeably less water to begin with; see Grinspoon, 1987.) Because of Venus' heat, any water in liquid form eventually evaporated, forming H_2O molecules in the air. The H_2O molecules were broken into H and O atoms by energetic solar radiation. The H atoms, being very light, tended to float to the top of the atmosphere and escape into space, a process well documented by atmospheric chemists. Now the question is: What happened to Venus' leftover oxygen? Soviet and American scientists, studying rock and soil data returned by the various Venus landers, believe that most of the oxygen combined with rock minerals (oxidized) and disappeared from the air.

If you could estimate the total amount of O_2 bound up in the oxidized minerals of Venus' rocks, and then imagine reconstituting it into the original water molecules, you could estimate how much water Venus originally had. Interestingly, some scientists have concluded Venus once had enough water to make "oceans" at least 10 m (30 ft) deep, though it is not known whether that much water ever actually collected on the surface at once. At any rate, the greenhouse effect and the resulting high temperatures destroyed any liquid oceans; Venus' CO_2 couldn't dissolve and was left as the major atmospheric gas.

If this theory is right, the sister planets, Earth and Venus, should have emitted similar total amounts of CO_2 from their volcanoes and the Earth's CO_2 should be traceable somewhere. It is. Inventories of Earth's carbonate rocks (rocks like limestone, formed by actions of sea creatures and by reaction of seafloor rocks with the ocean's dissolved CO_2) showed that those rocks contain about the same amount of CO_2 as Venus' atmosphere! Thus Venus and Earth *did* produce similar amounts of volcanic gases, but Earth's got trapped in its rocks through the mediation of the ocean. Earth's unusual N_2–O_2 atmosphere, which makes the higher life forms possible on Earth, thus seems to be a special consequence of our H_2O oceans, which prevented Earth from developing the "normal" CO_2 atmosphere of a terrestrial planet.

LANDSCAPES ON VENUS

Soviet space scientists have emphasized studies of Venus and have landed numerous probes on it. Their Venera 4 probe, which crashed on Venus in 1967, was the first human-built object to touch another planet. After many attempts, the Soviets successfully landed a number of probes that lasted in the surface heat for many minutes, making measurements of surface conditions. Veneras 9 and 10, in 1975, and Veneras 13 and 14, in 1982, made panoramic photos stretching from near the spacecraft to a bit of the horizon. Some of these photos are shown

in Figures 9-5 and 9-6. Their clarity shows that the surface is free from haze.

The photos revealed stark, dramatic landscapes with angular boulders, gravel, flat outcrops, and fine soil alternating at different sites. The varied states of erosion, with angular, young-looking rocks at some sites, suggest different degrees of geological activity. Although jet-stream winds (up to 185 mph) were detected at altitudes around 40 km, four lander probes found only gentle breezes of $\frac{1}{2}$ to 3 mph at the surface (Kerzhanovich and Marov, 1983).

As seen in Figure 9-6a, the rocks have orange-brown tones because they are bathed in the orangish light that filters through the clouds. Direct sunlight never falls on Venus' rocks, and the Sun is hidden beyond a high overcast. Studies of the photos reveal that the *intrinsic* color of the rocks is neutral gray. The orangish light would make the landscape look orangish to a (well-insulated!) astronaut looking at the daytime surface (just as your friend's gray shirt looks red in the light of a red neon sign). But the rocks of Venus would look gray if the astronaut turned a normal, white-light spotlight on them at night.[2] This "white-light" appearance is shown in Figure 9-6b.

Mapping Venus

The American 1978 Pioneer Venus mission included an orbiter that mapped altitudes all over Venus by means of radar. The resulting map of Venus, shown in Figure 9-7, yields exciting scientific information that clarifies Venus' sisterlike relation to Earth. The planet is 60% covered with rolling lowland plains, which are believed to be something like the basaltic-lava-covered plains of Earth's seafloors.

The radar maps also reveal Australia-sized areas raised about 2 to 5 km above the lowland plains. These are thought to be continents, or partially formed continents, indicating more evolution of Venus' lithosphere than occurred on the Moon, Mercury, or Mars.

---

[2]Curiously, the same studies revealed that if you brought the rocks inside the spaceship, they would slowly change to a reddish color as they cooled from the outdoor temperature of nearly 900°F to room temperature. This is because of optical properties of the oxidized (rusted) minerals, which are reddish at room temperature but turn gray when heated to Venus' temperatures (Pieters and others, 1986).

a

b

c

Figure 9-5 Three landscapes on Venus: views toward the horizon (top) from Russian Venera landers. Part of lander appears in the bottom of *a*. **a** First photo from the surface of Venus, showing loose boulders near Venera 9. **b** Rocks and gravel near Venera 13. **c** Platey rock surfaces near Venera 14. (Photos courtesy C. Florensky and A. Basilevsky, Vernadsky Institute, Moscow.)

a

b

Figure 9-6 A Venus landscape in color. **a** Under the natural, mustard-colored light filtering through Venus' atmosphere, the daytime landscape displays a yellowish-orange hue. **b** If the same scene were illuminated by ordinary white light (such as sunlight on Earth or a spotlight from a spaceship), the rocks would be seen to have a neutral gray color. Part of the Russian spacecraft Venera 13 and an attached color bar are seen at bottom. (Courtesy Vernadsky Institute, Moscow; image processing by Carlé Pieters and colleagues, Brown University.)

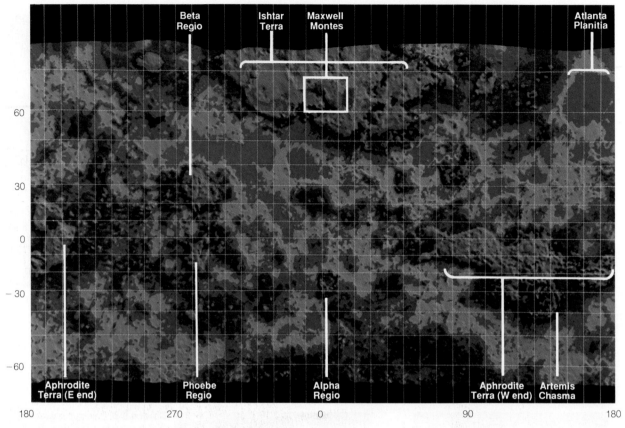

Figure 9-7 Radar map of the surface of Venus. Radar signals from the U.S. Pioneer Orbiter penetrated the clouds and measured altitudes of landforms underneath. Dark blue shows lowest terrain; red and pink, highest. Topography is reminiscent of Earth's, with the blue corresponding to seafloor interrupted by green-yellow continentlike blocks containing red mountain ranges. Latitude and longitude intervals of 10° are shown. The box shows the area of Figure 9-8. (NASA; courtesy M. Kobrick, Jet Propulsion Laboratory.)

In 1983, Russian scientists placed two radar-equipped probes in orbit around Venus and constructed more detailed radar maps of north polar areas. The American spacecraft Magellan reached Venus in 1990 and began producing even better radar maps of the whole planet, resolving details the size of a football stadium. This opened the way to geologic interpretation of the planet's development, and detailed comparisons with Venus' sister planet, Earth.

Because Venus is named for the goddess of femininity and is represented by the biological symbol for female, scientists named most of its features after real or mythical women. The largest continent, about half the size of Africa, is called Aphrodite Terra. Other features include Ishtar Terra with its volcanic peaks, and

a crater named Eve, which marks the zero meridian on maps.

Rock and Soil Compositions

Venera probes 8–10 and 13–14, as well as VEGA probes 1 and 2 (dropped by Russian spaceships on the way to Halley's comet), carried devices to measure soil and rock composition. The results indicate that the rolling plains are mostly covered by basaltic lava flows. This is the basis for comparing them with Earth's seafloor crust, which is generally a basaltic layer several kilometers thick. At one site was a more quartz-rich or granite-like rock, more typical of Earth's continents. This could be an indication of limited differentiation on

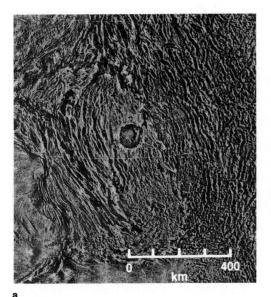
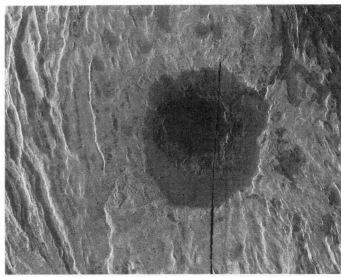

a   b

Figure 9-8 Radar images of the fractured slopes of the highest mountains on Venus, 11-km-high (36,000 ft) Maxwell Montes. **a** Broad view showing 100-km crater Cleopatra on eastern slopes of the mountain, surrounded by zones of parallel folds and fractures. (Soviet Venera 15 and 16 orbiter missions in 1983–1984; USSR Academy of Sciences.) **b** Detailed view of Cleopatra crater. Originally thought to be a volcanic caldera, it is now thought to be a meteorite impact crater. Folds and fractures at left may result from tectonic forces due to the weight of Maxwell Montes' gigantic pile of lava. (Magellan 1990 image; NASA.)

Venus. However, the site was not in the continent-like uplands, and we still do not know their composition.

WHAT THE RADAR MAPS TELL US

The radar maps and the identification of lavas both point toward a basic conclusion: The landscape of Venus is mostly volcanic. Giant volcanic peaks, higher than Everest, dot the planet. Often they are capped by volcanic calderas (volcanic craters formed by collapse of the surface to form a pit). The Magellan images in Figures 9-8 and 9-9 show additional spectacular volcanic features.

Some of Venus' volcanoes may be active. Certain measures of chemical changes in Venus' atmosphere suggest recent outbursts of volcanic gas. Also, fewer meteorite impact craters dot these surfaces than on the

Moon, but more than on Earth, and this means that Venus' average surface age is between that of Earth and Moon, perhaps a billion years. Some lava flows may have come from active volcanoes, as on Earth.

The main geologic differences between Venus and Earth seem to be twofold. (1) Venus lacks plate tectonics as active as Earth's, although wide regions are fractured by weaker tectonic stresses; (2) because of the lack of water, Venus lacks sediments and sedimentary rocks, and most of its surface is covered with volcanic lavas.

Venus gives an especially interesting clue about the beginnings of plate tectonics. Preliminary interpretations indicate that Venus has hot plumes rising in its mantle, but that they are not strong enough to drive full-fledged plate motions or continental drift. When they hit the underside of the surface layers, they have created lava-flooded, bull's-eye-like scars, several hundred kilometers across. These were discovered by the Russian probes and are called **coronas**. Study of them will probably clarify processes in our own mantle.

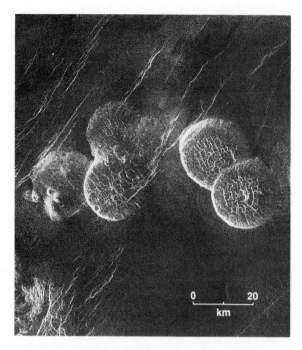

Figure 9-9 Magellan radar image revealed strange circular domes on east side of Alpha Regio, averaging about 25 km (15 mi) across and about 750 m (2500 ft) high. They are believed to be pancake-like eruptions of viscous lava. Cinder cones can be seen at left edge of left domes. Many tectonic fractures cross the area (bright lines). Vague shadings of light and dark at top mark different lava flows. Smallest details are about 150 m across. (NASA.)

A LESSON IN COMPARATIVE PLANETOLOGY: VENUS COMPARED WITH OTHER WORLDS

All these discoveries about Venus fit our general rule of thumb of greater geological evolution on more massive planets. Consider the sequence of worlds we have studied so far. The tiny Moon and the somewhat larger Mercury cooled rapidly. Their magma oceans allowed some lighter minerals to float to the surface and form anorthositic primordial crusts, later broken here and there by basaltic lava flows. On larger Venus, basaltic volcanism dominated and covered much of the planet with lava flows. Some small, protocontinent accumulations of more granitic rock may have formed, but mantle activity was inadequate to produce full-fledged plate tectonics. On Earth, any traces of a primeval anorthosite crust have been completely destroyed by the basalts that erupted to form much of the seafloor crust, and by plate tectonic activity that broke up and recycled

the most ancient surface layers. Earth's mantle has stayed hot and has been thoroughly churned by sluggish motions of molten material. This churning caused more complete differentiation and allowed low-density minerals to accumulate on the surface in large continental blocks.

ANOTHER LESSON IN COMPARATIVE PLANETOLOGY: WHY DO SOME PLANETS LACK ATMOSPHERES?

Our discussion of Earth in Chapter 6 showed that the planets probably formed with an initial gas concentration called a *primitive atmosphere,* which changed to a *secondary atmosphere* as new gases were added by outgassing. Why, then, do some planets lack atmospheres whereas other planets have dense ones? The explanation comes from three principles that govern the motions of gas molecules in atmospheres:

1. The higher the temperature, the higher the average speed of the molecules.

2. The lighter the molecules, the higher their average speeds. Light gases like hydrogen and helium have faster average speeds than heavier gases such as oxygen, nitrogen, carbon dioxide, or water vapor.

3. The larger the planet, the higher the speed needed for a molecule to escape into space.

If you could heat a planet's atmosphere, more and more molecules would move faster than escape velocity, and fast-moving molecules moving upward near the top of the atmosphere would shoot out into space, never to return. First hydrogen, then helium, and then heavier gases would leak away into space. Cold, massive planets are most likely to retain all the gases of their primitive and secondary atmospheres; hot, small planets (with weak gravity and low escape velocity) are most likely to lose all their gases. Calculations based on these principles show that planets as small as Mercury and the Moon have lost virtually all of their gases. Venus and Earth have lost most of their hydrogen and helium but have kept heavier gases.

These principles help explain why Venus could so rapidly have lost its water if water molecules broke into hydrogen and oxygen atoms: The light hydrogen atoms would quickly escape into space. Oxygen atoms react readily with minerals in the surface rocks, and thus disappear from the atmosphere as well.

SUMMARY

Venus is an Earth-sized planet, but it orbits nearer the Sun and shows striking differences from Earth. By comparing Venus and Earth we learn that an Earth without oceans could have retained much more CO_2 gas in its atmosphere, instead of in rocks. Venus also confirms that atmospheric CO_2 causes atmospheric heating through a strong greenhouse effect. Comparing the Moon, Mercury, and Venus shows that smaller planets apparently do not develop enough internal energy to drive the plate tectonic activity that has broken and reformed Earth's original, cratered crust.

From our studies of Earth, Moon, Mercury, and Venus, we can derive three general principles that will clarify phenomena of other planets as well:

1. *The larger the planet, the more internal geological activity there is likely to be.* Internal heat is the energy source that drives geological activity such as tectonic faulting, earthquakes, and volcanism; the larger a planet, the more radioactive minerals it contains and the more radioactivity there is to release heat. Also, the larger a planet, the better insulated the interior and the harder it is for the heat to escape. Small planets, on the other hand, cool rapidly and lose whatever heat they may have generated. Earth, unlike Mercury and the Moon, has enough internal energy to drive plate tectonics.

2. *The larger a planet is, the younger its surface features are likely to be.* This principle follows from the one above. The more internal heat, the thinner the lithosphere and the more likely it is for the lithosphere to be broken by recent geological activity. Small planets that cooled long ago retain very ancient surface features. Earth and probably Venus retain fewer ancient craters than Mercury and the Moon.

3. *The larger and cooler a planet is, the more likely it is to have an atmosphere, and the more likely this atmosphere is to have retained its original gases.*

CONCEPTS

prograde rotation

retrograde rotation

Venera 7

solar constant

greenhouse effect

PROBLEMS

1. Which is hotter, Mercury or Venus? Why?

2. Venus and Earth are about the same size and mass, and degassing volcanoes on each probably produced both

CO_2 and H_2O gas. Why is CO_2 a major constituent of the atmosphere only on Venus, whereas H_2O is not a *major* constituent of the atmosphere on either planet?

3. If astronauts are to walk on Venus, what sort of space suit design might be needed? Would it need to *contain* pressure or *resist* pressure?

4. Clouds of water vapor tend to absorb infrared radiation. Use this fact and the greenhouse effect to explain why desert climates on Earth have greater temperature extremes between day and night than moist climates.

5. State which of the following characteristics of Venus suggest a primitive surface (little disturbed since planet formation) and which suggest an evolved surface (affected by geological processes such as erosion, differentiation, and plate tectonics):
   a. Craters
   b. A large, rifted canyon
   c. The lack of long, folded mountain ranges
   d. Basaltic surface rocks
   e. Granitic surface rocks

6. Compare the state of Venus' geological evolution with those of the Moon, Mercury, and Earth.

7. What are the chances that life as we know it exists on Venus? Why?
   a. If Venus were to have a surface temperature of about 300 K and an abundance of H_2O in its clouds, how would you rate the chances for life? Why?
   b. If Venus were exactly like Earth, would life necessarily exist there?

8. How close is Venus during its nearest approach to Earth (see Table 8-1)? How many times farther is this than the distance to the Moon?

9. If a telescope shows Venus to be a thin crescent, where is Venus relative to Earth and the Sun?

PROJECTS

1. Determine whether Venus will be prominent in the evening or morning sky during this semester, and observe its motions and brightness from day to day. Observe on which date it is farthest from the Sun and estimate this angle.

2. Observe Venus in a telescope of at least 5-cm (2-in.) aperture on several dates a few weeks apart. Observe and sketch the changes in phase, and explain them in terms of Venus' motion relative to Earth and the Sun.

3. If a planetarium is convenient, arrange a demonstration of the motions of Venus.

CHAPTER 10

Mars

Historically, Mars has been the most exciting planet because it is the most Earth-like and also because it was long thought to harbor alien life. Indeed, for several decades, Mars was thought to be the home of an advanced civilization. Research has revealed a different Mars, but one still offering fascinating puzzles.

MARS AS SEEN WITH EARTH-BASED TELESCOPES

At its closest, Mars comes within about 56 million kilometers (35 million miles) of the Earth—closer than any other planet but Venus. When it is that close, a telescope of only 7 to 10 centimeters' aperture will show features on its reddish surface, including polar ice fields, clouds, and dusky markings (Figure 10-1)—features not unlike those of Earth. French–Italian observer Giovanni Domenico Cassini tracked the markings in the 1600s and determined that **Mars' rotation period** is 24^h37^m, only a bit longer than Earth's.

Seasonal Changes on Mars

Mars has seasons just like Earth, though each season lasts about twice as long because the Martian year is nearly twice ours. Telescopic observations in the 1700s and 1800s (Figures 10-1b and c) revealed **seasonal changes in the features of Mars**. In the Martian hemisphere experiencing summer, the bright, white polar cap shrinks away and may disappear from view, while the dusky markings darken and grow more prominent. Some early observers mistakenly thought the dark areas were oceans. Brighter, orange areas came to be called **deserts**.

While retaining roughly constant shapes, the markings also change slightly from year to year, as shown in

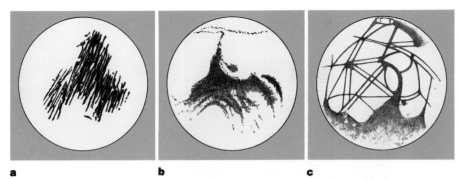

Figure 10-1 Drawings of Mars through telescopes over a three-century span. The north-extending dark triangle on all three drawings is a region known as Syrtis Major. **a** One of the earliest known sketches, by Christian Huygens in 1659. **b** Drawing by English observer W. R. Dawes during the 1864–1865 approach of Mars. Dawes recorded a streaky extension of Syrtis Major—the type of feature later called a "canal." **c** Italian observer Giovanni Schiaparelli first popularized the conception of thin, straight "canals," as in this 1888 sketch. North polar ice cap is at the top. (After Huygens, Dawes, and Schiaparelli.)

Figure 10-2. Many observers thought that the changing dark areas were regions of vegetation, perhaps losing their leaves in winter and turning dark and lush in summer, as on Earth. Today, scientists are confident that the changing markings are merely windblown dust, but many of those observers, especially in the late 1800s, erroneously believed that Mars had climate and vegetation like Earth. This opinion evolved even further as a result of the celebrated affair of the Martian "canals."

Canals on Mars?

In 1869 Father Angelo Secchi in Rome mapped streaky markings he called *canali*, maintaining the convention of naming dark areas after bodies of water. In 1877, Giovanni Schiaparelli, director of an observatory in Milan, popularized the term and drew the streaks much narrower and more linear than earlier observers had. He showed them forming a network of lines on Mars, as seen in Figure 10-1c. These features came to be called **canals**.

Many other observers did not see such features. The controversy grew hotter in 1895 with a vivid description of the canals by Percival Lowell. Lowell, a wealthy Bostonian who founded his own observatory in the exceptionally clear air of Flagstaff, Arizona, said the canals were very sharp lines:

It is the systematic network of the whole that is most amazing. Each line not only goes with wonderful directness from one point to another, but at this latter spot it contrives to meet, exactly, another line which has come with like directness from quite another direction.

Lowell concluded that the features really were canals—artificial ditches built by intelligent creatures to carry water. He pointed out that spectroscopic measurements had revealed Mars to be a dry place. He hypothesized that a once moist climate was becoming desertlike as water evaporated from the thin Martian atmosphere into space, and that a Martian civilization had turned to massive irrigation canals to carry water from their polar snow fields to the dry, warm equator.

This exciting hypothesis sparked raging debate for several decades. After spacecraft visits to Mars, however, we now know that no network of canals exists. How, then, could some astronomers have observed nonexistent features? Modern evidence indicates two reasons. First, streaky markings, including faulted canyons and dust deposits, do exist on Mars; seen through the shimmery atmosphere of Earth, these markings may resemble patterns of lines. If you squint at Figure 10-3, for instance, you will note the prominent horizontal linear canyon (lower center) and stubby vertical streaks (right center edge); these could be mistaken for lines by Earth-based observers. Second, and more important, some people are more likely than others to perceive streaky patches as straight lines, especially if they already believe the lines are there. Lowell was one of these; he even drew lines on Venus.

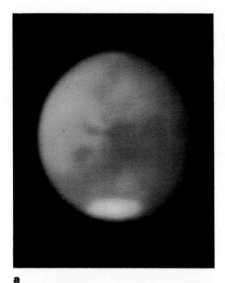

a

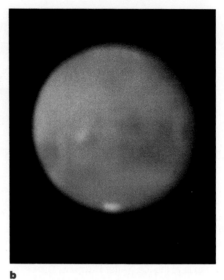

b

c

Figure 10-2 Three images of the same side of Mars. **a** 1971 photo shows south polar winter ice cap and dark markings. **b** 1973 photo shows a partially melted ice cap. Dark markings are washed out by a Martian dust storm, which reduces contrast. Bright yellow cloud (left center) is the core of the dust storm. **c** In 1988, sensitive new CCD technology allowed a shorter, sharper exposure. Special image processing techniques exaggerate contrast and detail. Stubby projections of dark areas northward into bright areas are the kinds of features that were once interpreted as "canals." (Catalina Observatory photos, courtesy S. M. Larson, University of Arizona.)

Names of Martian Features

Schiaparelli, a classical scholar as well as astronomer, named the larger Martian dark and light regions (as well as the illusory canals) after historical, mythological, and geographic features of his native Mediterranean area. These names, such as Hades, Arabia, and Libya, are still used for the major dark and light areas. In 1971, when the Mariner 9 spacecraft revealed actual geological structures such as craters, mountains, and canyons—all too small to be seen from Earth—these were assigned names as well. As on the Moon, craters were named after scientists. The largest canyon complex (Figure 10-3), big enough to stretch across the United States from coast to coast, was named Valles Marineris (Valleys of Mariner). A modern map with some of the markings and geological structures is shown in Figure 10-4.

THE LURE OF MARS

On October 30, 1938, Orson Welles broadcast a realistic radio play in which listeners heard "newsmen" report-

Figure 10-3 A portion of Mars showing the vast canyon Valles Marineris (bottom center), whose length is comparable to the width of the entire United States. Three large volcanic mountains with summit craters are prominent at the left. Meteorite impact craters, like those of the Moon but with more erosion, dot other parts of the picture, as in right center. In this processed image the colors are nearly natural, although the clouds are somewhat whiter than the true yellowish Martian cloud colors. As on Earth, the clouds tend to form over or near mountains; they nearly surround the three volcanoes. (NASA Viking composite image; courtesy A. S. McEwen, U.S. Geological Survey.)

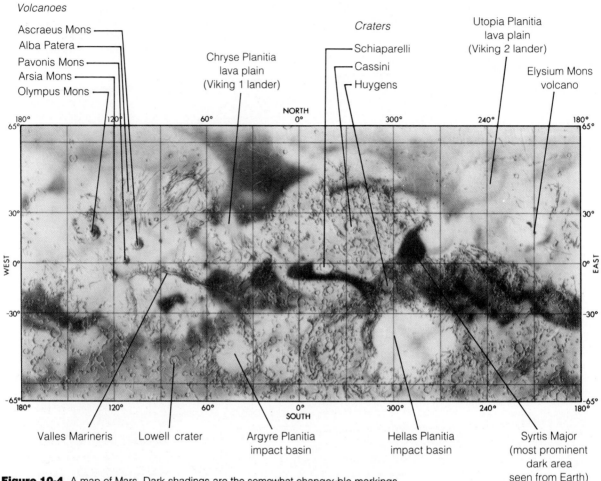

Figure 10-4 A map of Mars. Dark shadings are the somewhat changeable markings visible from Earth, probably associated with windblown dust. Topographic features such as craters and volcanoes are too small to see from Earth. (Base map courtesy R. M. Batson, U.S. Geological Survey.)

ing that Martians had invaded and were laying waste to New Jersey. Thousands believed it and the resulting panic caused a national scandal. In July 1965, Mariner 4 sent back the first close-up pictures of Mars, which were widely published on front pages of newspapers; the reporting rate of UFOs shot up by a factor of 6 for several weeks.

The groundwork for public fascination with Mars had been laid decades earlier. If Copernicus knocked the Earth out of the center of the universe, Lowell and the Victorians made people realize that Earth's civilization might not be the only one around. Earlier thinkers in the 1600s and 1700s had already suggested that other worlds might be inhabited—an idea known as the

plurality of worlds. And Darwin's theory of evolution, published in 1859, made more plausible the idea that other planets might produce totally alien species specially adapted to their environments.

The idea fascinated Victorian intellectuals. Tennyson and other poets wrote about it. In 1898, H. G. Wells published *The War of the Worlds*, in which Martians invade Earth to escape their dying planet. Wells commented:

The Tasmanians, in spite of their human likeness, were entirely swept out of existence in a war of extermination waged by European immigrants. . . . Are we such apostles of mercy as to complain if the Martians warred in the same spirit?

Figure 10-5 Under late afternoon Sun, rocks near the Viking 1 landing site cast picturesque shadows across the dunes of Mars. When this picture was taken, the atmosphere was particularly dusty, giving a hazy light with low-contrast shadows. Reddish dust settling out of the atmosphere has coated the tops of some rocks, such as the large boulder at left, 2 m (6 ft) wide and 9 m (30 ft) from the spacecraft. Parts of the landing craft containing the camera can be seen at the lower edges. (NASA.)

There was even a UFO scare in the 1890s that produced reports of Martian spaceships (described as looking like the first dirigibles, which were then flying).

Between the days of Lowell and the first space flights to Mars, several generations of readers grew up on stories by Edgar Rice Burroughs, Ray Bradbury, and others, which pictured a Lowellian Mars with remnants of a dying civilization holding out in nearly deserted cities on a dying, drying planet. Although incorrect in many details, these blends of theory and fancy fostered valid excitement about whether life might really exist on Mars and helped to encourage interest in actual voyages to the red planet. Yet, by the time of the first Martian voyages, Earth-based measurements had already revealed that Mars had colder, thinner, drier air than had been thought. Could such a planet support advanced life forms? Microbes? Or any life at all?

CONDITIONS ON THE SURFACE OF MARS

To find out if life exists on Mars, we sent space probes. Three Russian probes were first to reach the surface of Mars in 1971 and 1974, but all failed and none sent back useful data.

The first successful landing on Mars was by the **Viking 1** spacecraft, which touched down July 20, 1976 (7 y to the day after the first human landing on the Moon). It was followed on September 3, 1976, by a duplicate spacecraft, **Viking 2.** Both landings were on plains that looked relatively smooth from orbit but turned out to be rock strewn, as seen in Figures 10-5 and 10-6. Boulders as wide as 3 m were photographed among dunes near Lander 1 (Figure 10-5). Missing were the Martians, deserted cities, canals, and strange vegetation imagined by early writers.

Mars turned out to be a desolate, cold, yet beautiful desert. Its reddish color was vividly shown by color photos from Viking landers. Though some rocks appeared dark gray, like terrestrial lavas, most rocks and soil particles were covered with a coating of rustlike, reddish iron oxide minerals. Similar iron minerals give terrestrial deserts their familiar red-to-yellow coloration, especially when moisture is present only occasionally. Viking scientists were surprised to find the daytime sky of Mars reddish-tan instead of blue. The sky color is caused by much fine red dust stirred from the surface into the air by winds and deposited even on rocks as it settles out of the air (Figure 10-5).

Atmospheric Composition

Among the data analyzed by the Viking landers were samples of the **Martian atmosphere.** Its composi-

Figure 10-6 Winter on Mars. At the Viking 2 site near 48°N latitude, a thin layer of mixed CO_2 frost and H_2O frost covers the ground at night. During the day the CO_2 frost sublimes, leaving frozen H_2O covering much of this frosty scene adjacent to the previous pictures. On this late winter afternoon, the Sun is low in the sky behind us and the shadow of the Viking 2 spacecraft is seen in the foreground. (NASA.)

TABLE 10-1

Composition of Martian Atmosphere

Gas	Percent Volume
CO_2 (carbon dioxide)	95
N_2 (nitrogen)	2.7
Ar (argon)	1.6
CO (carbon monoxide)	0.6
O_2 (oxygen)	0.15
H_2O (water vapor)	0.03
Kr (krypton)	Trace
Xe (xenon)	Trace
O_3 (ozone)	0.000003

Note: Amounts of gases vary slightly with season and time of day. H_2O is especially variable. Some CO_2 condenses out of the atmosphere into the winter polar cap; changing cap sizes cause small changes in the total Martian atmospheric pressure.

tion, shown in Table 10-1, gives several clues about the planet's history. Like Venus, Mars has an atmosphere that is mostly carbon dioxide, probably generated chiefly by planetary degassing through volcanic activity. As noted in our discussions of Earth and Venus, volcanic gases, generated from melting of the interior rocky matter, are rich in carbon dioxide. Unlike Venus, Mars has only a very thin atmosphere. Various lines of evidence, which we will mention below, indicate that the present-day thin CO_2 is a remnant of a much thicker primeval CO_2 atmosphere.

The Martian Climate

Mars is very cold. The Viking probes landed during the Martian summer, but **air temperatures** at the two sites ranged from nighttime lows around 187 K (−123°F) to afternoon highs around 244 K (−20°F). Temperatures of the soil, which absorbs more sunlight than the air, exceed freezing (273 K, or 32°F) on some summer afternoons, so that any frost formed at night near the surface can melt and produce moisture or water vapor. Winds at the two sites were usually less than 17 kph, with gusts exceeding 50 kph. Much higher winds are believed to occur at certain seasons, however, raising clouds of dust that can be observed from Earth. **Air pressure** at each site was only about 0.7% that on Earth.

The Martian polar caps give vivid evidence that important gases freeze out of the Martian atmosphere (Figure 10-7). Even on the hemisphere that is having summer there is a small, permanent cap of frozen water. Many scientists believe the permanent ice cap at each pole is several kilometers thick. During Martian winter, temperatures plunge below the 146 K (−197°F) level at which carbon dioxide clouds form. Carbon dioxide snow, or "dry ice," accumulates on the polar ground to form a still larger winter cap. This transient winter cap is only a few meters thick and eventually shrinks during Martian spring. A thin, scenic layer of carbon dioxide frost accumulated on the ground around the Viking 2 lander at latitude 48°N during Martian winter (see Figure 10-8), but not at the Viking 1 site at 22°N.

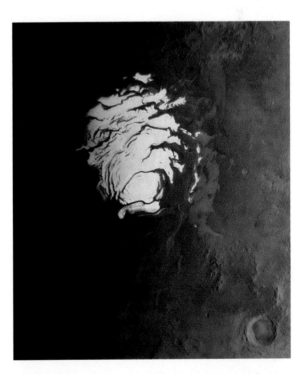

Figure 10-7 The south polar ice cap of Mars seen from orbit during the southern summer. It is roughly 360 km across and consists of CO_2 ice (possibly mixed with H_2O ice). Spiral breaks mark warmer, Sun-facing ridges in the layered sedimentary deposits around the pole. In winter, condensation of CO_2 frost and snow makes the cap expand to 10 times this size. Compare with Earth-based views of cap in Figure 10-2. Night side of the planet is at the bottom. (NASA Viking orbiter photo; courtesy Tammy Ruck and Larry Soderblom, U.S. Geological Survey.)

a

b

Figure 10-8 Onset of Martian winter at the Viking 2 site. **a** The boulder-strewn plain, showing a trench dug by a remote-controlled sampler arm. At right lies a protective hood ejected from the sampler. **b** Later, as winter sets in, the same scene shows deposits of CO_2 frost in the cold shadows of rocks. In the meantime, the sampler has dug another trench near the first one. (NASA.)

Rock Types

Martian rocks at the two Viking sites appear to be fragments of lava flows, as judged by their textures and colors and by the chemistry of the associated soil. Proportions of elements measured in the soil resemble those for soils derived from basaltic lavas on Earth and the Moon. A minor component in the soil is water, about 1%. This water exists not as a liquid or ice but as H_2O molecules in the crystal structure of the rock particles. It may be a remnant of more abundant water in the past. Many investigators believe that ice may be present among soil particles a few meters below the surface. This would form a layer of **permafrost**, permanently frozen soil similar to that in arctic tundra regions of Earth.

The bright deserts have more weathered soils and fine dust than the dark areas, which probably have more fresh, dark gravels and rock outcrops. The most Mars-like landscapes, rock types, and soil types on Earth are found in extremely dry volcanic deserts, as shown in Figure 10-9.

MAJOR GEOLOGICAL STRUCTURES

In addition to data from the surface, the mapping of large-scale geological structures from orbiting spacecraft has yielded information just as intriguing.

Deposits of windblown dust explain most of the patchy, variable markings visible from Earth. Comparison of close-up spacecraft photos from different seasons and different years reveals changes as the wind redistributes the dust in patchy and streaky markings, often in crater floors or downwind from crater walls or other topographic features. This effect can be seen at the right center edge of Figure 10-3.

Martian meteorite impact craters, such as seen in Figure 10-10, are interesting for three reasons. First, they show that impacts have been frequent on Mars, as on the other planets. Second, the craters come in more varied states of degradation than on the Moon or Mercury, and this shows that erosive processes have been more common on Mars than on the Moon or Mercury. The erosive processes include obliteration by lava flows, filling by windblown sediments, and probably erosion by flowing water. Third, the craters provide a means of estimating the ages of Martian surface features. The longer a surface has been exposed, the more meteorites have fallen on it to make craters. The old, heavily cratered regions of Mars are believed to be several billion years old, like the older uplands of the Moon. But the youngest, sparsely cratered volcanoes, lava flows, and eroded surfaces may be a few hundred million years old, or less.

The huge **Martian volcanoes** are especially notable. These were discovered by Mariner 9 in 1971. The highest, Olympus Mons (Figure 10-11), rises 24 km (78,000 ft) above the surrounding desert, forming a broad, dome-shaped mountain. Its huge base, about 500 km across, would cover Missouri. Three other sim-

a

b

c

d

Figure 10-9 Close terrestrial analogs to Martian landscapes occur in very arid regions with a history of weathering. **a** Glacier-dropped boulders in volcanic plains, Iceland. **b** Basaltic lava at 10,000-ft altitude on Mauna Loa volcano, Hawaii. **c** Windswept coastal desert of Peru. **d** Death Valley, California. Most of these areas are less weathered and the iron-bearing minerals are less oxidized, or rusted, than on Mars. Hence those areas are less red than Mars, but the rock structure and landscape is similar. (Photos by author.)

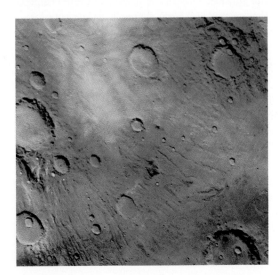

Figure 10-10 Orbital view of one of the older, cratered portions of Mars. On the one hand, we see lunarlike craters caused by ancient meteorite impacts; but on the other hand, the landscape shows cloudy haze and evidence of erosion unlike the Moon. (NASA Viking orbiter photo.)

Figure 10-11 The top of the mightiest Martian volcano, Olympus Mons, protrudes through morning clouds. The lava-furrowed flanks and the 65-km summit caldera complex can be seen. (NASA hand-tinted image derived from oblique Viking orbiter photo.)

ilar-sized volcanoes lie nearby, and numerous smaller ones dot the planet. Some of the volcanoes may still be active, but this is uncertain. The vast volcanic plains surrounding the four largest volcanoes look fresh. Some regions near the large volcanoes are intensely fractured, as if subjected to tectonic forces strong enough to split the lithosphere, but not strong enough to drive full-fledged plate tectonics, such as occurs on Earth. For example, the giant canyon of Valles Marineris (Figure 10-3), much larger than our Grand Canyon, appears to be a tectonic fracture system similar to the Red Sea on Earth, and about the same size.

THE MYSTERY OF
THE ANCIENT MARTIAN
CLIMATE

So far Mars might sound like the Moon or Mercury with a few extra volcanoes and a little air to blow the dust around. Mariner 9 shattered this conception by photographing **channels** that look like dry riverbeds, as seen in Figures 10-10 and 10-12. These channels meander in sinuous curves and often have tributaries. They get wider and deeper in the downslope direction and have sedimentary deposits on their floors. In short, they have all the features of arroyos cut by water or ice-clogged streams. Other theories of their origin—for instance,

that they might have been lava flow channels—do not explain all their features.[1]

The greatest surprise of Martian exploration was to confirm that Mars is very arid, and yet at the same time to discover what appear to be riverbeds! Evidently, Mars once had flowing rivers of liquid water. The number of recent impact craters interrupting the channels indicates that the channels are perhaps 1 to 3 billion years old. The duration of flow episodes is unknown. The channels are younger than the most ancient cratered regions, but older than most volcanoes. Some channels emanate from chaotic collapsed areas believed to have formed when ice deposits melted and released water onto the surface. Others—a network of fine channels near the equator and large channels with tributaries—suggest that some water came from other sources, possibly rainfall from a once denser atmosphere, as visualized in Figure 10-13. Other signs of ancient erosion, such as degradation of craters and the buildup of layered sedimentary deposits, also suggest much more atmospheric and erosive activity in the past.

Consistent with this, Viking scientists found chemical evidence that the atmosphere of Mars was 10 to

---

[1]Although a few of the major riverbeds lie near reported positions of the once popular "canals," there is little correspondence in general. The channels do not explain the canals.

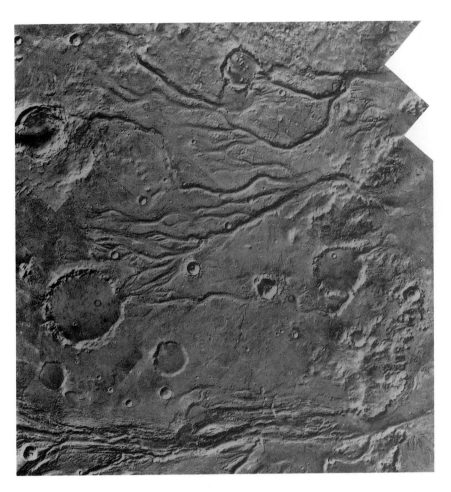

Figure 10-12 Dramatic evidence of ancient riverbeds on Mars is found in these channels and tributary systems. The area is about 180 km (110 mi) wide and drops about 3 km in the direction of flow, from the west (left) to the east. Water apparently cut into some old craters, but predated others. (NASA Viking photo from orbit.)

100 times denser in the past, making the early Martian atmosphere nearly as dense as Earth's atmosphere today (Owen and others, 1977; Haberle, 1986). Similarly, geologists studying the amount of lavas and their probable H_2O content have estimated that the total water vapor injected into Mars' atmosphere by all the volcanoes during all of Mars' history would have been enough to make a water or ice layer 50 m deep over most of Mars (Greeley, 1987). Other workers studying orbital photos point to features that resemble ancient shorelines, and there actually may have been ancient temporary lakes or oceans up to hundreds of meters deep in some locations (Kerr, 1986).

Where have the water and atmospheric gas gone? Many Mars analysts believe that much of the water is still on Mars. We know that much water is frozen in the permanent polar caps and that about 1% of the surface soil is chemically bound water. But much more water may still be frozen as permafrost below the surface. Some other gases may have escaped into space.

A more provocative question is: What could have caused the transition from the earlier conditions—with more liquid water, more air, and higher temperature—to the present arid, freezing conditions? Astronomers and meteorologists have combined forces to discover what factors could change planetary climates within intervals of a hundred million or a billion years.

Under present conditions, water is very unlikely to flow on Mars. It is usually frozen, and even if it warmed enough to melt, it would very rapidly evaporate into the thin Martian air. (In many regions the air pressure is so low that the water would spontaneously boil away into the air.) But calculations indicate that if Mars ever had more air, liquid water might have been more stable on the surface.

So the question is: How can we explain the depletion

a   b

Figure 10-13 Imaginary views of a Martian channel ("now" and "then"). **a** A typical channel floor under present conditions. (Painting by author.) **b** The same channel being inundated by a catastrophic flow of water. This might have occurred under ancient climatic conditions when the atmosphere was denser. Whether such scenes really occurred on Mars is still unknown. (Painting by Ron Miller.)

of an early, denser atmosphere to its present, thin state? Some evidence indicates that the atmospheric conditions depend strongly on the conditions at the polar caps. Cornell scientists have found that if solar radiation striking the caps were only 10 to 15% higher in the past, the ice caps would have been much smaller. Much of the now-frozen carbon dioxide would have existed as gas, raising the atmospheric density and pressure considerably. In other words, small changes in the polar climate could result in large changes in global climate (Sagan, Toon, and Gierasch, 1973).

Further research revealed that during formation of the giant Martian volcanoes (Olympus Mons and its neighbors), the massive lava accumulations on one side of Mars altered the planet's axial tilt, changing the total amount of solar heating on the cap during the summer. Thus, the massive volcanism of Mars may have triggered a climate change, perhaps in the last 1 or 2 billion years. Before the volcanism, the polar ice caps may have trapped less carbon dioxide and water, the air may have been denser, and water may have flowed occasionally on the surface. This climate change agrees with evidence that the channels are generally older than the volcanic lavas because channels rarely cut across young

lava surfaces. This also agrees with evidence from crater erosion that the change in Martian climate was not slow and gradual throughout Martian history, but was fairly abrupt, and possibly in the last half of Martian time.

Another process that probably depleted Mars' early atmosphere is the gradual trapping of carbon dioxide into carbonate rock minerals (Warren, 1987). Telescopic observations during the close approach of Mars in 1988 led to discovery of carbonate-bearing minerals on Mars, supporting this idea. Similarly, the water that was once important as a fluid has now been trapped as polar ice, underground ice, and water-bearing minerals in rocks.

In summary, the water and carbon dioxide that once made Mars more Earth-like are still there, but they are trapped in solid forms and cannot play much of a role in making the atmosphere and climate pleasant today.

These ideas about Martian climate history show how astronomical exploration of other planets can illuminate conditions on Earth. The Martian evidence forces us to realize that planetary climates may not be stable, and that Earth's climate, too, may be subject to astronomically caused changes.

ROCK SAMPLES FROM MARS?

To clarify Mars' history once and for all, we need samples of rock and soil from different geologic features on Mars. From them, we could measure the planet's chemistry and, more importantly, the ages of the different features. Ideally, the samples should come from carefully identified sites. Of course, no such samples are yet available, but, oddly enough, a few rocks from Mars may be among meteorites already in our museums!

Meteorites are rocks that fall from space. Most are 4.5-billion-year-old leftover debris from the formation of the planets. But about a dozen meteorites have puzzled scientists because they are basaltic lavas formed only about 1.3 billion years ago. Where could they have come from?

By the late 1970s, researchers knew that a few meteorites were rocks that had been blown off the Moon by ancient impacts and had later landed on Earth; however, the same researchers knew enough about lunar rocks to rule out the Moon as the source of the strange lavas. By the 1980s, most researchers concluded that they were rocks blown off Mars by ancient meteorite impacts on that planet. Among the lines of evidence: (1) gases trapped in the rocks had just the same composition as Viking measured for the air on Mars; (2) mineral evidence indicated the rocks had been shocked by explosive forces strong enough to blow them off Mars; (3) the ages of 1.3 billion years were consistent with the best guesses for the ages of common Martian lavas.

Assuming that they *did* come from Mars, the rocks reveal some things about that planet: (1) Mars has a mantle roughly similar to Earth's; (2) Martian volcanoes have erupted within the last 1.3 billion years; (3) Martian subsurface materials contain large amounts of water or ice (McSween, 1985). All this is consistent with our picture of Mars constructed from spacecraft data: for example, that Mars should be big enough to have melted and differentiated into a mantle and core; that it has large areas of lavas; and that it may have large amounts of permafrost in its frozen soil. The "accidental" Mars samples provided by the meteorites are a scientific bonus, but we do not know what locations they came from on Mars, and thus they are limited in what they can tell us about the red planet. For that reason, scientists are still anxious to get geologically documented samples from a variety of geologic features on Mars.

WHERE ARE THE MARTIANS?

Much of the motivation and excitement in exploring Mars has been in the search for extraterrestrial life. Discovery of life on Mars would be of major cultural importance. Just as the discoveries of the Copernican revolution showed that the Earth was not the center of the solar system, **life on Mars** (or elsewhere) would show that humans are not necessarily the lords of creation. On the other hand, proof that life never evolved on Mars in spite of favorable conditions would present an exciting challenge to our present concepts of the nature of life, because biological experiments suggest (but do not prove) that life should evolve whenever conditions are suitable.

The Viking mission was specifically designed to look for life on Mars.[2] Of the five Viking experiments involved, two gave strong negative results, but three gave ambiguous results. In the first experiment, the cameras showed no signs of life. The second experiment, a soil analysis, revealed no organic molecules in the soil at either Viking site, at a sensitivity of a few parts per billion. Because **organic molecules**, or massive molecules containing carbon, are essential building blocks of life, this test strongly indicates that living organisms do not now exist in Martian soil and have not existed in the recent past.

The other three experiments were designed to look for ongoing biological processes, such as metabolism and photosynthesis, by taking soil samples, putting them in special chambers (some with nutrients), and watching for chemical changes that would indicate microscopic organisms processing the material in the chamber. Although these experiments detected limited chemical changes, most scientists think the changes involved only chemical reactions in the soil, not life.

Thus, Viking has shown that Mars "is not teeming with life from pole to pole" (to use the memorable understatement by astronomer Carl Sagan). It certainly has no large animals or plants and probably has no present or recent microbial life. Why?

The most probable answer is that ultraviolet light from the Sun irradiates the Martian surface. This is because Mars lacks an ozone layer. Ultraviolet light

---

[2] By *life* scientists generally mean "life as we know it"—the ability of carbon atoms to combine with other atoms and form very complex molecules, which in turn form organisms that grow and reproduce.

Figure 10-14 Astronaut expeditions to the surface of Mars may eventually clarify Mars' secrets. Researchers could set up a base similar to Antarctic research stations and probe Martian geology and chemistry more effectively than robot probes. Such expeditions are within our technical capability; Russian cosmonauts have occupied their Salyut space station for durations equivalent to a flight to Mars. (Painting by Paul Hudson.)

breaks down organic molecules, sterilizing the rock surfaces and the soil particles that have been blown around in the atmosphere.

This theory has led to the idea that Mars might have evolved simple life forms in the ancient past, when the atmosphere was thicker, acting as a better shield. Some scientists have noted that an exciting goal for a future Martian expedition would be to try to study ancient, buried strata for possible fossils or other signs of ancient organic activity (Figure 10-14). This could be done through drilling or studying outcrops exposed by landslides in the sides of Martian cliffs. Perhaps the dusty Martian plains have not disclosed their last secrets about the origins of life in the solar system. In any case, we could learn more about ourselves if we could learn what prevented the flourishing of life on such an Earthlike planet.

MARTIAN SATELLITES: PHOBOS AND DEIMOS

Not all of the mysteries of Mars are on its surface. In 1877, the American astronomer Asaph Hall became the first human to see a satellite of Mars. Shortly afterward, he charted the positions of two Martian moons, naming the inner satellite Phobos ("fear") and the outer one Deimos ("terror") after the chariot horses of Mars in Greek mythology. Close-approaching spacecraft, including Viking and the 1989 Soviet Phobos-2 probe, revealed these moons to be strange, black, potato-shaped, cratered chunks of rock (Figures 10-15 and 10-16). Phobos is the larger moon, at 20 × 28 km (12 × 17 mi). Deimos is only 10 × 16 km (6 × 10 mi) across.

The craters of Phobos and Deimos were caused by collisions with small bits of meteoritic debris. The larg-

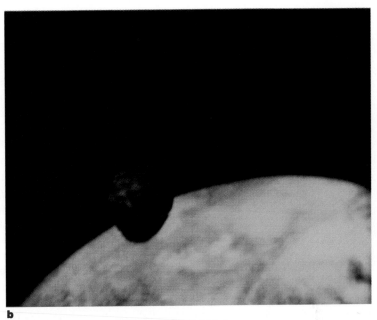

a **b**

Figure 10-15 Mars' larger moon, 28-km-long Phobos, is shown here in photos from two different space missions. **a** Photo from American Viking mission in 1976 shows the largest crater, Stickney (top), caused by an impact that almost shattered Phobos. Grooves run roughly radial to Stickney, perhaps marking fractures from the impact. (NASA.) **b** Color view from Soviet Phobos-2 mission in 1989 shows the moon hanging in front of Mars. This unusual view dramatically shows Phobos' very dark color. (Courtesy B. Zhukov, IKI [Institute for Cosmic Investigations], USSR.)

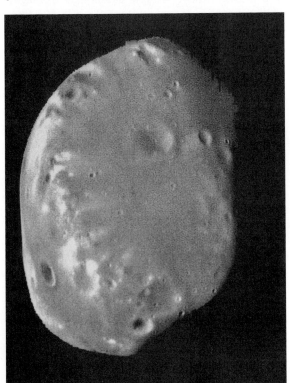

est crater on Phobos is 8 km (5 mi) across and can be seen at the top of Figure 10-15. It is named Stickney, the maiden name of Mrs. Hall, who encouraged her husband's successful search for the satellite.

A peculiar feature of Phobos was revealed by close-up photos from Viking orbiters. Networks of grooves reach widths of around 100 m and lengths of as much as 10 km. Some are rows of adjoining craters. They may mark positions of subsurface fractures (somewhat masked by surface dust) caused by the mighty Stickney impact.

Soviet scientists deduced from their Phobos-2 probe that Phobos has an unusually low density, only 1950 ± 100 kg/m^3, which is lower than that of any known meteorites or common rock types. From this they concluded that Phobos may have a porous aggregate structure, or may contain ice. Phobos' structure and resources

Figure 10-16 Deimos, a 16-km-long moon of Mars. Its surface texture is smoother than that of Phobos, perhaps because of a different duration of microcratering since the most recent large impacts. Streaky markings are believed to be due to slipping of loose material in locally downhill directions in Deimos' weak, asymmetric gravity field. (NASA.)

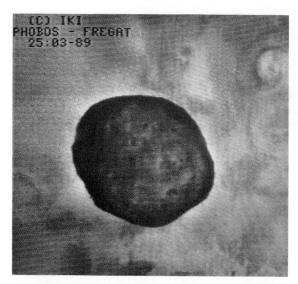

Figure 10-17 An eerie view of dark Phobos hanging in front of the enigmatic craters and dusky markings of Mars, from the Soviet Phobos-2 probe, symbolizes continuing interest in international exploration of Mars and its moons. (Courtesy B. Zhukov, IKI [Institute for Cosmic Investigations], USSR.)

(frozen water?) will be of interest to future explorers of the red planet.

The origins of Phobos and Deimos are unclear. Their composition apparently resembles that of a certain type of carbon-rich black asteroid known to be common in the nearby asteroid belt. For these reasons, many researchers believe Phobos and Deimos originated as asteroids and were later captured into orbit around Mars. This could have happened 4.5 billion years ago, when primordial Mars had a more extensive atmosphere that could slow asteroids happening to pass through its outer fringes. This slowing could have caused a passing asteroid to be captured into Martian orbit. Phobos and Deimos, according to some astronomers, might be pieces of a single asteroid broken during such a capture process. Further drag from the primitive extended atmosphere may have altered the moons' orbits, but the early atmosphere soon dissipated, leaving the moons stranded in their present orbits.

The Soviet Union launched two robot probes to Phobos in 1988. Phobos-1 was lost along the way due to programming errors, but Phobos-2 rendezvoused with Phobos in 1989 and acquired useful new data about Phobos and Mars. However, as it was being maneuvered close enough to drop a package of instruments onto Phobos, contact was lost before it could complete its mission.

EXPEDITIONS TO MARS

Interest in the comparative planetology of Mars and Earth, as well as the potential habitability of Mars, have led to emphasis on Mars as the major current target of planetary exploration. In the United States, the president's National Commission on Space in 1986 urged a permanently inhabited Martian base as a long-term space goal.[3] In 1989, President Bush set a goal of a human expedition to Mars. The project could be international—serving as a model for international technical cooperation as well as an impetus for scientific and engineering growth. The Commission report targets the base for the decade of the 2020s.

NASA plans a "Mars Observer" that would orbit Mars and gather new data on surface mineral composition, atmospheric dust, and geological structures, but its launch has been delayed from 1990 to 1992. The Soviets have proposed a more ambitious program. As mentioned in Chapter 4, they are preparing robotic surface exploration missions for the 1990s. A proposed Soviet 1994 mission would include balloons that would touch down on the surface at night and travel through the air by day. A joint American–Soviet sample-return mission in 1998 has been discussed. The samples would help us understand the dates of ancient climate changes, the fate of the early atmosphere, the role of flowing water, and the factors that retarded life from evolving on Mars. American and Soviet researchers plan to coordinate the work of all space probes near Mars in the 1990s. Meanwhile, the Soviets have gained the experience needed for interplanetary piloted flights by having their cosmonauts live in the Mir space station for more than a year, and they have discussed sending cosmonauts to Mars by around 2010. Future explorers may make Phobos a base of operations before proceeding on to Mars' surface (Figure 10-17).

A LESSON IN COMPARATIVE PLANETOLOGY: THE TOPOGRAPHY OF EARTH, VENUS, AND MARS

Orbital mapping of the planets has allowed researchers to prepare topographic maps showing the altitudes of

---

[3]The commission included Neil Armstrong, astronaut Kathryn Sullivan, test pilot Chuck Yeager, U.N. ambassador Jeane Kirkpatrick, as well as several well-known scientists and engineers.

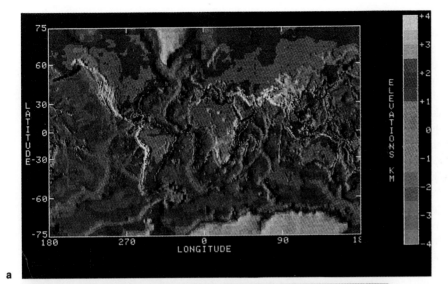

a

b

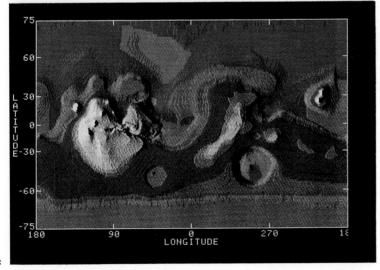

c

Figure 10-18 Topographic maps of the three largest terrestrial planets; altitude scale (shown by color bar) is the same in all three cases. **a** Earth, showing continental blocks with arc-shaped mountain chains formed by plate collisions. **b** Venus, showing lesser development of "continental" masses. **c** Mars, showing a more primitive surface with circular impact basins and highest spots created by giant volcanic peaks. (Computer-generated topographic maps from orbital radar and other data; courtesy M. Kobrick, Jet Propulsion Lab.)

points across the surfaces of Earth, Venus, and Mars. Figure 10-18 compares three such maps. The surface data have been digitized, with a pixel for each point of specified latitude and longitude. The color assigned to that pixel represents the altitude, shown by the color altitude scale, which is the same for all three maps.

A comparison of the maps reveals interesting differences in the geological "styles" of the planets related to their sizes. Earth, the largest, is dominated by rolling seafloor plains interrupted by continental blocks (Figure 10-18a). The same could be said for Venus, although there is not as much elevated "continental" land on Venus, perhaps because smaller Venus did not have as well-developed plate tectonic motions (Figure 10-18b). Indeed, we can see that Earth's major mountains are arc-shaped ranges that develop when plates collide. Because a smaller planet loses its internal heat faster, Venus did not have as much internal energy to drive such plate motions.

These ideas are affirmed as we turn to the map of still-smaller Mars (Figure 10-18c). Here the map shows a quite different style. This planet was too small to generate even enough tectonic energy to destroy all its original cratered topography. Thus we see that some of the deepest depressions are well-defined circular impact basins, not rolling seafloor plains. The Hellas basin (lower right center) is a prominent example. Most of the uplands rendered in khaki green are heavily cratered. There is a link with Venus, however. The highest Martian mountain areas, such as the broad Tharsis dome (tan, left center), are simply piles of volcanic lavas surmounted by the mighty Olympus Mons volcano (Figure 10-11) and other volcanic peaks, not unlike Venus' Maxwell Montes (Figure 10-18b, top center; Figure 9-8).

In other words, the maps confirm our rule of thumb at the end of Chapter 9: Worlds smaller than Mars preserve ancient surfaces dominated by the *external* forces of cratering that shaped the planet. For worlds around the size of the Moon and Mars, internal energy is significant enough that volcanic forces break through the lithosphere and resurface parts of the planet. Worlds larger than Mars have surfaces dominated by these *internal* forces, including volcanism and tectonic restructuring.

SUMMARY

Mars, the planet once thought to have fields of vegetation or even a dying civilization, has been revealed by spacecraft to be a barren but beautiful desert lacking any advanced life forms. Orbital photos reveal a wide variety of landscapes including lava flows, grand canyons, landslides, polar snowfields, eroded strata, dunes, impact craters, and arroyos. Biological experiments aboard Viking landers indicate that interesting chemical reactions take place in Martian soil, and the seemingly lifeless Martian environment may yet yield clues about the evolution of biochemical reactions and their dependence on climate.

Evidence about Mars' ancient climatic history seems to contradict its present-day barrenness. Nearly all water on Mars today is locked in polar ice, frozen in the soil, or chemically bound in the soil; virtually no liquid water exists. But liquid water apparently once flowed and eroded the surface, indicating different past climates. Current research indicates that the climates of both Mars and Earth may have varied significantly during their histories.

CONCEPTS

Mars' rotation period

seasonal changes on Mars

desert

canal

Viking 1

Viking 2

Martian atmosphere

Martian air temperature

Martian air pressure

permafrost

Martian meteorite impact craters

Martian volcanoes

channel

life on Mars

organic molecules

PROBLEMS

1. Describe a day on Mars as a future astronaut might experience it. Include the length of the day, the appearance of the landscape, possible clouds and winds, possible hazards, and objects visible in the sky.

2. What scientific opportunities for long-term exploration does Mars offer compared with the Moon? What qualities of the environment might make operating a long-term base or colony easier on Mars than on the Moon once initial materials were delivered to the site?

3. Compare photos of craters on Mars and on the Moon.
   a. Assuming that all craters had similar sharp rims when fresh, which craters have suffered most from erosion?
   b. What does this say about lunar versus Martian environments?

4. Give examples of how the Martian environment and geology are midway between those of the smaller planet Mercury and the larger planet Earth. Comment on atmosphere, craters, volcanism, and plate tectonics.

5. What scientific knowledge might be gained from close-up investigation of Phobos and Deimos? What measurements would be of interest if rocks from Phobos and Deimos were available for study?

6. Imagine you are a visitor from outer space exploring the solar system.
   a. If two Viking-type spacecraft landed at random places on Earth, took photos, measured the climate, and took soil samples, what might they reveal about Earth?
   b. How many landings might be needed to characterize Earth adequately?
   c. To characterize Mars to the same degree, how many might be needed?

PROJECTS

1. Observe Mars with a telescope, preferably within a few weeks of an opposition and with a telescope having an aperture of at least 15 cm (6 in.). Magnification around 250 to 300 is useful. Sketch the planet. Can you see any surface details? Usually the most prominent detail is one of the polar caps, a small, brilliant white area at the north or south limb contrasting with the orangish disk. Can you see any dark regions? Compare the view on different nights and at different times of the night. (Because Mars turns about once in 24 h, the same side of Mars will be turned toward Earth on successive evenings at about the same hour.) If no markings can be seen, three explanations are possible: Observing conditions are too poor; the hemisphere of Mars with very few markings may be turned toward Earth; a major dust storm may be raging on Mars, obscuring the markings.

2. For the previous observations, determine which side of Mars you were looking at. (Your instructor may need to assist you.) First determine the date and Universal Time of your observations (UT = EST + 5 h = PST + 8 h. Thus 10 P.M. EST on April 2 = 03^h00^m on April 3, Universal Time). In *Astronomical Almanac*, the table "Mars: Ephemeris for Physical Observations" gives the central meridian (or longitude on Mars of the center of the side facing Earth) at 0^h00^m Universal Time on each date. From these tables you can find the Martian central meridian for the time of your observation. (Mars turns about 14.7°/h.) Compare your observations with a map of Mars, locating the part of Mars that you observed.

CHAPTER 11

Jupiter and Its Moons

The terrestrial planets, which we have been studying, are huddled relatively close to the Sun. Now we leave them behind and move to *the outer solar system*—the part of the solar system beyond the asteroid belt.

INTRODUCING THE OUTER SOLAR SYSTEM

The outer solar system contains four **giant planets**—Jupiter, Saturn, Uranus, and Neptune—and a small planet, Pluto. The name of the monarch of the Roman gods is fitting for Jupiter. The biggest planet, it contains 71% of the total planetary mass—nearly $2\frac{1}{2}$ times as much as all other planets combined. All four of the giant planets also have large families of satellites—at least 50 in all. The four giant planets together contain $99\frac{1}{2}$% of the total planetary mass and harbor about 91% of the known satellites.

The four giant planets have much lower mean densities than the terrestrial planets—700 to 1600 kg/m^3 as compared with 3900 to 5500 kg/m^3. Saturn, at 700 kg/m^3, would float like an ice cube if we could find a big enough ocean. (Water's density is 1000 kg/m^3.) This simile is significant. The giant planets evidently *are* made largely of ices, as well as low-density liquids such as liquid hydrogen.

Jupiter's diameter measures a little over 10 times Earth's diameter, and Saturn just under 10 times. Uranus and Neptune have diameters about four times the Earth's. Placed on the face of Jupiter, Earth would look like a dime on a dinner plate (see Figure 8-2).

Perhaps the most important principle to remember about the outer solar system is that because it is further from the Sun and much colder than the inner solar system, it contains much more ice than the inner solar system. Because the gases that formed the Sun and its

a   b

Figure 11-1 Comparison of (**a**) an example of the best Earth-based telescopic imagery of Jupiter and (**b**) a close-up view from a spacecraft. Jupiter is entirely covered by multicolored clouds arranged in lacy belts swirled by wind motions. (Photo *a*: Catalina Observatory 61-in. telescope, courtesy S. M. Larson, University of Arizona; photo *b*: NASA Voyager photo, processed by A. S. McEwen, U.S. Geological Survey.)

planet-spawning surroundings were mostly hydrogen, the ices that formed in the outer solar system are frozen compounds of hydrogen, such as water (H_2O), methane (CH_4), and ammonia (NH_3). In the warm, inner solar system, these substances remained in gaseous form and did not add to the mass of the solid planets. Thus, instead of forming worlds of rock like the terrestrial planets, the outer solar system formed worlds of rock *plus* ice, often with 50% ice or more. This explains many properties of worlds in the outer solar system. The giant planets are giant for two reasons: (1) They had ices in addition to rock, and (2) once they reached a large size during their formation, their gravity was so great that they began to pull in the surrounding hydrogen-rich gases. (The gravity of Earth and smaller planets was too weak to hold these light gases.) Hence the giants are huge, hydrogen-rich worlds. Their surfaces are hidden by colored clouds. Their thick atmospheres may bear some resemblance to the primordial atmosphere of Earth.

Similarly, most of the moons in the outer solar system have icy surfaces or dirty-ice surfaces. On some of them, geological processes tended to evaporate the ice off the surface, leaving darker, soil-rich surfaces. On others, internal heating melted the ice, producing watery "lava" that erupted and formed bright ice patches. These moons are fascinating worlds—some larger than the planets Mercury and Pluto, as seen in Figure 8-1.

THE PLANET JUPITER

Even a week's observations with a backyard telescope reveal the swirling cloud patterns of Jupiter, prominent in Figure 11-1. The most obvious pattern is the system of dark and light cloud bands parallel to Jupiter's equator, as sketched in Figure 11-2. The dark ones are called **belts**; the bright ones are **zones**. Within these bands, wispy spots and streaks arise, develop, and die out. These features look small, but some of them are larger than Earth! Though the smaller ones evolve in days, the larger ones may last for months or years. Dark clouds may grow and darken an entire bright zone for months or years at a time (Figure 11-3). The clouds are believed to be composed primarily of ice crystals of ammonia, ammonium hydrosulfide, and frozen water.

N

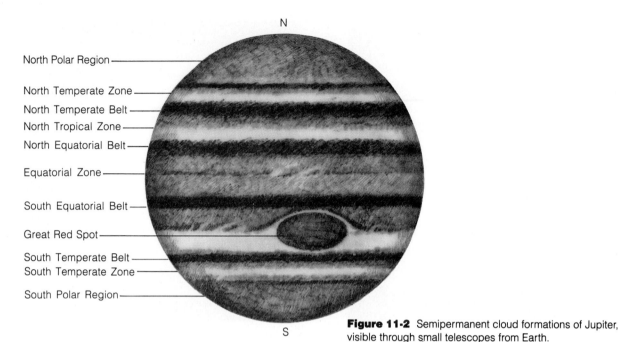

North Polar Region

North Temperate Zone
North Temperate Belt
North Tropical Zone
North Equatorial Belt

Equatorial Zone

South Equatorial Belt

Great Red Spot

South Temperate Belt
South Temperate Zone

South Polar Region

S

Figure 11-2 Semipermanent cloud formations of Jupiter, visible through small telescopes from Earth.

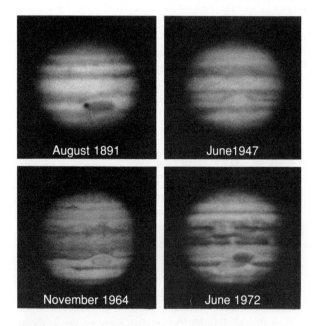

August 1891

June 1947

November 1964

June 1972

Figure 11-3 Photographs of Jupiter spanning 81 y, showing the changing array of Jupiter's belts and zones. The dark north temperate belt is relatively permanent, but the equatorial zone changes from bright (1891) to dark (1964). These images show the Great Red Spot. (Lowell Observatory.)

The cloud belts and zones have distinct colors, shown vividly in Figure 11-1. Usually belts are brown, reddish, or even greenish, whereas zones are light tan, whitish, or yellowish. The variegated colors persist even among small-scale clouds, as seen in the beautiful close-up of Figure 11-4. The colors are caused by photochemical reactions much like those that produce orangish-brown smog layers from pollutants on Earth. On Jupiter, the colorful minor constituents produced by these reactions may include hydrogen sulfide, organic particles, or metallic sodium particles.

Jupiter's Atmosphere

Just as on Earth, the cloud materials are only minor constituents of a much more extensive atmosphere of clear gas. On Earth, the gas is mostly nitrogen and oxygen, and the clouds of water droplets and ice crystals condense from the small amounts of water vapor in the air. On Jupiter, the gas is mostly hydrogen and helium, whereas the clouds condense from ammonia, water vapor, and other minor compounds (see Table 12-1). **Jupiter's atmospheric composition** is about four-fifths hydrogen and one-fifth helium by mass. It is very different from our nitrogen/oxygen atmosphere.

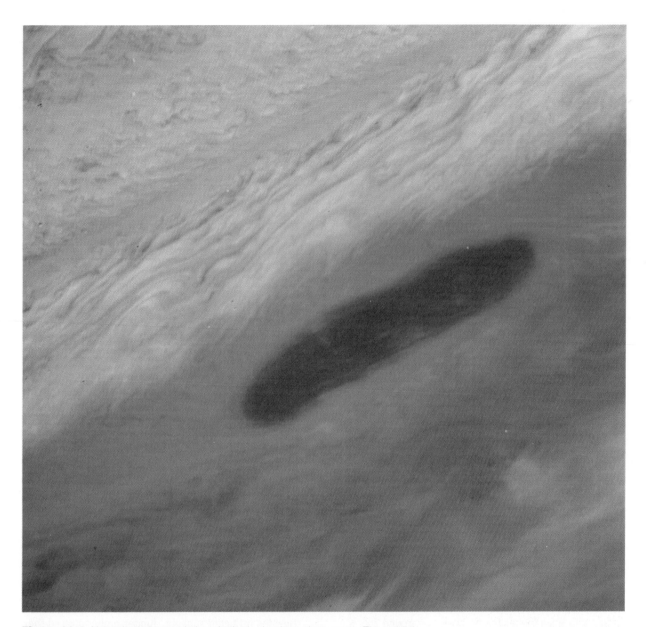

Figure 11-4 Nature's abstract painting: Jupiter's clouds at close range. The reddish-brown oval, more than half the diameter of Earth, may be a clearing in the upper cloud deck, allowing a view of deeper, redder clouds. Winds among the white clouds, which mark the north temperate zone, were measured at 260 mph. Smallest visible cloud features are about 80 km (50 mi) wide. (NASA Voyager 1 photo.)

The hydrogen and helium are thought to be a "fossil" atmosphere of the gas that surrounded all the planets as they formed. The Sun is made from similar gas. Water, ammonia, and methane are also present in Jupiter's atmosphere. Voyagers 1 and 2 in 1979 revealed other trace gases.

Infrared radiation reveals **Jupiter's temperature** in the upper atmosphere to be very cold because of the planet's great distance from the Sun—about 133 K (−220°F), on both the sunlit and nighttime sides. At a lower level, the poisonous clouds are warmer.

Gaps in the clouds have revealed still lower haze

Figure 11-5 Enormous cloudscapes would greet a visitor to the upper atmosphere of Jupiter or the other giant planets. This imaginary view shows cloud decks at several levels. In the distance, one of the moons partially eclipses the Sun. (Painting by Ron Miller.)

layers with even higher temperatures of around 250 K (−9°F). The lower regions may resemble the hydrogen-compound-rich primordial atmosphere of Earth when terrestrial life originated.

A few scientists have speculated that at these "comfortable" low levels complex organic molecules might have evolved into simple hydrogen-breathing organisms that could float in the atmosphere. A recent model of Jupiter's atmosphere calls for temperatures similar to those at Earth's surface at a level of about 60 km below the Jovian cloud tops, where the pressure would be about 10 times Earth's surface pressure. Such condi-

tions might be hospitable to primitive life, but most scientists doubt that any life forms exist on Jupiter. Future space probes to Jupiter may clarify this.

Voyagers' cameras revealed mighty blasts of lightning playing among the clouds and enormous auroral displays flickering high above the clouds in the polar regions.

For at least 300 y a storm three times bigger than the Earth has been raging on Jupiter. Through a telescope it appears as an enormous reddish oval in the south tropical zone. First studied by G. D. Cassini in 1665, it was rediscovered in 1887 and called the **Great Red Spot** (see Figure 11-1, 11-2, and 11-3). The Red Spot has reached diameters of 40,000 km. It and other, smaller transient spots are probably vast, hurricanelike storm systems in Jupiter's atmosphere. Small clouds approaching the Red Spot get caught in a counterclockwise circulation like leaves in a great whirlpool.

Figure 11-5 shows an imaginary view of the awesome cloud vistas near the top of the main cloud deck of this stormy world.

Jupiter's Rotation

Cloud belts and zones have a rotation period of about 9^h50^m near the equator, but several belts and zones at higher latitudes average 9^h56^m. The Red Spot has its own rate, sometimes lagging behind or drifting ahead of nearby clouds. The best estimate of the underlying planet's rotation is $9^h55\frac{1}{2}^m$, based on radio radiations from deep-atmosphere electrical storms. However, no one is sure whether any of these periods marks the true rotation of a well-defined solid or liquid surface beneath the clouds—or whether such a surface exists.

Jupiter's Infrared Radiation

Besides reflected sunlight, Jupiter emits infrared thermal radiation. **Jupiter's infrared thermal radiation** is generated by the heat of the planet itself. From measurements of the total amount of this radiation, scientists know the total amount of energy being radiated by Jupiter. Surprisingly, this figure turns out to be about twice as much energy as Jupiter absorbs from the Sun! This is very different from the case of Earth or other terrestrial planets, where the heat radiated from inside is negligible compared to the heat received from the Sun. Jupiter's extra internal heat must be coming from somewhere, but where?

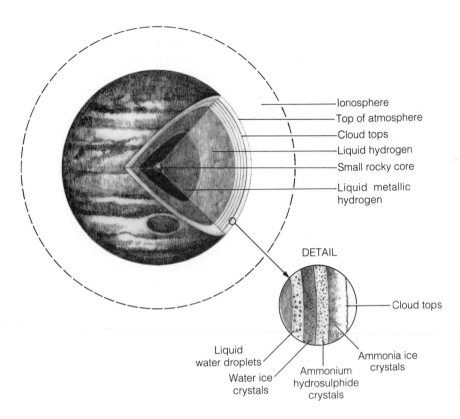

Ionosphere
Top of atmosphere
Cloud tops
Liquid hydrogen
Small rocky core
Liquid metallic hydrogen

DETAIL

Cloud tops

Liquid water droplets
Water ice crystals
Ammonium hydrosulphide crystals
Ammonia ice crystals

Figure 11-6 A view of the interior and atmosphere of Jupiter based on telescopic and spacecraft data. The atmospheric depth to the liquid zone is hundreds of kilometers. (NASA.)

Theorists believe Jupiter is slowly contracting, releasing gravitational energy as heat and radiation. This radiation was most intense when Jupiter formed and has declined ever since to the low level observed today.

Although Jupiter is radiating its own energy, it is not a true star, because its energy is not produced by thermonuclear fusion, the way a star's energy is. Jupiter's mass is not great enough to create the central pressure and heat necessary for starlike fusion reactions in its interior.

Jupiter's Internal Structure and Surface

What is Jupiter like under its clouds? Its mean density is too low for it to be a rock-and-iron planet like Earth. Instead, **Jupiter's interior** is believed to consist of 60% hydrogen, the rest being helium with small amounts of silicates and other "impurities." The heavier elements have sunk, so that a core of silicate or iron–silicate material—resembling terrestrial planets—may exist near the center. According to some theoretical models, the core may be over twice the size of Earth and have a temperature around 30,000 K.

Under the high pressure of Jupiter's interior, hydrogen takes on an unfamiliar form called **metallic hydrogen,** which can conduct electric currents. As shown in Figure 11-6, much of Jupiter's inner half may be liquid metallic hydrogen, in which convection currents carry heat to the surface and electric currents generate Jupiter's magnetic field. Much of the outer half may be a sea of ordinary liquid molecular hydrogen, H_2. There may be no solid surface at all, only a slushy mixture of liquid hydrogen and crystals of various compounds.

The gravity at this ill-defined "surface" is stronger than that of any other planet. A person weighing 150 lb on Earth would weigh about 400 lb on Jupiter! Similarly, the pressure of the thick atmosphere is roughly 100 times the air pressure on Earth. Various data suggest that the cloud-obscured, high-pressure ocean of liquid hydrogen begins several hundred kilometers below the cloud tops.

JUPITER'S SATELLITES

Jupiter has an impressive array of at least 16 moons ranging from 8-km bodies to four large worlds, including two slightly larger than the planet Mercury. The moons are numbered in order of discovery. The four largest, shown together in Figure 8-2, were discovered independently by Galileo and the German astronomer Marius on two consecutive nights in 1610 as they viewed Jupiter through the newly invented telescope. These bodies are called the **Galilean satellites.** You will recall their historical importance: When Galileo discovered that they orbit Jupiter, he cited this as proof that not all bodies orbit Earth—thus refuting the Ptolemaic theory and supporting the Copernican theory. Now we know that those moons are geologically, as well as historically, interesting.

Until the 1979 flights of Voyagers 1 and 2 through the satellite system, we knew the large satellites only as pinhead disks in even the largest telescopes. Some observers noted vague dusky markings. Because they are in the size range of the Moon and Mercury, these moons were assumed to be cratered, dead worlds with little geological activity. Spectra showed some differences among their surface materials, but no one realized how amazingly varied these moons are.

Two Basic Materials: Bright Ice and Black Soot

A simple but useful picture of Jupiter's moons can be obtained by thinking of *two* main types of materials that form at the very low temperatures of the outer solar system. The first is ice, or, more accurately, a mixture of various types of ice. Much of the ice is familiar H_2O. Other ices may include frozen CO_2 (what we call "dry ice"), frozen methane, and other frozen material. This ice mixture has a bright, whitish color.

The second main type of material is very black rocky minerals believed to be carbon and carbon compounds. This **carbonaceous material** is much like soot. Sometimes it is very dark chocolaty brown, probably due to colored organic compounds formed from the carbon. When it is not strongly diluted by being mixed with the bright ice, it gives an extremely dark coloration, as black as black velvet.

Thus we can think of a sort of "salt and pepper" model of material on Jupiter's moons: a mixture of white stuff and black stuff. The white material is ice, and the

black is sooty soil. As we will see, on some bodies' surfaces the ice has sublimed off into space, leaving the sooty material behind and giving very black color. On other bodies, water has erupted and coated the surface with fresh ice, making a cleaner, whiter color than the Moon or Mars. Still other bodies have a grayish-tan coloration from a mixture of the two materials.

These concepts help explain why each of Jupiter's major moons has a distinct "personality" and how the personalities shift as we move from one moon to another. Because certain heating effects were strongest near Jupiter, the materials of the nearer moons were most modified from their initial cold, ice-and-soot compositions. Therefore, we will start our survey of the satellites with the outer moons and work our way inward toward more and more altered moons. As we will see, these altered inner moons are truly strange, with properties hardly imagined until the flight of the Voyager probes. During our discussion, you can refer to Table 8-1 (page 122) for specific physical properties of these bodies.

Three Classes of Moons

Jupiter's system of moons reveals a pattern that we will encounter again with Saturn. There are three general classes of satellites of the giant planets. On the outermost edges of the system are **outermost, captured moons,** which seem to be asteroids captured by the planet's gravity. At intermediate distance are **large and intermediate-sized moons**—the Galilean satellites, in Jupiter's case. Close to the planet, on the outskirts of the ring system, are **nearby fragmented moonlets.** Figure 11-7 shows the Jupiter system to scale. The orbits of Jupiter's outermost moons are divided into two groups of four each, as seen in the right half of the diagram. The orbits of the four large, Galilean moons can be seen much closer to the planet. The small, fragmental moons are so close to Jupiter that they do not show up at all in this figure.

The Outermost Moons: Captured Black Asteroids

The eight outermost moons are so far from Jupiter that if they were much further, the Sun's gravity would be more important than Jupiter in controlling their motions. In other words, they are barely part of Jupiter's satellite system. Physically they have black surfaces that resem-

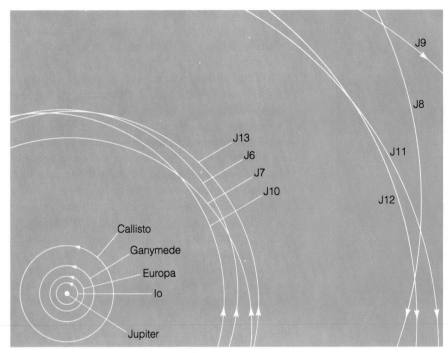

Figure 11-7 The "miniature solar system" of Jupiter and its satellites. Orbits of the four large Galilean moons are shown close to Jupiter (lower left). Still closer, smaller moons and the ring are too close to show on this scale. Two groups of outer moons lie in closely clustered orbits. Note that the outer group moves in retrograde direction—opposite to the usual sense of orbital motion in the solar system. See text for further discussion.

ble surfaces of many asteroids. Any ice that may originally have been exposed on these surfaces was sublimed into space during exposure to the Sun, leaving black, sooty surfaces. For these reasons, astronomers believe these moons are not native to Jupiter but are asteroids that originally orbited around the Sun, came close to Jupiter, and were captured into orbit. This could have happened early in the history of the solar system when Jupiter may have had a more extensive atmosphere; asteroids passing through the fringes of that atmosphere could have been slowed in a way that resulted in capture.

Satellites J6, J7, J10, and J13 have orbits around 12 million kilometers from Jupiter and inclined 27° to Jupiter's equator. All these moons, as well as the other moons mentioned so far, orbit in a **prograde** direction like our own Moon. (This is the "normal" direction of orbital motion in the solar system, counterclockwise as seen from the north celestial pole. See Figure 11-7.) But satellites J8, J9, J11, and J12 all orbit in a **retrograde** direction (clockwise as seen from the north celestial pole), 23 million kilometers from Jupiter with inclinations of around 52°. How did the outer satellites come to be clumped into two such groups? Analysts have suggested that during the capture, when the aster-

oids passed through the fringes of an early atmosphere, they broke into pieces as they were decelerated. Thus there may have been only two capture events, each contributing four large pieces. One object approached Jupiter in a prograde direction and fragmented to produce prograde moons in similar orbits; the other approached in a retrograde direction and produced retrograde moons. Perhaps smaller fragments in the two orbital groups remain to be discovered.

Because there are no close-up photos of any of these moons, we don't know their surface features. Probably they are cratered like Mars' moons, Phobos and Deimos. Indeed, they may be quite similar to Phobos and Deimos in appearance, size, and origin: Recall that Phobos and Deimos are also believed to be dark asteroids captured into satellite orbits.

The Four Big, Galilean Moons

Moving in toward Jupiter, we come to the four large, Galilean moons. These are not captured and must have formed as part of Jupiter's system. The outermost of these is **Callisto** (Figure 11-8). Second-largest of Jupiter's moons, it is 4800 km across, just 2% smaller than the planet Mercury.

Figure 11-8 Callisto, the outermost of Jupiter's four, large Galilean moons. The surface is a dark soil with some ice mixed in. Bright spots are craters where surface soil has apparently been blasted away, exposing cleaner ice. This picture uses an ultraviolet image for the blue-color component, enhancing color contrast between ice and the background soil. (NASA Voyager 2 photo.)

Figure 11-9 Part of the globe of Jupiter's largest moon Ganymede. Much of the surface consists of old, dark, cratered terrain like the surface of Callisto. The dark terrain is broken, however, by swaths of brighter, younger, more ice-rich material. At the upper left corner we can see part of a white polar cap of frost. (NASA Voyager 2 photo.)

In terms of surface geology, Callisto is a primitive world with little sign of internal geological activity or surface heating. Craters caused by meteorite impacts are the main landform. Compared with water's density of 1000 kg/m^3 and rock densities of about 2500 kg/m^3, Callisto's average density of 1800 kg/m^3 indicates a composition of about half ice and half rocky material. The surface is mostly a tannish-gray material that is probably a mixture of ice and carbonaceous matter. Dark soil may have concentrated on the surface as repeated impacts melted and vaporized some of the ice. As seen in Figure 11-8, each meteorite impact appears to have blown away the surface soil and exposed brighter, cleaner ice underneath. Judging from the high number of craters, the surface has changed little since its formative era 4 to 4.5 billion years ago.

Ganymede, with a 5262-km diameter, is the largest moon of Jupiter and the largest moon in the solar system. It is 10% larger than Callisto and 8% larger than the planet Mercury. Its average density of 1900 kg/m^3 is barely higher than Callisto's, implying a similar ice–rock composition. A thin polar cap, visible in Figure 11-9, is believed to be water frost, formed from water vapor that has been emitted from the interior

Ganymede has had somewhat more geologic evolution than Callisto. The ancient parts of Ganymede's surface are very similar to Callisto's surface, being heavily cratered and covered with dark, dusty soil (Figure 11-10). Unlike Callisto, though, these areas are divided by swaths of younger, light-toned, grooved terrain. The ancient dark crust seems to have split, allowing fresher icy material to erupt. Such splitting may imply internal heating that melted the icy component and caused expansion and cracking. Water erupted, forming ice flows—the Galilean satellites' equivalent of Earth's lava flows. The bright swaths appear to mark fractures along which the water erupted (Figure 11-11).

Ganymede is telling us an important story that may help us understand Earth better. Ganymede offers a case of incipient plate tectonics, because its brittle lithosphere of ice split into platelike blocks as it floated on a warmer, watery subsurface. Although Ganymede's interior was not active enough (hot enough?) to cause full-scale continental drift among the plates, offsets do show that the plates have moved slightly. Ganymede thus provides an important transitional example from cold, inactive planets to larger, geologically active worlds.

Europa, the next inner moon, is smaller than Callisto and Ganymede and much different. It is about the size of our own Moon. So much water has erupted and

Figure 11-10 A closer view of the central region of Figure 11-9 reveals that the dark region on Ganymede is cut by concentric arc-shaped fractures—probably part of an ancient multiring system that predated the breakup of the crust. Impact craters dot the area but are scarcer on the younger, bright icy areas. The large bright ovals appear to be ancient craters so large that they penetrated into liquid water beneath the dark crust, filled with water, and thus left nearly rimless icy white scars. (NASA Voyager 2 photo.)

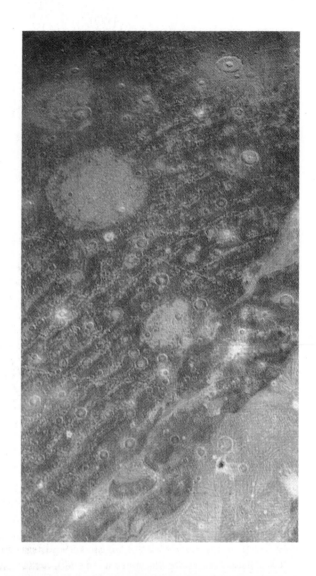

frozen on its surface that any ancient, dark, cratered surface it may have had, such as Callisto's or Ganymede's, has been covered over with smooth ice (Figure 11-12). The ice is creamy white, as bright as the paper of this book. The only markings are delicately shaded, shallow grooves crisscrossing the surface. The relatively high mean density of 2970 kg/m³ for Europa as a whole shows that the moon's interior consists mostly of rock, not ice. The surface ice is so young that virtually no impact craters have accumulated on it. The ice layer may be 100 km (60 mi) thick, overlying an ocean of liquid water, which in turn covers the rocky interior. The ice surface may be on average only a few hundred million years old, and may be continually forming. A

Figure 11-11 Imaginary panorama of Jupiter and the icy surface of its moon Ganymede. Ganymede's surface is split by parallel fissures, where water may have erupted and refrozen. This view shows a ridge of jagged ice blocks thrust up during movements of Ganymede's crust. In the sky are Jupiter's inner moons, Europa (left) and Io (in front of Jupiter). Jupiter would look about 15 times as big as our Moon in our sky, and Europa would look slightly smaller than our Moon. (Painting by Ron Miller.)

Figure 11-12 Jupiter's icy billiard ball moon Europa. This world is about the size of our Moon but is covered by nearly smooth, whitish ice. Pale dark streaks appear to mark fractures. (NASA Voyager photo.)

bright cloud seen on the final Voyager 2 photo of Europa may mark an erupting geyser! Some source of heating apparently melted the water component of Europa, causing the water eruptions that created the young ice flows.

Still greater heating affects the inner Galilean moon, **Io,** which is the most bizarre world in the solar system. Like Europa, it is about the size of our Moon. Voyagers 1 and 2 discovered active volcanoes all over Io, and a fantastic mottled surface of pale yellow, tan, orange, and white sulfur deposits (Figure 11-13). Voyager researchers joked that it should be called the "Pizza Moon," complete with cheese, tomato, and pepperoni strata.

In a realm of ice and cold carbonaceous compounds, what source of heat could possibly explain the melting of Europa and the erupting volcanoes and sulfur flows of Io? This would have been a total mystery when Voyager arrived except for some brilliant detective work by California dynamicists S. J. Peale, P. Cassen, and R. Reynolds (1979). While Voyager was on its way, they calculated that although Io's orbit is nearly circular, gravitational pulls of neighboring satellites have caused Io's distance from Jupiter to vary slightly throughout its history. As the distance varies, gravitational forces, called

tidal forces, act in a special way to stretch and compress Io. This flexing heats Io's interior, just as rapid flexing of a tennis ball makes it heat up from friction. The calculations showed that this heating effect is stronger in Io than in any other satellite and causes Io's interior to be molten. It may act to a weaker degree on Europa. In a beautiful example of the scientific method at work, Peale and his colleagues published a prediction, that volcanoes might be found on Io, in a journal that came out just a few days *before* Voyager 1 reached Io. A few days later, Voyager discovered the solar system's most active volcanoes, spewing 100-km-high plumes of debris.

The heating effects on Io were apparently so strong that any initial water (liquid or ice) was completely melted and evaporated or sublimed off the moon. The "salt-and-pepper" ice–soot mixture of other moons was thus destroyed. Even the carbonaceous rocky component was at least partly melted, allowing sulfur-dominated lavas to spread out in overlapping layers and form a colored, sulfurous surface (Figure 11-14).

Fragmental Moonlets on the Outskirts of the Ring

At least four additional moonlets exist inside the orbit of Io, near the edge of a faint ring discovered by the Voyagers. All are much smaller than the Galilean moons. The largest, designated J5 Amalthea, was discovered in 1892. It is a cratered lump 270 km long and 155 km wide (roughly the scale of New Jersey). Amalthea has an orangish color attributed to sulfur atoms ejected from Io and coating Amalthea. The other three moons, all discovered by Voyagers, are around 40 km across. One orbits between Amalthea and Io; the other two are nearly on the outer edge of Jupiter's ring. The small, nearby moonlets appear to be fragments of one or more original moons formed near Jupiter, but smashed by collisions with meteorites falling toward Jupiter. The impact rate is very high among these moonlets because incoming meteorites are concentrated there by Jupiter's gravity. That is why the nearest moonlets have a high probability of being fragmented.

THE RING OF JUPITER

The Voyagers' discovery of Jupiter's ring came as a surprise. It is too faint and too nearly edge-on to be seen from Earth (Figure 11-15). It probably consists

Figure 11-13 The strange volcanic moon Io. The surface is mottled by yellow, orange, and white volcanic deposits of sulfur and sulfur compounds. Darkest spots are probably active volcanic vents or pools of molten sulfur. Arrow marks one of the largest of Io's volcanoes, Pele. It is ejecting dark, ashy debris in direction of the arrow. Debris forms a huge, dome-like cloud, showing as a dark arch between the volcano vent and the edge of the planet. This fan of debris could cover most of the U.S. east of the Mississippi! A more irregular cloud of gas and debris, ejected by a volcano just over the horizon, can be seen at left edge as a bright, bluish glow. Special processing has been used to enhance the brightness and color of that cloud, which would be much fainter if seen by the naked eye from a spaceship over Io. (NASA Voyager 1 photo; courtesy A. S. McEwen, U.S. Geological Survey.)

mostly of microscopic dark particles being knocked off the inner moonlets near its edge. Forces acting on these particles make them spiral inward toward Jupiter, thus creating a ring through which material flows slowly over millions of years.

FUTURE STUDIES OF JUPITER

NASA's **Galileo program** is the only planned mission to Jupiter and its satellites announced by any of the spacefaring nations. Galileo is a large and complex robotic spacecraft designed to eject a small probe into Jupiter's atmosphere. The mother ship will go into orbit around Jupiter in order to visit and map one satellite after another. Launch is scheduled for late 1989 or 1990.

WHY GIANT PLANETS HAVE MASSIVE ATMOSPHERES

At the end of Chapter 9 we gave three principles that explain why some planets lack atmospheres. Let us apply them here to explain in more detail why giant planets have massive atmospheres rich in hydrogen gas. The first principle says that the higher the temperature, the higher the speed of molecules forming the atmosphere. This means that the atmospheres of the cold, outer planets will generally have slow-moving molecules. The second principle says that the lighter the molecule, the higher its speed. This means that on any given planet, hydrogen will be the fastest-moving molecule. The third principle says that the bigger the planet, the higher the speed needed for a molecule to escape the planet's grav-

Figure 11-14 Volcanic terrain near the south pole of Io. Low Sun angle reveals several depressed calderas and a volcanic peak (lower right corner). Blackest areas may be hot or molten sulfur, which grow darker at higher temperatures. Several vents (left center) show radiating spiderlike patterns of sulfurous lava flows. Plateaulike thin layers of lava have built up the surface in several areas. (NASA Voyager 1 photo.)

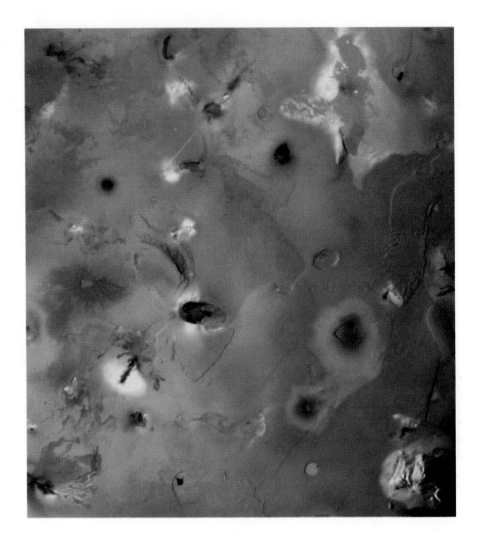

ity and shoot off into space. This means that giant planets will hold whatever gases they have much better than smaller planets like Earth.

Combining these principles, we can compare the Earth and a giant planet. On small, warm Earth, the fastest-moving hydrogen molecules will move more than fast enough to escape into space. Thus Earth has lost virtually all its original hydrogen, which was by far the most abundant primordial gas. The slower nitrogen and oxygen did not escape. Conversely, on a cold giant like Jupiter, the fast-moving hydrogen and all other gases have been retained. Because the original hydrogen was more abundant than all other gases put together, today Jupiter's hydrogen-rich atmosphere is very massive and very deep.

SUMMARY

Jupiter is the largest planet in the solar system. Its massive atmosphere is about four-fifths hydrogen and one-fifth helium by mass and is probably a remnant of gas from which the Sun and planets formed. The planet is covered by dense colored clouds arranged in bands parallel to the equator.

Jupiter's system of moons provides us with a "miniature solar system" to compare with the system of planets. The four largest moons are called the Galilean moons. Each of these satellites is unique. Callisto is cratered and has a gray ice/soil surface. Ganymede has a similar surface, except that it is broken by swaths of younger, brighter ice. Europa is completely surfaced by clean, bright, barely cratered ice. Strangest of all is Io,

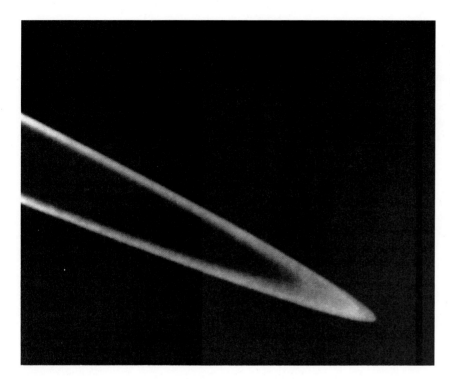

Figure 11-15 Jupiter's ring is believed to be fine material spiraling toward Jupiter from small moons near the outer edge of the ring. Jupiter is out of the picture at left. (NASA Voyager photo.)

which has been heated by tidal forces, creating the only known satellite with active volcanoes. Indeed, Io is the most volcanically active body in the solar system and is covered with yellowish sulfur-rich lavas. Inward from the large moons are a number of small moonlets on the outskirts of a thin ring. The situation is reminiscent of the solar system's four giant planets circling beyond the small terrestrial planets. On the outskirts of Jupiter's system are eight small, black moons that seem to be captured asteroids.

CONCEPTS

giant planet

belts

zones

Jupiter's atmospheric composition

Jupiter's temperature

Great Red Spot

Jupiter's infrared thermal radiation

Jupiter's interior

metallic hydrogen

Galilean satellites

carbonaceous material

outermost, captured moons

large and intermediate-sized moons

nearby, fragmented moonlets

prograde and retrograde satellite orbits

Callisto

Ganymede

Europa

Io

Galileo program

PROBLEMS

1. Answer the following problems:
   a. How might studying cloud patterns on Jupiter, Mars, and Venus help us understand terrestrial meteorological theory?
   b. What planetary or environmental characteristics might figure in such theory?
   c. How do these factors vary among these planets?
   d. Why are other planets not included on the list?

2. Which planet has:
   a. The highest surface gravity?
   b. The lowest surface temperature?
   c. The highest atmospheric pressure?
   d. The most atmospheric hydrogen?
   e. The highest percentage of atmospheric oxygen?

3. List places in the solar system where you might expect to find droplets, pools, or oceans of liquid water.

4. How would gravity on the Galilean satellites compare with gravity on the Moon (see data in Table 8-1)? Which three bodies in the solar system would you expect to have general environments most like the Moon's?

5. Describe a landscape on Io.

6. Two large planets have the same size and mass but orbit at different distances from their sun.

    a. Which would you expect to have more hydrogen? Why?

    b. If they have had different amounts of volcanism, which would you expect to have more carbon dioxide? Why?

PROJECTS

1. Observe Jupiter using a telescope with at least an 8-cm (3-in.) aperture. Sketch the pattern of belts and zones. Which are the most prominent belts? Which zones are brightest? Compare these results with photos in this book. Is the Red Spot or other dark or bright spots visible on the side of Jupiter being observed?

2. Observe Jupiter using large binoculars or a telescope with at least a $2\frac{1}{2}$-cm (1-in.) aperture. How many satellites are visible? Observe the satellite system at different hours over a period of several days and try to identify the satellites. (This could be done as a class project, with different students making sketches at different hours.)

3. With a telescope having at least a 15-cm (6-in.) aperture, determine the rotation period of Jupiter by recording the time when the Red Spot is centered on the disk. Note that intervals between appearances must be an integral number (1, 2, 3, and so on) of rotation periods.

CHAPTER 12

The Outermost Planets and Their Moons

Jupiter is only the first of four giant planets that occupy the outer solar system. Beyond Jupiter are the giants Saturn, Uranus, and Neptune. Each of the four giants has a massive gaseous atmosphere, an extensive system of interesting satellites, and probably a unique ring system. (The status of a ring, or segments of rings, around Neptune is unclear, as we will see.) The existence of four such systems offers us good opportunities for comparative planetology. By comparing the giants, the systematics of the different moon systems, and the rings, we can gain insights into the evolution of worlds, the common themes of planetary evolution, and the circumstances that produce unique properties, such as the volcanoes of Io.

Overlapping the orbit of Neptune is the orbit of the tiny world Pluto. Pluto is conventionally listed as the ninth planet, but it has distinctive properties, and is smaller than our own Moon. As we will see in this chapter and the next, labeling Pluto as an independent ninth planet may not be the best way to look at it.

THE PLANET SATURN

Saturn is most famous for its rings, the only ring system that can be easily seen in Earth-based telescopes (Figure 12-1). The globe itself is less interesting than Jupiter and lacks colorful features such as the Great Red Spot. Saturn's clouds are organized into bright zones and dark belts, sketched in Figure 12-2. The colors are yellowish and tan, and, as indicated in Figure 12-1, the contrast is less than on Jupiter. Because Saturn is nearly twice as far from the Sun as Jupiter, it is colder. This difference explains the lower degree of color, because at the cloudtop temperatures, around 100 K ($-279°F$), there is less formation of colored organic compounds.

TABLE 12·1		
Atmospheric Compositions of Jupiter and Saturn		
	Estimated Percentage by Mass	
Gas	Jupiter	Saturn
H_2 (hydrogen)	79	88
He (helium)	19	11
Ne (neon)	1?	?
H_2O (water)	Trace	?
NH_3 (ammonia)	0.5?	0.2
Ar (argon)	0.3?	?
CH_4 (methane)	0.2?	0.6
C_2H_6 (ethane)	Trace	0.02
PH_3 (phosphine)	Trace	Trace
C_2H_2 (acetylene)	Trace	Trace

Note: Data derived mainly from Voyager 1 and 2 infrared observations (*Science 204:*972; *206:*952; *212:*192). "Trace" indicates a small fraction detected; "?" indicates a likely constituent that has not been measured. The hydrogen and helium abundances are very close to those in the Sun.

Figure 12-1 An example of the best photography of Saturn from Earth-based telescopes. The yellowish color of the clouds contrasts with the whitish tones of the ice composing the ring particles. (Catalina Observatory 61-in. telescope; courtesy S. M. Larson, University of Arizona.)

In spite of the blander colors, the overall composition of Saturn's atmosphere is similar to Jupiter's—mostly hydrogen (H_2) and helium (He), with minor amounts of methane (CH_4) and other gases (Table 12-1).

As on Jupiter, atmospheric circulation varies with latitude. Rotation periods near the equator are about

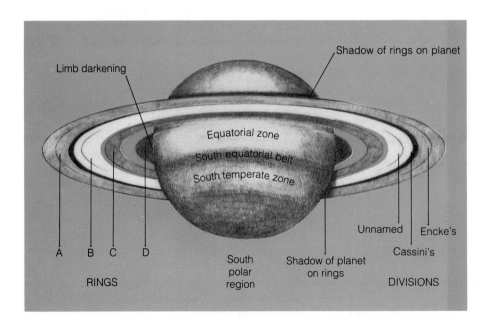

Figure 12-2 Features of Saturn. Telescopes with apertures as small as 5 cm (2 in.) will show the rings; telescopes larger than about 25 cm (10 in.) will sometimes show all these features.

Figure 12-3 This extraordinary view of Saturn in crescent phase was obtained by Voyager 2 as it left Saturn's system: Such backlighting of Saturn can never be seen from Earth. The rings' shadow makes a black band on the globe, and the shadow of the globe cuts a black swath through the rings. Note that the globe of Saturn can be seen *through* most parts of the rings where they pass in front of the globe, proving that the rings are not solid. (NASA.)

10^h14^m, but cloud formations near the poles show longer periods, such as 10^h40^m.

High layers of methane haze, which absorb some wavelengths of light, help reduce the contrast among Saturn's cloud markings. Nonetheless, cloud patterns were seen when Voyager space probes made dramatic close-up photos of Saturn (Figures 12-3 and 12-4). Photos such as Figure 12-4 revealed turbulent structures and stronger eastward jet streams than on Jupiter, with wind speeds up to 450 m/s (1000 mph) eastward along the equatorial zone. Only the larger disturbances can be seen from Earth; they can be tracked for many weeks.

The mean density of Saturn is the lowest of any planet and less than that of water, indicating that Saturn has only a small core of rocky material. Its layer of metallic hydrogen is smaller than Jupiter's. Most of its bulk is

Figure 12-4 Contrast-enhanced close-up of Saturn's cloud belts reveals structure similar to that in Jupiter's clouds. Two dark ovals are visible at upper right; each is about half the size of Jupiter's Great Red Spot. They may have origins similar to the Red Spot. The north equatorial belt is at the bottom (dark); north temperate zone is in the middle (bright). (NASA Voyager 1 photo.)

molecular hydrogen in either liquid or highly compressed gaseous form. A core of rocky compounds, something like a terrestrial planet but with 15 times the mass of Earth, exists at the center of Saturn.

SATURN'S RINGS

Galileo's first sight of Saturn, in 1610, was a strange one: a blurry disk with blurry objects on either side. He drew it as a triple planet. Not until 1655 did Christian Huygens discover that a ring system encircled the planet.[1] A small modern telescope easily shows the astonishingly beautiful rings, largest in the solar system. The rings' proportions are surprising. They stretch about 274,000 km (171,000 mi) tip to tip, but they are probably less than 100 m thick! The system is so thin that terrestrial observers lose sight of it altogether when

it appears edge-on for a few days once every 15 y as Earth passes through the ring system's plane (see Figure 12-5).

Saturn's rings are composed of billions of separate particles, too small to be seen individually from Earth or in Voyager photos. This was proved by a stepwise combination of theory and observation. In 1895, Scottish physicist James Clerk Maxwell showed that the rings could not be a solid disk because they are closer to Saturn than a certain distance called Roche's limit. Within this limiting distance, a disk would be pulled apart into myriads of particles. In 1895, American astronomer James Keeler confirmed spectroscopically that different parts of the rings move at different speeds, each in its own orbit around Saturn. Modern proof that the rings are not solid comes from spacecraft photos showing that we can see through parts of them, as in Figure 12-3. In 1970, American astronomers proved spectroscopically that the ring particles are composed of, or covered by, frozen water (Lebofsky, Johnson, and McCord, 1970). Various measurements from Earth and from the Voyagers show that common particles in the rings range from ping pong ball to house sizes. Some particles of larger or smaller size may also exist. The remarkable view from within the rings is suggested by Figure 12-6.

The rings are divided by several gaps. The biggest is called **Cassini's division**, after G. D. Cassini who discovered it in 1675 (see Figures 12-1, 12-2, and

[1]Huygens announced this discovery in the form of a famous anagram (succession of apparently meaningless letters). Huygens later explained that the anagram was to represent the sentence "Annulo cingitur, tenui, plano, nusquam cohaerente, ad eclipticam inclinato" ("It is surrounded by a thin, flat ring, nowhere touching, inclined to the ecliptic"). In the days before copyrights, such anagrams were a common form in which scientists published initial results, thus establishing priority while giving themselves time for more observations. Not until about 1661 were most observers convinced of Huygens' discovery (Alexander, 1962).

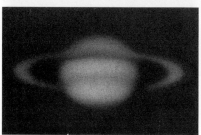

Figure 12-5 Varying aspects of Saturn during its 29-y orbit around the Sun, as photographed with Earth-based telescopes. Because the rings maintain a fixed relation to the ecliptic, the Earth-based observer sometimes sees from "above" and sometimes "below" the ring plane. The series shows half the 29-y cycle. (Lowell Observatory.)

Figure 12-6 In this imaginary view we are floating among the icy particles within Saturn's rings. Saturn's globe is at left, and the distant Sun (right) backlights the scene. The rings, perhaps less than 100 m thick, stretch off into the distance, lower center. (Painting by author.)

12-3). A dusky ring region outside Cassini's division is called ring A; the brighter region just inside it is called ring B. Other rings and finer divisions have been mapped by telescopic observers, as seen in Figure 12-2.

Scientists were astonished when Voyager 1 and 2 photos showed not just the few major divisions from Figure 12-2 but thousands of fine-scale divisions and ringlets, as glimpsed in Figure 12-7. Wavy ring-edges, gentle twists in ringlets, and transient radial shadings

were also found. What maintains such an intricate structure? One theory was that unseen moonlets clear swaths among the ring particles, create wavy edges, and maintain ringlets by their gravitational effects. Supporting this theory was the Voyager 1 discovery of two 200-km moonlets straddling the narrow F ring just outside ring A, confining ring F to a narrow zone. Moons of this type, which confine rings into narrow zones, are called **shepherd satellites.** Voyager 2 searches failed to turn

Figure 12-7 Close-up photos of Saturn's rings show an unexpectedly intricate structure—including thousands of individual ringlets. Overexposed Saturn can be seen in the background, along with the shadow of the rings (top). The dimmest ring (top) is ring C; ring A is in the foreground. (NASA Voyager photo.)

up other shepherd moons among the rings, down to a diameter of a few kilometers. Nonetheless, ring structure may result from a combination of poorly understood gravitational effects caused by known moons and still-unseen moonlets.

How did such a remarkable ring system form? A clue is that any swarm of particles released into the space closely surrounding a rotating, flattened planet, no matter how chaotic they were initially, would soon reorganize into a flat ring over the planet's equator. This is due to certain forces acting on the particles from the planet's flattened shape, together with redirection of the particles' orbits due to collisions among the particles themselves. Therefore, the question of ring origin comes down to a question of how a large number of ice particles might have been created near Saturn. Most theorists think this happened when one or more primeval icy moons, orbiting very near Saturn, were shattered by collisions with passing interplanetary bodies, called asteroids and comets. The ejected fragments could have provided most of the mass of the rings.

By contrast, Jupiter's very thin ring is much less massive. Recall that its tiny particles are a transient stream, eroding off small moons on the margins of the ring. Saturn also has small moons on the margins of its rings (fragments of the older, broken moon?) and may

receive additional particles from them, but most of the particles of Saturn's rings probably date back to a large fragmentation event. When was that event? Probably, it happened soon after Saturn formed, when impact rates were high. The rings may be nearly as old as Saturn itself, but may have had minor changes as the nearby inner moonlets underwent further collisions with small asteroids or comets.

SATURN'S SATELLITES

Saturn's satellite system, like Jupiter's, is extensive and includes exotic worlds. There are at least 17 moons (see Table 8-1, page 122). As with the Jupiter system, we find three classes of satellites. Saturn has one outermost, captured moon, a host of large and intermediate-sized moons at intermediate distances, and five small, nearby fragmental moons on the outskirts of the rings.

Saturn's Nearby, Fragmental Moonlets

Figure 12-8 shows one of the five small moons orbiting Saturn on the outskirts of its ring system. It is heavily cratered. As mentioned above, even more violent cra-

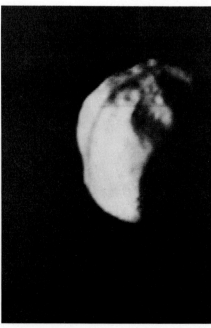

Figure 12-8 Saturn's satellite S11 is a 70 × 135 km object, probably eroded into its irregular shape by intense cratering. These two views, taken 13 min apart, show the shadow of Saturn's nearby rings moving across the satellite. (NASA Voyager 1 photo.)

tering events probably shattered the parent body or bodies of this and the other nearby moonlets, creating debris that formed Saturn's rings. Thus, the nearby moonlets are "fossil remnants" of the rings' parents.

Saturn's Intermediate-Sized Moons

At intermediate distances are seven intermediate-sized moons and one giant moon. Most of the intermediate-sized moons are bright, cratered, ice-surfaced worlds. Figure 12-9 shows an example—1118-km (680-mi) **Dione**—silhouetted against the orangish clouds of Saturn. These moons' densities indicate that they are mostly composed of ice, with only a little bit of rocky material.

Several of these moons have unique personalities. For instance, although most of them have ancient, cratered surfaces, 500-km (310-mile) **Enceladus** has some fresh, sparsely cratered plains (Figure 12-10), reminiscent of Jupiter's resurfaced moon Europa. This is the brightest surface in the solar system. It suggests that Enceladus may have had some internal heating, causing water to erupt and coat parts of the surface with fresh ice.

Another unusual moon is 1436-km (900-mi) **Iapetus**. Like our Moon and most other moons of Saturn and Jupiter, Iapetus keeps one face toward Saturn. Therefore, one hemisphere always faces forward as Iapetus moves in its orbit, and the other hemisphere trails. Strangely, although the trailing face is bright ice like the other intermediate-sized moons, the leading face is black! Researchers believe this peculiarity has arisen because black soil is blasted off Saturn's outermost, captured moon, Phoebe, by meteorite impacts; the soil particles then spiral inward, coating Iapetus' leading face. It would be interesting to study Iapetus' two hemispheres and their boundary at close range in order to clarify this puzzle; but no very sharp, close-up photos of Iapetus have yet been made.

Phoebe: Saturn's Captured Moon

Several clues indicate that **Phoebe** is different from Saturn's other moons and not native to Saturn's system. First, it is on the outskirts of the system, more than three times farther from Saturn than its nearest neighbor Iapetus. This is the region in which it would be easiest to capture a passing asteroid into orbit around Saturn, because the change in velocity to go from a sun-centered to a planet-centered orbit is least at great

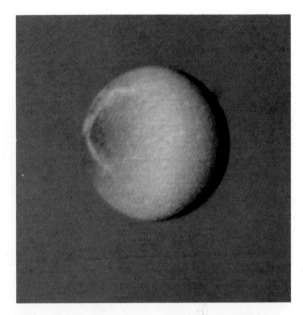

Figure 12-9 An unusual view of Dione silhouetted not against space, as in most views, but against the orangish clouds of Saturn in the background. The color contrast emphasizes Dione's whitish, icy surface. The trailing hemisphere (left) has more dusky mottling and bright swaths of uncertain origin. (NASA Voyager photo.)

distance from the planet. Second, it has the only retrograde orbit in the system, moving in the opposite direction from all other moons. Third, Phoebe is very black, and its colors and spectral properties resemble those of certain outer solar system carbonaceous asteroids; they are different from the rest of Saturn's moons, which are bright and icy.

At 220-km diameter (140 mi), Phoebe is moderate in size. It was probably an interplanetary body that approached primordial Saturn with a retrograde sense of motion around the planet. Each giant planet is believed to have been surrounded by an extended gas cloud at that time, and Phoebe was probably captured when it was slowed by this gas cloud.

Titan: A Huge Moon with a Thick Atmosphere

We called Io the most bizarre moon in the solar system, but Saturn's largest moon, **Titan**, is a close runner-up. With a diameter of 5150 km (3200 mi) it is probably the second-largest moon in the solar system and the only

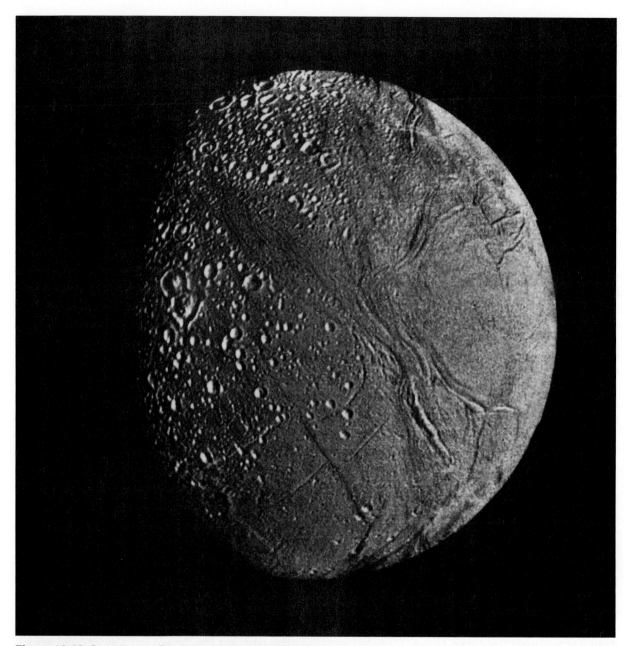

Figure 12-10 Saturn's moon Enceladus, 502 km across, offers a missing link between the "cracked billiard ball" surface of Jupiter's moon Europa and the cratered, fissured surface of Jupiter's moon Ganymede—though Enceladus is only one-sixth the size of Europa. Portions of the ancient cratered surface seem to have been cracked as water erupted and froze into smooth ice plains, obliterating older craters—for example, bottom central region. This suggests that Enceladus may have been heated enough to melt portions of its icy interior at some time in its history. (NASA Voyager 2 photo.)

moon with a thick atmosphere. The atmosphere was discovered in 1944, when spectra of Titan revealed the gas methane (CH_4). Observations in 1973 showed that Titan's atmosphere is not clear but filled with reddish haze, seen in Figure 12-11 (bottom right). This haze is photochemical smog, produced by reactions of the methane and other compounds when they are exposed to sunlight—like the smog produced by the interaction of sunlight and hydrocarbons over Los Angeles. Titan is the smoggiest world in the solar system.

The Voyagers showed that methane and smog are no more than 10% of the atmosphere; the main constituent is nitrogen. The atmosphere is denser than realized before the Voyager flights, by far the densest of any satellite. In fact, Voyager 1 discovered that the surface air pressure is 1.6 times that on Earth! Because Earth's air is also mostly nitrogen, Titan's atmosphere offers fascinating comparisons to present or primeval conditions on Earth. The main difference is that Titan is very cold, around 93 K (−292°F).

Based on their measurements of temperature and pressure at Titan's surface, researchers visualize dramatic weather conditions, illustrated in Figure 12-12.

Methane may exist not only as a gas but may also be able to rain out of the clouds and exist as snow or ice, playing the same triple role of gas, liquid, and icy solid that water does on Earth. In addition to methane, meteorologists conclude that ethane (C_2H_6) would form in the atmosphere, slowly drizzle out, and accumulate on the ground. Thus the surface of Titan may be largely covered by a cold ocean of liquid methane and ethane estimated to be a kilometer deep. Still more complicated, gasolinelike compounds including acetylene, ethylene and hydrogen cyanide may form in the smog and rain out of the hazy clouds. As these facts unfolded during the Voyager mission, one Voyager scientist characterized Titan as "a bizarre murky swamp," shown in Figure 12-12b. More interesting to scientists are the possible biochemical reactions among these organic molecules. Life is unlikely in the cold temperatures of Titan, but biochemists would be very excited to learn what complex organic compounds have formed on Titan. Indeed, Titan may be a natural laboratory for studying primordial biochemical evolution on Earth! Titan remains a fascinating target for future exploration. The United States and Europe are pursuing a joint project, Cassini,

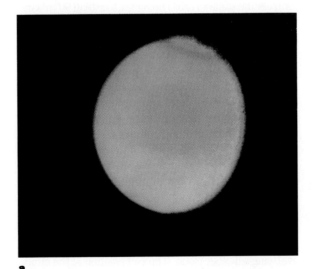
a

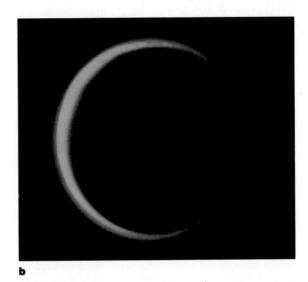
b

Figure 12-11 Saturn's giant moon Titan, about as big as Mercury. Titan has a smoggy atmosphere of nitrogen, colored orangish by organic compounds. **a** This gibbous lighting view shows a dark cloud band around the north polar region. The southern hemisphere haze is brighter than the northern hemisphere. **b** Under crescent lighting, the illuminated atmosphere can be traced all the way around the moon. The color is bluer where sunlight has been scattered further through the atmosphere—for the same reason that scattered sunlight colors our own sky blue. (NASA Voyager photos.)

Figure 12-12 Scenes on Saturn's giant moon Titan. **a** From above the smoggy layer of orangish haze, visitors would see Saturn hanging in a blue sky. (Painting by author.) **b** Below the clouds it is quite dark. Measurements of surface temperature and pressure suggest snowstorms and rainstorms of solid and liquid methane compounds—with possible rivers, "waterfalls," and oceans of liquid methane and ethane. (Painting by Ron Miller.)

to parachute probes beneath the mysterious clouds to see what is really on the surface.

THE PLANET URANUS

In 1781 William Herschel became the first known person to discover a new planet when he detected **Uranus** during an ambitious star-mapping project. Uranus is so far from Earth that an ordinary telescope reveals almost no markings on it. But in 1986, Voyager 2 flew past Uranus, giving us our first close-ups of its bluish haze (Figure 12-13), its dark rings, and its unexpectedly varied moons.

Herschel found several satellites of Uranus as well. The orbits of these satellites lie in the plane of the planet's equator but are steeply inclined to the planet's *orbital plane* by 98°. As the French dynamicist Laplace noted in 1829, steep inclination to the orbital plane can arise only if the planet has a substantial equatorial bulge, whose gravitational attraction holds the satellites over the planet's equator. Thus Uranus itself must have an equatorial bulge, and its equator must be inclined 98° to its orbital plane, an unusual planetary configuration shown in Figure 12-14. This **obliquity** (deviation of the equa-

tor from the plane of orbit) being greater than 90° means that the rotation of Uranus is *retrograde*, as illustrated in Figure 12-14. The rotation period of the planet, under the clouds, was measured by Voyager to be 17.2 h.

Thus Uranus has a unique seasonal sequence. When the north pole points almost directly toward the Sun, the southern hemisphere is plunged into a long, dark winter, lasting for about a quarter of the planet's 84-y revolution. After this 21-y south polar winter and north polar summer, the Sun shines on the equatorial regions. Each point on the planet now goes from day to night during the planet's 17-h rotation. After 21 more years the south pole points approximately toward the Sun and the southern hemisphere experiences a 21-y summer.

The Voyager flyby revealed that Uranus' atmosphere has a deep layer of almost featureless hazy gas overlying a deeper cloud. The composition of the atmosphere is similar to those of Jupiter and Saturn—about three-fourths hydrogen (H_2) and one-fourth helium (He) by mass, with small amounts of methane (CH_4) and other gases. The haze layer is very cold: Voyager recorded a minimum temperature of 51 K ($-368°F$). The temperature rises, however, deeper into the clouds. Perhaps because of the cold, Uranus' visible haze layer

a

Figure 12-13 The hazy sky-blue planet Uranus. **a** As
Voyager 2 approached Uranus from the inner solar system
and the Sun's direction, it took this photo with Uranus at a
gibbous phase under nearly full sun. The view looks down
on the pole, which is pointed nearly toward the Sun due to
Uranus' unusual axial tilt. Cloud features are almost
invisible, though a faint dark band can be seen along the
upper right edge. **b** As Voyager left, it captured this view of
crescent Uranus looking back toward the Sun. (NASA
Voyager 2 photos.)

b

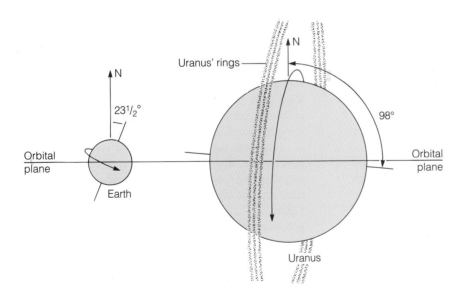

Figure 12-14 Comparison of the
sizes and rotations of Earth and
Uranus. The Earth has an obliq-
uity (or axial tilt) of $23\frac{1}{2}°$ and a
prograde (west-to-east) rotation.
Uranus has much steeper obliquity
and retrograde rotation.

lacks the orange and tan-colored compounds that give rich color to the clouds of Jupiter and Saturn. Instead, the traces of methane in the high haze absorb red light and scatter blue light, lending an ethereal sky-blue color to the disk as seen in Figure 12-13. Only extreme processing of relatively featureless images reveals cloud structures showing through the high haze.

Calculations suggest that the inner quarter of the planet's diameter is occupied by a central rocky core, perhaps molten, similar to Earth in size. This core is surrounded by a deep ocean of hot, high-pressure, liquid water. Above this is the deep atmosphere comprising the outer half of the planet.

URANUS' RINGS

Uranus has an unusual system of very thin, distinct rings. **Uranus' rings** were discovered in 1977, but were not well understood until Voyager 2 obtained the first clear pictures. Shepherd satellites, like those on each side of Saturn's narrow F ring, are believed to force the particles into the narrow ring formations. Voyager cameras discovered two shepherd moons straddling the thickest Uranian ring (Figure 12-15). Gravitational forces of undiscovered moonlets may help confine the other rings.

URANUS' SATELLITES

Uranus has at least two of the three types of satellites found among the giant planets: 5 intermediate-sized moons and 10 more nearby fragmental moonlets. The moons are named after characters in Shakespeare's plays and an Alexander Pope poem. Before Voyager, only the first five were known. The 10 small moonlets were discovered as Voyager 2 flew by the planet. Prior to the Voyager flight, scientists had assumed that Uranus' moons would be little more interesting than inert iceballs because they were so far from the Sun, so cold, and of such modest size. However, Voyager revealed that **Uranus' satellites** have much more variety than had been predicted.

The nearby fragmental moons are small, dark bodies on the edges of the rings; they are similar to but darker than Saturn's nearby moonlets. They are heavily cratered, and probably are fragments of one or more originally larger moons.

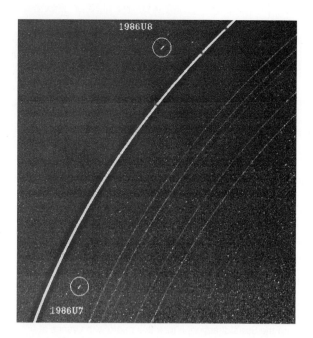

Figure 12-15 Portion of the rings of Uranus and two shepherd moons (circled), whose gravitational pulls help confine the ring particles to narrow widths. Width of widest, outer ring is only 100 km (60 mi). All nine rings appear; the inner three may be glimpsed by holding up the book and sighting along the line of the rings. (NASA Voyager 2 photo.)

Of the five intermediate-sized moons, the outer four are the largest, about a third the size of our Moon. True to our rule of thumb that small worlds show primarily external cratering without internally generated geologic disturbances, the moons are fairly heavily cratered, as seen in Figure 12-16. But as can be seen in the figure, there are complications. Several of the outer four moons are more strongly fractured by canyons than is our Moon, suggesting internal heating by unknown sources, or perhaps dynamic forces arising from interactions with the other satellites.

The most surprising example of complex surface geology was found on the smallest of the five intermediate-sized moons—480-km **Miranda.** It is the most fractured and resurfaced of all the moons. As seen in Figure 12-17, it is laced by discolored patches, complex grooves, and cliffs. One cliff, seen in Figure 12-18, has a sheer, faulted face sloping to a height of 5 km (16,000 ft) above its base. How could a moon of this size violate our general rule of little geological activity on small worlds? How did it generate enough internal energy to cause such faulting and fracturing? One theory is that

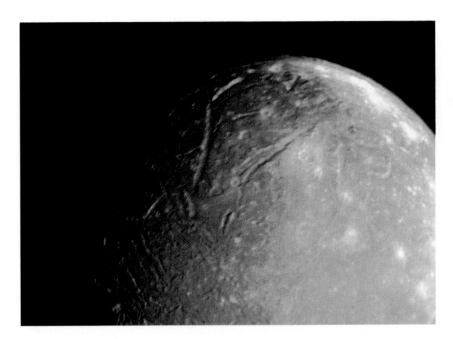

Figure 12-16 Uranus' 1160-km-diameter moon Ariel. This view shows the prominent system of fractures and canyons that have broken the surface of this moon. Brightest spots are impact craters that have apparently blown away the darker soil and exposed cleaner ice. (NASA Voyager 2 photo.)

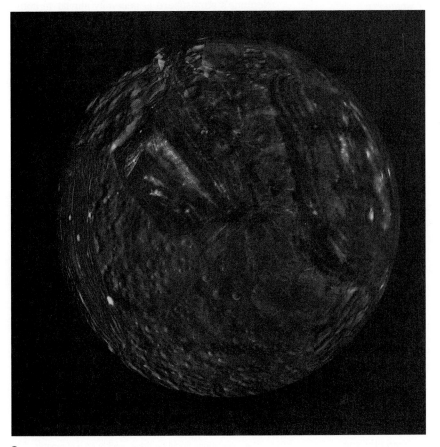

a

b

Figure 12-17 Uranus' strange, 484-km-diameter moon Miranda. Old crater surfaces (bottom) are broken by swaths of striped, younger terrain, which have accumulated fewer impacts. Deep fractured canyons can be seen at the top. **a** Mosaic made by computer-processed combination of high-resolution photos. **b** Lower-resolution color view of part of the same hemisphere. (NASA Voyager 2 photos; mosaic processed by U.S. Geological Survey; courtesy Jet Propulsion Laboratory.)

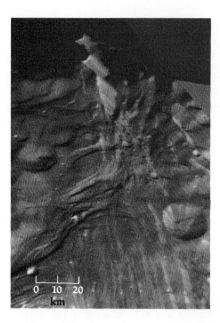

Figure 12-18 Detailed view of a system of fractures, faults, and valleys on Uranus' moon Miranda. Compare with the full-disk view of Miranda (Figure 12-17), where this valley appears near the top. At the top, extending onto the night side, is an enormous cliff with an estimated height of 5 km (16,000 ft) and a slope of about 30°. Much of the region to its right, and in the foreground, has slumped along parallel fractures. (NASA Voyager 2 photo; upper right corner, outside original frame, printed as gray here.)

because of certain gravitational forces among the moons, Miranda may have been forced into a period of chaotic rotation, causing stresses that resulted in fracturing. From our descriptions of Io, Europa, Enceladus, and Miranda, we can see that the forces acting on a satellite orbiting among other satellites around a giant planet are more complicated than had been imagined a few years ago. Researchers are still trying to understand their consequences. Whatever Miranda's history, its fractured face is the cause of much bewilderment.

THE PLANET NEPTUNE

The discovery of Uranus led to the discovery of Neptune. After 1800, theorists tried unsuccessfully to fit observations of Uranus' position into Kepler's laws of planetary motion. Uranus seemed to have its own somewhat irregular motions. A few scientists thought this might signal a breakdown of Newton's law of gravity at great distances from the Sun. Others correctly suggested that Uranus was being attracted by a still more distant planet. In the 1840s an English astronomer and a French mathematician set out independently to predict where the new planet should be.

The English astronomer J.C. Adams finished his prediction in 1845 after two years of calculations,[2] but he had trouble getting his senior professors at Cambridge interested in searching for the new planet with a telescope. In 1846 English astronomers began a desultory search. Several times they actually charted the new planet but failed to realize what they had seen.

Meanwhile, the French mathematician U. J. Leverrier, who had finished his work soon after Adams, interested two young German astronomers in the search. Armed with Leverrier's predictions, they located the new planet within half an hour of starting their search on September 12, 1846. It was given the name **Neptune**. Although Adams and Leverrier are both now credited with the discovery, the incident became an international scandal at the time because of the Britons' failure to grasp their opportunity.

Neptune's atmosphere is generally similar in composition to that of Uranus, but it has much more prominent and active cloud features. These include oval stormlike spots (Figure 12-19), similar to the Great Red Spot of Jupiter. These clouds change structure over a period of a day or so. Their activity level surprised Voyager scientists who photographed them in 1989. The activity may be related to the fact that Neptune radiates more internal heat than Uranus, but the cause of this internal heat is uncertain.

Earth-based operations showed that the clouds rotate in about 18 hours, but Voyager measurements of the magnetic field showed that the bulk of the planet, under the clouds, rotates in about 16 hours. Neptune has a magnetic field, but the center of its symmetry is offset from the center of Neptune by about 0.4 of Neptune's radius, suggesting an unusual internal structure.

NEPTUNE'S ARC-RINGS

Like the other three giant planets, Neptune also has a ring system. Observations from Earth during the 1980s showed that the rings are unique in having uneven density around their circumference. They appeared to have

---

[2] Today the calculations could be done quickly on a computer.

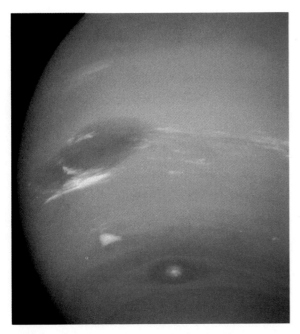

Figure 12-19 Color view of a portion of Neptune shows the bluish color of its atmosphere. Dark oval clouds, such as the prominent "Great Dark Oval," mark circulating storm systems, capped by white condensed clouds higher in the atmosphere. (NASA Voyager 2 photo.)

dense arclike segments. Voyager photos in 1989 confirmed the denser arcs in some rings (Figure 12-13), but detected ring particles all around the planet in each ring. The concentration of particles into certain arcs is probably caused by gravitational forces of nearby satellites, but remains a challenge for dynamicists to explain.

NEPTUNE'S SATELLITES

Neptune has a peculiar **satellite system:** The largest satellite is moving in a circular but highly inclined retrograde orbit, instead of the low-inclination prograde orbit common for large moons. This satellite, **Triton,** has an estimated diameter of about 2760 km, making it the seventh-largest moon in the solar system and bigger than the planet Pluto. As with several other moons we have discussed, the more observations that are obtained, the more unique Triton becomes. Spectroscopic observations in the late 1970s indicate that Triton has a thin atmosphere containing methane (CH_4) gas (Cruikshank and Silvaggio, 1979).

Voyager cameras confirmed the atmosphere by detecting thin clouds and a thin haze layer 5–10 km

above the surface. Other Voyager instruments indicated that Triton's thin atmosphere is mostly nitrogen, with some methane. With a surface pressure of only about 0.01 millibar, it is barely a thousandth as dense as Mars' thin air. Many of the bright and dark surface markings are believed to result from the contrast between bright nitrogen and methane frosts, condensed from the atmosphere in the coldest regions, and darker soils exposed where the frost has sublimed.

As seen in Figure 12-21, Triton has a unique surface. Hardly any impact craters appear. Instead, as with the peculiar moons Europa (of Jupiter) and Enceladus (of Saturn), Triton's surface is covered by very bright ice and frost and seems to have undergone fracturing and eruptions of liquids that froze into ice. Unique to Triton are uplands of lumpy, icy hills, and flat "lakes" of ice that mimic the lava-flooded plains of our Moon. The scarcity of impact craters shows that this surface is geologically young. Stubby, parallel, dark streaks (upper part of Figure 12-21) emanate from dark spots a few kilometers across; researchers think they are ash deposits from volcanic vents, blown in parallel directions by winds of Triton's thin air. Close examination of Voyager 2 photographs after the mission revealed a volcanic eruption in progress, thus making Triton only the third object in the solar system, after Earth and Jupiter's moon Io, known to have active volcanoes.

The source of the internal energy that powered Triton's eruptions may be related to its orbital history. The

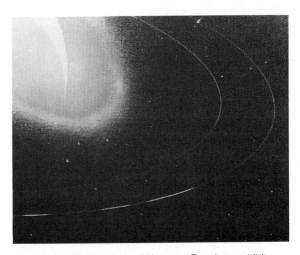

Figure 12-20 The rings of Neptune. Two rings exhibit denser arclike segments (bottom), but are continuous around Neptune. Fainter broad rings were also found inside each of these rings. (NASA Voyager 2 photo.)

strange orbit, being retrograde and inclined, suggests that Triton is not in its original path. It may have had its orbit altered by some passing interplanetary body, or it may itself be an interplanetary body that was captured into orbit around Neptune. Triton resembles the planet Pluto in size, surface composition, and low den-

Figure 12-21 Neptune's largest satellite, Triton, shows a strangely mottled surface of bright ice and frost deposits. Few impact craters are found; fractures cross the central region. The top, brighter area is around the south pole and is believed to mark a polar frost cap. (NASA Voyager 2 mosaic.)

Figure 12-22 Ice "lakes" of Triton. In the midst of amorphous hilly terrain, two depressions (top and center), which may mark ancient impact basins, have been flooded with liquid that froze into a smooth, ice surface. The prominent pit in the center is probably a meteorite impact crater. (NASA Voyager 2 photo.)

sity ($\sim$2000 kg/m^3). Both have originated as small interplanetary bodies. Once Triton was placed in an irregular orbit around Neptune, tidal gravitational forces may have caused both a slow evolution of the orbit to its present circular shape, and also a heating of the interior, allowing volcanism. The same tidal gravitational forces are presently causing Triton to spiral inward toward Neptune; it will crash into the planet in the distant future.

Neptune's outermost moon, 170-km Nereid, may also be captured. It moves in a highly inclined prograde orbit. It is fairly dark (reflectivity about 10 percent, similar to the interplanetary body Chiron). The peculiarities of both Triton's and Nereid's orbits support the idea of some major ancient disturbance to the Neptune system, probably the passage or capture of an interplanetary body.

As with Jupiter, Saturn, and Uranus, Voyager discovered new, small moons close to the planet. Six moons, discovered by the time this book went to press, are listed in Table 8-1 on page 122. Others may still be found from the Voyager pictures.

THE PLANET PLUTO AND ITS MOON

Scientists had expected that gravitational pulls by Neptune would explain irregularities in Uranus' motions. But Neptune's gravity did not account for all of them, and Neptune itself has some unexplained irregular motions. These irregularities suggested at least one more planet beyond Neptune. For this reason, Percival Lowell began a search for a ninth planet in 1905. As shown in Figure 12-23, this search led to the discovery of **Pluto** in 1930, when Clyde Tombaugh found it on Lowell Observatory photos. After it was named for a god of the underworld, a new planetary symbol (℞) was created from the first two letters, which were also Lowell's initials.

Though Pluto was hailed as the ninth and final planet, its status as a planet is currently being questioned, for at least five reasons:

1. It is much smaller than any other full-fledged planet. At an estimated diameter of 2300 km, it is smaller than our Moon and only 47% as big as Mercury.

2. It fails to continue the trend of the giant planets in size; among them, it is an anomaly.

3. Its orbit is not close to twice the size of Neptune's, in keeping with other planet spacings; instead, it overlaps Neptune's orbit.

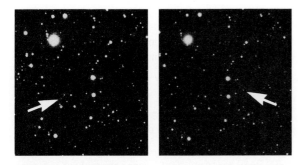

Figure 12-23 Portions of the photographs on which Pluto (arrow) was discovered. The planet was distinguished from stars by its motion. (Lowell Observatory.)

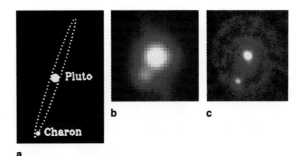

Figure 12-24 Pluto, the double planet, as seen by the Hubble Space Telescope. **a** Map of Pluto, Charon, and Charon's orbit as they appeared at the time of the photography. **b** The approximate best level of detail that can be photographed from Earth-based telescopes. **c** Image of Pluto and Charon as photographed with the Hubble Space Telescope, using computer corrections to compensate for the telescope's flawed optics. This was the first long-exposure photo made with the Hubble Telescope on a moving object; the telescope had to track Pluto during the exposure. (NASA.)

4. An interplanetary body (named Chiron) was discovered in 1977 between Saturn and Uranus, raising the possibility that the outermost solar system has many interplanetary bodies of various sizes. Moreover, we know that many comet nuclei exist beyond Neptune's orbit (see next chapter). Pluto might just be the largest of such interplanetary bodies.

5. Pluto's moon is much larger in diameter with respect to its planet (57%) than are most moons.

For these reasons astronomers are beginning to think of Pluto (and its moon) as more like a giant double asteroid rather than a normal planet. The distinction may be more than just semantic: Pluto may indeed be more closely related to asteroids and comets than to full-fledged planets. But most writers list Pluto as the ninth planet.

Pluto's light variations and eclipses of its moon establish that it rotates every 6.39 d on an axis tipped about 120° to its orbit. Its density is close to 2000 kg/m³, composed of about 70% rock and 30% ice (Tholen and others, 1987; Tholen and Buie, 1988). Spectroscopic observations indicate methane ice on the surface and perhaps a thin atmosphere of methane gas, probably derived from methane ice deposits on the surface. The surface reflects some 60% of the light falling on it, characteristic of bright, icy material (Binzel, 1990).

Pluto's satellite, named **Charon**, was discovered in 1978. The name comes from a mythological figure associated with Pluto. Charon (pronounced KEHR-on, not to be confused with interplanetary body Chiron, mentioned in item 4 above) circles Pluto in 6.39 d, keeping one face toward Pluto. Charon appears somewhat darker than Pluto, reflecting about 40% of the incident light—a figure characteristic of slightly dirty ice.

Spectroscopic studies show interesting differences between the two bodies. Charon is not only darker than

Pluto, but it has less methane ice and more water ice. This difference may involve a loss of lightweight methane ice molecules from Charon, due to its lower gravity, leaving more of the dark soil behind (Marcialis and others, 1987). Clearly, scientists would like to know more about the history of this distant pair of worlds.

On cold, remote Pluto and Charon, the surface temperature is only about 50 K (−369°F), and the Sun would be too far away to be perceived as a disk. From this outpost of the solar system, the Sun would look like an intensely bright streetlight across the street, reminding us that it is one star out of many.

"PLANET X?"

Several astronomers have sought dynamical or photographic evidence of a planet beyond Pluto, sometimes called **"Planet X."** Clyde Tombaugh, the discoverer of Pluto, conducted a long search of the region beyond Pluto and ruled out any planet as large as Neptune near the plane of the solar system out to a distance of around 100 AU.

However, the 1978 discovery of Pluto's satellite allowed a better mass determination than had been possible earlier for Pluto, and the mass turned out to be too small to account for all of the irregularities of the motions of Neptune. This finding has prompted a number of astronomers to suspect that other objects may exist, perhaps well out of the plane of the solar system. Astronomers have searched for such objects without success (Luu and Jewitt, 1988; Littmann, 1989), but the search is incomplete. "Planet X" might be in a highly inclined eccentric orbit. There is growing suspicion that the outer solar system may harbor additional small bodies.

SUMMARY

The four giant planets—Jupiter, Saturn, Uranus, Neptune—and their satellite systems show some systematic similarities as well as differences. All four of the giant planets have dense, cloudy atmospheres, averaging around four-fifths hydrogen and one-fifth helium by mass. All have rings composed of billions of small particles ranging in scale from microscopic to house-sized. All have systems of satellites. And most of the moons contain abundant ice as well as black soil, because this part of the solar system is so cold that ices were a main constituent of the material available for building moons and planets.

Among differences, the colder, more remote planets have more hazy, less colorful cloud patterns. Jupiter has reddish and tan clouds; Uranus and Neptune have greenish-blue clouds from methane-rich haze. The ring systems have different amounts of material; Saturn's is most prominent, and Neptune's is oddly concentrated in dense arcs. Uranus and its satellite system are tipped so that the equator and satellite orbits lie nearly perpendicular to the plane of the solar system.

Pluto is an "oddball"—much smaller than the other planets and having a moon half as big as itself. It may be more closely related to other small bodies (such as asteroids and comets—see the next chapter) than to the true planets.

CONCEPTS

Saturn	Uranus' rings
Saturn's rings	Uranus' satellites
Cassini's division	Miranda
shepherd satellites	Neptune
Saturn's satellite system	Neptune's arc-rings
Dione	Neptune's satellite system
Enceladus	Triton
Iapetus	Nereid
Phoebe	Pluto
Titan	Charon
Uranus	"Planet X"
obliquity	

PROBLEMS

1. Why do Uranus and Neptune have more Earth-like colors (bluish) than Jupiter and Saturn?

2. Discuss how rings may have formed around giant planets.

3. Why doesn't Earth have a hydrogen/helium atmosphere like the giant planets do?

4. Describe the appearance of the sky to an observer flying above the clouds in the upper atmosphere of Saturn at dusk at low latitudes. Do the same for Saturn's polar regions.

5. Discuss why a probe to the surface of Titan might be interesting to biologists trying to understand how life started on Earth.

6. What difficulties might be met in sending a spacecraft through the rings of Saturn to examine the ring environment? Consider a pass *perpendicular* to the ring plane versus a pass *in* the ring plane. (Note that without elaborate retrorockets, such a spacecraft would probably travel about 20 to 30 km/s relative to the rings.)

7. Describe the seasons and other effects that would occur if Earth had the same obliquity as Uranus.

8. Do orbits of planets ever cross, as seen from far north or south of the plane of the solar system?

9. Where would Bode's rule, if extended, predict a tenth planet? Would it be easier to detect if covered with snow or with rock? Why?

10. Explain why observers on Earth can never see Saturn appearing as it does in Figure 12-3.

PROJECTS

1. With a telescope having at least an 8-cm (3-in.) aperture, observe Saturn. Sketch the rings. Estimate the angle by which the rings are tilted toward Earth during your observation. Does this angle change much from day to day? From year to year? (Your teacher might save drawings by students from past years for comparison.) Can you see any belts or zones? (Usually they are less prominent than on Jupiter.) Can you see Cassini's division? Sketch nearby starlike objects that may be satellites and track them from night to night. Identify Titan, the brightest satellite.

2. With a telescope having at least 15-cm (6-in.) aperture, observe Uranus and Neptune. What color are they? Can you see any detail? Any satellites?

CHAPTER 13

Comets, Meteors, Asteroids, and Meteorites

On June 30, 1908, a mysterious explosion occurred in Siberia. English observatories, 3600 km (2200 mi) away, noted unusual air pressure waves. Seismic vibrations were recorded 1000 km (600 mi) away. At 500 km, observers reported "deafening bangs" and a fiery cloud. The explosion was caused by an unknown object that struck the Earth's atmosphere from space.

Some 200 km (120 mi) from the explosion, the object was seen as "an irregularly shaped, brilliantly white, somewhat elongated mass . . . with [angular] diameter far greater than the Moon's." Carpenters were thrown from a building and crockery knocked off shelves. An eyewitness 110 km from the blast reported that

the whole northern part of the sky appeared to be covered with fire. . . . I felt great heat as if my shirt had caught fire . . . there was a . . . mighty crash. . . . I was thrown onto the ground about [7 m] from the porch. . . . A hot wind, as from a cannon, blew past the huts from the north. . . . Many panes in the windows were blown out, and the iron hasp in the door of the barn was broken.

Probably the closest observers were some reindeer herders asleep in their tents about 80 km (49 mi) from the site. They and their tents were blown into the air and several of the herders lost consciousness momentarily. "Everything around was shrouded in smoke and fog from the burning fallen trees."

The cause of this remarkable explosion was a collision between Earth and a relatively modest bit of interplanetary debris. Several types of such debris exist.

Comets are icy worldlets a few kilometers across. **Asteroids** are rocky–metallic worldlets ranging from a few hundred meters or less up to 1000 km in diameter. Still smaller bits of debris, probably dislodged from comets and asteroids, are sometimes called meteoroids. Table 13-1 shows orbital and other data on selected examples of these classes.

TABLE 13·1

Properties of Selected Small Bodies in the Solar System

Class	Example	Diameter (km)	Orbit Semimajor Axis (AU)	Eccentricity	Inclination	Remarks
Comets						
Short-period	Encke	6?	2.2	0.85	12°	Probably dirty ice
	Halley	8 × 15	18	0.97	162°	Probably dirty ice
Long-period	Kohoutek	Few?	Very large	1.0	14°	Probably dirty ice
Meteors						
Shower	Perseid	10^{-6}	40	0.97	114°	Cometary debris
	Taurid	10^{-6}	2.2	0.80	2°	Cometary debris
Fireballs	July 31, 1966	10^{-4}?	32	0.98	42°	May be related to comets or asteroids
	May 31, 1966	10^{-4}?	3.0	0.80	9°	
Asteroids						
Belt	1 Ceres	1020	2.8	0.08	11°	Carbonaceous rock surface
	2 Pallas	538	2.8	0.23	35°	Rocky surface
	3 Juno	248	2.7	0.26	13°	Rocky surface
	4 Vesta	549	2.4	0.09	7°	Lavalike surface
	14 Irene	170	2.6	0.16	9°	Rocky surface
Trojan	624 Hektor	100 × 300	5.1	0.02	18°	Unusual shape
Apollo	433 Eros	7 × 19 × 30	1.5	0.22	11°	Elongated
	Apollo	Few?	1.5	0.56	6°	Lost before orbit precisely measured
Asteroid/comet	2060 Chiron	100–320?	13.7	0.38	7°	Cataloged as asteroid; developed comet activity
Meteorites						
	Pribram	10^{-4}	2.5	0.68	10° ⎫	Rocky fragments
	Lost City	10^{-4}	1.7	0.42	12° ⎬	recovered
	Leutkirch[a]	10^{-4}?	1.6	0.40	2.5°	Not collected

Source: Data from Chapman and Morrison (1974); Hartmann (1972, 1975); Gehrels (1972). The diameter of Comet Encke was measured with radar by Kamoun and others (1982).

[a]An object photographed over Europe in 1974. No fragments recovered as of late 1974, but believed to be a stone meteorite.

Some of these bodies have planetlike orbits that are nearly circular, prograde, and lying in the plane of the solar system. Others have very different orbits—highly elliptical, retrograde, or inclined at steep angles to the solar system's plane. Some of these orbits are shown in Figure 13-1, which emphasizes the diversity of interplanetary objects.

Objects whose orbits cross the orbits of the Earth

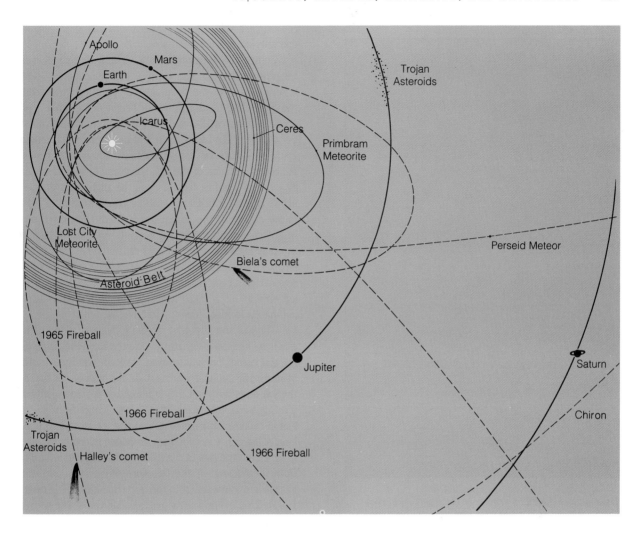

Figure 13-1 A portion of the solar system showing orbits of a few selected
interplanetary bodies including comets, belt asteroids, Trojan asteroids, Apollo asteroids,
meteorites, and large meteors, or fireballs. Orbits of comets and probable cometary
debris are dashed. Orbits of planets, asteroids, and probable asteroid debris are solid.

and other planets can collide with the planets. This is
rare for bodies as big as comets and asteroids but common for tiny meteoroids. When they collide with Earth,
they are heated by friction with the atmosphere (just
as returning space vehicles are). Bodies hitting Earth
are given one of two names. **Meteors** are the smaller
ones (usually up to a few centimeters in diameter) that
burn out completely before striking the ground. They
are mostly debris from comets. **Meteorites** are the
larger ones that survive and fall to the ground: Made
of rock or metal, they are free samples of other planetary bodies—most likely asteroids. Their properties
testify that they broke off of larger bodies, called **par-**

ent bodies, in the remote past. One unsolved mystery
is the question of what were the parent bodies of different meteorite groups.

Comets, asteroids, and meteoroids of every size
are debris left over from the origin of the solar system
4.6 billion years ago. As the present planets formed,
the solar system was filled with innumerable small, preplanetary bodies, ranging up to 1000 km across, known
as **planetismals.**

Comets, asteroids, and meteorites are all leftover
planetismals and planetismal fragments. Thus they provide a direct link to the conditions under which the
planets and the Sun itself formed. By implication, they

Figure 13-2 Comet Ikeya-Seki in the dawn sky (1965). The comet's head is the sharply defined tip nearest the horizon; the tail extends diffusely upward. (35-mm camera; 15-min exposure guided on comet). (Photo by S. M. Larson, University of Arizona.)

provide clues to the formative conditions of other stars and hence provide a link with the most distant stellar regions, a topic of much of the rest of this book.

COMETS

Comets are the most spectacular of the small bodies in the solar system. When they pass through the inner solar system near Earth, they can be seen drifting slowly from night to night among the stars. (Writers sometimes incorrectly describe comets as "flashing across the sky" like shooting stars. They do not. They seem to hang motionless and ghostly among the stars. Their motion relative to the stars can be detected by the naked eye only after a few hours.)

Comets have several parts. The brightest part is the **comet head.** The **comet tail** is a fainter glow extending out of the head, usually pointing away from the Sun. The head and tail of bright comets are well shown in Figures 13-2 and 13-3. Although a typical comet tail can be traced for only a few degrees by the naked eye, binoculars or long-exposure photos may reveal fainter extensions of the tail extending tens of degrees or even extending clear across the night sky. A telescope reveals a brilliant, starlike concentration of light at the center of the comet head. At the center of this bright concentration is the **comet nucleus,** which is the only substantial, solid part of the comet, but is too small to be resolved by telescopes on Earth. Studies reveal that a typical comet nucleus is a worldlet of dirty ice only about 1 to 20 km (a few miles) across—tiny compared to most planets and moons!

The gas and dust that make up the rest of the comet's head and tail are material emitted from the nucleus. As the comet nucleus moves through the inner solar system, the sunlight warms it. In the vacuum of space, the ice does not melt into a liquid but instead changes directly from solid ice to gas. This change from ice to gas is called **sublimation,** and the ice is said to sublime. This gas, together with dislodged microscopic dust grains from the dirt in the nucleus, is then carried away from the nucleus by the pressure of radiation and thin gas rushing outward from the Sun (Brandt and Niedner, 1986). This moving solar gas is called the **solar wind.** Only microscopic grains get blown outward by it; larger grains are too heavy to be caught in the solar wind. (Similarly, a handful of dust gets blown by the wind but stones do not.) The gas and dust from the nucleus form the comet's tail, often stretching more than an astronomical unit. Caught in the solar wind, the tail streams out behind the comet as the comet approaches the Sun, but leads as the comet recedes from the Sun (Figure 13-4). It is like the long hair of a woman walking in a strong wind, streaming out behind as the woman walks into the wind, but in front of her if she reverses direction.

Comets as Omens

Before comets' true natures were known, they were often regarded as evil omens. For example, the appearance of Halley's comet in A.D. 66 was said to have heralded the destruction of Jerusalem in A.D. 70. Five circuits later it was said to mark the defeat of Attila the

Hun in 451. In 1066 it presided over the Norman conquest of England. In 1456, its appearance coincided with a threatened invasion of Europe by the Turks, who had already taken Constantinople three years before. Pope Calixtus III ordered prayers for deliverance "from the devil, the Turk, and the comet." Of course, we now recognize that some noteworthy events are likely in most cultures every decade or so, and comets are no longer regarded as ominous portents by most people.

Discovery of Comets' Orbits

In 1704 the English astronomer Edmond Halley applied newly developed methods of computing orbits and discovered that comets travel on *long, elliptical orbits* around the Sun and that certain comets reappear. Calculating the orbits of 24 well-recorded comets by Newton's methods, Halley found that four comets (seen in 1456, 1531, 1607, and 1682) had the same orbit and a periodicity near 75 y. Halley correctly inferred that these appearances were by a single comet and that the slight irregularities in periodicity were caused by gravitational disturbances from the planets, especially Jupiter. Halley predicted that the comet would return about 1758. It did so on Christmas night. It was named Halley's comet.[1] The discovery that comets are visitors on ordinary elliptical orbits, with predictable motions, helped dispel the superstition that comets are evil omens.

Halley's comet became very famous due to its prominence in 1910, when Earth passed through its tail. It was not so prominent in 1986 during its most recent approach, because it did not pass as close to Earth (Figure 13-4). Even so, the 1986 approach marked the first close-up study of a comet by space probes.

The Physical Nature of the Comet Nucleus

After naturalists discovered that comets are not wispy atmospheric phenomena or evil visitors, but objects orbiting the Sun, the next question became: What is the detailed nature of a comet nucleus?

The answer came in two major steps. First, around 1950, Harvard astronomer Fred Whipple recognized that the comet nucleus is a solid body composed of ice

---

[1]Comets discovered today are first named for their discoverers but are later given scientific names in order of their passage around the Sun—for example, Comet 1982 I.

Figure 13-3 A view of Halley's comet during its approach in 1986. In that year, it was in Southern Hemisphere skies and not well placed for northern observers. The photo shows the pinkish color of the diffuse dust tail and the bluish color of the narrower, straighter gaseous tail. (Photo by William Liller from Easter Island for NASA's International Halley Watch; 4-min exposure on Fujichrome 400 film with 8-in. Schmidt telescopic camera at f1.5.)

and dust grains. This came to be called the **dirty iceberg model**. The evidence came from the kinds of gases streaming out of the nucleus into the comet tail. These include atoms of carbon (C), hydrogen (H), oxygen (O), ions of nitrogen (N_2^+), carbon dioxide (CO_2^+), and complex organic molecules such as CH_3CN, and other compounds. Whipple recognized that such atoms, ions, and molecules would be exactly the types released when common ices of the outer solar system—such as H_2O, CO_2, CH_4, and NH_3—turn into gas. So he inferred that the nucleus contains these ices, as well as dust grains that form the dust in the tail. This model explained observed comet phenomena and fit the fact that ices such as frozen H_2O, CO_2, CH_4, and NH_3 are common

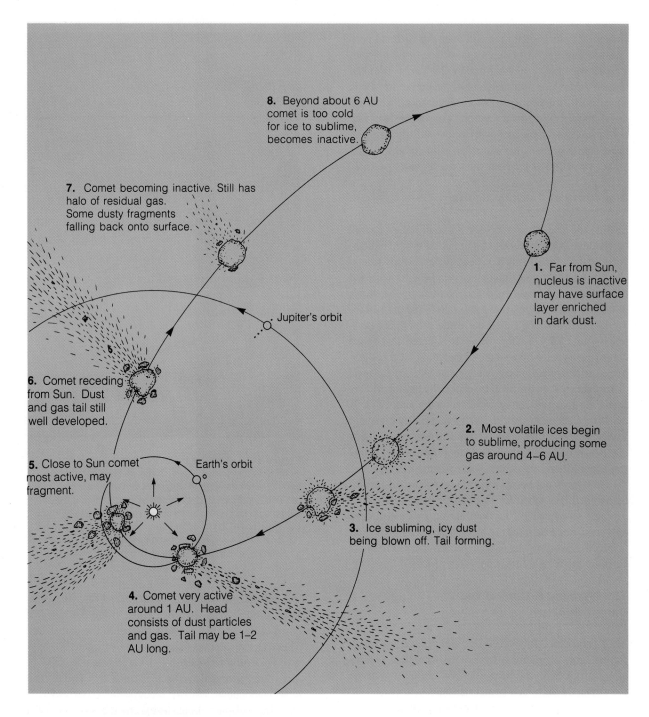

8. Beyond about 6 AU comet is too cold for ice to sublime, becomes inactive.

1. Far from Sun, nucleus is inactive may have surface layer enriched in dark dust.

7. Comet becoming inactive. Still has halo of residual gas. Some dusty fragments falling back onto surface.

Jupiter's orbit

6. Comet receding from Sun. Dust and gas tail still well developed.

5. Close to Sun comet most active, may fragment.

Earth's orbit

2. Most volatile ices begin to sublime, producing some gas around 4–6 AU.

3. Ice subliming, icy dust being blown off. Tail forming.

4. Comet very active around 1 AU. Head consists of dust particles and gas. Tail may be 1–2 AU long.

Figure 13-4 Stages in the development of a typical comet as it travels from the outer solar system through the inner solar system, where it loops rapidly around the Sun and moves outward again. The tail develops only within a few astronomical units of the Sun, where sunlight is warm enough to sublime the ice in the nucleus. In order to show comet detail, the figure is not drawn to scale.

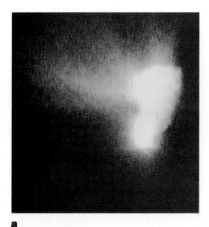 

a   **b**

Figure 13-5 The nucleus of Halley's comet at moderate distance from two different space probes in March 1986. Both views are hazy because of the dusty environment, but they show luminous jets issuing from a peanut-shaped nucleus. Sunlight is from the left in both views, and jets come from the sunlit side, where gas is subliming rapidly. **a** View from Russian Vega 2 probe at 8000-km distance. Lighting at gibbous phase illuminates much of the nucleus surface in this view. (USSR Academy of Sciences.) **b** View from European Giotto probe at about 20,000 km distance. Crescent lighting silhouettes the dark side of the nucleus against the luminous head. (Copyright Max-Planck-Institut für Astronomie.)

in the cold depths of the outer solar system beyond Jupiter, where comets come from. When a comet enters the inner solar system and approaches the Sun, these ices are heated. In the vacuum of space, they sublime into atoms, ions, and molecules of their constituents. As the ice erodes under solar heating, trapped dust grains work loose and are carried off in the out-rushing gas.

The second major step in understanding the comet nucleus was the opportunity to visit one. In 1986, as Halley's comet sailed through the inner solar system, an international fleet of five spacecraft (two Japanese, two Russian, and one European) flew to meet it. After the first four reconnoitered the environment at moderate distance, the European probe flew by at very close range, about 600 km. This probe was named Giotto, after the early Italian artist whose 1304 painting may have been the first rendering of Halley's comet. As seen in Figures 13-5 and 13-6, Giotto's close-up photos revealed that the nucleus of Halley's comet is a solid, peanut-shaped body about 15 × 8 km in size, and black in color. The gas and dust shoot off the nucleus in jets emanating from active spots that may be vents or exposures of fresher ice. The gas was found to be about 80% water vapor by volume, implying a large amount of ordi-

nary frozen water in the nucleus. The dust was found to be rich in C, H, O, and N, somewhat resembling carbon-rich meteoritic particles collected on Earth. This is different material from terrestrial soil, which is richer in silicon, iron, other metals, and their oxides. The black, carbon-rich dust grains are what gives the nucleus its black surface, which reflects only 4% of the light striking it. It is as dark as black velvet! Scientific interest has grown in the comet dust, which apparently contains an abundant supply of the building blocks needed for life. No life is expected to have evolved in frozen comets, but perhaps comets can teach us about the primitive stages of carbon-based organic chemistry in the solar system.

The Origin of Comets

The most important step in unraveling the origin of comets was the work of Dutch astronomer Jan Oort, around 1950. He studied the statistical distribution of comets' orbits and discovered that comets spend most of their time on the extreme outskirts of the solar system, around 50,000 AU from the Sun. These comets are far beyond Pluto, which orbits only 39 AU from the Sun. This swarm of comets surrounding the solar sys-

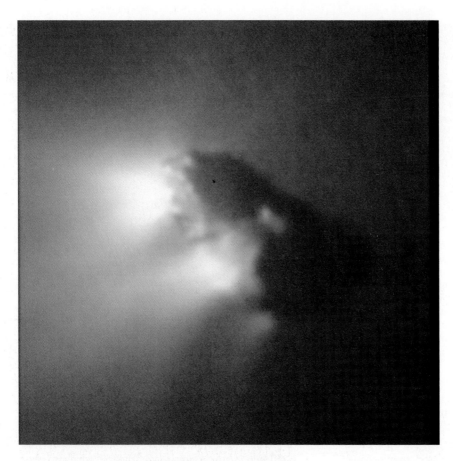

Figure 13-6 In the weeks after Giotto's flight past the nucleus of Halley's comet, the hazy images were highly processed to improve clarity and combine the best parts of different images. This resulting picture is the best available. It shows bright jets of gas and dust shooting from localized active vents on the sunlit side of the nucleus. The smallest details are about the size of a football field. There is little color to be seen in the nucleus, which is surprisingly irregular in shape. The unlit "night" side is silhouetted against the softly glowing head on the right. (Giotto photo courtesy Harold Reitsema and Alan Delamere, Ball Aerospace; copyright 1986 MPAE.)

tem is called the **Oort cloud.** It contains billions of comets. Here, comets are so far from the Sun that the gravitational attractions of nearby stars, as well as the Sun's, are important. Following Kepler's and Newton's laws, the comets drift slowly in their orbits, and their motions are disturbed by nearby stars' gravitational forces, until they eventually fall back into the inner solar system; there they may be near enough to the Sun for only a few months before they are warmed up, emit gas, develop a glowing head and tail, and become visible. Away from the Sun, these comets are only cold, icy bodies without a head or tail.

The Oort cloud is only a "holding bin" for comets. The important question about comets' origins is: How did they get into the Oort cloud? Some may have formed on the inner edge of the Oort cloud, beyond Pluto, where there was not enough material to form a full-fledged planet. But many of them are believed to have formed among the giant planets of the outer solar system. In this cold region, icy bodies aggregated into the 1-, 10-, and 20-km dimensions of comets. But consider the his-

tory of such a body. As a nearby giant planet, such as Uranus, grew, the comet's orbit was eventually disturbed by the planet's strong gravity. Eventually, the comet was likely to make a near pass by the planet, and its orbit would have been strongly altered, throwing the comet far out of the solar system, just as Pioneer and Voyager space probes were thrown out of the solar system by their near passes by Jupiter, Saturn, Uranus, and/or Neptune. Thus, comets appear to be survivors of the earliest so-called planetesimals, which were the building blocks of planets. (We will discuss these planetesimals in more detail in Chapter 14.) Because comets preserve these icy primitive bodies in the "deep freeze" of the cold Oort cloud region, comets are of growing interest to scientists for the samples they provide of the earliest solar system materials.

Future Work on Comets

Scientists would like to learn more about the elements and organic compounds in comets in order to compare

Figure 13-7 A proposed future comet mission would park a probe in orbit alongside a comet and watch it "turn on" as it approached the Sun. This painting, made two months before the Halley encounters, attempted to predict some features of the nucleus. It shows jetting similar to that seen in spacecraft photos such as Figure 13-5. (Painting by author.)

them with terrestrial data on the origins of life. They would also like to understand the physical processes that occur during comet activity. Is the black dirt mixed through the ice of the nucleus? Or does a layer of black surface gravel mask cleaner ice below? Could future astronauts use comet ices for water supplies?

American scientists have proposed a "CRAF" (Comet Rendezvous, Asteroid Flyby) mission, and Soviet scientists have proposed a similar mission. As shown in Figure 13-7, this probe would "park" in a position near a comet far from the Sun and then observe the turn-on processes as the comet becomes active and begins to blow off surface materials. A torpedolike "penetrator" would be fired into the comet to measure the composition of its dirty ice. On the way to the comet, the probe would pass by one or more asteroids, thus getting data for comparing asteroids and comets.

METEORS AND METEOR SHOWERS

Meteors that flash momentarily across the sky might seem totally unrelated to comets. However, a study of their frequency on different nights reveals a direct connection. On an average night you may see about 3 meteors per hour before midnight and about 15 meteors per hour after midnight. (You see more meteors after midnight because you are then located on the leading edge of the Earth as it moves forward in its orbit, sweeping

up interplanetary debris.) But on certain dates each year you may see **meteor showers** of 60 meteors or more per hour, with all the meteors in the shower radiating from the same part of the sky (Table 13-2).

The best-known example is the Perseid shower, which occurs every year around August 12, when bright meteors streak across the sky every few minutes from the direction of the constellation Perseus. (A shower is named for the constellation most prominent in the area of the sky from which the shower radiates.) Occasionally the showers are so intense that meteors fall too fast to count. During the Leonid shower of November 17, 1966, meteors fell for some minutes like snowflakes in a blizzard at an estimated rate of more than 2000 meteors per minute (see Figure 13-8).

Most meteors are far too small to reach the ground, "burning" at altitudes around 75 to 100 km. Occasional large ones, called **fireballs,** are very bright and spec-

Figure 13-8 An unusually intense meteor shower; the Leonids of November 17, 1966. The rate of meteors visible to the naked eye was estimated to exceed 2000 per minute. Such an intense shower happens only a few times per century. The brightest star (upper left) is Rigel, in the constellation Orion. This exposure of a few minutes' duration was made with a 35-mm camera. (D. R. McLean.)

Dates of Prominent Meteor Showers

Shower Name (After Source Constellation)	Date of Maximum Activity[a]	Associated Comet
Lyrid	April 21, morning	1861 I
Perseid	August 12, morning	1862 III
Draconid[b]	October 10, evening	Giacobini–Zinner
Orionid	October 21, morning	Halley
Taurid	November 7, midnight	Encke
Leonid	November 16, morning	1866 I
Geminid	December 12, morning	"Asteroid" 1983 TB[c]

[a]Showers can last several days before and after the peak activity on the listed date. Observations are best when the constellation in question is high above the horizon, usually just before dawn.

[b]The Draconids are now weak because their orbits have been disturbed by the gravity of planets, but further disturbances may again strengthen the shower in the future.

[c]This object was discovered in 1983 by the IRAS satellite, and may be a "burnt-out" comet nucleus.

tacular. They generally explode in the air instead of hitting the ground, again indicating that they are too fragile to survive atmospheric entry.

What is the connection between the showers and the comets? In 1866 G. V. Schiaparelli (of Martian canal fame) discovered that the Perseid meteor shower occurred whenever Earth crossed the orbit of Comet 1862 III. The Perseids, then, must be spread out along the orbit of that comet. Other relationships were soon found between specific meteor showers and specific comets, as Table 13-2 shows. *Therefore, most meteors must be small bits of debris scattered from comets.*

Since the 1960s, rockets and balloons in the high atmosphere have collected microscopic fragments believed to have come from comets and asteroids. These fragments include carbonaceous materials similar to those found in Halley's comet and in certain types of meteorites.

In the 1980s, scientists discovered the "missing link" between comets and meteor showers. The Infrared Astronomical Satellite (IRAS) had mapped the sky in infrared light coming from all sorts of celestial materials. Among the asteroids, nebulae, and other sources, researchers found trails of dust spread along the orbits of active comets (Sykes and others, 1986). The solar system is ringed with these tenuous **dust trails** of microscopic particles blown off the comets' surfaces when they are active. Visible meteors may be just the

larger particles in these trails. As we will see in later chapters, these rings of dust may relate to rings of dust found around other stars.

ASTEROIDS

Asteroids are the largest of the interplanetary bodies, ranging in size from Ceres (1020 km in diameter), down to a few hundred meters across and probably less. Spectral studies show that they are rocky, often with metal embedded in them. They appear in various parts of the solar system but are most abundant in the **asteroid belt**—a region between Mars and Jupiter.

Discovery of Asteroids

Normally too small to be seen by the naked eye, asteroids were unknown before 1800. Astronomers discovered the first asteroid, Ceres, in 1801 while searching for a new planet in the large space between Mars and Jupiter.[2]

---

[2]As noted by Arthur C. Clarke, this discovery came just when the philosopher Hegel had "proved" philosophically that there could be no more than the seven then-known planetary bodies, Mercury through Uranus—which suggests that going out and looking is worth more than sitting at home speculating.

Ceres turned out to be smaller than a full-fledged planet, being less than a third the size of our Moon. Between 1802 and 1807, three more small, planetlike bodies turned up between 2.3 and 2.8 AU from the Sun. Because of their small size, they came to be called minor planets, or *asteroids,* a name that now applies to any interplanetary body that is not a comet—that is, is not known to emit gas. Thousands of asteroids have since been discovered by photographic searches.

Asteroids are known by numbers (assigned in order of discovery, but only after the orbit has been accurately identified) and a name (chosen by the discoverer)—1 Ceres and 2 Pallas, for example. The names cover a wide range of human interests, including cities, mythology (1915 Quetzalcoatl), politicians (1932 Hooveria!), spouses, and other lovers.[3] Over 2000 are now cataloged, and perhaps 100,000 observable ones remain uncharted.

Asteroids Outside the Belt

Several interesting subgroups of asteroids have orbits outside the main asteroid belt. **Apollo asteroids,** for example, come into the inner solar system and cross the Earth's orbit. They are named after Apollo, first of the group to be discovered. Dozens are known, including one as small as 800 m across. Most are irregularly shaped, and many approach much closer to Earth than any planet. Indeed, meteorites that hit the Earth can be thought of as the smallest examples of Apollo objects!

Apollo asteroids generally last only a few hundred million years before crashing into Earth or another planet, creating impact craters. New Apollo asteroids are constantly created—probably as asteroids are thrown out of the belt by the gravitational force of massive Jupiter. Other Apollos may be burnt-out stony remnants of short-period comets that have exhausted most of their ices after spending too much time in the inner solar system.

Trojan asteroids[4] lie in two swarms in Jupiter's orbit, 60° ahead of and 60° behind the planet. These two spots are called **Lagrangian points** after the astronomer Joseph Louis Lagrange, who discovered that

particles are held in the two swarms by Jovian and solar gravitational forces. At least 45 Trojans have been found at one Lagrangian point, and observers estimate about 700 observable Trojan asteroids. The largest Trojan is Hektor, estimated to be 100 km wide and 300 km long, tumbling end over end in Jupiter's orbit. Trojan asteroids and their neighbors, the outer satellites of Jupiter (J6 through J13), are similar in having a very dark color and a different composition (more carbon minerals) than that of belt asteroids. They probably formed near Jupiter's orbit. Jupiter's outer satellites may even be Trojan-like asteroids that were captured into orbits around the giant planet by Jupiter's strong gravity.

A unique body, cataloged as an asteroid, is even further from the Sun. It is called Chiron, and was discovered in 1977. Its orbit ranges from a point just inside Saturn's orbit to a point just inside Uranus's orbit. It is the only "asteroid" known beyond Saturn's orbit. Size estimates place it near 200 km in diameter. Its asteroid status was challenged in 1988 when D. J. Tholen, D. P. Cruikshank, and this author discovered that Chiron was experiencing a brightening well beyond its normal brightness. At that time, Chiron was moving closer to the Sun, toward a closest approach in 1996. The brightening probably heralds the onset of cometary activity. We suspect that Chiron may be rich in ices, which are warming and throwing off gas and dust, causing Chiron to brighten. If so, Chiron may be reclassified as the largest known comet. Chiron, the small planet Pluto, and Pluto's satellite Charon may be only the largest examples of a host of ice-rich worldlets on the outskirts of the solar system, with others waiting to be discovered.

The Rocky Material of Asteroids

As indicated in Chapter 5, the spectra of asteroids can be used to obtain rough estimates of their mineral properties. The asteroids divide into several classes with different spectra, and the spectra of these classes have been compared with spectra of minerals and rocks from Earth, the Moon, and many meteorites. The spectra of some asteroid classes closely match those of some types of meteorites, strong evidence that certain meteorites are asteroid fragments.

The different asteroid compositional classes are concentrated at different distances from the Sun, so that the asteroid belt has a rough zonal structure. Thus, in the inner and central parts of the belt,[5] we find many

[3]Recently I learned of a new astronomical amusement—making sentences using only asteroid names. My favorite (not my own creation, I'm sorry to say) is "Rockefellia Neva Edda McDonalda Hamburga."

[4]The name *Trojan* comes from the tradition of naming asteroids after heroes in the Homeric epics.

[5]"Inner" asteroid belt is the portion nearest Mars.

Figure 13-9 Beyond the Moon, the closest planetary bodies are Apollo asteroids. Small Apollos, a few hundred meters across, may approach close enough to the Earth–Moon system (lower right) to be visited relatively easily by spaceships. Some researchers believe their materials could be economically exploited for use on or near Earth. (Painting by author.)

asteroids with a rock or mixed rock-and-iron composition similar to two important classes of meteorites called stony and stony-iron (see the next section). In some asteroids, such as 1 Ceres, the largest, the surface materials include water-bearing minerals; the water molecules are chemically bound in the mineral grains (Lebofsky, 1981). Also in this region are some asteroids (such as 4 Vesta, the second largest, 549 km across) that did melt—they have surface rock similar to basaltic lavas of Earth and the Moon.

Beyond the middle of the asteroid belt, most asteroids are very black. They are probably composed of soot-like carbon-rich compounds and water-rich silicate minerals. Some are tinted reddish by organic compounds. All these materials (also found in certain meteorites) formed in a region of the solar system colder than Earth's environs. Many such asteroids probably contain ice, hidden in the blackish soil; they are transitions between rocky asteroids and the comet nuclei of the outermost solar system. Chiron, the asteroid that turned into a comet, is a good example.

Asteroids that come near the Earth are a mixture of all these various types, nudged by the gravitational pull of giant Jupiter from their original, distant orbits into new orbits that pass close to the Earth.

Being fragments, many asteroids have irregular shapes. Most may resemble Phobos and Deimos (Figures 10-15 and 10-16). One unusual asteroid, measured by radar in 1989, has a shape like two snowballs stuck together (Ostro and others, 1990). The Galileo spacecraft, on its way to Jupiter, will make the first closeup photos of an asteroid, named Gaspra, in late 1991. This should give important new data about asteroid compositions.

Mines in the Sky?

Because some asteroids probably contain metals and other useful minerals, and because some come very near Earth, asteroids may someday be economically exploited. Studies are already under way to examine the feasibility of flying to Apollo asteroids for scientific and economic exploration, as suggested by Figure 13-9. In terms of energy expenditure, some close asteroids are actually easier to reach and return from than the Moon! Asteroids of pure nickel–iron might be mined in space, brought to space stations for processing, or shaped into crude entry vehicles and landed on Earth in remote areas.

Based on meteorite samples, other Apollo asteroids are believed to contain good ores of economically important platinum-group metals. Masses worth billions of dollars—enough to supply Earth's need for certain metals for decades—might be obtainable (Gaffey and McCord, 1977; Hartmann, 1982; O'Leary, 1983). Other asteroid materials such as rock, ice, and hydrogen might be used for constructing and maintaining space stations. Use of such materials might ultimately relieve some of the environmental pressures on Earth, where we are currently mining ever deeper and releasing pollutants from ever-lower grade ores and fuels, because the easily accessible ores and fuels have already been used up.

Origin of Asteroids

Asteroids are probably planetesimals that never finished accumulating into a planet. Possibly disturbed by nearby Jupiter, the planetesimals in the belt region collided too fast to coalesce into a planet. Instead, they broke into thousands of fragments, as depicted in Figure 13-10. Most smaller asteroids in the main belt are such fragments. In 1983, the IRAS satellite discovered distinct rings of dust circling the Sun in the asteroid belt, each interpreted as debris from a relatively recent, individual asteroid collision. Thus the collisions continue, sporadically, even today.

METEORITES

Meteorites are rocks and chunks of metal that fall from the sky. Modern evidence suggests that they are fragments of Apollo-type asteroids that cross Earth's orbit and eventually collide with Earth.

Meteorites were, for many centuries, a source of awe and superstition. Numerous examples have been recorded of meteorites enshrined as holy objects in the temples of various religions, including Christianity and Islam.

In 1794, the German physicist E. F. F. Chladni studied some of these meteorites and was able to show that they were different from normal terrestrial stones. Even so, the French Academy—the scientific establishment of the day—dismissed as superstition the notion that stones could fall from the sky. As luck would have it, a meteorite exploded over a French town in 1803, pelting the area with stones. An official study by the Academy proved once and for all that these meteorites had really fallen out of the sky and were not ordinary rocks of the Earth.

Interplanetary debris collides with Earth at very high speeds, usually 11 to 60 km/s (24,000 to 134,000

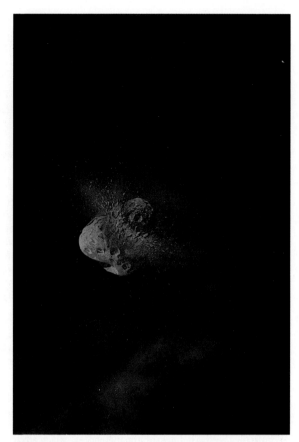

Figure 13-10 Imaginary view of two asteroids colliding. Fragments created by such a collision may ultimately fall on the Earth and planets as meteorites. Because the two asteroids are of different mineralogical types, some of the meteorites produced will be mixtures of different rock types. The larger body is about 30 km across. (Painting by author.)

mph). At such speeds, material is heated by friction with the air. Dust grains and pea-sized pieces burn up before striking the ground. Larger pieces are usually slowed by drag, although at least two grapefruit-scale specimens have punched through house roofs, one bruising a woman occupant! Because they pass through the atmosphere too fast for their interiors to be strongly heated, stories of meteorites remaining red-hot for hours after falling are untrue.

Large meteorites are rare. Only a few brick-sized meteorites are recovered each year. In 1972 an object weighing perhaps 1000 tons just missed Earth, skipping off the outer atmosphere; it was filmed from the ground and detected by Air Force reconnaissance satellites. Objects weighing 10,000 tons, like the Siberian object

TABLE 13·3

Types of Meteorites

Meteorite Type	Percentage of All Falls[a]	Remarks
Stony		
Carbonaceous chondrite	5.7	Most primitive, least altered material available from early solar system.
Chondrite	80.0	Commonest type. Defined by millimeter-scale spherical silicate inclusions, sometimes glassy, called *chondrules*.
Achondrite	7.1	Most nearly like terrestrial rocks. Defined by lack of chondrules, which have been destroyed by a heating process. Some resemblance to terrestrial lunar igneous rocks of basaltic type.
Stony–iron	1.5	Contain stony and metallic sections in contact with each other.
Iron	5.7	Nickel–iron material. Museum specimens are often cut, polished, and etched to show interlocking crystal structure.
Total	100	—

[a]This table is based on meteorites called *falls*—those actually seen to fall. Meteorites found by chance in the soil, called *finds,* are more numerous but less valuable statistically because they are biased toward iron meteorites, which attract attention whenever found in the ground, while stony meteorites eventually weather to resemble ordinary stones.

of 1908, which are large enough to cause nuclear-scale blasts, fall every few centuries. Larger blasts, thousands of years apart, may form craters many kilometers across, such as that in Figure 6-10.

Because meteorites are "free samples" of planetary matter, recovering and reporting a meteorite is a rare honor. Meteorite discoveries should be presented, or at least lent, to research institutions—for example, the Smithsonian Institution in Cambridge, Massachusetts, and Washington, D.C., or the Center for Meteorite Studies, Arizona State University, Tempe.

Origin and Types of Meteorites

After decades of study, scientists have begun to understand how meteorites formed and what they have to tell us about the ancient history of the solar system. The key to the story is that meteorites are fragments of asteroids (and sometimes mixtures of fragments of dis-

similar asteroids), created when asteroids collided and blew apart at various times during the history of the solar system. Their fragments were ejected in various directions, as indicated in Figure 13-10. Many were thrown into orbits that eventually intersected the Earth's. In fact, studies of minerals in several types of meteorites show that they formed inside parent bodies with diameters of tens to hundreds of kilometers—fitting the theory that they formed inside asteroids. The different types of meteorites thus tell us about conditions inside asteroids' parent bodies. See Table 13.3.

Meteorites are among the most complex rocks studied by geologists. Scientists have subdivided them into different rock types, and sometimes they appear as **brecciated meteorites**—meteorites made of mixed fragments of different meteorite types all jumbled into one rock! There are three broad classes: stony, iron, and stony-iron meteorites. **Stony meteorites** are by far the most common, comprising 94% of all meteorites that fall out of space. Stony meteorites are rocks from the surfaces, crusts, and mantles of asteroids. Some are lavalike rocks that have been melted, but others are very primitive rocks that have never been melted since their parent asteroid formed.

An important subclass of stony meteorites is the **carbonaceous meteorites,** which comprise about 5% of all meteorites. These are black, weak, carbon-rich stones. The mixture of oxygen isotopes in the oxygen of their minerals is very different from that in terrestrial rocks, lunar rocks, or other meteorites, showing that they formed far from the terrestrial planets or asteroid belt. Their composition is close to that measured by the Giotto probe in the dust of Halley's comet. Thus, they probably represent the kinds of black, carbon-rich minerals formed in the outer solar system and now found among comets and Trojan asteroids.

Iron meteorites are pure metal objects, composed of nickel-containing iron alloys. They seem to be fragments of the metal cores of asteroids. Like Earth, many asteroids apparently melted and formed iron-rich cores surrounded by rocky mantles and crusts. The rare **stony-iron meteorites,** only about 1% of all meteorites, are intricate mixes of iron alloy and rock, apparently representing the interface between cores and mantles of their parent asteroids.

Ages and Origins of Meteorites

One of the most important aspects of meteorites is that they can be dated by the techniques discussed in Chapter 6, and this gives us a history of events in the early solar system. Dating by various techniques shows that the most primitive meteorites formed during a "brief" interval of only 20 million years, 4.6 billion years ago. "Formed" means that they made the transition from dispersed dust grain and gas to solid, rocklike objects. The melting that formed the iron cores also happened at this time, although the heat source is uncertain. This interval, then, marked the birth of planetary material. Other types of dating show that many major collisions happened at this time, 4.6 billion years ago, and sporadic additional collisions smashed some meteorites' parent bodies at other times scattered through solar system history.

Several lines of evidence help prove that most meteorites originated as fragments of shattered asteroids. First is the rough match between spectra of meteorites and spectra of surviving asteroids. A second example is the study of asteroid minerals, which show by their structure that they formed inside bodies ranging up to a few hundred kilometers in diameter; this matches the sizes of observed asteroids in the asteroid belt. Third is the orbital analysis that demonstrates how Jupiter's gravity could disturb certain asteroids' orbits, kicking them out of the belt and onto orbits that could collide with Earth, thus explaining how asteroid fragments could arrive here.

The meteorites' evidence of early melting among the asteroid parent bodies is one good example of how important meteorites are in reconstructing the early histories of planetary bodies and the solar system.

ZODIACAL LIGHT

The smallest interplanetary particles are microscopic dust grains and individual molecules and atoms spread out along the plane of the solar system and concentrated toward the Sun. If you look west in a very clear rural sky as the last glow of evening twilight disappears (or look east before sunrise), you can detect the diffuse glow of sunlight reflecting off the cloud of these particles. It appears as the **zodiacal light**—a faint glowing band of light extending up from the horizon and along the ecliptic plane, shown in Figure 13-11. It is brightest at the horizon. Measurements made by astronauts indicate that it merges with the bright glow of the Sun's atmosphere.

The zodiacal light can be thought of as the visible effect of countless dust grains—once distributed in trails along comets' orbits but eventually spread out into a uniform disk-shaped cloud of dust around the Sun.

SIBERIA REVISITED: ASTEROID OR COMET IMPACT?

The mysterious Siberian explosion of 1908, described at the beginning of this chapter, is now easier for us to understand. Interplanetary space contains debris of many sizes. Rocky and icy chunks up to a few kilometers in size cross Earth's orbit and are likely to collide with it from time to time. One of these objects apparently struck the atmosphere over Siberia on June 30, 1908.

Other dramatic events in Russia at this time kept Russian scientists from visiting the site until 1927. During the 1927 expedition and later expeditions, scientists

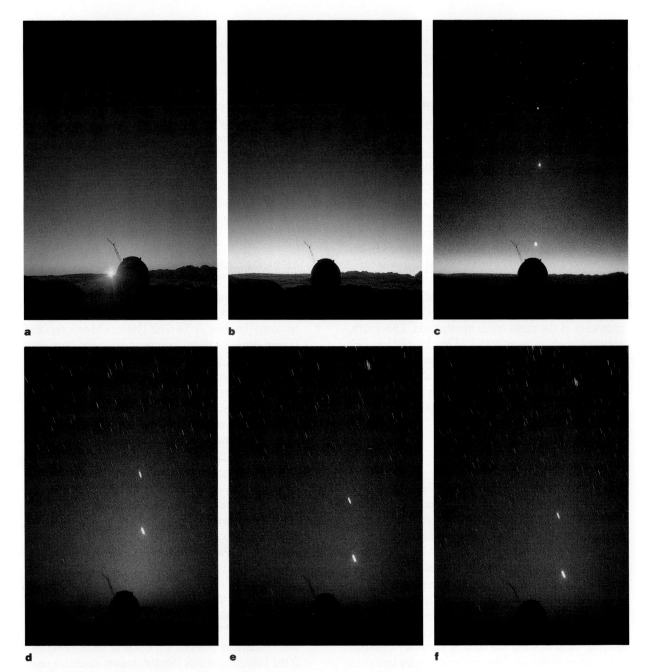

Figure 13-11 Sunset and the emergence of the zodiacal light. **a** Sunset. **b** Blue color still dominates the sky $\frac{1}{2}$ h later and obliterates the faint zodiacal light. The crescent moon, Venus, and Jupiter, in order up from the horizon, are emerging and define the position of the zodiac. **c** The zodiacal light begins to be visible along the zodiac 1 h after sunset. **d** The zodiacal light reaches its best visibility about $1\frac{1}{2}$ h after sunset, appearing as a diffuse band of light rising nearly vertically from the horizon and running near Venus and Jupiter. **e** and **f** The zodiacal light slowly sets 2 h and $2\frac{1}{2}$ h after sunset, respectively. Features in these photos can be seen only away from cities under dark, clear skies. (Exposures *a–b* on Kodachrome, *c–f* on 3M ASA 1000 film; 28-mm wide-angle lens with nearly 90° vertical field of view; exposures *c–f* 5 min at f 2.8. (Photos by author from Mauna Kea Observatory.)

Figure 13-12 Devastation caused by the great Tunguska meteorite fall of 1908. This view, more than a decade later, shows trees some miles from "ground zero" blown over by the force of the explosion. (Photo from E. L. Krinov, 1966.)

found that, surprisingly, the object did not reach the ground and form a crater. It apparently exploded in the atmosphere. Trees at "ground zero" were still standing but had their branches stripped by the blast forces in a downward direction. Trees had been knocked over by the blast out to 30 km from ground zero, as shown in Figure 13-12. A forest fire was started and trees were scorched by the blast out to about 14 km. Researchers found carbonaceous dust but no meteorites.

Most investigators believe the object was composed of weak material that disintegrated in the atmosphere. It may have been a small asteroid resembling a stony meteorite some 90 to 200 m across (Sekanina, 1983) or a similar-sized icy comet fragment. It injected much dust high into the stratosphere. On June 30 and July 1, sunlight shining over the North Pole illuminated this dust during the night, and newspapers could be read at midnight in western Siberia and Europe.

Such was the effect of a relatively small bit of interplanetary debris striking Earth. To put this event in perspective, if the same object had exploded over New York City, the scorched area would have reached nearly to Newark, New Jersey. Trees would have been felled beyond Newark and over a third of Long Island. The man knocked off his porch could have been in suburban Philadelphia. "Deafening bangs" might have been heard in Pittsburgh, Washington, D.C., and Montreal.

New reconnaissance satellites may provide early warning against such objects, which may hit Earth every century or so, allowing us to take action or even intercept the objects to use their materials for more constructive purposes.

An interesting sidelight on large ancient impacts is that the distortions of geological strata on Earth have been economically valuable in some cases. Rich iron ore deposits at Sudbury, Ontario, occur in a probable impact crater more than 100 km across and about 1.8 billion years old; the impact exposed native iron-bearing strata. Several oil deposits in North America and the Soviet Union occur in sediments collected in ancient craters.

SUMMARY

All of the solar system's small bodies—comets, meteors, asteroids, and meteorites—are examples of planetesimals or their fragments. Planetesimals are preplanetary bodies that formed in the solar system 4.6 billion years ago, during an interval of less than 100 million years.

Most asteroids are now in a belt between Mars and Jupiter; the planetesimals in other regions interacted with planets and either were ejected from the solar system after near misses or crashed onto planet and satellite surfaces, making craters. Asteroids in the inner belt are dominated

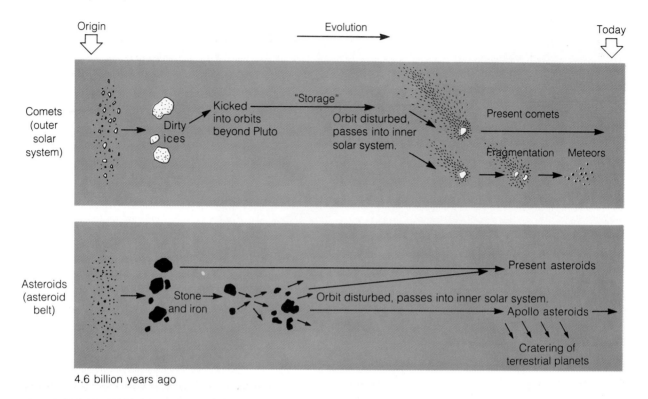

Figure 13-13 Schematic histories of comets and asteroids. Initial aggregation into multikilometer bodies (left) is followed by a subsequent history of orbit evolution. Disturbances by gravitational forces change the orbits. Asteroids in the asteroid belt undergo collisions and fragmentation.

by metal-rich rock materials; asteroids from the outer belt to Jupiter have carbon- and water-rich minerals. Planetesimals formed farther out were ice rich and evolved to what we know as comets. Many of them were thrown out of the planetary region into the Oort cloud, on the outskirts of the solar system. The relation between the various asteroids, comets, meteors, meteorites, and craters that we see today is summarized in Figure 13-13.

The most detailed information about these bodies comes from meteorites, whose chemistry reveals that they came from asteroidlike parent bodies a few hundred kilometers across. Many of these were melted, causing minerals to separate into iron and silicate phases. Others survive with only minimal heating, providing samples nearly unaltered since the solar system's earliest days.

Thus the small bodies of the solar system provide some of the best clues about the origin of planets, both in our solar system and perhaps near other stars. Recent studies also indicate that they may be exploited as a source of raw materials for use on Earth and for space exploration.

CONCEPTS

comet	fireball
asteroid	dust trails
meteor	asteroid belt
meteorite	Apollo asteroids
parent body	Trojan asteroids
planetesimal	Lagrangian points
comet head	brecciated meteorite
comet tail	stony meteorite
comet nucleus	carbonaceous meteorite
sublimation	iron meteorite
solar wind	stony–iron meteorite
dirty iceberg model	zodiacal light
Oort cloud	
meteor shower	

PROBLEMS

1. If a comet should happen to pass through Saturn's satellite system, why would it probably not be detected from Earth?

2. Kepler's third law states that $a^3 = P^2$, where a is the semimajor axis of a body orbiting the Sun (expressed in astronomical units) and P is the period (expressed in years). Should this result apply to comets? If a typical comet in Oort's cloud has a semimajor axis of about 100,000 AU (10^5 AU), how often would it return to the inner solar system?

3. In terms of measuring and reporting useful scientific information, what actions would be appropriate if you observed an extraordinarily bright meteor or fireball? In two columns, list examples of useful and nonuseful descriptions of the fireball's speed, brightness, and apparent size. What actions would be appropriate if you saw a meteorite strike the ground?

4. Summarize relations between the bodies and particles responsible for meteors, comets, the zodiacal light, asteroids, and meteorites.

5. Suppose future astronauts could match orbits with a comet and reach its nucleus. Describe the possible surface appearance of a comet. Consider gravity, surface materials, sky appearance, and so on. Which would be easier to match orbits with: a long-period or short-period comet?

6. Typical interplanetary material may move at about 15 km/s relative to the Earth–Moon system.
   a. If a kilometer-scale asteroid was discovered on a collision course with Earth when it was 15 million kilometers away, how much warning time would we have?
   b. What would be the potential dangers?
   c. If a much smaller asteroid was similarly discovered at the distance of the Moon, how much warning time would we have?
   d. How long would the objects take to pass through the 100-km thickness of the atmosphere?

7. Based on everyday experience, what is the danger of an event such as that in Problem 6 compared with the danger of other natural disasters, such as earthquakes? Is a large-scale meteorite disaster a plausible source of myths during the 10,000 y history of humanity? Defend your answer.

PROJECTS

1. Use a large piece of cardboard to make a model of the inner solar system out to the orbit of Jupiter. Assume that the planets travel approximately in the plane of the cardboard. Use orbital properties listed in Table 13-1 to cut out scale models of the orbits of various interplanetary bodies. (A slit through the cardboard could be used to show how comet or asteroid orbits penetrate through the ecliptic plane. Note that the Sun must always occupy one focus of each orbit.)
   a. Show how the geometry of passage of a comet (or other body) through the ecliptic plane, especially for highly eccentric orbits, depends on the angle of the orbit between the point nearest to the Sun in the object's orbit and the ecliptic plane. (This angle is fixed for each body but is omitted from Table 13-1 for simplicity.)
   b. Show how the prominence of a given comet may depend strongly on where the Earth is in its orbit as the comet passes through the inner solar system.

2. Visit Meteor Crater, Arizona. Why is this feature misnamed? Observe the blocks of ejecta and deformation of rock strata, as explained in museum signs and tapes. What would prehistoric observers, if any (estimated impact date was 20,000 y ago), have witnessed at various distances from the blast that formed this crater nearly a mile across? (Other impact sites are known in various states, but they are eroded or undeveloped.)

3. Examine meteorite specimens in a local museum. Compare the appearance of stones and irons. In iron samples that have been cut, etched, and polished, look for the crystal patterns that give information about cooling rate and environment when the meteorite formed inside its parent body.

CHAPTER 14

The Origin of the Solar System

The planets, satellites, meteoritic material, and Sun did not exist before about 4.6 billion years ago. Samples from the Earth, Moon, and meteorites suggest that the solar system formed from preexisting material within an interval of about 20 to 100 million years. Before that time, the atoms in the Earth, in this book, and in your own body were floating in clouds of thin gas in interstellar space. It is remarkable to realize that this interstellar material somehow aggregated not only into a dazzling star but also into a family of surrounding planets. How did the Sun and its planetary system form?[1]

FACTS TO BE EXPLAINED BY A THEORY OF ORIGIN

Nobel laureate Hannes Alfvén, who has spent years researching the solar system's origin, once said: "To trace the origin of the solar system is archaeology, not physics." He meant that our ignorance of the initial conditions forces us to work backward through time, reasoning from whatever clues we can find. The most important clues are *facts about the solar system that have no obvious explanation from present-day conditions* but must have arisen from initial conditions as the solar system formed. Table 14-1 lists some of these clues. In this chapter, we will account for them one by one. This process leads to an interesting realization. As we sift through clues found in meteorites, lunar rocks, and orbits in our solar system, we will learn about the formation of not only our own system but also the stars

---

[1]Note that this question is not addressing the origin of the whole universe, which probably occurred around 14 billion years ago, or the origin of our galaxy, which probably occurred around 12–14 billion years ago. These topics will be taken up in later chapters.

TABLE 14-1

Solar System Characteristics to Be Explained by a Theory of Origin

1. All the planets' orbits lie roughly in a single plane.
2. The Sun's rotational equator lies nearly in this plane.
3. Planetary orbits are nearly circular.
4. All planets' revolutions, and the Sun's rotation, are in the same west-to-east direction, called prograde (or direct).
5. Planets differ in composition.
6. The composition of planets varies roughly with distance from the Sun: Dense, metal-rich planets lie in the inner system, whereas giant, hydrogen-rich planets lie in the outer system.
7. Meteorites differ in chemical and geological properties from all known planetary and lunar rocks.
8. The Sun and all the planets except Venus and Uranus rotate on their axis in the same direction (prograde rotation) as well. Obliquity (tilt between equatorial and orbital planes) is generally small.
9. Planets and most asteroids rotate with rather similar periods, about 5 to 23 h, unless an obvious gravitational tidal force from a satellite or from the Sun slows them.
10. Distances between planets usually obey Bode's rule.
11. Planet–satellite systems resemble the solar system.
12. As a group, comets' orbits define a large, almost spherical cloud around the solar system.

themselves. Thus this chapter makes a link to the ensuing chapters.

THE PROTOSUN

Before the Sun existed, its material must have been distributed in interstellar space in a large cloud like the interstellar clouds we see today. The Sun began to form a little more than 4.6 billion years ago when this cloud (or part of it) became so dense that its own inward-pulling gravitational forces became stronger than the outward force of the pressure generated by the motions of its atoms and molecules. The cloud thus began to shrink. The term **protosun** is used for the early, contracting Sun when it was larger than its current size.

Early Contraction and Flattening

Even if gas in the cloud had been randomly circulating initially, the contracting cloud would have developed a net rotation in one direction or another. To show this, an experiment can be done with a cup of coffee or a pan of water. If you stir the liquid vigorously but as randomly as you can and then wait a moment and put a drop of cream in it, the cream will usually reveal a smooth rotation in one direction or the other, because the sum of the random motions will usually be a small net angular motion in one direction or the other.

The cloud would have rotated faster as its mass contracted toward its center, just as a figure skater spins faster when she pulls in her arms. Mathematical studies show that centrifugal effects associated with the faster spin caused the outer parts of the cloud to flatten into a disk. Material in the center contracted fastest, forming the Sun, as shown in Figure 14-1. The planets formed in this disk, accounting for fact 1 in Table 14-1: *the single plane of all planets' orbits.* Because the Sun itself was an integral part of this disk, fact 2, *the Sun's equator lying in the same plane,* is also explained.

At first, when the cloud was large, its atoms were far apart and could have fallen more or less freely toward the center of gravity in the middle of the cloud. This state is called **free-fall contraction.** If free-fall persisted, the cloud could have completely contracted into the Sun in only a few thousand years. But eventually, because of their growing concentration and random motions, the atoms in the cloud began to collide with each other. Atoms interacting in a gas cause outward *pressure.* A buildup of pressure in the cloud would have slowed the collapse. Thus the inward force of gravity

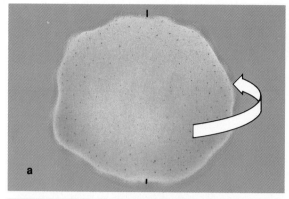

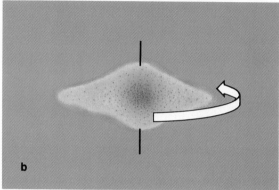

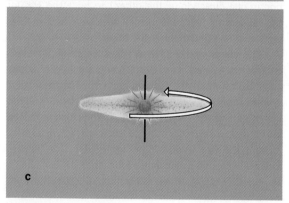

Figure 14-1 Three stages in the evolution of the protosun. **a** A slowly rotating interstellar gas cloud begins to contract by rapid, nearly free-fall contraction because of its own gravity. **b** A central condensation forms and the cloud rotates faster and flattens. **c** The Sun forms by slow, Helmholtz contraction in the cloud center, surrounded by a rotating disk of gas.

and the outward force of gas pressure competed. Because the properties of gases are well understood, astrophysicists can analyze the later contraction and evolution of the cloud under these competing forces.

Helmholtz Contraction

Gravitational contraction in which the shrinkage is slowed by outward pressure is called **Helmholtz contraction** (after Hermann von Helmholtz, the German astrophysicist who first studied it). In 1871, Helmholtz showed how contraction would have caused heat to accumulate in the protosolar cloud. As Helmholtz noted (quoted in Shapley and Howarth, 1929):

If a weight falls from a height and strikes the ground, its mass loses . . . the visible motion which it had as a whole—in fact, however, this is not lost; it is transferred to the smallest elementary particles of the mass, and this invisible vibration of the molecules is [what we call] heat.

In the contracting protosun, atoms or swarms of atoms would have fallen toward the center until they collided with other parts of the gas cloud. Temperature would have increased inside the cloud. Wien's law and other physical laws guarantee that the protosun would have radiated energy and warmed the surrounding cloud.

Calculations based on the Helmholtz theory indicate the conditions: Eventually the cloud's central temperature rose to 10 million Kelvin or more, starting the nuclear reactions that made it a star and not just a ball of inert gas. Meanwhile, and more important for the formation of planets, the outer parts of the cloud, shown in Figure 14-1, formed a disk of gas as big as the solar system, at temperatures of a few thousand Kelvin. *These theorized steps have been confirmed by actual observations of newly formed stars surrounded by clouds of gas and dust.*

THE SOLAR NEBULA

A cloud of gas and dust in space is called a **nebula** (plural: *nebulae*) from the Latin term for mist. The disk-shaped nebula that surrounded the contracting Sun is called the **solar nebula.** Molecules of gas or grains of dust must have moved in circular orbits, because noncircular orbits would have crossed the paths of other particles, leading to collisions that would have damped out the noncircular motions—in the same way that noncircular eddies get damped, or canceled, in the cup of coffee mentioned earlier. Thus, neglecting small-scale eddies in the gas, broad-scale motions in the cloud were in parallel circular orbits, accounting for facts 3 and 4

TABLE 14-2

Condensation Sequence in the Solar Nebula

Approximate Temperature (K)	Elements Condensing	Form of Condensate (with Examples)	Comments
2000	None		Gaseous nebula
1600	Al, Ti, Ca	Oxide (Al_2O_3, CaO) grains	
1400	Fe, Ni	Nickel–iron grains	Parent material of planetary cores, iron meteorites?
1300	Si	Silicate and ferrosilicate grains [enstatite, $MgSiO_3$; pyroxene, $CaMgSi_2O_6$; olivine, $(Mg, Fe)_2 SiO_4$]	First stony material, combined to form meteorites; some still preserved in primitive meteorites
100–300	H, N, C	Ice particles (water, H_2O; ammonia, NH_3; methane, CH_4)	Large amounts of ice; still preserved in outer planets and comets

Sources: Data from Lewis (1974), Grossman (1975), and others.

in Table 14-1. As the nebula stabilized, its gas began to cool.

Condensation of Dust in the Solar Nebula

The solar nebula was initially gas that Helmholtz contraction heated to at least 2000 K. At such a temperature, virtually all elements were in gaseous form. As is the case with other cosmic gas, most solar nebula atoms were hydrogen, but a few percent were heavier atoms such as silicon, iron, and other planet-forming material.

How did solid particles form in this gas? The answer can be seen on Earth. When air masses cool, their condensable constituents form particles: snowflakes, raindrops, hailstones, or the ice crystals in cirrus clouds. Similarly, as the solar nebula cooled, condensable constituents formed tiny solid particles of dust. Various mineral compounds appeared in a sequence known as the **condensation sequence.**

Chemical studies show that as the temperature in any part of the nebula dropped toward 1600 K, certain metallic elements such as aluminum and titanium condensed to form metallic oxides in the form of **grains,** or microscopic solid particles, as shown in Table 14-2. At about 1400 K a more important constituent, iron, condensed. Microscopic bits of nickel–iron alloy formed as grains or perhaps coated existing grains. Still more important, at about 1300 K abundant silicates began to appear in solid form. For instance, a magnesium silicate

mineral, enstatite ($MgSiO_3$), formed at about 1200 K (Lewis, 1974). These silicate minerals are the common rock-forming materials, so the solar nebula at this point was acquiring a large quantity of fine dust with rocky composition.

Complex mixtures of magnesium-, calcium-, and iron-rich silicates condensed, depending on the temperature, pressure, and composition of the gas at various points in the solar nebula. Local conditions were determined by the distance from the newly formed Sun. Therefore, different compositions of mineral particles may have dominated at different locations, explaining facts 5 and 6 in Table 14-1. Similarly, the outer nebula must have been cooler than the inner nebula, being further from the Sun. In about the region of the middle asteroid belt and beyond, at about 300 K, water was trapped in many minerals. Beyond the outermost asteroid belt, snowflakes of H_2O ice condensed. At 100 to 200 K, in the outermost nebula, ammonia and methane ices also condensed. In the outer solar system, these ices survive even the direct, weak light of the distant Sun; various kinds of ice are found today on comets and icy satellites of the giant planets.

Meteorites as Evidence

The compositions of meteorites strongly support this theory of condensation. In particular, black carbonaceous meteorites contain inclusions of material believed to be among the earliest solid particles in the solar

Figure 14-2 A piece of the Allende carbonaceous meteorite, showing white inclusions. These demonstrate how material that formed in one environment was later trapped in other material. The black matrix is composed of microscopic dust grains condensed at a few hundred Kelvin. The white inclusions contain aluminum-rich minerals condensed at high temperatures. As recognized in the 1970s, the inclusions shed light on the earliest formative conditions in the solar system. (Courtesy R. S. Clarke, Smithsonian Institution.)

system. These inclusions, shown in Figure 14-2, are rich in elements that would have condensed first (at the highest temperatures), such as osmium and tungsten. Their minerals formed at temperatures of about 1450–1840 K, consistent with fact 7 in Table 14-1. Yet the bulk of carbonaceous meteorites contain microscopic grains formed at low temperatures. Consistent with this, water is a common constituent in minerals in parts of carbonaceous meteorites.

A PRESOLAR EXPLOSION?

As meteorite compositions were being studied in the 1970s, an interesting mystery emerged. The light-colored inclusions of high-temperature minerals in certain carbonaceous meteorites (Figure 14-2) were found to have queer abundances of certain isotopes. Most elements occur in one main, stable isotopic form and several other forms, both unstable (radioactive) and stable. In the light-colored inclusions, researchers found certain isotopes that could only have been created very shortly before the solar system itself.

For instance, they found xenon-129, a form of xenon that arises from the decay of radioactive iodine-129. This decay process is very fast, geologically speaking, once the iodine is created—iodine-129's half-life is only 17 million years. Therefore, the parent iodine must have

been trapped in the meteorite inclusions in the brief interval (1 to 20 million years?) between the creation of the iodine and its decay into xenon. Remember that the inclusions were among the first minerals condensed in the solar nebula, and they are the main ones that got a dose of the mysterious iodine.

The mystery is how a batch of radioactive iodine and other isotopes was created just *before* the first planetary material formed and how it got trapped in that material. The types of isotopes involved are now believed to be created by nuclear reactions inside certain massive stars, which burn their nuclear fuel quickly and then explode. Thus the evidence indicates that such a star exploded near the presolar nebula spewing short-lived radioactive isotopes and other debris into the cloud that was becoming the solar nebula. Indeed, the blast from the explosion probably helped compress the presolar cloud, helping to initiate the collapse that produced the Sun!

FROM PLANETESIMALS TO PLANETS

Although the preplanetary particles may have formed as microscopic grains, they clearly grew larger. (Otherwise there would be no planets!) The hypothetical intermediate bodies, from millimeters to many kilometers in size, are usually called **planetesimals,** as noted in Chapter 13. Evidence that they existed includes the following:

1. Craters on planets and satellites indicate impacts of planetesimals with diameters of at least 100 km (Figure 14-3).

2. Meteorites are their surviving fragments, and their microstructures reveal how the dust grains clumped together.

3. Asteroids and comet nuclei reaching more than 100 km in diameter survive today throughout the solar system.

But exactly how did microscopic grains aggregate to produce 100-km-sized planetesimals? If they had circled the Sun in paths comparable to present asteroid and comet orbits, they would have collided with one another at speeds much faster than rifle bullets. They would have shattered (as seen in Figure 13-10) and the solar system would still be a nebula of dust and grit.

But did dust particles have such high speeds in the early solar nebula? According to dynamical analyses, they collected in a swarm of particles with nearly parallel, circular orbits in the central plane of the disk. Because the orbits were nearly parallel, planetesimals approached each other gently and their collision velocities were low. In low-velocity collisions, some dust grains simply stuck together, held by weak adhesive forces such as gravity and electrostatic attraction. Mutual gravity probably caused groups of planetesimals to clump together, growing to many kilometers in size during only a few thousand years—a mere moment in cosmic time. At this point, the solar nebula would have resembled the scene in Figure 14-4. Rocky and icy planetesimals, such as shown in Figure 14-5, orbited in dusty clouds that dimmed the central Sun.

Dynamical studies indicate that coalescing particles tend to form bodies rotating in a prograde motion (fact 8, Table 14-1) with similar rotation periods (fact 9). No one is sure why the planetary spacings are regular (fact 10), but studies suggest that gravitational forces tended to divide the solar nebula into ring-shaped zones, each favoring the formation of one planet.

The larger planetesimals collided with smaller ones, knocking off debris. Some small particles fell back on their surfaces, forming a powdery soil layer that, like the lunar regolith, was effective in trapping other small fragments. Thus the largest planetesimals grew fastest,

sweeping up the others. These bodies may have grown to 100-km size in a few million years. Some planetesimals, the *parent bodies* of meteorites, were heated, melted, and differentiated into metal and rock portions. Some were shattered by collisions with other large or fast neighbors, freeing iron, stony–iron, and rocky meteorites from their interiors, as illustrated in Figure 13-10. But as some shattered, others grew to replace them.

Terrestrial Planets

In the inner part of the solar system, these collisions between planetesimals simply continued until Mercury-sized to Earth-sized planets were formed. Most of the planetesimals in this region were composed of silicate rocky material, familiar to us from the rocks of Earth and the Moon. Thus all the terrestrial planets grew from this material until most of it was swept up. Any remaining gas and fine dust was blown away by outrushing gas and radiation from the newly-heated Sun.

Giant Planets

The giant planets started out forming in the same way as the terrestrial planets, accreting planetesimals. More material was apparently available in the giant planet zone, however, perhaps because ices as well as rocky material had condensed in this cold region, augmenting local planetesimal masses. Thus the "embryo planets" that were to become Jupiter, Saturn, Uranus, and Neptune grew to Earth's size and beyond.

By the time they reached about 15 times the mass of present-day Earth, an interesting thing happened. At 15 Earth masses, they had such strong gravity that they began to pull in gas from the surrounding solar nebula. Thus they accreted not only planetesimals to make a solid/liquid planet, but they also accreted massive atmospheres of gas whose composition approximately equaled that of the nebular gas. Hence the giant planets can be thought of as two-phase planets. Their cores are "giant terrestrial planets" averaging around 15 times more massive than Earth; and these cores are surrounded by giant atmospheres. The terrestrial planets never accumulated these giant atmospheres of hydrogen-rich nebular gas because they never attained enough mass to pull in the nebular gas.

To support this scenario, we can compare the compositions of the terrestrial planets with those of the

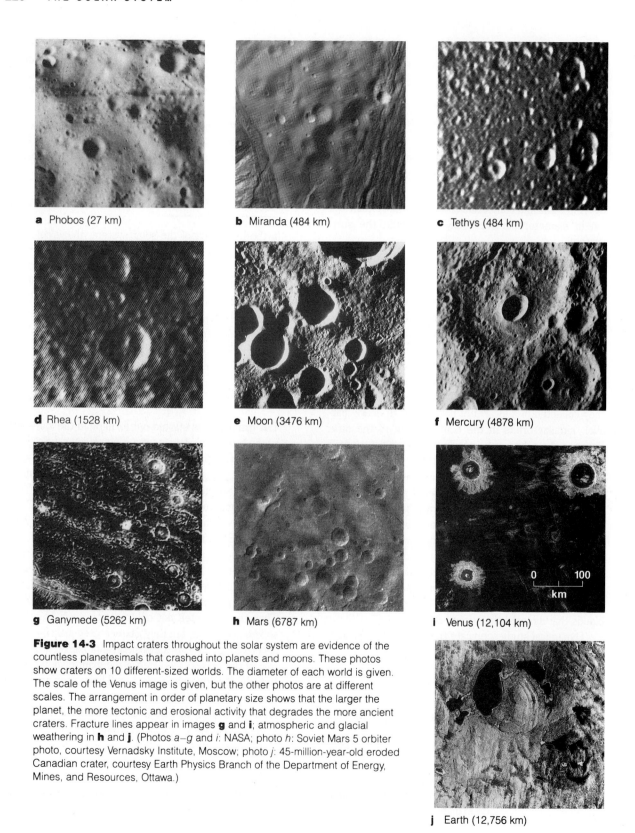

a Phobos (27 km)

b Miranda (484 km)

c Tethys (484 km)

d Rhea (1528 km)

e Moon (3476 km)

f Mercury (4878 km)

g Ganymede (5262 km)

h Mars (6787 km)

i Venus (12,104 km)

j Earth (12,756 km)

Figure 14-3 Impact craters throughout the solar system are evidence of the countless planetesimals that crashed into planets and moons. These photos show craters on 10 different-sized worlds. The diameter of each world is given. The scale of the Venus image is given, but the other photos are at different scales. The arrangement in order of planetary size shows that the larger the planet, the more tectonic and erosional activity that degrades the more ancient craters. Fracture lines appear in images **g** and **i**; atmospheric and glacial weathering in **h** and **j**. (Photos *a–g* and *i*: NASA; photo *h*: Soviet Mars 5 orbiter photo, courtesy Vernadsky Institute, Moscow; photo *j*: 45-million-year-old eroded Canadian crater, courtesy Earth Physics Branch of the Department of Energy, Mines, and Resources, Ottawa.)

Figure 14-4 A scene in the early solar nebula. Planetesimals of rocky and icy materials orbit in the foreground. The Sun is partially obscured and reddened by dust and gas in the inner nebula. (Painting by author.)

Figure 14-5 Artist's conception of a planetesimal in the primeval solar system. It has grown by aggregation of innumerable smaller grains and fragments, has been fractured and cratered by larger collisions, and has been rounded by sandblasting effects of smaller collisions. (Painting by Michael Carroll.)

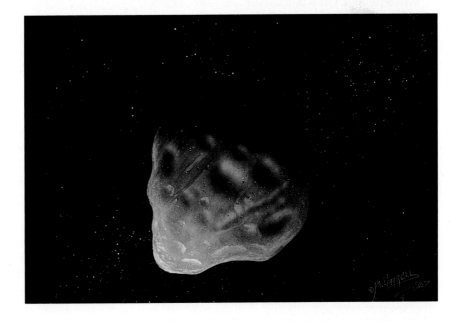

giants and with that of the nebular gas. Although we don't have samples of the nebular gas, we *can* measure the Sun's present-day composition, which is believed to be about the same as that of the nebular gas from which the Sun formed. Table 14-3 makes this comparison. Here we see that the gases from which the Sun and nebula formed were almost three-fourths hydrogen and about one-fourth helium. In the giant planets we see a

similar situation. Recall that the terrestrial planets' atmospheres are mostly carbon dioxide degassed from inside the planet; they have virtually no trace of massive hydrogen–helium concentrations. Given that the uncertainties in the Table 14-3 data are a few percent, we see that the hydrogen–helium concentrations of the three measured giant planets are remarkably like the composition of the solar nebula. These measurements sup-

TABLE 14·3

Comparison of Solar Nebula and Giant Planet Atmospheres

| | Composition Percentages by Mass | | | |
Gas	Solar Nebula (Sun's Present Composition)	Jupiter	Saturn	Uranus
H₂ (hydrogen)	71	79	88	76
He (helium)	27	19	11	23
Total	98	98	99	99

Source: Grevesse (1988); Anders and Ebihara (1982, solar); Voyager reports (Stone and others, 1979, 1982, 1986); Hubbard and Stevenson (1986).

Note: Uncertainties are a few percent. Results are basically similar; scientists debate significance of lower helium content for Saturn.

port the theory of giant planet formation with nebular atmospheres added to the planets by their gravity.

As the nebular gas was attracted into each of the four giant planets, it formed a miniature "solar nebula" around each giant. And in this disk-shaped cloud of gas and dust, the accretionary growth process was repeated all over again. That is, each giant planet became an analog of the Sun, and moons grew around it analogous to planets. In these miniature planetary systems, the most abundant building materials were ice and the black carbonaceous dirt common to the outer solar system. Thus giant planets spawned systems of dirty-ice satellites in prograde, circular orbits, all lying close to the planet's equatorial plane, accounting for fact 11 in Table 14-1.

Asteroids

The asteroids are easy to understand in this scenario. They are planetesimals that never made it all the way to planethood. There is probably a specific reason why an asteroid belt was left stranded between Mars and Jupiter, where a planet normally should have grown. Ceres, the largest asteroid, had already grown to 1000-km diameter by the time the growth process stopped in that zone. Probably the reason it stopped was that Jupiter had grown so huge that its gravity disturbed the motions of the asteroids in the zone we now know as the asteroid belt. This disturbance increased the collision velocities of the asteroids, causing them to smash

into innumerable fragments during each collision instead of coalescing into an even-larger body.

Comets

Our general picture of planet formation also allows us to explain comets. There must have been asteroidlike planetesimals that formed from dirty ice in the present region of the giant planets. Most of them were consumed in growing the rocky-icy cores of the giant planets, but millions more were left in interplanetary space. The gravitational forces of the giant planets were so strong that whenever an icy planetesimal experienced a near-miss with a giant planet, the planetesimal was thrown off into a new orbit. Such orbits typically took the icy planetesimals out to distances as far as thousands of astronomical units from the Sun. That is, the icy planetesimals were thrown into the Oort cloud! Thus we can account for fact 12 in Table 14-1. The Oort cloud of comet nuclei is the storehouse for icy planetesimals that were ejected from the outer solar system during the final stages of giant planet growth.

What Became of the Remaining Planetesimals and the Solar Nebula?

Most of the planetesimals were eventually accumulated by the planets, scattered into the Oort cloud, or left stranded in the asteroid belt, probably within 100 million years of the solar system's beginning. Each time a planetesimal crashed into a planet to add to the planet's mass, a crater was formed. For another 500 million years, as lunar rocks have taught us, the last remaining interplanetary planetesimals rained down onto planetary surfaces, creating still more craters on all planets and satellites, as shown in Figure 14-3. A few planetesimals were captured into orbits around the newly formed planets, explaining the dozen or so satellites with irregular orbits that seem to have been captured instead of growing in a nebular disk around the planet.

GRADUAL EVOLUTION— PLUS A FEW CATASTROPHES

We have stressed that this growth of planets by innumerable small planetesimals produced certain regularities of the solar system: planetary orbits that lie in the plane of the Sun's equator, the regular Bode law spacings of the planets' orbits, prograde orbital revolutions, prograde rotations, the mostly small obliquities of planets, and the regular systems of prograde satellites lying in equatorial planes of their planets' equators. These regularities of the solar system require smooth evolutionary growth from a system of many small planetesimals, not a catastrophic creation in some chaotic system.

Nonetheless, one of the beauties of the modern picture of planet formation is that it requires a few catastrophic events to explain the nonregularities of the solar system (Wetherill, 1985). These catastrophic events would involve the collisions or near-misses of the planets with the largest of their nearby planetesimal neighbors as the planetesimals were being swept up. As each planet grew, it experienced collisions of different sizes. If a collision was large enough, it could have had important effects.

We have already seen one example of this type of thinking: the theory that a Mars-sized planetesimal hit Earth late in its growth and blew off material from which the Moon formed. Another example is that Uranus was probably hit by a relatively large planetesimal, tipping its rotation axis to lie nearly *in* the plane of the solar system instead of nearly perpendicular to it, as is the case for most other planets. Other properties of the solar system—such as the different styles of ring systems and the geological differences between hemispheres of some planets, such as Mars—may trace back to large impacts. The largest impacts were not large enough to randomize the characteristics of planets and their orbits, but they were large enough to give planets individuality of character!

THE CHEMICAL COMPOSITIONS OF PLANETS

Why do the planets vary in chemical composition? Why do meteorites differ from lunar and terrestrial materials? The answers to these questions now become quite clear. The condensation sequence shows that different groups of minerals existed at different *temperatures* in the nebula. In the cold outer parts of the nebula, hydrogen-rich ices formed, but in the hot inner parts, only metals and silicates. Massive bodies such as the Sun and Jupiter retained their original gases; studies of these bodies show that the nebula was originally mostly hydrogen and helium with only a small fraction of heavier atoms such as silicon, oxygen, and iron.

Virtually none of the hydrogen or helium was trapped in the solid materials of the inner solar system, however. Unmelted meteorites show that primitive silicates were composed mostly of oxygen, iron, silicon, and magnesium. When these materials melted, probably from the heat of radioactivity in certain minerals, the iron drained to form central cores resembling iron meteorites. This differentiation process meant that the surface materials of planets became more iron-poor and silicon-rich than the primitive materials. This is why rocks on the surface of Earth are made mostly out of silicon-oxide-related minerals.

STELLAR EVIDENCE FOR OTHER PLANETARY SYSTEMS

Because no unusually rare processes were necessary to form the solar system, other stars may also have spawned planetary systems. All stars should form from gravitationally unstable, collapsing nebulae. All stars should have some angular momentum and thus produce rotating, flattened clouds around themselves as they form. Such clouds should cool as their heat radiates away into space, and dust grains of varied compositions should therefore appear. Many observations of stars indicate that these processes do occur elsewhere.

But would the dust always accumulate into planet-sized bodies? Probably not, according to simple observations of stars. Many if not most stars are double or triple star systems. In such a system the second largest object is not a planet, like Jupiter, but a full-fledged star with nuclear reactions and an incandescent gaseous surface. Such massive companions may interfere with planets forming in the system.

However, one of the exciting areas of astronomy in the 1980s has been the discovery of star systems with features related to those of our planetary system. First, there have been numerous detections of ever-smaller companions to apparently single stars. None may be quite as small as Jupiter, but smaller ones are reported each year, and the new discoveries are approaching Jupiter in size. Second, astronomers were surprised to discover with infrared satellites that many seemingly single, Sun-like stars have thin dust clouds around them. These clouds may be analogs of the dust trails from comets or the zodiacal light. They may indicate that planetary objects—comets, asteroids, or even full-fledged planets—lurk still unseen near these stars.

Thus there is a stimulating new merger between studies of our own planetary system and the study of other stars. A new interdisciplinary field is emerging: the study of planetary material around other stars. If only 0.1% of all stars had planets on which liquid water could exist, then 100 million habitable planets could plausibly exist in our galaxy! We will come back to this subject—and its provocative consequences for possible alien life—in later chapters.

SUMMARY

Information about the origin of the solar system has been culled from meteorites, lunar samples, the oldest terrestrial rocks, and chemical and dynamical analysis of planets and satellites. This information indicates that the Sun formed from a contracting cloud of gas about 4.6 billion years ago. As outer parts of this cloud cooled, solid grains of various minerals and ices condensed and accumulated into planetesimals during a relatively brief interval, lasting from a few million to 100 million years.

During this interval, neighboring planetesimals collided, often gently enough to allow them to hold together by gravity. In this way small planetesimals aggregated into a few larger bodies. Sometimes these bodies collided at high enough speeds to shatter each other, producing meteoritelike fragments. The largest bodies survived and grew into planets. Moons formed near planets by a similar process. Many planetesimals crashed into planets and moons, forming craters. When icy planetesimals in the outer solar system made near-miss encounters with giant planets, the

planets' gravity flung many of them almost out of the solar system, forming the Oort cloud reservoir of comets (as described in Chapter 13).

The Sun and the outer planets, with their high masses and strong gravity, retained the light, hydrogen-rich gases of the original cloud. Lower-gravity terrestrial planets lost these gases.

No data about the early solar system restrict these processes to the solar system. Thus many scientists suspect that planetary debris or full-fledged planets may have formed near some other stars besides our Sun.

CONCEPTS

protosun	solar nebula
free-fall contraction	condensation sequence
Helmholtz contraction	grain
nebula	planetesimal

PROBLEMS

1. The gas in the early solar system was about 76% hydrogen (Table 14-3). Considering the theory of thermal escape of gases from planetary atmospheres (see Chapter 9), explain the absence of abundant hydrogen in the terrestrial planets' atmospheres. Why does this not need to be listed as a fact to be explained by theories of solar system *origin* in Table 14-1?

2. Because all planetary material condensed from the same nebula, why do meteorites have different chemical and geological properties than rocks you might find in your own yard?

3. Because of heating by the Sun and by the contraction process, gases in the inner solar system were probably warmer than gases in the outer solar system when planetary solid matter formed. In terms of the condensation sequence, relate this to the estimated or observed composition of the planets.

4. Judging from planetary composition, where was the inner boundary of the part of the solar nebula where ices condensed?

5. If planets orbited their sun in randomly inclined prograde and retrograde orbits, how might theories of such a solar system's origin differ from the theory described in this chapter?

6. List some observations that support the theory of solar system origin described in this chapter.

7. If a 5-year-old member of your family asked where the world came from, how would you answer?

Stars and Their Evolution

Star cycles. The bright reddish object on the left is a cloud of gas and dust where stars form. The brightest star on the right is a supernova, or exploding star, that appeared in 1987. Such an explosion marks a massive star's death.
(National Optical Astronomy Observatories.)

CHAPTER 15

The Sun:
The Nature of
the Nearest Star

Much of modern astronomy deals with stars—how they generate their energy, what kinds of light they radiate, how they form, and how they evolve. How can we study such remote objects as the stars? We can study the closest one, which is nearer to Earth than five of the nine planets. Light from its surface reaches us in only eight minutes. Our eyes are dazzled by it. Earth is bathed in its flow of radiation, washed by the winds of its outer atmosphere, blasted by seething swarms of atoms blown out of it, bombarded by bursts of X rays and radio waves emitted by it. It is our Sun, a million-kilometer ball of hydrogen and helium in the center of the solar system, 10 times Jupiter's size and roughly 100 times the Earth's size.

Humans pondered the stars for many centuries before they realized that the Sun is just another star, and the stars are suns. The Sun is the only star on which we can see surface details. It is so close that we can watch storms develop on its surface and track them as they are carried around it by **solar rotation.** This rotation takes about 25.4 d relative to the stars (and 27.3 d relative to Earth, because Earth's orbital motion is in the same direction as the solar rotation and must be added in). As in Jupiter's atmosphere, the equatorial region rotates faster than the polar regions—proof that the Sun has a gaseous, not solid, surface.

SPECTROSCOPIC DISCOVERIES

We cannot send space vehicles to land on the Sun, as we have done for planets. To study the Sun or stars we must rely on interpreting their light, sampled by using telescopes.

Chapter 5 described certain properties of light, such as the relation of color of radiated light to the temperature of the radiating material. Here we will touch on a few of these basic concepts again as we explain how they apply to the Sun.

The Solar Spectrum

As Newton showed, sunlight is a mixture of all colors. Light with this specific mixture of colors is called **white light,** because it looks white to our eyes. Sunlight passing through Newton's prism revealed a **spectrum,** the array of these colors in order of wavelength, as shown in Figure 5-1. Review pages 72–73 (especially Figure 5-10) to recall how parts of the ultraviolet and infrared spectrum of sunlight and starlight are absorbed by our atmosphere before reaching the ground.

In 1817, German physicist Joseph Fraunhofer found that certain wavelengths were missing from the Sun's spectrum, so that the spectrum appeared to be crossed by narrow, dark lines.[1] As explained in Chapter 5, these are called **absorption lines** (Figure 5-19). Fraunhofer named them A, B, C, and so on, from red to blue.

What were these lines? Let us review and expand on what we learned in Chapter 5. By the mid-1800s, scientists discovered that when a given element is burned, it emits glows of certain colors and no others. These very narrow wavelength intervals, unique to each element and as unmistakable as a set of fingerprints, are called **emission lines.** Researchers soon found that some of the emission lines exactly matched the position of Fraunhofer's solar absorption lines. For instance, Fraunhofer's D absorption line (actually a close pair) matched an emission line from sodium; his H and K lines, calcium. Did this mean that the D line in the solar spectrum was caused by sodium in the Sun, and the H and K lines by calcium?

Kirchhoff's Laws

The answer proved to be yes. The proof? In the 1850s, German physicist Gustav Kirchhoff discovered in the laboratory the conditions that produce the three different kinds of spectra described in Chapter 5: the glow consisting of all colors, called the **continuum** (illustrated in Figure 5-1); absorption lines; and emission lines. For instance, when Kirchhoff looked through his

spectroscope toward a sodium flame against a dark background, he saw the sodium D line in emission, like the emission lines in Figure 5-6. But when he changed the background to a brilliant beam of sunlight passing through the same flame, he saw a strong D absorption line similar to the absorption lines shown in Figure 5-9. In each case, the D lines came from the gaseous sodium atoms in the flame.

Thus, in simple terms, the absorption lines found by Fraunhofer in the solar spectrum occur when photons of solar white light pass outward through the cooler gas of the solar atmosphere. Photons of certain wavelengths (such as the D line), when striking certain atoms (such as sodium), are absorbed from the outgoing beam, causing that color to be missing from the beam.

Kirchhoff reduced such observations to three statements called **Kirchhoff's laws:**

> **1. A gas at high pressure, a liquid, or a solid, if heated to incandescence, will glow with a continuous spectrum, or continuum.**
>
> **2. A hot gas under low pressure will produce only certain bright colors, called emission lines.**
>
> **3. A cool gas at low pressure, if placed between the observer and a hot continuous-spectrum source, absorbs certain colors, causing absorption lines in the observed spectrum.**

The continuum arises when free electrons are available; this occurs in high-pressure gases, liquids, or solids. Emission lines arise from electrons inside atoms in an excited state, as in a hot gas. Absorption lines arise when atoms are in, or near, the ground state, as in a cooler gas. If these principles seem unclear, you should review Chapter 5, especially pages 68–72.

Kirchhoff himself found that the absorption lines and emission lines of a given gas have identical wavelengths, as seen by comparing Figures 5-6 and 5-9. What is seen depends on the temperature and density of the gas relative to the radiation coming from behind it, as indicated in the third law. Later an important modification was made to Kirchhoff's laws: An absorption spectrum need not originate *in front of,* or in a cooler gas than, the continuous spectrum. It can arise within the same gas as the continuous spectrum. This is because within a single gas, electrons may be jumping upward in some atoms (making absorption lines) and downward from the

---

[1]For a good review of the early discoveries about solar radiation and its effects on Earth, see Meadows (1984).

TABLE 15-1

Composition of the Solar System

Element	Percent Mass of the Sun
Hydrogen (H)	70.8
Helium (He)	27.4
Oxygen (O)	1.0
Carbon (C)	0.3
Neon (Ne)	0.2
Iron (Fe)	0.1
Nitrogen (N)	0.1
Silicon (Si)	0.07
Magnesium (Mg)	0.06
Sulfur (S)	0.04
Nickel (Ni)	0.04
Total	100.0

Source: Adapted from Anders and Grevesse (1988).
Note: Based on spectroscopic measurements of the Sun and measurements of meteorites and other samples. This represents approximately the bulk composition of the early Sun and solar nebula.

free state into other atoms (forming the continuum). Indeed, in the Sun some absorption lines originate in the same surface layers that produce the continuous spectrum we call sunlight. These layers form the well-defined visible surface of the Sun, called the **photosphere.**

In summary, spectroscopy allows us to determine many things about the properties of the Sun. Most important, *identification of the spectral lines allows identification of the elements in the Sun.* Application of Kirchhoff's laws proves that the photosphere is a layer of hot gas. Application of Wien's law (page 68) allows us to use the wavelength of the strongest solar radiation to determine the photosphere temperature. Other spectroscopic principles enable astronomers to measure temperatures and pressures at different depths in the gas near the Sun's surface.

COMPOSITION OF THE SUN

After more than a century of spectroscopic study, the **Sun's composition** is accurately known. It is almost three-quarters hydrogen and one-fourth helium by mass—roughly the same H/He proportions we found in the giant planets' atmospheres. The heavy elements common in Earth comprise only 2% of the Sun by mass. The most abundant elements are listed in Table 15-1, which is believed to reflect the composition of material from which the solar system formed.

An interesting episode occurred in 1868, when the French astronomer Pierre Janssen and the English astronomer Norman Lockyer independently found solar spectral lines corresponding to a previously unknown element. *Helium* (named from the Greek *helios,* "sun") was the first element to be discovered in space instead of on Earth (where it was not observed until 1891).

IMAGES OF THE SUN

Normal photographs of the Sun (or any other scene) are made from light of many colors, because normal photographic films are sensitive to many wavelengths. However, color filters and other devices can restrict the wavelength range. By using a narrow enough range, we can see the Sun in only the light emitted by a chosen gas (hydrogen, for example) in a specific atomic state, thus tracing the distribution of that gas. The **spectroheliograph** is an ingenious instrument that allows this to be done. It spreads the light into different colors, from which one color alone is selected to form an image. For example, in certain states of temperature and pressure, hydrogen emits an especially useful red color called the **hydrogen alpha line,** or *Hα emission,* shown in Figure 5-6. In an image using Hα light, as in Figure 15-1 b and c, the only bright regions are those where hydrogen is emitting this specific red glow. Hα emission thus becomes an indicator of hydrogen's presence at certain temperatures and pressures. Dark parts of the image are regions where little of the hydrogen is in this Hα-emitting condition. Thus images in Hα light, or light emitted by other gaseous atoms, are very useful in mapping the conditions of that particular gas in various regions of the surface.

SOLAR ENERGY FROM NUCLEAR REACTIONS

Hermann von Helmholtz showed in 1871 that the energy output of the Sun corresponds to the burning of 1500 lb of coal every hour on every *square foot* of the Sun's surface. No ordinary chemical reactions can produce

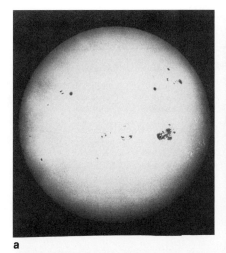

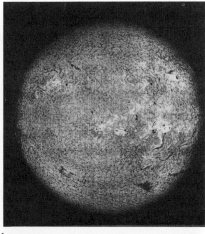

a   b

Figure 15-1 a The Sun in normal visible light, with sunspots. Shading around edge, called limb darkening, is caused by solar atmosphere's absorption of light. **b** The Sun on the same date, photographed by a spectroheliograph in red light emitted by hydrogen. Bright areas involve intense emission of light by excited hydrogen atoms. **c** An unusual spectroheliograph photo. Researchers and hiker cooperated to silhouette the hiker against the setting Sun. (Images *a* and *b* from Hale Observatories; *c* from National Oceanic and Atmospheric Administration.)

c

energy at this rate! Thus, Helmholtz realized, the Sun is not "burning" in the normal sense.

Then what *is* the source of the Sun's heat and light? In the 1920s, astrophysicists realized that the energy of the Sun and other stars comes from **nuclear reactions**—interactions of atomic nuclei near the star's center. Normally, nuclei are protected from interacting by their surrounding clouds of electrons. Familiar **chemical reactions,** such as coal burning, involve interactions only between *electrons* of different atoms, far outside the central nucleus. If the temperature and pressure are high enough, however, atoms collide fast enough to knock away electrons, allowing the nuclei to interact. This happens in the Sun's central core.

States of Matter in Stars and in the Universe

When we deal with nuclear reactions and the material inside stars, we are no longer dealing with the familiar forms of matter that make up the solids, liquids, and gases of our planet and our bodies. We are about to

TABLE 15·2

States of Matter in the Universe

Approximate Temperature Region	Velocity of Typical Particles		State of Matter	Typical Location	Typical Radiation Emitted
60 billion K	Relativistic (appreciable fraction of speed of light)		NUCLEAR FRAGMENTATION (nuclear particles collide hard enough to shatter)	Accretion disk near a black hole	Gamma rays, X rays
10 million K	500 km/s		BARE NUCLEI, FUSION (nuclear particles collide and fuse)	Core of a star	X rays Ultraviolet light
5000 K	10 km/s		IONIZED GAS (electrons knocked free)	Atmosphere of a star	Visible light
500 K	1 km/s		GAS (separate atoms)	Atmosphere of a planet	Infrared light
300 K	$\frac{1}{2}$ km/s		LIQUID (some atoms linked in chains)	Water	Far infrared light
0 K	0		SOLID (atoms linked in lattice)	Rock	Radio waves

make the leap from cold planetary matter to hot stellar matter. This is a good moment to pause and describe the way that matter behaves under a variety of conditions.

There is a simple way to arrange the states of matter—by temperature as in Table 15-2. Although this arrangement oversimplifies certain effects of pressure and other variables, it is very useful in understanding the forms of material we are dealing with in this book. The table is arranged from the bottom up in order of increasing temperature.

At the lowest temperatures, matter is "frozen." It exists as a solid. This means that the atoms are bonded together, often in a lattice pattern. The bonds are formed by the sharing of electrons (tiny dots) between nuclei. The nuclei consist of protons (blue) and neutrons (white). Most of the mass of each planet is in this solid state—indeed, the crystals that form rocks are good examples of atoms bound together in lattice patterns.

Now recall from page 67 that our conception of temperature is merely a way of measuring the rate of motions of the atoms. The faster the motions, the higher the temperature and vice versa. At absolute zero temperature, 0 K, atomic particles would have no motion. But as we raise the temperature of a solid, its atoms vibrate faster and faster.

At room temperature, around 300 K, the typical atoms in typical substances are moving at around $\frac{1}{2}$ km/s. This activity is sufficient to break many of the atoms loose from their lattice. We perceive this as the substance melting, or turning into a liquid. The liquid oceans of planet Earth are a good example. In the liquid state, chains or groups of atoms may move among each other.

By the time we reach 500 K, however, the atoms are moving at around 1 km/s and the chains are broken. Now we have created a gas, in which the individual atoms (or molecules, such as H_2O in water vapor) move freely. This is the state of matter in the air we breathe. But all these forms of matter are cooler than matter in the Sun or in most stars.

If we keep heating the gas, the speeds of the atoms increase. They hit harder and harder. At a temperature of a few thousand degrees, the atoms are hitting each other so hard that they break the electrons free from their orbits. The Sun's surface, for instance, has a temperature a little over 5000 K. There the hydrogen nuclei, many of them stripped of their electrons, move at speeds around 10 km/s. Gas that has had its electrons knocked off is called **ionized gas**. This is the form of matter not only in the Sun's surface layers but also in a familiar flame. Because the electrons are constantly being

bumped off atoms and rejoining them, they are constantly changing energy states and giving off light. Following Kirchhoff's second law, the light of a candle flame is in the form of emission lines. But as we noted above, in accordance with Kirchhoff's first law, the light given off from the high-pressure gas in the surface of the Sun or a star is in the form of a continuum.

The insides of stars are even hotter than their surfaces. The center of the Sun, for instance, is at about 15 million Kelvin. What happens if we confine the gas (in a container or inside a star) and keep raising the temperature toward such a value? The electrons have all been stripped off and the bare nuclei of atoms are approaching each other. Their mutual positive charges tend to make them repel each other, and at first they just bounce apart. But, as always, when the temperature increases, the particles move faster and collide. Atomic nuclei might be compared to spitballs: If they just brush together at low speed, they merely bound apart; but if they hit hard enough they mush together. Among nuclei, it takes temperatures of the order 10 million Kelvin before the fastest nuclei begin to fuse. Typical hydrogen nuclei at this temperature move at around 500 km/s, and the fastest ones in a sample of the ionized gas will be moving even faster. This merger of nuclei at high temperatures is called *fusion*. The fusion may be among individual protons (blue in Table 15-2) and neutrons (white), or it may be between nuclei to build even larger nuclei with many protons and neutrons.

Fusion of atoms inside stars is one of the most important processes in the universe. We will be dealing with its consequences throughout the rest of this book. One of its most important effects is that the fusion process gives off energy (shown in Table 15-2 as wavy lines representing radiation). This energy heats the gas, maintains the temperature at the center of the Sun at around 15 million Kelvin, and thus keeps the reactions going.

But what happens if we force the temperature still higher? Eventually, at temperatures of billions of degrees, the ionized gas particles are moving at nearly the speed of light. Then the nuclei hit so hard that they shatter in a shower of radiation and tiny subatomic particles. In later chapters, we will encounter such environments—for example, near black holes.

The Nuclear Reactions Inside the Sun

Nuclear reactions inside stars do two important things:

The Proton–Proton Chain: Energy Source of the Sun

Step 1

$$^1H + {}^1H \rightarrow {}^2H + e^+ + \text{neutrino}$$

Step 2

$$^2H + {}^1H \rightarrow {}^3He + \text{photon}$$

Step 3

$$^3He + {}^3He \rightarrow {}^4He + {}^1H + {}^1H + \text{photon}$$

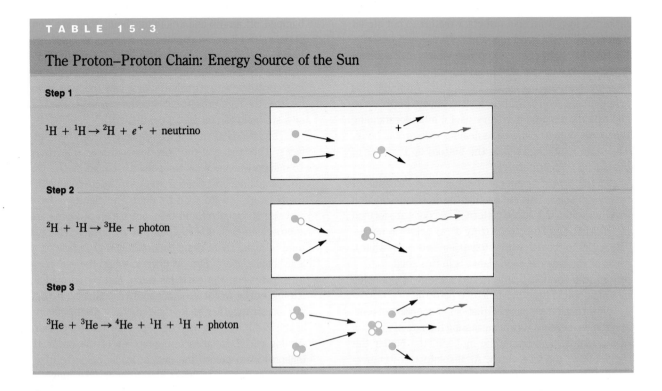

They generate energy and they gradually change the star's composition because they destroy light nuclei and create more and more heavy nuclei in their place. The principal reactions inside the Sun are believed to be a three-part sequence that fuses four hydrogen atoms into a helium atom. Because this chain of reactions starts with two hydrogen nuclei—that is, two single protons—it is called the **proton–proton chain.** To see how it works, let us follow it in detail in Table 15-3. The reaction occurs in three steps, shown both in diagrammatic form and in the form of the nuclear reaction equations written by physicists. In step 1, two protons collide and fuse. The fusion produces a form of hydrogen nucleus designated ^2H. A tiny positive particle called a positron is given off, together with a massless (or virtually massless) particle called a neutrino. In step 2, the hydrogen nucleus hits another proton and fuses into a form of helium known as helium-3, designated ^3He. A photon of radiation is emitted. In step 3, two of the ^3He nuclei collide and fuse into the normal form of helium, helium-4, designated ^4He. Two protons are left over and another photon is emitted.

Note that the reaction has done two important things: It has given off energy in the form of radiation of photons, and it has created helium out of the lighter element hydrogen.

The total amount of mass left at the end of the three-step chain is slightly less than the mass of the initial hydrogen atoms. During the fusion, a small amount of mass m is converted to an amount of energy E, according to Einstein's famous equation $E = mc^2$. (The constant c is the velocity of light. In the Sun's fusion sequence, about 0.007 kg of matter is converted into energy for each kilogram of hydrogen processed. This liberates 4×10^{26} J/s inside the Sun, and the Sun radiates this much energy every second to maintain its equilibrium. This corresponds to 400 trillion trillion watts—which equals a lot of light bulbs!

Every second, the Sun converts 4 million tons of hydrogen into energy and radiates it into space. Long before the Sun can use up all its mass, the solar core will have converted so much hydrogen to helium that there will not be enough hydrogen left in the core to fuel further reactions and the reactions will stop. According to a recent calculation, the Sun won't run out of hydrogen for about 4 billion years. The consumption of hydrogen inside stars proves the important point that stars are not permanent, but must evolve and run down.

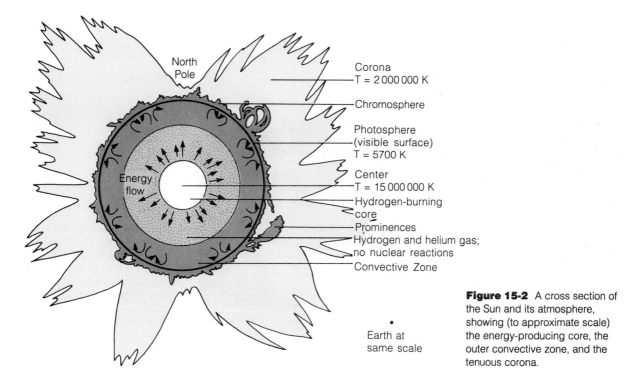

North Pole

Corona
T = 2 000 000 K

Chromosphere

Photosphere
(visible surface)
T = 5700 K

Center
T = 15 000 000 K

Hydrogen-burning core

Prominences

Hydrogen and helium gas;
no nuclear reactions

Convective Zone

Energy flow

•
Earth at same scale

Figure 15-2 A cross section of the Sun and its atmosphere, showing (to approximate scale) the energy-producing core, the outer convective zone, and the tenuous corona.

THE SUN'S INTERIOR STRUCTURE

As a result of Helmholtz contraction (see Chapter 14), the temperature at the Sun's center reached 15 million Kelvin when the Sun formed. This is when the proton–proton reactions became established. This remains the current temperature in the **solar core,** where nuclear energy is generated.

Approximate conditions in the layers between the core and the surface can be calculated by using equations that describe the pressure at any depth, the properties of gas under different pressures, and energy generation rates at points inside the Sun. The results appear in Figure 15-2.

These calculations indicate that the gas pressure at the Sun's core is about 250 billion times the air pressure at Earth's surface. This high pressure compresses the gas in the core to a density of about 158,000 kg/m³— 158 times denser than water and about 20 times denser than iron. One cubic inch of this gas would weigh nearly 6 lb! The core of the Sun occupies about the inner quarter of the Sun's radius.

How Energy Gets from the Core to the Surface

As heat energy always flows from hot to cool regions, solar energy travels outward from the hot core, through a cooler zone of mixed hydrogen and helium, toward the surface. Throughout most of the Sun's volume, this energy moves primarily by **radiation.** That is, the energy radiates through the gas in the form of light, just as light travels through our atmosphere from a light bulb. Very little energy moves through the Sun by **conduction,** the mechanism by which a pan on a stove becomes hot.

In the outer part of the Sun we find a third mechanism of energy transport—**convection.** Convection occurs when the temperature difference per unit length between the hot and cold regions is so great that neither radiation nor conduction can carry off the outward-bound energy fast enough. So-called "cells" of gas, having become heated enough to expand, become less dense than their surroundings and rise toward the surface, move across the Sun at about 20 m/s, cool by radiating their energy into space (sunlight!), and sink (Thomsen,

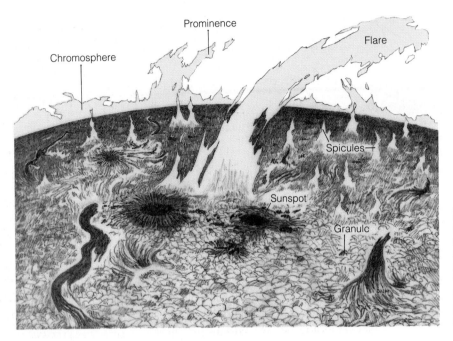

Figure 15-3 A schematic, oblique view of the solar surface showing features mentioned in the text. Granules are the result of the convective "boiling" of the surface. Flamelike spicules, prominences, and flares shoot up above the surface.

1985). Convection is essentially like the process of boiling, where moving blobs of hot material carry the heat upward and disturb the surface. The slight mottling of the solar surface, visible in photos, is due to this disturbance by convecting cells of gas, called *granules* (see Figure 15-3).

The Solar Neutrino Puzzle

Step 1 of the proton–proton chain (Table 15-3, p. 232) produces a neutrino, a virtually massless subatomic particle. From the Sun's total energy production, astronomers have calculated the number of neutrinos that ought to come from the reactions in the Sun's interior. Though neutrinos are hard to detect, detectors to measure neutrino numbers went into action in the 1970s. To scientists' amazement, the experiments detect only about one-third as many solar neutrinos as predicted!

This is a serious problem for solar physicists! Is our understanding of solar nuclear reactions fatally flawed? By the mid-1980s, several theories had been advanced to explain the missing neutrinos. In one theory, certain other hitherto undetected particles (charmingly called WIMPs, for *weakly interacting massive particles*) were

postulated to carry off some of the Sun's energy, thus cooling it enough to retard neutrino production rates. Alternatively, a Russian theory suggested that certain types of neutrinos transmute into undetectable particles. These theories are still being tested, and show the close connection between modern astronomy and nuclear physics (Bahcall, 1990).

THE PHOTOSPHERE: THE SOLAR SURFACE

Energy ascending from inside the Sun heats the photosphere—the bright surface layer of gas that radiates the visible light of the Sun—to a temperature of about 5700 K.

If the Sun is a giant ball of gas, why does it appear to have a sharply defined surface? The answer involves the **opacity** of the gas—its ability to obscure light passing through it. Air, for example, has low opacity. But ionization of solar gas makes it opaque like a flame. The opaque layer gives the appearance of a sharply defined surface, but a solid probe dropped into the sun—if it survived the 5700-K temperature of the surface layer—could plunge right through it like an airplane passing

Figure 15-4 The Sun's inner atmosphere during an eclipse. Pink flames of hot hydrogen, colored by the red Hα emission of excited hydrogen atoms, protrude from the Sun's surface at the top and at several other points around the Moon's black silhouette. The pearly diffuse glow is the inner corona. (NASA photo by astronauts using solar telescope in Skylab space station.)

through the surface of a cloud. (Other flamelike features of the photosphere are shown in Figure 15-3.)

CHROMOSPHERE AND CORONA: THE SOLAR ATMOSPHERE

The clear region above the photosphere is called the solar atmosphere. Just above the photosphere is a red-glowing region called the **chromosphere** (which means "color layer"), indicated in Figure 15-3. Its light is mainly the red Hα emission described in Figure 5-6. During an eclipse of the Sun, when the Moon blots out the bright light of the solar surface, the chromosphere can be seen by the naked eye as a ring of small intense red flames, just visible in Figure 15-4.

Above the chromosphere is the rarefied, hot gas of the **corona**. Gas in the corona reaches the amazing temperature of 2 million K, due to heating by violent convective motion in the photosphere and chromosphere. As a result of the extreme heat, it expands rapidly into space. Whereas the gas density is about

0.001 kg/m^3 within a few hundred kilometers of the photosphere, it drops to 10^{-7} kg/m^3 in the middle chromosphere and to less than 10^{-11} kg/m^3 in the outer corona. Clearly the corona is only the outermost, tenuous atmosphere of the Sun. During eclipses, the corona is visible to the naked eye as a pearly, glowing gas around the Sun, as seen in Figures 15-4 and 1-14.

SUNSPOTS AND SUNSPOT ACTIVITY

Although sunspots can sometimes be seen with the naked eye when the Sun is dimmed by fog or a dark glass,[2] their nature was not realized until 1613, when Galileo studied them and concluded that they are located on the solar surface and are carried around the Sun by solar rotation.

The Nature of Sunspots

A **sunspot** is a magnetically disturbed region that is cooler than its surroundings. A sunspot looks dark only because its gases, at 4000–4500 K, radiate less than the surrounding gas at about 5700 K. Motions of solar gas near sunspots are controlled not by atmospheric forces, as with terrestrial storms, but by magnetic fields of the Sun. *Ions* (charged atoms or molecules), which are common in the Sun, cannot move freely in a magnetic field but must stream in the direction of the field—for example, from the north magnetic pole to the south.

Unlike a neutral gas, the motions of ionized gas are thus strongly influenced by magnetic fields. For this reason, ionized gas in the sunspots and elsewhere in the solar atmosphere moves in peculiar patterns that indicate the twists of the solar magnetic field.

Huge clouds of ionized gas, larger than the whole Earth, erupt from the disturbed regions of sunspots. These *prominences* can be seen when silhouetted above the solar limb, or edge, as seen in Figure 15-5. The largest blasts of material and their very active sunspot sites are called *flares*. Figure 15-6 shows the detailed structure of a sunspot.

---

[2]**Never point any telescope or binoculars at the Sun without professional aid!** Unmodified telescope optics collect enough solar light and heat to blind the observer. Even pointing a telescope *near* the Sun may concentrate light inside the instrument in such a way as to damage it, as well as risking an observer's eyesight.

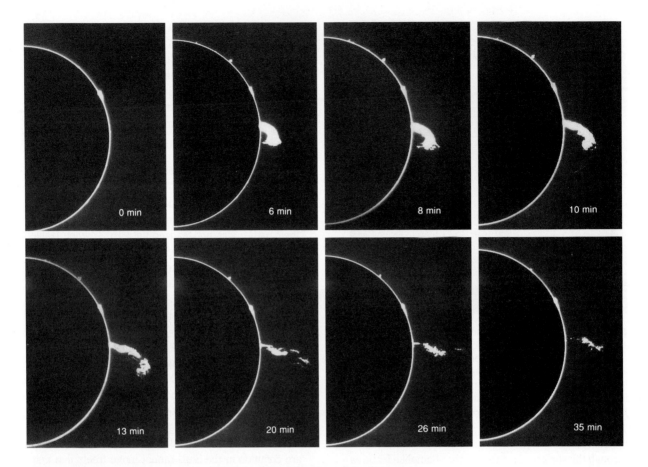

Figure 15-5 This sequence of photos shows a jet of gas blasting off the Sun over a period of 35 min. The photos were made with a coronograph, which obscures the bright solar disk and allows solar atmospheric activity to be monitored. (National Center for Atmospheric Research.)

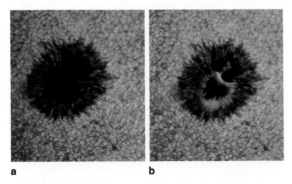

Figure 15-6 Vertical views of a sunspot surrounded by bright photosphere with typical granules, each about 1000–2000 km across. **a** Normal photographic image. **b** Montage of normal image and longer exposure showing filaments of gas and other details within the darkest, coolest part of the spot. (Photos by W. Livingston, National Solar Observatory.)

The 22-y Solar Cycle

Around 1830 an obscure German amateur astronomer, H. Schwabe, began observing sunspots as a hobby. After years of tabulating his counts, he announced in 1851 a **solar cycle:** The *number* and *positions* of sunspots vary in a cycle, as shown in Figure 15-7. This discovery, followed a year later by the realization that terrestrial magnetic compass deviations follow exactly the same cycle, was a key step in understanding the Sun and its effects on the Earth.

The solar cycle's duration averages 22 y and consists of two 11-y subcycles, as shown in Figure 15-8. At a time of "sunspot minimum" (when there is a minimum number of sunspots), the few visible spots are grouped within about 10° of the solar equator. When a new cycle begins in a year or so, groups of new spots

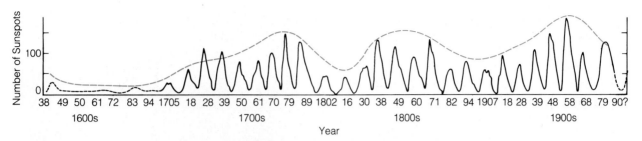

Figure 15-7 Sunspot counts since the 1600s show the cycle averaging 11 y for sunspot numbers (half the 22-y magnetic cycle), with evidence for a longer 80-y cycle (white dashed line). The extended period of low sunspot activity in the 1600s is believed to correlate with climate changes at that time. Dates give years of maximum sunspot numbers. (After data of M. Waldmeier in Gibson, 1973; Pasachoff, 1980.)

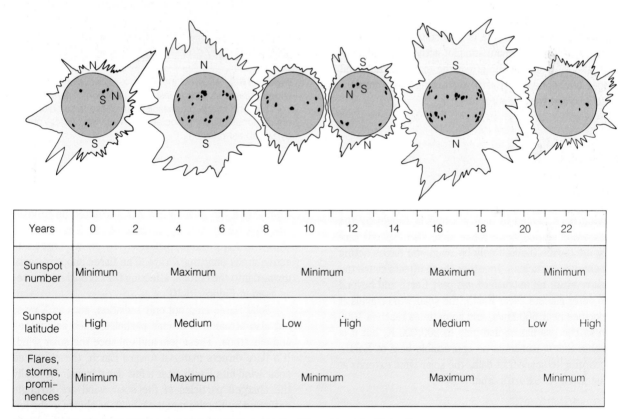

Years	0	2	4	6	8	10	12	14	16	18	20	22
Sunspot number	Minimum		Maximum			Minimum			Maximum		Minimum	
Sunspot latitude	High		Medium		Low		High		Medium		Low	High
Flares, storms, prominences	Minimum		Maximum			Minimum			Maximum		Minimum	

Figure 15-8 A schematic sequence of changes during the 22-y solar cycle. Note also the different coronal appearances that are typical of the minimum and maximum of the cycle. (See text.)

appear at high latitudes, about 30° from the solar equator. The spots often appear in pairs, and the eastern spot in each northern hemisphere pair is of specific polarity—for example, a north magnetic pole. In the southern hemisphere the polarity is reversed. After a few years, the sunspot number reaches a maximum and the spots are at intermediate latitudes, about 20° from the solar equator. After about 11 y, the spots appear mostly about 10° from the equator, and a sunspot minimum occurs again.

Now the cycle begins to repeat, with one noticeable difference. The new spots forming at ±30° have reversed polarity. The eastern spots in the northern hemisphere are now south magnetic poles! Thus it takes another 11 y to complete the full cycle, when all features resume their initial pattern. A sunspot maximum should occur around 1990.

Still more remarkable is the fact that the magnetic field of *the entire Sun* reverses during each 11-y subcycle; thus the entire Sun participates in the full 22-y cycle. Imaginary observers on the Sun would find their compasses pointing north in one direction for 11 y (subject to disturbances by frequent magnetic storms) and in exactly the opposite direction for the next 11 y. This behavior is not entirely unique: The Earth's field reverses every few hundred thousand to few million years. Patterns of reversal on both the Sun and Earth may involve cyclic flow patterns in the deep fluid cores of the two bodies. The sunspot cycle is important to us because during years of maximum sunspot activity, solar particles shooting off the Sun affect the magnetic field and upper atmosphere of Earth, disturbing radio communications and causing aurorae.

SOLAR WIND

Particles blasted out of flares and spots rush outward through interplanetary space. The solar coronal plasma, having been heated to nearly 2 million Kelvin by the violence of photospheric convection, also expands rapidly into space (limited only by magnetic forces acting on charged particles). Together these effects cause the **solar wind:** an outrush of gas past Earth and beyond the outer planets. Near Earth, the solar wind travels at velocities near 600 km/s, and sometimes reaches 1000 km/s. The gas has cooled only to 200,000 K, but it is so thin that it transmits no appreciable heat to Earth. According to spacecraft data, the solar wind extends at least as far as Saturn's orbit.

AURORAE AND SOLAR–TERRESTRIAL RELATIONS

During a solar flare, the total visible radiation from the Sun changes by much less than 1%, but the X-ray radiation may increase by a hundredfold. The X-ray photons are energetic because they have very short wavelength. When they strike Earth's upper atmosphere, they change its distribution of ions and affect radio transmission on

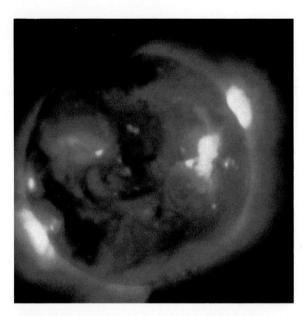

Figure 15-9 A view of the Sun in X-ray wavelengths showing flares as sites of intense X radiation. X rays are invisible, and image processors here chose false-color orange tones usually associated with sunlight. (NASA photo by Skylab astronauts.)

the ground. Imaging of the Sun at X-ray wavelengths from above the atmosphere has permitted us to visualize the remarkable appearance of the X-ray Sun, as shown in Figure 15-9, which vividly shows flares and active areas emitting X rays. The flares shoot material upward into the corona, affecting the coronal structure, as seen in Figure 15-10.

Solar flares emit not only radiation, such as X rays, but also streams of atomic particles, such as protons and electrons. These join and enhance the solar wind. If a flare directs material toward Earth, the enhanced solar wind hits Earth after a few days' travel. Normally the charged particles of the solar wind are strongly deflected by Earth's magnetic field, as shown in the left side of Figure 15-11. But during solar flares, the surge in the solar wind is often so strong that Earth's magnetic field is seriously distorted, affecting distributions and motions of charged particles throughout Earth's vicinity.

The ions near Earth are concentrated into doughnut-shaped regions encircling Earth, approximately over the equator. These **Van Allen belts** of radiation were discovered in 1958 by the first artificial satellites. They are shown by the x's in Figure 15-11. They are particularly affected by disturbances in the solar wind. Under

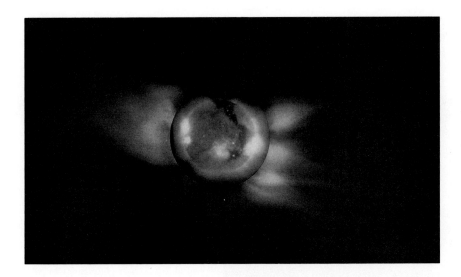

Figure 15-10 Surface and atmospheric activity of the Sun. During the June 30, 1973, solar eclipse in Kenya, the outer corona of the Sun was photographed in white light (outer image). About an hour earlier, astronauts in the orbiting Skylab photographed the solar surface (circular inset) using X radiation, showing centers of flare activity. Major streamers in the outer corona are shown to be aligned with X-ray flares on the surface. (National Center for Atmospheric Research; American Science and Engineering, Inc.; NASA.)

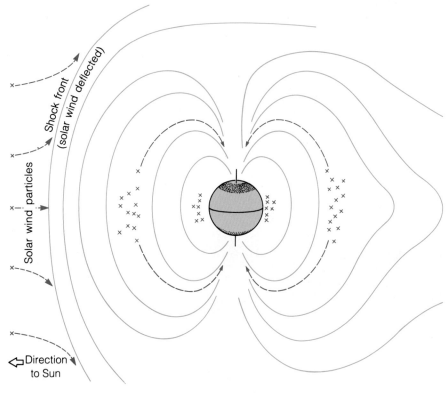

Figure 15-11 Interaction of solar wind ions (from left) with Earth's magnetic field. The shock front is analogous to the bow wave cut by a moving boat. Dashed lines show the typical paths of solar ions (x's). Those ions that penetrate Earth's field accumulate in doughnut-shaped Van Allen radiation belts around Earth. Concentrations of x's mark their positions. Ions in the belts eventually empty into Earth's polar atmosphere, colliding with air molecules and forming auroral zones near the north and south magnetic poles.

normal conditions, as the solar wind sweeps around Earth, it builds up voltages of 100,000 volts or more between the outer regions of the magnetic field and the atmosphere, driving some charged particles along the magnetic field lines toward the poles (dashed lines in the middle of Figure 15-11). These particles crash into

the upper atmosphere, excite the gas atoms there, and cause them to emit light. This glow can be seen from the ground as a weird, colored glow in the sky (Figures 15-12 and 15-13). It is called the **aurora** (plural: *aurorae*). The ions crash into the atmosphere only near the magnetic poles, causing intermittent aurorae in the Arctic

a

b

Figure 15-12 Various forms of the aurora. **a** Part of a typical curtain display seen from Alaska. (Photo by Nancy Simmerman.) **b** An unusual ring of vertical rays over Alaska. (Photo by Michio Hoshino.)

and Antarctica. These aurorae are also called the northern lights and southern lights (or aurora borealis and aurora australis, respectively). Although the northern lights are normally visible only in Alaska and Canada, large solar flares disrupt the situation so much that ions may be dumped into the atmosphere over lower latitudes, including much of the United States. During an aurora display, the sky may be filled with a stationary red glow, moving bands of red, white, and greenish light, or huge displays that look like colorful draperies rippling silently in the wind. During the greatest flare-related disturbances, aurorae may be seen from latitudes as low as Mexico and India, compass needles may deviate by a degree from normal, and electromagnetic effects may overload electric power lines, burn out transformers, and disrupt telephone, telegraph, radio, and power transmission. Because these effects are related to mighty solar flares, they are most common

Figure 15-13 An auroral display from *above*. This spectacular scene was photographed by astronauts in Spacelab 3 orbiting in a space shuttle halfway between Antarctica and Australia. Faint moonlit clouds can be seen on Earth, as bright vertical curtains of aurora play across the upper atmosphere. (NASA.)

around the years of sunspot maximum, such as around 1990.

SOLAR ENERGY: A COSMIC FUEL

Solar radiation reaches Earth at a rate of 1.37 kilowatts/ m^2. This value is called the **solar constant**. Actually, it may not be perfectly constant. Careful measures of sunlight from space probes show slight variations, less than a tenth of a percent per year. During the few days of solar flares, the total energy rate does not change much, but the rate at certain wavelengths, such as X rays, can change drastically, affecting Earth's upper atmosphere structure.

Some researchers suspect that longer-term changes affect climatic conditions. For example, during the period 1645–1715, sunspot numbers were unusually low. Tree ring structures and other evidence suggest that Earth's climate conditions were altered during this period. The study of solar effects on long-term climate remains an active area of research.

Some 1.37 kilowatts of solar energy hits each daylit square meter of Earth; the roof area of an average home intercepts roughly 1000 to 2000 kilowatt-hours of energy on each sunny day, equalling the average daily energy consumption of an American household (Snell and others, 1976). Thus, solar energy is abundant. It is a cosmic energy source. As *fossil* fuels such as coal and oil are consumed over the next several generations, *cosmic* sources of energy will become increasingly important.

One giant test of solar technology is the power station Solar One, near Barstow, California. It uses 1818 mirrors to focus sunlight on a central collector, generating 10,000 kilowatts of energy (Kreith and Meyer, 1983). Rooftop solar cell tiles, still under research, might make homes more self-sufficient and relieve home energy expenses, which average around $1000 per year for many households. In the same way, development of solar energy technology (plus other cosmic energy sources such as geothermal energy) might make Third World, as well as developed nations, energy independent, reducing the global tensions that result as nations struggle to control the last, dwindling, fossil fuel reserves, such as the oil of the Persian Gulf. Understanding and harnessing the Sun's power thus goes far beyond an academic problem; it could affect our daily lives and the world's social future.

SUMMARY

The Sun can be studied both observationally and theoretically. Observational studies have yielded information about solar gas primarily by studying the solar spectrum. These studies reveal, among other things, that the Sun is almost three-quarters hydrogen; most of the rest is helium, and a few percent is composed of heavier elements.

Theoretical studies reveal that as the Sun formed by contraction of an interstellar gas cloud, it got so hot that atoms at the center collided at high speeds. These collisions cause nuclear reactions in which hydrogen atoms are fused into helium atoms, releasing energy. This nuclear fusion is the source of the Sun's light and heat. The most important reactions in the present-day Sun are a series called the proton–proton cycle.

Transport of this energy from the Sun's center to the outer layers violently disturbs the surface, producing phenomena such as sunspots and flamelike prominences. Particles are shot off the Sun in outward-moving gas called the solar wind, which interacts with the Earth, causing aurorae and other phenomena.

Comparing the Sun with other energy sources used on Earth shows that we are rapidly consuming our planetary budget of fossil fuels and will soon have to convert to cosmic energy sources, such as solar energy, to maintain our present rates of energy consumption.

CONCEPTS

solar rotation	proton–proton chain
white light	solar core
spectrum	radiation
absorption line	conduction
emission line	convection
continuum	opacity
Kirchhoff's laws	chromosphere
photosphere	corona
Sun's composition	sunspot
spectroheliograph	solar cycle
hydrogen alpha line	solar wind
nuclear reaction	Van Allen belts
chemical reaction	aurora
ionized gas	solar constant

PROBLEMS

1. By what logic do astronomers infer that the Sun's energy comes from nuclear fusion reactions of the proton–proton cycle? How do we know it does not come from chemical burning?

2. If you see a cluster of sunspots in the center of the Sun's disk, how long would the spots take to reach the limb, carried by the Sun's rotation? How long would it take for the cluster to appear again at the center of the disk?

3. Why is the solar cycle said to be 22 y long, even though the number of sunspots rises and falls every 11 y?

4. Suppose you could make detailed comparisons of the appearance of the Sun at different moments of time. What variations in appearance would you see if the intervals were:
  a. 10 min?
  b. 1 wk?
  c. 5 y?
  d. 10 y?
  e. 100 million years?
  f. 9 billion years?

5. Why is a radio disturbance on Earth likely to occur within minutes of a solar flare near the center of the Sun's disk, whereas an aurora occurs a day or two later, if at all?

6. How much more massive is the Sun than the total of all planetary mass (see Table 8-1 for data)?

7. What is the most abundant element in the solar system? The second most abundant element?

8. If Earth formed in the same gas cloud as the Sun, why is Earth made from different material than the Sun?

9. Why will the Sun change drastically in several billion years?

10. Why is the Sun's energy generated mostly at its center, and not near its surface?

PROJECTS

1. According to the principle of the pinhole camera, light passing through a small hole will cast an image if projected onto a screen many hole diameters away. Confirm this by cutting a 1-cm hole of any shape in a large cardboard sheet and allowing sunlight to pass through the hole onto a white sheet in a dark room or enclosure several feet away (3 m— more if possible). Confirm that the projected image is round, an actual image of the Sun's disk. Are any sunspots visible?

2. Cut a 1-cm hole in a sheet of cardboard and use it as a mask over the end of a small telescope. *After* masking the telescope, point it toward the Sun and project an image of the Sun through an eyepiece onto a white card. **Under no circumstances should anyone ever look through the eyepiece, since all the light entering the telescope is concentrated at that point and can burn the retina!** Professional supervision of the telescope is suggested. Note also that an unmasked large telescope may concentrate enough light and heat to crack eyepiece lenses.

Are sunspots visible? If so, trace the image on a piece of paper and reobserve on the next day. Confirm the rotation of the Sun by following sunspot positions for several days. Class records kept from year to year can be used to record the cyclic variations of the numbers of sunspots.

CHAPTER 16

Measuring the Basic Properties of Stars

Now we are ready to take the great leap out of the solar system into the realm of the stars. It is quite a leap—the next star beyond the Sun is 260,000 times as far from Earth as the Sun and 6800 times as far as Pluto. If we make a model solar system the size of a half dollar, the nearby stars would be dots smaller than a period in this book, scattered about half a block apart.

Because of optical limitations and turbulence in Earth's atmosphere, the world's largest telescope cannot distinguish details smaller than a few hundredths of a second of arc (about 10 millionths of a degree). But the stars with the largest apparent angular size are no larger than about 0.04 seconds. So the surface details of nearly all distant stars are hidden from us, in contrast to the great detail we can study on the Sun.

The apparent large size of images of bright stars in photos such as Figure 16-1 has nothing to do with the star's shape itself, but is produced merely by overexposure of the brightest stars. Similarly, features such as the rings of light and "cross hairs" around the brightest images are optical effects arising in the telescope system.

In spite of this, we know there are giant stars larger than the whole orbit of Mars, stars the size of Earth, and stars the size of an asteroid. There are red stars and blue stars. There are stars of gas so thin you can see through parts of them, and stars with rocklike crusts that may contain diamonds. There are stars that are isolated spheres, stars with disklike rings around them, and stars that are exploding.

All of these bodies fit the definition of a star: A **star** is an object with so much central heat and pressure that energy is (or has been) generated in its interior by nuclear reactions. The most familiar stars visible in the night sky are balls of gas with solar composition and sizes usually a few times smaller or larger than the Sun.

How can we know all these details about stars if we

Figure 16-1 Telescopic image of the bluish star Alcyone—brightest star in the Pleiades—and adjacent stars. The true disk of the star would be a microscopic point in the center of the overexposed image. The four spikes and the surrounding circle are not stellar phenomena, but only optical effects produced inside the reflecting telescope and on the photographic plate used to make the picture. The fainter stars in the picture illustrate the great range of apparent brightness among stars. Faint blue wisps around Alcyone are portions of a gas cloud associated with the star—a feature rarely visible among stars. (Copyright by California Institute of Technology and Carnegie Institution of Washington, by permission from the Hale Observatories.)

cannot even see the disks of stars in telescopes? In the next chapters we will describe the details of familiar and unfamiliar types of stars, but first we will describe *how* we learn about them.

NAMES OF STARS

Stars are named and cataloged by several systems. Because Ptolemy's *Almagest* was passed on by Arab astronomers, many of the brightest stars ended up with Arabic names. Because *al-* is the common Arabic article, many star names start with *al:* Algol, Aldebaran, Altair, Alcor. (Other scientific "*al* words" also have Arabic origins: *algebra, alchemy, alkali,* and *almanac.*) Stars in constellations are cataloged in approximate order of brightness using Greek letters. Thus the brightest star

in the constellation of the Centaur is called Alpha Centauri (α Centauri). Fainter stars or stars with unusual properties are often known by English letters followed by constellation names or by catalog numbers, such as T Tauri or B.D. 4° + 4048.

DEFINING A STELLAR DISTANCE SCALE: LIGHT-YEARS AND PARSECS

The vast distances that separate stars and make them so hard to observe are awkward to express in ordinary units. Therefore, astronomers use units appropriate to these distances: the light-year and the parsec. The **light-year** (abbreviated "ly") is the distance light travels in 1 y, which is about 6 million million miles, or 10^{16} m. Notice that the light-year is a unit of *distance,* not time. The common mistake of using light-year as if it were a unit of time is like saying that the ball game lasted for 2 miles.[1]

The nearest star beyond the Sun, Proxima Centauri (which is in orbit around Alpha Centauri), is about 4.3 light-years away. The Sun could be said to be 8 light-minutes away. The North Star, Polaris, is about 650 ly away. Polaris' light takes about 650 y to reach us, so we are seeing it now as it was in the 1300s! If Polaris had suddenly exploded in 1950, we would not know it until about A.D. 2600.

Astronomers more commonly use a still larger unit of distance called the **parsec** (abbreviated "pc"):

$$1 \text{ parsec} = 3.26 \text{ light-years}$$
$$= 3 \times 10^{16} \text{ m}$$

In this book we will use the parsec and its multiples (kiloparsecs and megaparsecs) to express cosmic distances as we move to more remote parts of the universe. You can convert parsecs to light-years approximately by multiplying by 3. Another convenient fact to remember is that near the Sun, stars are roughly a parsec apart. For instance, Alpha and Proxima Centauri, the closest stars to the solar system, are about 1.3 parsecs away.

---

[1]Even noted space pilot Han Solo confused distance and time units when he claimed in *Star Wars* that the Millennium Falcon could "make the Kessel run" in less than 12 parsecs.

EXPRESSING THE BRIGHTNESS OF STARS: MAGNITUDE SCALES

A nearby candle may *appear* to be brighter than a distant streetlight, but in *absolute* terms the candle is much dimmer. This statement contains the essence of the problem of stellar brightness. A casual glance at a star does not reveal whether it is a nearby glowing ember or a distant great beacon. Hence astronomers distinguish between **apparent brightness** (the brightness perceived by an observer on Earth) and **absolute brightness** (the brightness that would be perceived if all stars were magically placed at a standard distance).

Apparent Magnitude

Let us look first at apparent brightness. Because scientists generally use the decimal system, we might expect astronomers to use a brightness scale in which each unit of brightness difference corresponds to a factor of 10. No such luck. Astronomers, too, are victims of history. When Hipparchus cataloged 1000 stars in about 130 B.C., he ranked their **apparent magnitude,** or apparent brightness, on a scale of 1 to 6, with 1st-magnitude stars the brightest and 6th-magnitude stars the faintest visible to the naked eye. This system stuck. As the famous astrophysicist Arthur Eddington remarked: "You have to remember that stellar magnitude is like a golfer's handicap—the bigger the number, the worse the performance."

Nineteenth- and twentieth-century astronomers extended that scale to cover objects brighter than 1st magnitude and dimmer than 6th magnitude. Because Hipparchus' 1st-magnitude stars are about 100 times brighter than his 6th-magnitude stars, they used this as the definition of the brightness system. A difference of five magnitudes is *defined* as a factor of 100 in brightness. Any given star is about $2\frac{1}{2}$ times brighter than a star of the next fainter magnitude.[2]

The system is illustrated in Figure 16-2, which shows a photo of the constellation Orion the Hunter, with some stars labeled with their magnitudes (0 to 9) on the right and, in the case of brighter stars, their Greek letter names on the left.

---

[2]The actual factor is $\sqrt[5]{100}$, or 2.512. This unusual factor as a unit of brightness difference is used in no other science, but it is so ingrained in astronomy that we continue to use it here.

TABLE 16-1	
Apparent Magnitudes of Selected Objects	
Object	**Apparent Magnitude**
Sun	-26.5
Full moon	-12.5
Venus (at brightest)	-4.4
Mars (at brightest)	-2.7
Jupiter (at brightest)	-2.6
Sirius (brightest star)	-1.4
Canopus (second brightest star)[a]	-0.7
Vega	0.0
Spica	1.0
Naked eye limit in urban areas	3–4
Uranus	5.5
Bright asteroid	6
Naked eye limit in rural areas	6–6.5
Neptune	7.8
Limit for typical binoculars	9–10
Limit for 15-cm (6-in.) telescope	13
Pluto	15
Limit for visual observation with largest telescopes	19.5
Limit for photographs with largest telescopes	23.5
Expected limit for Hubble Space Telescope	28±

Source: Data from Allen (1973).
[a]This lesser-known Southern Hemisphere star is used as a prime orientation point for spacecraft. A small light detector on spacecraft is called the "Canopus sensor."

Table 16-1 shows how the apparent magnitude scale has been extended, including fractional magnitudes. Extending the scale beyond the brightest objects included by Hipparchus meant using negative numbers. Thus the planet Venus at its brightest ranks more than minus 4th magnitude (-4.4), whereas the Sun reaches beyond minus 26th (-26.5).

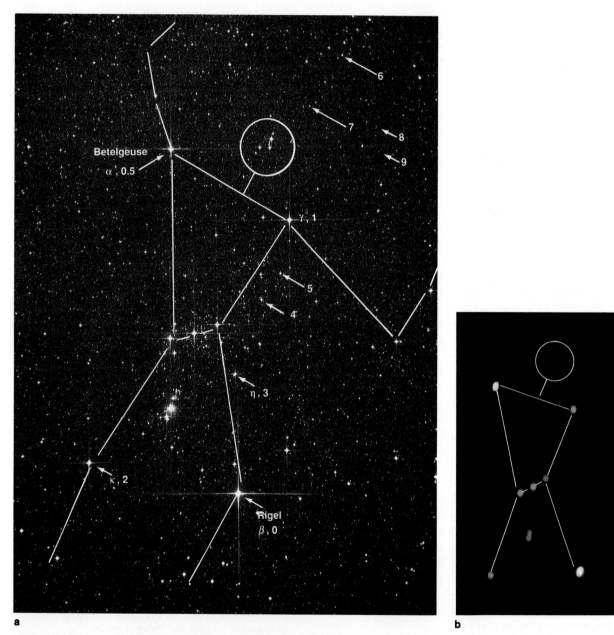

a

b

Figure 16-2 The constellation of Orion the Hunter (stick figure) showing Greek letter designations and approximate visual apparent magnitudes of certain stars. This exposure records stars down to about 10th magnitude. Different colors of stars are evident; most stars in this region are hot, blue stars, but Betelgeuse is a prominent cool, orangish-red star. (These color differences can be seen by the naked eye.) The three stars below Orion's Belt mark the sword. The middle "star" of the sword is the Orion Nebula (see Chapter 20), marked by red-glowing hydrogen gas. (Photo by author; 11-min guided exposure on 3M ISO 1000 film, 55-mm lens, f2.8, "star filter" emphasizes bright stars with diffraction spikes.) **b** An image of the same region, deliberately out of focus, expands the images of the stars into blobs that emphasize their color. Note the blue colors of most stars, due to their high temperatures, and note the difference in reddish hue between the orange thermal radiation of Betelgeuse and the red nonthermal Hα emission from the Orion Nebula. (Photo by author, 12-min guided exposure on Ektachrome ISO 1600 film, 55-mm lens, f2.8.)

Because various light detectors (the eye, photographic films, and photoelectric devices) have different sensitivities to different colors, astronomers specify exactly what color (such as blue light, red light, or infrared light) any measurements of magnitudes refer to. Here we will usually be referring to a system having the same color sensitivity as the human eye, sometimes called **visual apparent magnitude** (often abbreviated m_v).

Absolute Magnitude

Because the apparent magnitude of each star depends on the star's distance, it doesn't express the star's true energy output. This true, or intrinsic, energy output can be expressed using a magnitude system called the star's **absolute magnitude,** which is simply the apparent magnitude the star would have if it were moved to a standard distance of 10 pc. In this system, a star with the same energy output as the Sun has an absolute magnitude of about +5. A star with 100 times the energy output of the Sun is said to have an absolute magnitude of 0. A star 100 times fainter has an absolute magnitude of +10.

BASIC PRINCIPLES OF STELLAR SPECTRA

The last chapter showed that the Sun's **spectrum,** or distribution of light into different colors (**wavelengths),** gives information about the Sun's atmosphere, surface layers, and interior. The same is true of other stars. So critical is **spectroscopy**—the study of spectra—to astronomy that many astronomers devote their entire careers to it. Chapter 5 gives more details on the principles of spectra, and you may wish to review that material while studying this chapter. Let us briefly review the nature of spectra to see how they can be used to reveal the nature of stars.

Spectra can be presented in two ways—photographically or as a chart, depending on the instrument used (Figure 16-3). In a photographic image of the spectrum, the intensities of the photographic print represent intensities of light, or radiant energy. The vertical width of the spectrum is arbitrary, and the blue end is usually to the left, as in Figure 16-3.

In a chart, as in the lower portion of Figure 16-3, the vertical direction represents the intensity of the

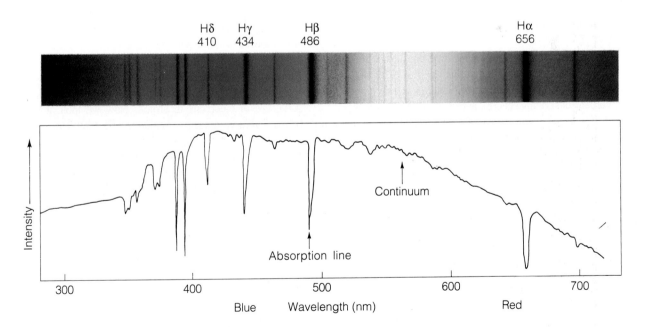

Figure 16-3 The visual portion of a typical stellar spectrogram is shown here in schematic form as it would appear on a color photographic plate (top) and as a graph (bottom). The spectral lines appear as dark vertical streaks in the photographic spectrum and as deep valleys in the scan.

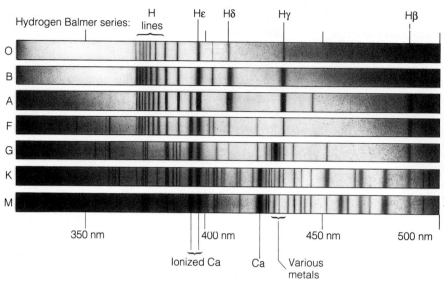

Figure 16-4 Representations of stellar spectra in the seven spectral classes. Classes A and B show the Balmer hydrogen lines especially well. (Compare with Table 16-2.)

light. Wavelength is shown along the bottom, again with red usually to the right. **Absorption lines** appear as dark vertical lines on a photographic spectrum and as notches or valleys on a graph. **Emission lines** appear as bright vertical lines or as sharp peaks. The general level of brightness between absorption or emission lines is the **continuum**.

Spectra of Stars

In 1872, Henry Draper, a pioneer in astronomical photography, first photographed stellar spectra. This represented a tremendous advance. Instead of sketching or verbally describing spectra, astronomers could directly record, compare, and measure them. Spectra of thousands of stars became available for precise analysis.

Such massive amounts of data required a classification scheme. This was begun in the 1880s by Harvard astronomer Edward C. Pickering and completed by Annie J. Cannon and a group of young assistants who invented a system of **spectral classes** based on the number and appearance of spectral lines. The classes—A, B, C, and so on—started with spectra with strong hydrogen lines. When Cannon published the Henry Draper Catalog (1918–1924), containing spectral data on 225,320 stars, it became the basis for all modern astronomical spectroscopy.

Further work showed that the classes had to be rearranged to bring them into a true physical sequence based on temperature, as shown in Figure 16-4. The sequence finally adopted begins with the hottest stars—class O—which show ionized helium lines in their spectra. The sequence of classes is O, B, A, F, G, K, and M.[3] The M stars are the coolest. About 99% of all stars can be classified into these groups. For finer discrimination the classes are sometimes divided from 0 to 9; the Sun, for example, is said to be a G2 star.

Spectral Lines: Indicators of Atomic Structure

In Chapter 5 we gave a simplified description of how emission and absorption lines depend on the orbital structure of electrons in atoms. Now we look at the process in more detail by considering the simplest, most abundant, and most important type of atom in the universe: hydrogen. As shown in Figure 16-5, hydrogen atoms' orbits can be numbered. Because a neutral hydrogen atom has just one electron, an atom in the ground state would have one electron in the orbit $n = 1$. If the atom had been bumped by other atoms or if it had

---

[3]As Annie Cannon noted, the rearrangement of letters made the sequence harder to remember. In the 1920s, Harvard astronomer Henry Norris Russell solved this problem with Yankee ingenuity by proposing a sentence as a memory aid: Oh, Be A Fine Girl, Kiss Me! Two later classes, R and N, were easily accommodated by adding the words *Right Now*. However, when class S was added, a controversy broke out between Harvard and Cal Tech over whether it signified the word *Sweetie* or *Smack*.

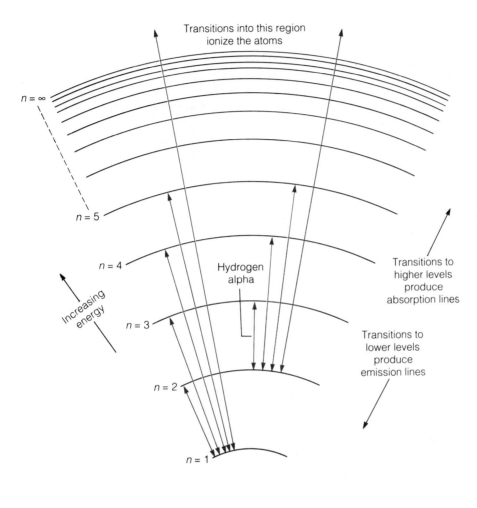

Transitions into this region
ionize the atoms

$n = \infty$

$n = 5$

$n = 4$

Hydrogen
alpha

Increasing
energy

$n = 3$

$n = 2$

$n = 1$

Transitions to
higher levels
produce
absorption lines

Transitions to
lower levels
produce
emission lines

Nucleus

Figure 16-5 Schematic diagram of a hydrogen atom showing different possible electron orbits or energy levels ($n = 1, 2, 3, \ldots$). Each change of orbit by an electron produces a spectral line, because energy is removed from or added to the light beam. The series of lines starting from or ending on $n = 1$ occurs in the ultraviolet part of the spectrum and is called the Lyman series. Lines starting from or ending on $n = 2$ occur in the visible part of the spectrum and are called the Balmer series. The well-known red Hα line of the Balmer series involves transitions between $n = 2$ and $n = 3$.

absorbed radiation, the electron might be in the orbit $n = 2$, 3, and so on. Further absorption of energy might cause it to jump from $n = 3$ to $n = 4$, creating an absorption line, or it might spontaneously revert from $n = 3$ to $n = 2$, creating an emission line. Each possible transition (1 to 2, 2 to 3, 4 to 2, and so on) creates a different line. As Figure 16-5 shows, transitions between the $n = 2$ level and higher levels create the lines prominent in the visible part of the spectrum, called the **hydrogen Balmer series** of lines. The famous red line called **hydrogen alpha** (Hα, shown in Figure 16-3, top) or **Balmer alpha,** involves transitions between $n = 2$ and $n = 3$ in the hydrogen atom. This line lends its brilliant red color to many astronomical gases, including the solar chromosphere and many nebulae. At various temperatures and pressures, the different lines of a given element will have different relative strengths.

Note the power of the knowledge we have discussed so far. *Identification* of the spectral lines tells us what elements are present in a star, because different elements display different patterns of lines, as unique as fingerprints. *Measurement of the relative line strengths* of a given element tells us about the temperature, density, and pressure conditions of the star's photospheric (surface) layer. This method supplements the temperature information gained from other techniques, such as Wien's law, discussed in Chapter 5.

Spectral Classes as a Temperature Sequence

Each spectral class corresponds to a different *temperature* in the gases in the light-emitting layers of the stars. **Temperature** is merely a measure of the average velocity of the atoms or molecules of a substance. The hotter a gas, the faster its atoms and molecules move. If the temperature quadruples, the velocities of the atoms double. The faster the atoms collide, the more they disturb or dislodge each other's electrons. Furthermore, the hotter the gas, the more radiation it emits; the resulting photons also disturb electron structures of atoms. Because spectral lines depend on electron structures, and because spectral classes depend on spectral lines, spectral classes therefore form a temperature sequence.

Consider what happens as we reduce the temperature of matter. The hottest stars, the O stars, have temperatures of 40,000 K or more, as shown in Table 16-2. Here the atoms and radiation are so energetic that even the most tightly bound atoms, such as helium, have had their electrons knocked off, as the table indicates. By class B, the temperature has dropped to around 18,000 K, and helium atoms keep their electrons. By classes A and F, even relatively weakly bound hydrogen keeps its electrons. Many metals have at least one electron that can be very easily knocked off the outer part of the atom, and in G-type stars like the Sun, at around 5500 K, lines of ionized metals are prominent along with Balmer lines of hydrogen. By class K, at 4000 K, even many metals are neutral. By class M, at 3000 K, energies are so low that different atoms can stick together into molecules; even water molecules have been identified in spectra of such cool stars.

Two Important Laws of Spectroscopy

We now introduce two important effects that will help us understand many astronomical phenomena: the Doppler effect and the Stefan-Boltzmann law.

> The *Doppler effect* is a shift in wavelength of a spectral absorption or emission line away from its normal wavelength, caused by motion of the light source toward or away from the observer.

If the source approaches, there is a **blue shift** toward shorter wavelengths, or bluer light. If the source recedes,

a **red shift** occurs toward longer wavelengths, or redder light. Thus, if a star moves away from us at high speed, all of its spectral lines will appear at slightly redder wavelengths than normal. This shift can be measured, and thus tells us the velocity of the star's motion away from us. The amount of the shift is just proportional to the approach or recession speed of the source (as long as that speed is well below the speed of light). In the shorthand of mathematics,

$$\frac{\text{Shift in wavelength}}{\text{Normal wavelength}} = \frac{\text{approach or recession speed}}{\text{velocity of light}}$$

If the light source is receding at 10% the speed of light, the light will be red-shifted by 10% of its normal wavelength; a line normally found at wavelength 500 nm would appear at 550 nm. By measuring such wavelengths, velocities of stars toward us or away from us can be studied.[4]

> The *Stefan–Boltzmann law,* discovered by two Austrian physicists about 1880, states that the higher the temperature of a surface, the more energy radiated by each square meter in each second.

MEASURING NINE IMPORTANT STELLAR PROPERTIES

We now can apply the Stefan–Boltzmann law, the Doppler effect, and other principles to explain how nine important physical properties of remote stars can actually be measured. These applications, to be discussed in the following sections, are summarized in Table 16-3.

How to Measure a Star's Distance

From what you have read so far, can you think of a way to measure the distance of a star? The question is not farfetched, because star distances were first measured more than a century ago. A crude method may be applied to the information in Table 16-1. The stars that look brightest have *apparent* magnitudes about 25 magni-

---

[4]The same shift occurs for sound. This explains why the noise of a rapidly traveling automobile, truck, or train changes from higher pitch (shorter wavelength) to lower pitch (longer wavelength) as the vehicle rushes by. The sound a child makes to imitate a race car rushing by is a representation of the Doppler shift.

TABLE 16·2

Principal Spectral Classes of Stars

Type	Spectral Class	Typical Temperature (K)	Source of Prominent Spectral Lines	Representative Stars
Hottest, bluest	O	40,000	Ionized helium atoms	Alnitak (ζ Orionis)
Bluish	B	18,000	Neutral helium atoms	Spica (α Virginis)
Bluish-white	A	10,000	Neutral hydrogen atoms	Sirius (α Canis Majoris)
White	F	7000	Neutral hydrogen atoms	Procyon (α Canis Minoris)
Yellowish-white	G	5500	Neutral hydrogen, ionized calcium	Sun
Orangish	K	4000	Neutral metal atoms	Arcturus (α Bootes)
Coolest, reddest	M	3000	Molecules and neutral metals	Antares (α Scorpii)

TABLE 16·3

Nine Basic Properties of Stars and How They Are Measured

Property	Method of Measurement
Distance	Trigonometric parallax
Luminosity (absolute magnitude)	Distance combined with apparent brightness
Temperature	Color or spectra
Diameter	Luminosity and temperature
Mass	Measures of binary stars and use of Kepler's laws
Composition	Spectra
Rotation	Spectra, using Doppler effect
Circumstellar material	Spectra, using absorption lines and Doppler effect
Motion	Astrometry or spectra, using Doppler effect

tudes fainter than the Sun. In 1829, the English scientist William Wollaston used this simple fact to estimate that most typical stars must be at least 100,000 times more distant than the Sun, because dimming by 25 magnitudes corresponds to increasing distance 100,000 times.

Because this gives only a typical distance, we need a better technique to measure distances of individual stars. The most important such technique is **parallax** measurement. You will recall that parallax is the apparent shift in the position of an object caused by a shift in the observer's position (page 32).

The principle is easy to understand, as shown in Figure 16-6. Hold your finger in front of your face and look past it toward some distant objects. Your finger represents a nearby star; the objects, distant stars. Your right eye represents the view on one side of the Sun. Your left eye represents the view 3 mo later, after the Earth has traveled to a point lined up with the Sun (a shift in the Earth's position by 1 AU as seen from the star). First wink one eye and then the other. Your finger (the nearby star) seems to shift back and forth. Hold your finger only a few centimeters from your eyes; the

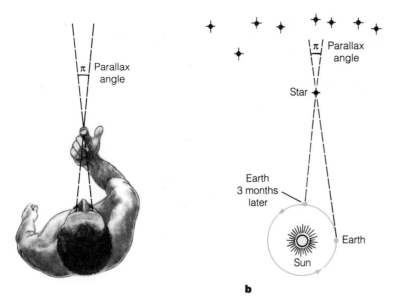

Figure 16-6 The principle of parallax determination applied to one's finger **a** and to stars **b**. Nearby objects observed from two positions appear to shift by a parallactic angle π compared with background objects; measurement of π gives the distance of the object, once the separation of the two positions is known. (By astronomical convention, the angle cataloged as a star's parallax corresponds to the angle diagrammed here; but astronomers actually observe from two positions 6 mo apart. This yields twice as large a parallactic shift, which is easier to measure.)

shift is large. Hold your finger at arm's length; the shift is smaller. Likewise, the farther away the nearby star, the smaller the parallax. The parallax in this experiment may be measured in degrees, but the parallaxes of actual stars are all less than a second of arc.

Credit for first measuring the parallax of a star is usually given to the German astronomer Friedrich Bessel, who in 1838 published the parallax and distance of 61 Cygni (star 61 in the constellation of Cygnus the Swan). Bessel found the parallax of 61 Cygni to be $\frac{1}{3}$ second of arc and the distance to be about 3 pc. Most stars are farther than 3 pc away, and their parallaxes are correspondingly smaller. Parallaxes as small as $\frac{1}{20}$ second of arc have been reliably measured, and the distance of such stars is 20 pc.

As the preceding figures show, the distance of a star in parsecs is simply the inverse of its parallax angle in seconds of arc. This explains the origin of the term parsec: It is the distance corresponding to a PARallax of one SECond of arc.

The more distant a star, the smaller its parallax. Thus if a star is too distant, its parallax is too small to be measured. Parallaxes smaller than about $\frac{1}{20}$ second of arc are difficult to measure accurately, though improved techniques are being developed. *Therefore stars farther away than about 20 pc are beyond the* **distance limit for very reliable parallaxes.** Distances from 20 pc to 200 pc can be estimated by new, refined parallax techniques.

Knowing accurate distances is the prime requirement of measuring most other properties of stars. Thus the estimated 1000 to 2000 stars that lie within 20 pc are our main statistical sample for measuring stellar properties. Other techniques have been devised for estimating distances to more remote stars, but they all depend ultimately on the accuracy of the parallax measures of nearby stars.

How to Measure a Star's Luminosity

The **luminosity** of a star is its *absolute brightness*—the total amount of energy it radiates each second. Unlike apparent brightness (discussed earlier in this chapter), luminosity is intrinsic to a star. The most useful concept of luminosity is **bolometric luminosity**—*the total amount of energy radiated each second in all forms at all wavelengths.*

The bolometric luminosity of the Sun ($L_\odot$) is 4 × 10^{26} watts. (Note: one watt = 1 joule of energy radiated each second. The subscript $\odot$ is a symbol for the Sun.) Many stars, of course, are much brighter than the Sun. A star twice as luminous as the Sun, radiating 8 × 10^{26} watts, would be said to have a luminosity of 2 $L_\odot$.

The most basic method of estimating luminosity derives from the measurement of distance. A faint light in the night may be a candle a hundred meters away, a streetlight a few kilometers away, or a brilliant lighthouse beacon 100 km away. Once we know the distance

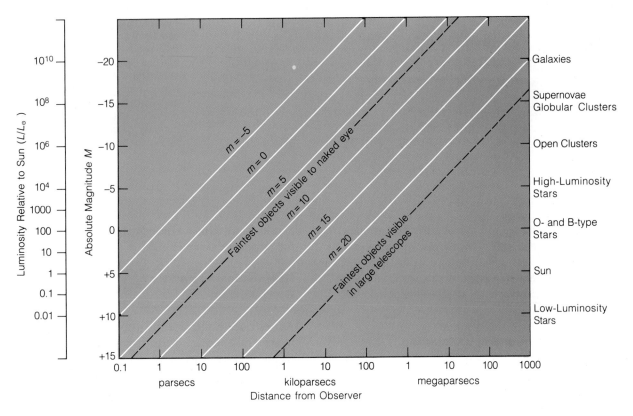

Figure 16-7 If a certain star has a known absolute magnitude M (vertical scale, possibly determined from spectra or other means), astronomers can measure its apparent magnitude m and estimate its distance from a chart like this one. For example, a high-luminosity star of $M = -5$ could be visible to the naked eye with visual magnitude $m = 5$ at a distance of around 1000 pc. Note that galaxies (see right end of graph) are so luminous that some can be visible to the naked eye at distances of about 1 Mpc, or 1,000,000 pc. The values shown here assume clear space; see Chapter 20 for a discussion of dimming due to interstellar "smog" in many regions of space.

to an object, we can determine its absolute brightness. For example, Figure 16-7 shows that a 1st-magnitude star (that is, apparent magnitude $m = 1$) 10 pc away would have an absolute magnitude of $+1$ and a luminosity of about 40 $L_\odot$.

This method becomes less accurate for very distant stars. For one thing, as mentioned earlier, parallax distance measures become unreliable beyond about 20 pc. More important, there is some interstellar haze that dims starlight over larger distances. Just as a fog bank might keep you from telling whether a distant light was a streetlight or a more distant lighthouse, variable interstellar haze in uncertain amounts complicates estimates of stars' luminosities.

However, other methods are available. For exam-

ple, very luminous stars have slightly different spectral characteristics from faint stars of the same spectral class. Spectra can therefore be used to measure luminosities as shown in Figure 16-8.

Note that the luminosity, the distance, and the apparent magnitude of an object are all interrelated. As shown in Figure 16-7, if we know any two of these quantities, we can estimate the third.

How to Measure a Star's Temperature

In measuring stellar temperatures, spectra again come to our aid. Spectral classes offer a rough guide to temperature. At a more precise level of detail, we can mea-

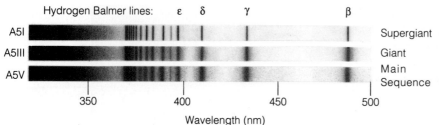

Figure 16-8 Comparison of spectra of a supergiant, giant, and main-sequence star of constant spectral class. In the expanded atmosphere of the larger, more luminous stars, pressures are less and atoms less disturbed by collisions, resulting in slightly narrower, better-defined absorption lines.

sure the spectrum, see which color is the most strongly radiated, and then apply Wien's law to derive the temperature.

Stellar temperatures range all the way from about 2500 K for cool M-type stars to 40,000 K or more for hot O-type stars.

How to Measure a Star's Diameter

Most stellar diameters are measured from temperature and luminosity by applying the Stefan–Boltzmann law. By this law, a star radiating a certain number of watts and having a certain temperature must have a certain area. The star's diameter can be found from its area. Such measures indicate a vast range of stellar diameters, from less than Earth size, through Sun size, to huge stars larger than the diameter of Mars' orbit!

How to Measure a Star's Mass

According to Kepler's laws as modified by Newton, when any two cosmic objects are in orbit around each other, the period of revolution (the time it takes to complete an orbit) increases as the distance between the objects increases and as the sum of the masses of the two objects decreases. Many pairs of stars orbit around each other. For many of these *binary stars* we can measure both the period of revolution and the distance between the two stars. Thus we can calculate the sum of the masses. By applying other Newtonian laws, astronomers are able to use additional orbital properties to derive the mass of each individual star in the binary pair.

Study of many stars by this technique reveals an important fact: *Stars with nearly identical spectra usually have nearly identical masses.* This fact allows the masses

of many stars to be estimated roughly from their spectral properties alone.

How to Measure a Star's Composition

As described earlier, stellar compositions are revealed by spectra. There are two problems: detecting the *presence* of an element and measuring its *amount*. The presence of an element is detected by identifying at least one—or preferably several—of its absorption lines or emission lines in the spectrum of the star.

The amount of the element is indicated by the appearance of the spectral line. Generally, the wider and darker the absorption lines, the more atoms of the element are present. Likewise, the wider and brighter the emission lines, the more atoms. Studies of this type around 1925 by Cecilia Payne-Gaposchkin and others showed that most nearby stars are approximately Sun-like in composition.

How to Measure a Star's Rotation

If a rotating star is seen from any direction except along the rotation axis, one edge will be approaching the observer and one edge will be receding (Figure 16-9). Light emitted or absorbed at the approaching edge will be blue-shifted. Light from the other edge will be red-shifted. Consider an absorption line being formed in all parts of the star's atmosphere. If the star were not rotating, neither shift would occur and the line would be very narrow. Because the star rotates, the line is broadened. The faster the rotation and the closer our line of sight to the equatorial plane, the more line broadening occurs.

The rotations of stars can thus be inferred to some

extent from measurements of the broadening of spectral lines. In general, the fastest rotators are the hot, bluish, massive stars of class O, which have equatorial rotation speeds of around 300 km/s. Stars cooler than F stars rotate much more slowly. The Sun's equatorial speed is only 2 km/s.

How to Detect Circumstellar Material

Unstable stars, such as newly forming or dying stars, may throw out more massive clouds of gas and dust than the thin, flamelike prominences ejected by the Sun (see Figure 16-10). These surround the stars and are called **circumstellar nebulae.** Because many of these clouds are nearly transparent, we can see light arising from both the near side and far side of the cloud, as shown in Figure 16-11. As such a cloud expands away from the star, absorption lines arising on the near side (where the starlight passes through the gas) are blue-shifted compared with lines in the star because the gas is approaching the observer. Emission lines from glowing gas on the near side are also blue-shifted, but emission lines from the far side of the cloud are red-shifted.[5] These shifted lines are telltale clues that a star has an expanding circumstellar nebula.

In certain newly formed stars the circumstellar material is not glowing gas but dust grains. These grains are warmed by the star and detected by the thermal infrared radiation they give off.

How to Detect a Star's Motion

The Doppler shift is a very simple way to detect *part* of a star's motion—its **radial velocity,** or *motion along the line of sight.* A consistent blue or red shift of all of a star's spectral lines proves that the star is moving toward or away from us. Among the 50 nearest stars, about half the radial velocities measured are more than 20 km/s toward or away from us.

The other component of a star's motion is its **tangential velocity**—the *motion perpendicular to the line of sight.* It cannot be measured as simply as radial velocity. In order to measure tangential velocity, we must measure the distance of the star and its rate of angular

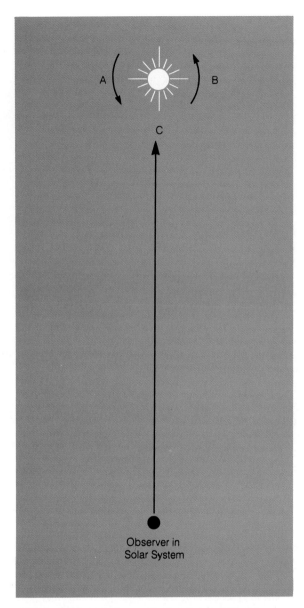

Figure 16-9 Light from a rotating star would be blue-shifted if from side *A,* red-shifted if from side *B,* or unshifted if from the central region *C.* The net result is a broadening of spectral lines.

motion across the sky, called **proper motion.** (Values of a few seconds of arc per year are common for nearby stars.) These measurements are often lumped together in a special branch of astronomy called **astrometry.**

If both radial and tangential velocities are known, they can be combined to give the star's **space velocity:** its true speed and direction of motion in three-

---

[5]For a discussion of how such spectral lines are produced, see Kirchhoff's third law (Chapter 15).

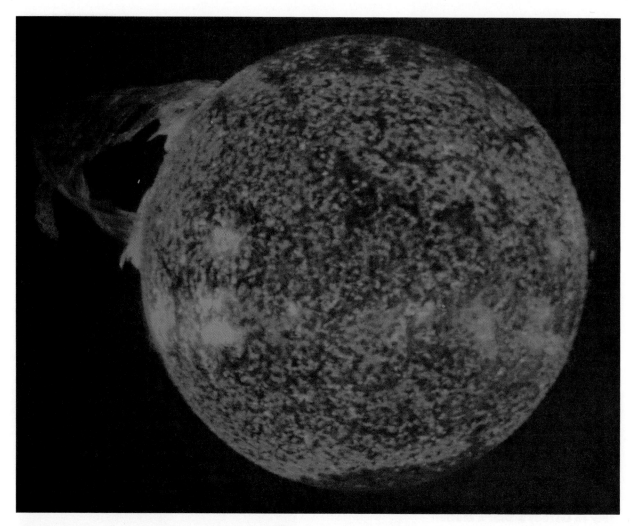

Figure 16-10 A star blowing off a cloud of gas; one of the largest solar flares ever observed shoots glowing gas off the Sun. This is a false-color image of a photo made at extreme ultraviolet wavelengths (30 nm)—light emitted by excited helium atoms but invisible to the human eye. Interestingly, the photo shows that the Sun's poles emit little light at this wavelength. (NASA photo by Skylab astronauts, 1973.)

dimensional space *relative to the Sun* (Figure 16-12). Space velocities of most stars near the Sun are a few tens of kilometers per second and are nearly random in direction.

SUMMARY

This chapter emphasizes methods of studying stars, rather than the nature of the stars themselves. The chapter has two basic sections. The first section describes concepts:

the system of naming stars; the definitions of *light-year* and *parsec;* the system of magnitudes for measuring brightness; stellar spectra; the Doppler effect; and the Stefan–Boltzmann law, which describes the amount of stellar radiation. The second section describes methods for determining nine basic stellar properties, as summarized in Table 16-3.

Clearly, our journey outward from Earth has taken us far beyond the places that we can investigate in the near future by explorations with space crews or instruments. Even at the speed of light (which is 10,000 times greater than speeds our instruments have achieved), it would take

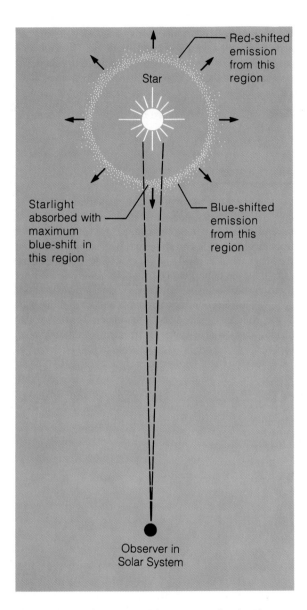

Figure 16-11 Gaseous envelopes expanding (as shown here) or rotating around stars produce characteristic spectral effects, allowing their detection even if they cannot be seen separately.

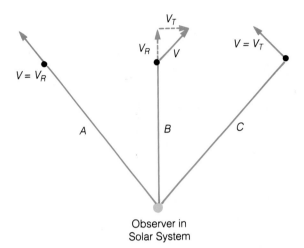

Figure 16-12 Three examples of a star's space velocity *(V)* compared with its radial velocity *(V$_R$)* and tangential velocity *(V$_T$)*. In *A*, the star's space velocity is aligned with the line of sight; no proper motion would be seen. *B* is the most common case, in which both *V$_R$* and *V$_T$* are appreciable. In *C*, the space velocity is perpendicular to the line of sight; no radial velocity or Doppler shift would be seen.

CONCEPTS

star	temperature
light-year	Doppler effect
parsec	blue shift
apparent brightness	red shift
absolute brightness	Stefan–Boltzmann law
apparent magnitude	parallax
visual apparent magnitude	distance limit for very reliable parallaxes
absolute magnitude	
spectrum	luminosity
wavelength	bolometric luminosity
spectroscopy	circumstellar nebula
absorption line	radial velocity
emission line	tangential velocity
continuum	proper motion
spectral class	astrometry
hydrogen Balmer series	space velocity
hydrogen alpha (Balmer alpha)	

years to reach the nearest stars. Thus, instead of sending instruments to the stars, we must rely on the messages in the starlight coming to us. Our abilities to interpret starlight are rapidly increasing, not only with construction of new large telescopes but also with the launching of such telescopes into orbit.

PROBLEMS

1. If a telescope of 25-m (1000-in.) aperture could be put in orbit or on the Moon, it could resolve an angle of about 0.005 second of arc. Compare this resolution with the maximum angular size known for stars (0.03″). Why would this telescope perform better on the Moon than on Earth? Why do astronomers never use the highest magnifications theoretically possible with earthbound telescopes?

2. A star is 20 pc away. How many years has its light taken to reach us?

3. What is the chemical composition of most stars?

4. A reddish star and a bluish star have the same radius. Which is hotter? Which has higher luminosity? Describe your reasoning.

5. A reddish star and a bluish star have the same luminosity. Which is larger? Describe your reasoning.

6. How does the magnitude scale differ from the scales by which we measure lengths (centimeters or inches), temperature (degrees Celsius or Kelvin), weight (kilograms or pounds), or other familiar properties?

7. A certain star has exactly the same spectrum as the Sun but is 30 magnitudes fainter in apparent magnitude. List some conclusions you would draw about it and describe your reasoning.

8. A star has a parallax of 0.05 second of arc. How far away is it?

9. The spectral lines of metals, such as calcium, are prominent in the solar spectrum. Hydrogen lines are less prominent. Why does this not indicate that the Sun consists mostly of these elements instead of hydrogen?

10. Each of a certain star's spectral lines is found to be spread out over a wide range of wavelength. What might cause this?

PROJECT

1. Using a telescope of fairly high magnification (such as $300\times$ or $400\times$), examine the image of a bright star on several different nights and sketch it. Are there differences from night to night? Can you identify telescopic effects on the image, such as small rings or cross-hair–shaped spikes caused by the optical effect called diffraction? If so, label them on your sketches. Note any shimmering due to atmospheric turbulence or air currents in and around the telescope. Run the eyepiece inside or outside the focus point, making the image a round blob. Shimmering and other turbulent effects are often more evident in this way.

Use a piece of cardboard with a 1- or 2-in. hole to reduce the size of the aperture of the telescope. What changes do you see? Examine a double star (two stars very close together) and perform this experiment. Is the resolution better or worse at small aperture?

CHAPTER 17

The Systematics of Nearby Stars: The H–R Diagram

As soon as astronomers could begin to measure different properties of stars, they began to categorize the stars. This chapter will discuss three interesting discoveries that resulted. First, stars come in a wide variety of forms: massive and not so massive, large and small, bright and faint. Second, one reason for the different forms is that stars evolve from one form to another. Third, another reason is that stars of different initial mass evolve at different rates and into different final forms.

In this sense, the population of stars can be likened to that of people. If you travel from country to country and observe people, you tend to see mostly ordinary people—youngsters, middle-aged people, and elderly people. The creation of new people—the birth of babies—is secluded in special places such as hospitals and is not readily visible to the tourist. In addition, the tourist may see tombstones, but the actual deaths of people are usually brief events not commonly encountered. This metaphor explains the approach of this chapter and the next two chapters. In this chapter we will tour the local population of stars. We will find that the great majority of these stars are ordinary stars—that is, youthful, middle-aged, or growing elderly.

CLASSIFYING STAR TYPES: THE H–R DIAGRAM

When astronomers began to determine properties of stars by using the techniques described in the last chapter, they found a puzzling variety of star forms. Masses range from about 0.1 to 60 $M_\odot$. One star in a billion is a dazzling object even more massive than 60 $M_\odot$. Luminosities range from around one-millionth to a million times that of the Sun. Surface temperatures range from

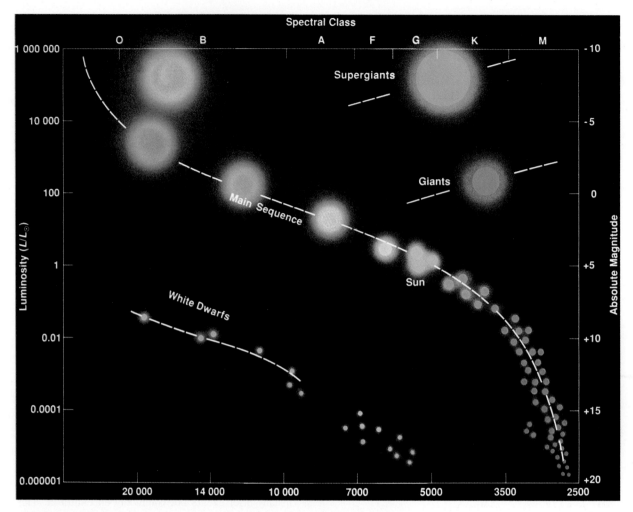

Figure 17-1 An H–R diagram for a selection of different types of stars. Schematic drawings give colors and relative sizes, but not to true scale (which would require much larger red giants). The main sequence, running diagonally across the diagram, comprises the "normal" hydrogen-burning stars.

about one-third to nearly 10 times the solar temperature. Diameters range from less than 1/100 to more than 100 times the Sun's size.

When astronomers realized that there is such a variety of star forms, their first step was to devise a sensible way to arrange and study the data about them, in hopes of finding relationships among the various forms of stars. This could have been done in several ways, but one method has become traditional. This method was introduced around 1905 to 1915 by Danish astronomer Ejnar Hertzsprung and American astronomer Henry Norris Russell. They constructed diagrams plotting the spectral class of stars versus their luminosity. This type of plot is usually called the **H–R diagram** in

honor of Hertzsprung and Russell.[1] An example is seen in Figure 17-1, where schematic color images are used to show stars of different types.

As we saw in the last chapter, temperature is the principal factor that governs the differences among spectra of O, B, A, F, G, K, and M stars. For this reason, many astrophysicists make H–R diagrams with a temperature scale instead of a spectral class scale. In this book, we identify temperature, spectral class, lumi-

---

[1]Sometimes this graph is called the spectrum–luminosity diagram or color–magnitude diagram, because plots of spectral type versus luminosity or color versus absolute magnitude produce equivalent information.

nosity, and absolute magnitude along the edges of the diagram. (Following Russell's original version, the spectral classes are arranged with *cooler* stars to the right.)

Different locations on the H–R diagram correspond to different types of stars. It is important to understand why this is true. First, remember from Wien's law that stars of different temperature have different color. Thus, as shown schematically in Figure 17-1, the right part of the diagram corresponds to redder stars and the left part to bluer stars. Similarly, the upper part corresponds to more luminous stars, and the lower part to less luminous stars.

Note that the Sun is near the middle of the main sequence (luminosity $1\,L_\odot$, spectral type G). If we move upward from the Sun's position, we encounter stars more luminous than the Sun, even though they have the same temperature. Thus these stars must be larger than the Sun. Similarly, a star directly below the Sun's position on the diagram must be smaller than the Sun.

Do stars appear *throughout* the diagram? Or are there only certain types such as main-sequence stars, giants, and dwarfs? The answer is that only certain parts of the H–R diagram are crowded with stars. Other types of stars, which would fall in other parts of the diagram, are rare or nonexistent. Using the H–R diagram, astronomers began to investigate why this is so.

For example, even the earliest H–R diagram revealed an important discovery about stars: *Among most stars there is a smooth relation between spectral class and luminosity.* This band came to be called the *main sequence* of stars. Hertzsprung called stars obeying this relation (falling on the diagonal band of the H–R diagram) **main-sequence stars.** The main sequence runs from hot luminous blue stars (upper left) to cool faint red stars (lower right). It turns out that these main-sequence stars all have something in common: They generate their energy by consuming hydrogen in nuclear reactions, just as the Sun does. We will return to that in a moment, but first let's consider some other aspects of the H–R diagram.

THE NEARBY STARS AS A REPRESENTATIVE SAMPLE OF ALL STARS

Figure 17-1 shows an H–R diagram plotting various classes of discovered stars. Some of these classes are brighter and larger than the Sun and some are fainter and smaller. But what kinds of stars are representative of the stars in the broad region of our galaxy near the Sun? Are they mostly Sun-like? Or brighter? Or fainter?

Figure 17-2 answers these questions with an H–R diagram plotted for the 100 stars closest to the Sun.

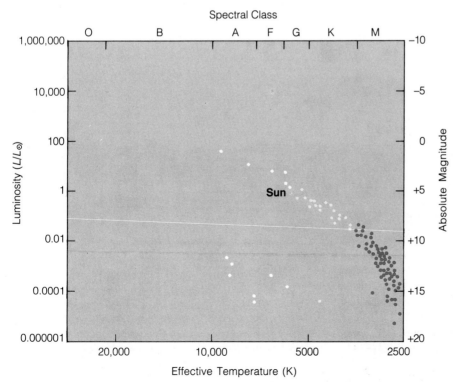

Figure 17-2 An H–R diagram for the 100 nearest stars—a representative sample from a volume of space around the Sun. This diagram shows that most stars are on the main sequence. Note also that most are fainter and cooler than the Sun. Therefore, they fall in the lower right corner.

TABLE 17-1

The 27 Nearest Stellar Systems (Stars Within 4 Parsecs)

Distance (pc)	Star Name	Component	Apparent Magnitude (m_v)	Absolute Magnitude (M_v)	Spectral Type[a]	Mass ($M_\odot$)	Radius ($R_\odot$)	Semimajor Axis in Multiple Systems (AU)
0.0	Sun (and Jupiter)	A	−27	5	G2	1.0	1.0	5.2
		B	—	—	—	0.001	0.1	
1.3	Alpha Centauri	A	0	4	G2	1.1	1.2	23.6 (AB)
		B	1	6	K0	0.9	0.9	
		C[b]	11	15	M5	0.1	?	13,000 (AC)
1.8	Barnard's Star	—	10	13	M5	?	?	—
2.3	Wolf 359	—	14	17	M8	?	?	—
2.5	BD + 36°2147	A	8	10	M2	0.35	?	0.07
		B	?	?	?	0.02	?	
2.7	L726-8 (= UR Ceti)	A	12	15	M6	0.044	?	10.9
		B	13	16	M6	0.035	?	
2.9	Sirius	A	−2	1	A1	2.3	1.8	19.9
		B	8	11	Wh. dw., A4	1.0	0.02	
2.9	Ross 154 (= Gliese 729)	—	11	13	M5	?	?	—
3.1	Ross 248	—	12	15	M6	?	?	—
3.2	L789-6	—	12	15	M6	?	?	—
3.3	Epsilon Eridani	—	4	6	K2	0.9	?	—
3.3	Ross 128	—	11	14	M5	?	?	—
3.3	61 Cygni	A	5	8	K5	0.63	?	85 (AB)
		B	6	8	K7	0.6	?	
		C	?	?	?	0.008	?	
3.4	Epsilon Indi	—	5	7	K5	?	?	—
3.5	Procyon	A	0	3	F5	1.8	1.7	15.7
		B	11	13	Wh. dw.	0.6	0.01	

These are **representative stars** in our part of space. The Sun is plotted near the center, in spectral class G at luminosity 1 $L_\odot$. The diagram reveals the interesting fact that *most representative stars are fainter, cooler, and smaller than the sun.* Figure 17-2 also shows that *about 93% of representative stars fall on the main sequence.*

Table 17-1 lists more detailed properties of the 27 stellar systems within 4 pc of the Sun. This table reveals another interesting fact about representative stars: *Most stars are members of systems in which two or three stars orbit around one another.* Stars without stellar companions, such as the Sun, are in the minority.

The 27 Nearest Stellar Systems (Stars Within 4 Parsecs), *continued*

Distance (pc)	Star Name	Component	Apparent Magnitude (m_v)	Absolute Magnitude (M_v)	Spectral Type[a]	Mass $(M_\odot)$	Radius $(R_\odot)$	Semimajor Axis in Multiple Systems (AU)
3.5	Σ 2398 (= Gliese 725 = Groombridge 34)	A B	9 10	11 12	M4 M5	0.4 0.4	? ?	60
3.5	BD + 43°44 (= Gliese 15)	A B C	8 11 ?	10 13 ?	M1 M6 K?	? ?	?	156 (AB)
3.6	Tau Ceti	—	4	6	G8	?	1.0	—
3.6	CD − 36°15693 = Lacaille 9352	—	7	10	M2	?	?	—
3.7	BD + 5°1668 (= Luyten's Star)	A B	10 ?	12 ?	M4 ?	? ?	? ?	?
3.7	G51-15	—	15	17	?	?	?	—
3.8	L725-32	—	12	14	M5	?	?	—
3.8	CD − 39°14192 = Lacaille 8760	—	7	9	M0	?	?	—
3.9	Kapteyn's Star	—	9	11	M0	?	?	—
4.0	Krüger 60 (= DO Cep = Gliese 860)	A B C	10 11 ?	12 13 ?	M4 M6 ?	0.27 0.16 0.01	0.51 ? ?	9.5 (AB)
4.0	Ross 614	A B	11 15	13 17	M5 ?	0.14 0.08	? ?	3.9
4.0	BD − 12°4523 (= Gliese 628)	—	10	12	M5	?	?	—

Source: Data from van de Kamp (1981) and Baum (1986). A companion to Barnard's Star listed by van de Kamp is regarded as dubious and deleted here.

[a]The numeral represents a subdivision of each spectral class from 0 (hottest) to 9 (coolest). A G5 is halfway between the F and K classes, and a G9 would be closest to a K0.

[b]Proxima Centauri, the closest known star beyond the Sun.

The closest star system to the Sun is the triple system of **Alpha Centauri,** brightest star in the southern constellation of Centaurus. Shown in Figure 17-3, Alpha Centauri is easily visible from Hawaii, but it is too far south to be prominent from most of the rest of the United States. Because Alpha Centauri is so close and so bright, we can accurately measure the distances, spectral classes, temperatures, luminosities, and sizes of its components. By measuring the orbital size and period, we can use Kepler's laws of orbital motion and Newton's law of universal gravitation to deduce the masses of the components. The main two components

Figure 17-3 The nearest star system, Alpha Centauri, and the adjacent region of the sky. Neither of two companion stars orbiting around Alpha is seen here—one (Alpha B) is too close to Alpha A, and the other (Proxima) is too faint. Sketch map in **b** identifies prominent features, including the famous constellation called the Southern Cross; a dust cloud known as the Coal Sack, which blots out part of the background Milky Way; and nebulae glowing with the red color of Hα emission. The Coal Sack is easily visible to the naked eye, but the red glows are too faint for the eye to perceive as colored. (Photo by author; 18-min exposure, 55-mm lens, f2.8, Ektachrome ISO 1600 film, Mauna Kea, Hawaii.)

a

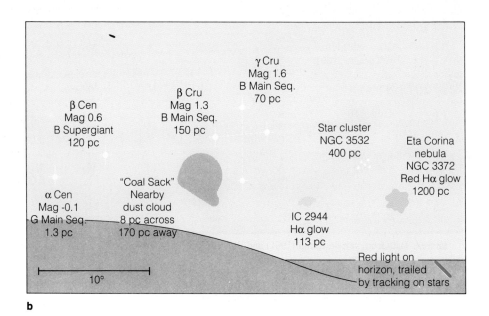

b

are Alpha Centauri A and B, which are very similar to the Sun and located 1.33 ± 0.01 pc away. These two components orbit around each other in 80 y and are easily visible as a pair—yellowish and orange—in a small telescope. In the 1800s, telescopic observer Sir John Herschel called them "beyond all comparison the most striking object of the kind in the heavens."

Orbiting around A and B once every 1½ million

Figure 17-4 Imaginary view of the red M-type supergiant star Antares as seen from a hypothetical planet 9½ AU away—the same distance Saturn is from our Sun. Although the Sun would resemble only a brilliant dot from such a planet, Antares is so big that it would cover an angle of almost 40°. It would look 80 times larger than our Sun in Earth's sky and have a reddish-orange color. The fainter star in the upper right is a bluish B-type star that orbits around Antares at a distance of several hundred AU. (Painting by author.)

years is a small faint star with only one-tenth the Sun's mass. It happens to be on our side of the system at a distance of only 1.29 pc (Kamper and Wesselink, 1978). Because it is the Sun's closest neighboring star, it is called Proxima Centauri (from the same root as *proximity*).

REPRESENTATIVE STARS VERSUS PROMINENT STARS

If we try to expand our statistical sample of stars by tabulating more distant ones, we run into a problem. At great distances the less luminous stars are too faint to see. The only distant stars we see are unusual, luminous ones. In fact, the **prominent stars** in our night sky are mostly distant, unusually luminous stars. As we learned from Figure 17-2, these stars are quite rare in three-dimensional space. Hertzsprung aptly called them the "whales among the fishes." Thus the statistics of prominent stars do not give us a sample of representative stars. Rather, they are biased toward the "whales." Nonetheless, a tabulation of such stars is important because it reveals that some of the "whales" don't belong to the main sequence.

Non-Main-Sequence Stars

If we combine a sampling of the most prominent stars in our sky (as ranked by apparent magnitude) with an equal-sized sampling of the nearby stars, we get an

H-R diagram similar to Figure 17-1. Whereas the nearby stars are the small red stars at the bottom of the main sequence, the prominent stars turn out to be mostly "whales" in the upper part of the diagram.

Now let's examine more closely some of these different types of non-main-sequence stars.

As shown in Figure 17-1, some red K and M stars have much higher luminosity than the faint, main-sequence K and M stars. That is, they radiate more light than the fainter stars of the same temperature. According to the Stefan–Boltzmann law, they must have more area. In other words, they must be larger. Hertzsprung named them **giant stars.** Most are called *red giants,* from their reddish color. Other stars were found to be brighter than the giants or the main-sequence O stars; they came to be called **supergiant stars.** The enormous size of the red supergiant Antares, for example, can be visualized in the imaginary close-up view shown in Figure 17-4. Figure 17-1 schematically dramatizes the relative brightnesses and sizes of these stars.

As more data came in, still other stars appeared on the H-R diagram below the main sequence. By the same logic used for giants, these stars were recognized as unusually small. They came to be called **white dwarf stars,**[2] because their colors were intermediate, like the Sun's, being neither strongly blue nor red.

[2]Unfortunately, main-sequence stars are sometimes called dwarf stars to distinguish them from giants, but for clarity we reserve the term *dwarf* for white dwarfs.

TABLE 17·2

The 17 Stars of Greatest Apparent Brightness

Star Name	Apparent Magnitude (m_v)	Bolometric Luminosity $(L_\odot)$	Type[a]	Radius $(R_\odot)$	Distance (pc)
Sun	−26.7	1.0	Main sequence	1.0	0.0
Sirius (α Canis Majoris)	−1.4	23	Main sequence (primary)	1.8	2.7
Canopus (α Carinae)	−0.7	~1400	Supergiant	30	34
Arcturus (α Boötis)	−0.1	115	Red giant	~25	11
Rigel Kent (α Centauri)	0.0	1.5	Main sequence (primary)	1.1	1.33
Vega (α Lyrae)	0.0	~58	Main sequence	~3	8.3
Capella (α Aurigae)	0.1	~90	Red giant (primary)	13	14
Rigel (β Orionis)	0.1	~60,000	Supergiant (primary)	~40	~280
Procyon (α Canis Minoris)	0.4	6	Main sequence (primary)	2.2	3.5

The work of Hertzsprung, Russell, and their followers was important because it showed systematic groupings of stars: one group fitting on the main sequence, a distinctly different group of giants, a group of supergiants, and a group of dwarfs. Further work showed that these groups occur in all known regions of space. This meant that universal physical processes could be sought to explain the groups.

Table 17-2 tabulates the 17 most prominent stars in our sky and reveals that nearly half are giants or supergiants. Although the whales among the fishes are rare, when we tabulate all stars in three-dimensional space, they account for many of the familiar stars.

Occasionally, we find groups of stars at about the same distance from us and formed at about the same time. These "star clusters," such as the one in Figure 17-5, offer us, so to speak, living H–R diagrams in the sky. Just by studying the colors and brightness of stars in Figure 17-5, we can see the pattern of the H–R diagram. Most of the stars are on the main sequence; the brightest ones are blue-white, and the faintest ones are red. This particular scene, known as the Jewel Box cluster, is all the more attractive because the fourth brightest star is an orangish-red giant or supergiant, offering a startling color contrast.

EXPLAINING THE TYPES OF STARS: DIFFERENT MASSES AND DIFFERENT AGES

What causes the distinctive types of stars found on the H–R diagram, such as main-sequence stars, the giants, and the white dwarfs? First, stars have different masses. *High-mass* main-sequence stars are hotter and brighter and larger than *low-mass* main-sequence stars. Second, stars evolve from one type to another, and because we

The 17 Stars of Greatest Apparent Brightness, *continued*

Star Name	Apparent Magnitude (m_v)	Bolometric Luminosity ($L_\odot$)	Type[a]	Radius ($R_\odot$)	Distance (pc)
Achernar (α Eridani)	0.5	~650	Main sequence	~7	37
Hadar (β Centauri)	0.7	~10,000	Supergiant (primary)	~10	150
Betelgeuse (α Oriðnis)	0.7	10,000	Supergiant	800	160
Altair (α Aquilae)	0.8	~9	Main sequence	1.5	5
Aldebaran (α Tauri)	0.9	125	Red giant (primary)	~40	21
Acrux (α Crucis)	0.9	~2500	Main sequence (primary)	~3	~110
Antares (α Scorpii)	0.9	~9000	Supergiant (primary)	~600	~160
Spica (α Virginis)	1.0	~2300	Main sequence (primary)	8	84

Source: Data from Burnham (1978).
Note: These are stars brighter than the first apparent visual magnitude.
[a]The designation "primary" indicates data for the brighter companion in binary pairs.

see stars with different ages, we see stars with different forms.

How did we discover the principles that control the structure and evolution of stars? Much of this work was begun by the English astrophysicist Arthur Eddington in the 1920s, and was carried on by other scientists through the 1950s. Eddington showed that the same two major, opposing influences which we saw at work in the formation of the Sun are competing in any star: Gravity pulls *inward* on stellar gas while gas pressure and radiation pressure push *outward*.[3] In a stable star, these forces are just balanced.

As we have already seen in the discussion of Helmholtz contraction (page 216), heat is produced during gravitational contraction of a star. A stable main-sequence star is one that has contracted until the inside is hot enough to start nuclear reactions among hydrogen atoms. At this point the interior becomes a stable heat source, radiating light and creating enough outward pressure to counterbalance the inward force of gravity. What had been a contracting ball of gas now becomes a star with a stable size, governed by the interior energy-generating conditions.

Explaining the Main Sequence

The source of energy in a main-sequence star's interior is nuclear reactions in which hydrogen is consumed. Because stars form with a huge supply of hydrogen, they remain stable on the main sequence for a relatively long time at a fixed size. If a stable star were magically expanded, its gas would cool and the reactions would

[3]As mentioned in Chapter 15, radiation exerts a distinct pressure on material through which it passes. In stars, especially massive ones, this pressure is important.

Figure 17-5 The Jewel Box cluster, cataloged as NGC 4755, displays different-colored stars that demonstrate the different star types charted on the H–R diagram. The cluster is named from Sir John Herschel's remark that it resembles a superb piece of jewelry. It is about 8 pc across and 2400 pc away. The brightest stars are mostly hot, bluish-white, young B stars; the faintest stars are red, low-mass stars. (Copyright Association of Universities for Research in Astronomy, Cerro Tololo Interamerican Observatory.)

decline, reducing the outward pressure, and the outer layers would fall back to their original state. If it were magically compressed, the inside would get denser and the reactions would increase, raising the outward pressure and expanding the star. The star has to stay in its stable state *as long as its internal chemistry and energy production rate stay the same.*

Calculations based on those ideas revealed a **mass–luminosity relation:** A hydrogen-burning star more massive than the Sun has higher luminosity and surface temperature than the Sun. Hence, on the H–R diagram it would lie to the upper left of the Sun. Similarly, a star of lower mass would lie to the lower right of the Sun. These stars therefore fall along a line in the H–R diagram—the main sequence. In other words: *The main sequence is explained as the group of stars of different masses that have reached stable configurations and are generating energy by consuming hydrogen in nuclear reactions.* Any hydrogen-burning[4] star with 1 $M_\odot$ and solar composition must look like the Sun. Any hydrogen-burning star with 2 $M_\odot$ and solar composition must be brighter and larger than the Sun. The O-type super-

giants at the upper left of the main sequence in Figure 17-1 may have 40 or even 100 times the Sun's mass.

Eddington's explanation of the main sequence also helped explain why some stars lie off the main sequence. The same assumptions simply do not apply to them. In other words, giants, supergiants, and dwarfs are stars that do *not* have the same composition or energy generation processes as the Sun.

Why Main-Sequence Stars Must Evolve Off the Main Sequence

Now we can explain the cause of **stellar evolution** from one form to another form: A star burning hydrogen converts it to helium and changes its composition; therefore, the star must change to a new equilibrium structure. All nuclear reactions cause changes in composition, and all changes in composition cause evolution to new structure.

AN OVERVIEW OF STELLAR EVOLUTION

Because of the work of Eddington and others, modern astrophysicists can calculate models of stars of any given mass and chemical composition. We can calculate not

---

[4]*Burning* technically refers to chemical reactions involving only electrons. But astrophysicists use the term informally for nuclear reactions that convert small fractions of the mass of atomic nuclei into energy.

only what the Sun's interior structure is like now, but what its future structure will be when 1% of the present hydrogen is gone, 2% is gone, and so forth. Each different composition corresponds to a certain (temporary) structure and to a certain position in the H–R diagram. Knowing the rates of nuclear reactions, we can calculate the time scale associated with this evolution from one state to another. Using computer modeling, we can estimate what happens during unstable transition states, such as the transition from mainsequence to giant and postgiant stages.

Because we can calculate a star's temperature, luminosity, and other properties for each stage of its evolution, these stages can be plotted on an H–R diagram. The sequence of such points is called an **evolutionary track**—the trail a star follows on an H–R diagram as it evolves.

Evolution of a One-Solar-Mass Star (Such as the Sun)

Figure 17-6 shows the approximate evolutionary track across the H–R diagram followed by the Sun or any star of solar mass. Some parts of the track are more certain than other parts. Consideration of different parts of the track shows clearly why different groups of stars appear on the H–R diagram.

To understand this, note first that Figure 17-6 is an interesting version of the H–R diagram because of the lines of constant radius. That is, any star 10 times as large as the Sun has to fall somewhere on the line marked $10 \, R_\odot$ because that line contains the only combination of temperature and luminosity that produces a star of $10 \, R_\odot$.

Now consider the Sun's history. As we saw in Chap-

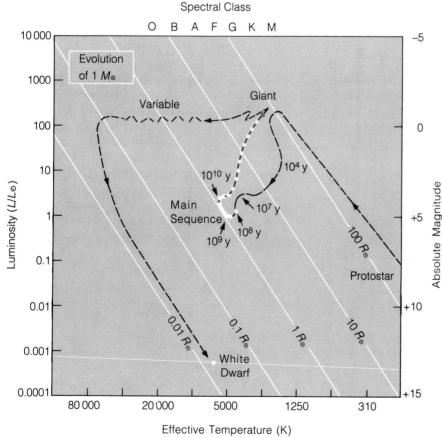

Figure 17-6 An H–R diagram showing the calculated evolutionary track for the Sun (or any star of 1 $M_\odot$). Dots are placed about 100 million years apart. The track shows the evolution from protostar state to main sequence, then to giant, variable, and white dwarf states. (Based on calculations by Westbrook and Tarter, 1975; Iben, 1967; Larson, 1969; and others.)

TABLE 17·3

Key Elements in Nuclear Reactions at Different Stages in Stellar Evolution

		Elements Being "Burned"	Comments
Time	Pre-main sequence	Lithium and other light elements	Relatively unimportant source of energy during formation of star before main-sequence stages (see Chap. 18)
	Main sequence	Hydrogen	Primary energy source; defines main sequence of stars; proton–proton chain; carbon cycle (see Chap. 18)
	Post-main sequence	Helium	Important after hydrogen is depleted; involved in reactions during rapid evolution off main sequence and into giant branch of H–R diagram (see Chap. 19)
		Heavy elements, including carbon and metals	Energy sources during final period on giant branch and during rapid evolution off giant branch, including supergiant states (see Chap. 19)

ter 14, the Sun began its life as a cool, dim interstellar cloud. Therefore, a Sun-like star first appears as a protostar on the H–R diagram at the right, with low temperature, low luminosity, and large radius, perhaps exceeding $100\,R_\odot$. It contracts rapidly, becoming quite bright, and then reaches a few solar radii in about a million years (only an instant of cosmic time). At this point it is approaching the main sequence and is called a pre-main-sequence star. Nuclear reactions begin in the interior, because the central temperature reaches some 10 million degrees or more (see Table 15-2 and discussion on page 231). This whole process of star formation is treated in more detail in the next chapter.

Once the nuclear reactions start, the huge reservoir of hydrogen begins to be converted into helium, and a star reaches a relatively stable, main-sequence configuration. A star takes only about 100 million years to reach the main sequence, but it is destined to sit on the main sequence for roughly 9 billion years.

Eventually, the supply of hydrogen begins to be so exhausted in the central core that changes become more rapid. The core is now mostly helium. The star's composition having changed, its structure must change. Because there is less hydrogen in the core to fuel the reactions, not enough energy is produced to maintain the outward pressure and keep the star in its main-

sequence state. The interior contracts, keeping the temperature high enough to maintain hydrogen burning in a shell around the helium core. Meanwhile, the outer layers of the star adjust by expanding. The star rapidly moves off the main sequence to a higher luminosity and much larger radius. It thus becomes a giant. All such states after the star leaves the main sequence are called post-main-sequence states.

Theoretical calculations can trace a star quite accurately to the giant stage, but subsequent states are less well understood, as indicated by the dashed line in Figure 17-6. Nuclear reactions begin to "burn" heavier elements, such as helium, fusing them into still heavier nuclei, as indicated in Table 17-3. Generally, after one element is consumed, slight core contractions raise the temperature to a point where still heavier elements begin to burn in fusion reactions. Evolution now takes the star somewhere to the left. Stars in this region are unstable, often variable, and rapidly evolving.

After each "combustible" element is used up by nuclear reactions in the core, the whole star contracts. Ultimately, it gets so dense it cannot contract any further. It has now reached the size of a planet—very tiny compared to its original size. A white dwarf has been formed. It is very hot and because of certain structural peculiarities, it cannot radiate its heat or cool very rapidly, even though it no longer produces any heat by nuclear reactions. The process of creating white dwarfs and other superdense stellar corpses is described in more detail in Chapter 19.

The following sequence summarizes the evolution of Sun-like stars according to the groups on the H–R diagram (multiple arrows indicate faster evolution):

Protostar
   ↘ ↘ ↘
   Pre-main sequence
      ↘ ↘
      Main sequence
         ↘
         Giant
            ↘ ↘
            Variable or unstable
               ↘ ↘ ↘
                  White dwarf or other small star

You will get a better idea of how these stages relate if you trace the dashed line in Figure 17-6 with your pencil. Start with the pencil point in the lower right and move forward one round dot each second. One second equals 100 million years. As you can see, the star almost reaches the main sequence in just one of these "cosmic seconds" but then spends nearly 100 cosmic seconds on the main sequence. Then it evolves rapidly to the giant stage and is transformed (in another cosmic second or so) to a small, faint, defunct state below the main sequence. No wonder we see so many main-sequence stars: Most stars spend nearly their whole lives in this state.

Massive Stars Evolve Fastest

Stars evolve at different rates. The more massive a star, the higher its interior temperatures and the faster it uses its nuclear fuel. A Sun-like star of 1 $M_{\odot}$ stays on the main sequence about 9 billion years (9×10^9 y), but a star of 10 $M_{\odot}$ stays there only about 20 million years (2×10^7 y). The entire history of a star of 1 $M_{\odot}$ (protostar to white dwarf) takes about 11 billion years, whereas a 10-$M_{\odot}$ star lasts only about 24 million years. In spite of the differences in total time, the largest fraction of each star's life is still spent on or near the main sequence.

The most massive stars stay on the main sequence for only the twinkling of a cosmic eye. Some of them

evolve into the supergiant region, and some less massive ones become ordinary giants. All of them quickly evolve to unstable configurations; the most massive ones explode; and all eventually fade from visual prominence.

Determining the Ages of Stars

Consider a group of stars that formed at the same time. After about 10 million years, stars larger than 20 $M_{\odot}$ will have disappeared from the main sequence. That is, the H–R diagram of the group will contain no main-sequence O stars. After about 100 million years, stars more massive than 4 $M_{\odot}$ will have evolved off the main sequence, and the H–R diagram will contain very few main-sequence B-class stars. The older the cluster, the more of the main-sequence stars will be gone. The missing stars will have been transformed into giants, white dwarfs, or even fainter terminal objects.

Thus we reach an important conclusion: *The H–R diagram can serve as a tool for dating groups of stars that formed together.* This principle will be applied often in later chapters as we probe the **ages of stars** in our galaxy. Isolated individual stars are harder to date.

SUMMARY

The nearby stars, within about a hundred parsecs of the Sun, display a range of masses, luminosities, temperatures, and compositions. These properties are not randomly distributed but grouped into distinct types. Some stars are on the main sequence, some are giants, some supergiants, and some dwarfs. Differences among these types are conveniently displayed on the H–R diagram, which plots stellar luminosity against temperature.

To explain these different types of stars, this chapter developed three important principles. First, stars form with a variety of masses covering a range from as low as 0.1 to 60 $M_{\odot}$ and more. As the stars form and begin to burn hydrogen in nuclear reactions, they settle onto the main sequence, but the more massive the star, the more luminous its main-sequence position in the H–R diagram.

Second, stars evolve from one form to another. They begin with pre-main-sequence configurations, settle onto the main sequence for a relatively long time to burn their hydrogen, and then evolve off the main sequence. Post-main-sequence forms include giants, variables, and dwarfs.

Third, stars evolve at different rates. The more massive a star, the faster it goes through its sequence of life stages. This is because more massive stars reach higher central temperatures and therefore consume their hydrogen faster.

The great majority of stars, including most of those in the Sun's neighborhood, have masses in the range of 0.1 to 1 $M_\odot$. They are relatively faint and cool; they lie in the lower right part of the H–R diagram. The most massive stars are much brighter and can be seen from much further away. They lie in the upper part of the H–R diagram. Thus many of the stars prominent in the night sky are prominent not because they are close but because they are the very luminous "whales among the fishes," far away among myriads of distant fainter stars.

CONCEPTS

H–R diagram

main-sequence star

representative star

Alpha Centauri

prominent star

giant star

supergiant star

white dwarf star

mass–luminosity relation

stellar evolution

evolutionary track

ages of stars

PROBLEMS

1. Suppose you could fly around interstellar space, encountering stars at random.
    a. Describe the stars you would encounter most often. Mention masses and H–R diagram positions.
    b. About what percentage of stars would be as massive as or more massive than the Sun? (*Hint:* Use statistics from either Table 17-1 or 17-2, as appropriate.)
    c. Would the stars encountered be similar to bright stars picked at random in our night sky?

2. Do giant stars necessarily have more mass than main-sequence stars? Why are they called giants?

3. Four stars occupy the four corners of an H–R diagram. Which one has:
    a. The highest temperature?
    b. The greatest luminosity?
    c. The greatest radius?

4. Two stars lie on the main sequence in different parts of the H–R diagram. One is bluer and one, redder. Which is:
    a. Larger?
    b. More luminous?
    c. More massive?
    d. Hotter?

5. Two stars form at the same time from the same cloud in interstellar space, but one is more massive than the other. Describe their differences or similarities at later moments in time.

6. What would an imaginary terrestrial observer see as the Sun runs out of hydrogen in the future? If life is confined to Earth when this happens, would life perish from heat or from cold?

7. Which part of Problem 4 could not be answered just from location in the H–R diagram if both stars were not on the main sequence?

8. Consider newly forming cosmic objects of stellar composition. What would you expect to happen in terms of stars' life cycles as the masses decrease from 1 $M_\odot$ toward values comparable to Jupiter's mass?

9. Show that astrophysical estimates of the Sun's lifetime are consistent with meteoritic and lunar data on the age of the solar system. Why would the data be inconsistent if astrophysicists calculated that solar-type stars stay on the main sequence only 1 billion years?

10. Why don't normal stars all collapse at once to the size of asteroids or tennis balls under their own weight, because gravity always pulls their material toward their center?

11. How was the chemical composition of the Sun 3 billion years ago different from what it is now?

12. Red giants have proceeded further in the evolutionary sequence than main-sequence stars, but is it correct to make the blanket statement that red giants are older than main-sequence stars? Explain. Which star is older, the Sun or a red giant of 10 $M_\odot$?

PROJECT

1. Locate or construct spherical objects that have the same proportional sizes as selected different types of stars, such as an Antares-type supergiant, the Sun, and a white dwarf.

CHAPTER 18

Stellar Evolution I:
Birth and Middle Age

Just as new living individuals are being formed from chemical materials against a background of slowly changing genetic pools, new stars are being formed from interstellar gas and dust against a background of slowly changing elemental abundances in the cosmic gas. Locally—and by this we now mean a substantial part of our galaxy—there is a constantly changing environment, with new stars forming and old stars blasting material into interstellar space. From a philosophical point of view, identifying and understanding newly forming stars are challenges in our long quest to comprehend the role of stars in our universe.

THREE PROOFS OF "PRESENT-DAY" STAR FORMATION

Here are three lines of **evidence for "present-day" star formation**—that is, star formation in the current part of astronomical history.

Youth of the Solar System The solar system is only about 4.6 billion years old, whereas the entire Milky Way galaxy—our system of 100 billion stars—is at least 10 or 12 billion years old. Thus our own Sun formed much more recently than the galaxy, and stars did not all form in one burst at the beginning.

Short-Lived Star Clusters Many young, massive stars are grouped in *open star clusters*. The stars in the cluster formed at about the same time. Because of the tendency of each star in the cluster to follow its own orbit around the galactic center, most clusters dissociate into isolated stars in only a few hundred million years. Few of these clusters are much older than this. The famous Pleiades cluster, for example, is only about 50 million years old. The existence of such short-lived clusters shows that star clusters did not all form at the beginning, but continue to form today.

Short-Lived Massive Stars Chapter 17 showed that massive stars evolve fastest. Calculations show that stars of 20 to 40 $M_\odot$ can last only a few million years in a visible state, yet we see them shining. They must have formed less than a few million years ago.

Thus star formation has been a continuing process during the whole history of the galaxy, including the last million years. There are stars in the sky younger than the species *Homo sapiens*.

THE PROTOSTAR STAGE

Stars form from tenuous material between other stars. The average conditions between stars are a near vacuum with very thin, cold gas. Atoms and molecules number only about 1 to 5 per cubic centimeter. About three-quarters of them are hydrogen (H atoms or H_2 molecules), and most of the rest are helium—as is the case with the atoms in most nearby stars. There are scattered dust particles. The temperature is typically around 100 K ($-279°$F).

Molecular Clouds

In certain regions of space, the density of gas and dust may be 10,000 times greater. With some 10,000 atoms per cubic centimeter, the atoms are close enough to collide and form molecules, rather than single atoms. These regions are called *molecular clouds*. They are rich in atoms, dust grains, and diverse molecules such as molecular hydrogen (H_2), water (H_2O), and more complex forms such as ethyl alcohol (C_2H_5OH) and the amino acid glycine (NH_2CH_2COOH). Such complex molecules are virtually absent in most parts of interstellar space. Although a molecular cloud is denser than most interstellar gas, it is still a near vacuum compared with ordinary room air, which has about 20 billion billion (2 × 10^{19}) molecules per cubic centimeter.

Molecular clouds are important because they are the most active regions of star formation. In these regions the atoms and molecules and dust grains are crowded close enough to begin to attract each other gravitationally—a key step in forming stars from gas (Scoville and Young, 1984). An example of the dramatic and chaotic

forms of gas and dust in such regions is shown in Figure 18-1. Many such nebulae are known. Most are cataloged in the "New General Catalog" of celestial objects and hence are usually designated by "NGC" numbers. Many also have popular names based on their appearance in telescopes; the "Eagle Nebula" in Figure 18-1b is an example.

Toward an Astrophysical Theory of Star Formation

Chapter 14 explained how the Sun formed when one of these diffuse interstellar clouds contracted and produced a central star surrounded by a dusty nebula. Now we need a more general theory of this process that can explain other stars with different masses.

When scientists use the word *theory,* they usually mean a well-tested body of related ideas often with a mathematical formulation that can be applied to a variety of cases and that is often backed up by observations.[1] Such a theory of star formation has been developed. This theory can be approached by asking: What causes some clouds to contract and not others? Why don't all interstellar clouds contract into stars and be done with it, once and for all? The answer involves the same two opposing forces we have considered before: gravity versus thermal pressure. Gravity pulls all the atoms in a cloud inward. But even at only 100 K the atoms are dancing, striking each other, and creating an outward pressure that opposes the tendency to collapse.

Gravity increases if the density of the cloud increases, because this gets more mass into the same amount of space. Outward pressure increases if the gas heats up, because this makes the atoms and molecules move faster. We can already say, then, that the reason not all clouds contract is that not all clouds are dense enough or cool enough. Furthermore, a noncontracting cloud can be turned into a contracting cloud if it suddenly experiences some turbulence (caused by a nearby stellar explosion, for example) that compresses it.

Astronomers want a more quantitative **theory of star formation** than this description. For interstellar material of any given density, they have calculated what

---

[1]A less complete or less tested idea is called a *hypothesis,* or a *working hypothesis.* An untested idea is often called *speculation.* Unfortunately, newspapers and magazines often publish speculations but label them with the more imposing term *theory.*

combination of mass and temperature a cloud would need to begin its **gravitational contraction,** or shrinkage toward stellar dimensions. The calculations show that in the denser interstellar medium, at about 100 K, masses contracting would be around 10^{34} kg, or several thou-

sand solar masses, and about 100 times the mass of the largest stars. But our theory was supposed to predict the formation of stars! Is something wrong? In fact, the theory agrees with observations, because, as we noted in the second of the three proofs of present-day star

a

Figure 18-1 The spectacular Eagle Nebula and the associated star cluster NGC 7711 dominate a chaotic region of star formation about 2000 pc away. The nebula is a cloud of gas and dust about 6 pc across. The red color comes from hydrogen atoms excited to glow with the red "hydrogen alpha" emission line. Dark clots, called Bok globules, are concentrations of dust. Hot, bluish-white, newly formed O stars are the brightest stars in the region. The complex is only about 3 million years old. **a** General view. (Anglo-Australian Telescope Board, courtesy David Malin.) **b** The heart of the Eagle Nebula with its star cluster. and silhouetted dust clouds. (Copyright Association of Universities for Research in Astronomy, NOAO.)

b

formation, stars apparently form in star clusters that have masses thousands of times greater than the Sun. In other words, conditions in the interstellar medium are such that *the gas subdivides into enormous concentrations containing enough mass for thousands of stars.* These huge clouds can be thought of as protoclusters.

But how do *individual* stars form? As a protocluster cloud continues to contract to still higher densities, smaller, star-sized condensations can form inside it. In other words, shrinking protoclusters divide into individual stars in a process called **subfragmentation.** The Canadian researcher A. Wright (1970), for example, followed the contraction of a 6 million $M_\odot$, 120-pc-wide cloud and a 500-$M_\odot$, 12-pc-wide cloud. He found that

both large and small density perturbations will grow in size, given the conditions found throughout most of a gas cloud. It would appear that fragmentation is an almost inevitable consequence of the cloud collapse process.

The new fragments become stars, and the whole mass turns into a star cluster. It is not hard to see why the interstellar gas in the disk of the galaxy might begin to contract. The material is not uniform; clots of dust and gas exist and are constantly being stirred by the galaxy's rotation, the expansion of hot nebulae, and other influences. Naturally, some clouds accumulate material or become compressed until their density exceeds the critical density that allows contraction to begin.

The Protostar's Collapse

The preceding discussion allows us to define a **protostar** as a cloud of interstellar dust and gas that is dense and cool enough to begin contracting gravitationally into a star. An interstellar cloud may hover on the verge of this state for millions of years. A nearby exploding star or some other disturbance might compress a gas cloud and thus trigger the collapse, which then proceeds very rapidly. For instance, typical protostars contract to stellar dimensions in a hundred thousand years—a wink of the cosmic eye. This explains why astronomers use the term **collapse:** The initial contraction is very rapid in terms of astronomical time. The collapse may not be smooth; the inner core of the cloud may collapse first, later absorbing the surrounding material; and under certain conditions a single rotating cloud may break up into two or more stars (Boss, 1985). Of course, once the cloud begins to reach stellar dimensions (a few

astronomical units, or 0.00001 pc), the atoms of gas bump into each other frequently enough to produce substantial outward pressure, so that the collapse is slowed. At this point we can speak of the object as a *pre-main-sequence star.*

THE PRE-MAIN-SEQUENCE STAGE

What are the forms of pre-main-sequence stars? Where do they lie on the H–R diagram? Can they actually be observed among the stars in space? Can we actually see stars being born?

The **pre-main-sequence star** stage covers the evolution from the end of the protostar stage to the main sequence. Astrophysicists such as the Japanese theorist C. Hayashi (1961) and his American colleague L. Henyey (Henyey, LeLevier, and Levée, 1955) pioneered in calculating the evolutionary tracks and appearance of stars contracting toward their main-sequence configurations.

The energy during most of the pre-main-sequence period does *not* come from nuclear reactions, which have not yet started inside the protostar; instead, heat is generated by the Helmholtz contraction (page **216**). After only a few thousand years of protostellar Helmholtz collapse, surface temperatures reach a few thousand Kelvin, thus causing visible radiation. Hayashi's work showed that convection would transport large amounts of energy from the interiors of most newly forming stars, making them very bright for short periods known as **high-luminosity phases** or *Hayashi phases.* A star of solar mass, for example, contracts in less than 1000 y from a huge cloud to a size about 20 times larger than the Sun with a luminosity about 100 times greater than the present Sun's.

These calculations have three important consequences. First, they show that stars have complicated, if short-lived, evolutionary histories even before nuclear reactions start. Second, they show that *newly forming stars must lie above and to the right of the main sequence.* (This is important in identifying new stars by direct observation.) Third, the calculations indicate that stars spend only a small fraction of their lifetimes in the pre-main-sequence stage.

More massive stars evolve faster to the main sequence than less massive stars. For instance, a 15-$M_\odot$ star reaches the main sequence in only 100,000 y. A 5-$M_\odot$ star takes about 1 million years, whereas a Sun-like star takes several million years.

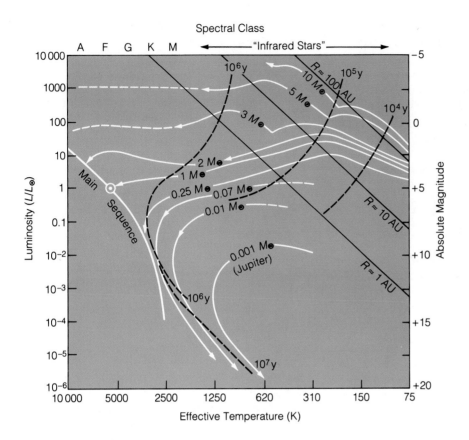

Figure 18-2 H–R diagram showing apparent tracks followed by pre-main-sequence protostars of various masses and their opaque cocoon nebulae, as viewed from outside. These configurations would be visible as infrared stars only for their first 10^5 y or so. Black dotted lines show time scales; black solid lines show radii of contracting nebulae. Stars smaller than 0.08 $M_\odot$ never settle onto the main sequence but continue to fade (bottom center).

Prediction of Cocoon Nebulae and Infrared Stars

While some theorists were calculating evolving conditions in newly forming stars, others, such as the Mexican astronomer A. Poveda, hypothesized that as young stars form from collapsing clouds, remnants of the clouds might surround and obscure these stars. As in the early solar nebula, dust would form in the cooling cloud and block the outgoing starlight. Only after a few hundred thousand years would the nebula eventually dissipate, revealing the star.

The term **cocoon nebula** was coined in 1967 to describe the nebula that hides a star from view during its earliest formative period (perhaps a few million years for a solar-sized star). The nebula is cast off later, just as a cocoon is cast off by an emerging butterfly.

What would the newly forming star and cocoon nebula look like to an observer before the nebula dissipated? Could examples ever be observed? The outermost dust of the opaque cocoon would have a temperature of only a few hundred Kelvin. Therefore the nebula would radiate

not visible light but rather infrared light (review Wien's law, page **68**), and the star–nebula complex, as seen from outside, would lie far to the right of the main sequence. Such an object is called an **infrared star** and can be detected at wavelengths of a few micrometers. As shown in Figure 18-2, it would evolve across this region of the H–R diagram in less than a million years and thus be detectable as an infrared star for a relatively brief period of its life. Stars of this type should therefore represent only a small fraction of all stars.

What Happens If the Mass Is Too Small?

If a cloud is dense and cool enough to contract but has less than about 8% of a solar mass, it will contract but never develop a high enough central pressure and temperature to reach a full-fledged main-sequence state. Protostars from about 1 to 8% $M_\odot$ may warm up temporarily due to their Helmholtz contraction, and may even develop a few feeble nuclear reactions, but eventually they fade without settling on the main sequence

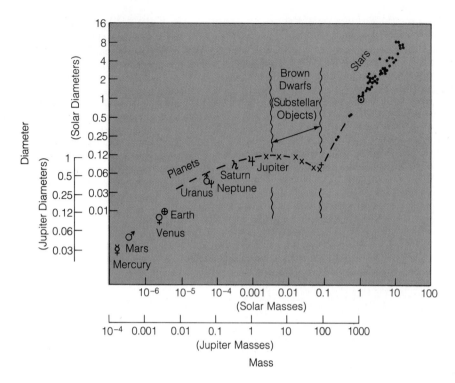

Figure 18-3 The transition from planets through brown dwarfs to stars. A combination of theoretical and observed data shows the changing diameter of objects as we add more mass. If we start with planetary masses (lower left) and add material of solar composition, the size levels off at about the diameter of Jupiter and then declines slightly due to compression by gravity. As additional mass pushes the total mass over 0.08 $M_\odot$, where nuclear reactions begin, the size again expands. [After data by D'Antona (1987), Lunine and others (1986), and Hubbard (1987, private communication).]

for any appreciable time. Smaller objects, including even Jupiter-sized planets, also warm up temporarily due to their Helmholtz contraction; they may glow for a while in the infrared, but they fade before any nuclear reactions can start. Even these nonstars therefore have evolutionary tracks that can be plotted on an H–R diagram, as shown at the bottom of Figure 18-2.

The smallest known luminous object in the sky, cataloged as RG 0050-2722, may be a temporarily glowing object of this type. It has an estimated mass of about 2.3% $M_\odot$ (only 23 × Jupiter's mass!) and a temperature of 2600 K. It is about 25 pc from us.

The Transition from Planet to Star: Brown Dwarfs

Objects with less than 8% $M_\odot$ (80 × Jupiter's mass), but larger than planets, are called **brown dwarfs,** or substellar objects, because, lacking nuclear reactions, they are not true stars. The name *brown dwarf* comes from the theory that the larger ones get red hot and glow with a dim, brownish-red light.

What would brown dwarfs really be like, if we could see them up close? Suppose we could do this experiment: Start with Jupiter, keep adding solar-composition

gas of hydrogen and helium, and go from planets through the brown dwarfs' size range to true stars. The result of this imaginary experiment is shown in Figure 18-3. As we add mass to Jupiter, its diameter gets only a little larger until the total mass reaches roughly 2 M_{Jupiter}, where it levels off. As we add even more mass, the object actually contracts a little, because it has so much gravity that it can compress the gas to a denser state. Then, as we approach 80 M_{Jupiter} where nuclear reactions begin, the object "turns on" as a star and again gets much larger as we add more hydrogen fuel. Thus conceptually there are fairly natural cutoff points in the transition from planets to brown dwarfs to stars. Although there has been some controversy about how to define the word *planet*, we will use the cutoff point at 2 M_{Jupiter} and call anything smaller a planet. Brown dwarfs or substellar objects range from about 2 to 80 M_{Jupiter}. They would get quite hot, and the larger ones would look red-hot and luminous to the eye, as suggested by the imaginary view in Figure 18-4. Although luminous, they would lack the nuclear reactions of a true star. Only objects above 80 M_{Jupiter} are true stars.

Astronomers have been searching for good examples of brown dwarfs, which would help us understand the relations between planets and stars. Brown dwarfs,

Figure 18-4 Imaginary view of a brown dwarf star from a hypothetical nearby planet. The brown dwarf emits a dull glow from its barely red-hot surface. In the distance is a true star, the "sun" of this imaginary system. (Painting by Ron Miller.)

of course, are hard to detect because of their faintness. One possible example was reported in 1987 by B. Zuckerman and E. Becklin of UCLA and Hawaii. Using infrared detectors they found an object with temperature 1200 K close to a white dwarf star known as Giclas 29-38. The object itself was not photographable in visible light, but analysis of its dynamics suggests that it is probably not a 1200-K dust cloud but a brown dwarf orbiting around the white dwarf. Because few other examples have been found, astronomers wonder whether there is a cutoff point in star formation at stars of a few percent of a solar mass. The search for brown dwarfs continues and the issue is unresolved.

What Happens If the Mass Is Too Big?

If the cloud is dense enough to contract but is more than about 100 $M_\odot$, the contraction is violent and produces an extremely high central temperature and pressure.

Under these conditions, so much energy is generated inside the new star that the star is very luminous and may blow itself apart almost immediately without spending much time on the main sequence. This rapid destruction is more appropriate to the topic of the next chapter, and will be taken up again there. We will see that explosions of massive stars explain many features of our starry surroundings.

EXAMPLES OF PRE-MAIN-SEQUENCE OBSERVED OBJECTS

Several types of observed objects are believed to relate to late protostellar stages or pre-main-sequence stages of star formation.

Bok Globules

Bok globules are dense clouds of dust and gas that may be contracting protostars. They are made visible

a

Figure 18-5 a A panorama of star formation in the midwinter evening sky, about 80° wide, around the constellation Orion. Many of the features are unusually young. Most of the brightest stars, such as Sirius and Orion's Belt, are recently formed bluish-white O and B stars. However, two prominent stars, Betelgeuse and Aldebaran, are red giants easily distinguished by their color (not only on this photo but also by naked eye). (11-min exposure with guided 35-mm camera, wide-angle 24-mm lens at f2.8. Star-pattern filter emphasizes bright stars and their colors by spilling some light into spike-shaped diffraction pattern; photo by author.) **b** Key map. Distances (parsecs) and ages (millions of years) are marked. Rings of radiating dashes mark associations of T Tauri stars, not prominent to the eye or on the photo. Compare with the enlarged view of Orion shown in Figure 16-2.

by being silhouetted against brighter nebulosity in the background; examples are the Coal Sack Nebula, silhouetted against the southern Milky Way as shown in Figure 17-3, and the small, dark blobs seen in Figure 18-1.

Infrared Stars and Cocoon Nebulae

Some of the best examples of infrared stars, cocoon nebulae, and other young stars are found in the **Orion star-forming region,** a large area of the sky around the constellation Orion. This region, which is about 400 to 700 pc away, is a hotbed of dense clouds and star-forming activity. When we look in this direction on a starry night, we can see many results of star formation, such as in Figures 18-5 and 18-6. Many newly forming objects lie hidden to our eyes inside dust clouds of interstellar space. But these are now being mapped by detectors sensitive to infrared light that passes through the clouds, even though the clouds block visible light.

Figure 18-7 shows an image of such a dark dust cloud, revealing the newly born infrared stars within it.

These and other objects exhibit many features that tie in well not only with the theory of star formation but also with evidence about conditions in the nebula that surrounded our own Sun as it formed. One of the most exciting implications of the observational work on newly formed stars is that we may be seeing dusty systems in which planets are forming now or might form in the "near" future—that is, in the next few million years.

T Tauri Stars

The most important pre-main-sequence stars are the **T Tauri stars,** named after the variable star T in the constellation Taurus. Although a great many of them are found in the region of Taurus and the neighboring constellation Orion, they can also be found in many other parts of the sky.

T Tauri stars may represent a transitional stage

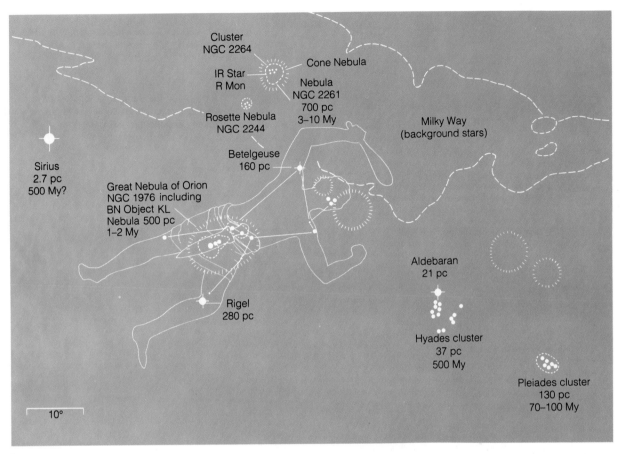

b

between infrared stars surrounded by opaque nebulae and stable stars that have lost their cocoons and settled on the main sequence. There are many signs of their youth and instability. The number of T Tauri stars alone in a star-forming cubic parsec may exceed the number of *all* stars per cubic parsec near the Sun by more than a factor of 10! (See Cohen and Kuhi, 1979.) T Tauri stars vary irregularly in brightness. Penn State astronomer Joan Schmeltz (1984) finds that the variations stem from changes both in the star's surface layers and in the enshrouding dust clouds. T Tauri stars lie to the right of the main sequence in the H–R diagram, where young stars are supposed to lie. Many have the infrared radiation characteristic of cocoon nebulae, as pointed out by the Mexican astronomer E. Mendoza (1968).

T Tauri stars are typically 20,000 to a million years old (Cohen and Kuhi, 1979). This is younger than the human species! Stars that have evolved beyond the T Tauri stage would probably be indistinguishable from main-sequence stars.

Disks of Dust Around Young and Middle-Aged Stars

Although there were many signs that cocoon nebulae might bear a connection to the proposed solar nebula in which the planets formed, there were no direct observations of disk-shaped nebulae around newly formed stars until the 1980s.

In 1983, however, the Infrared Astronomical Satellite (IRAS) was launched into orbit with the capability of mapping stars and material around them at the far-infrared wavelengths emitted by cool dust. IRAS promptly made a new discovery: some two dozen stars that had clouds of infrared-emitting dust extending hundreds of astronomical units from the star. The discovery led ground-based astronomers to use sophisticated techniques to make detailed images of these systems in order to discover their properties. Starting in 1984, astronomers began discovering flat, disk-shaped nebulae of dust near these stars. These discoveries seem

a

Figure 18-6 The Rosette Nebula and star cluster NGC 2244 are notable features of the skyscape in Figure 18-5. About 800 pc from Earth, the Rosette Nebula is a spheroidal cloud about 17 pc across. In its center is a cluster of young stars. **a** General view of the region. **b** Enlarged view showing blue colors of the brightest, hottest stars in the cluster; the dark dust globules may be contracting to form still more stars. (Copyright Anglo-Australian Telescope Board, courtesy D. Malin.)

b

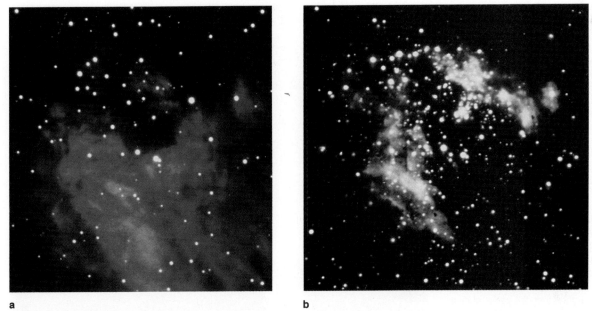

a   b

Figure 18-7 Piercing a veil of dust in the star-forming nebula M17 (NGC 6618). The same area of the sky is shown in **a** visible light and **b** infrared light. The visible image shows a red-glowing cloud. Dark dust clouds in the top half of the picture blot out part of the nebula and enclose a group of newly formed infrared stars. The infrared light from the stars penetrates through the dust cloud, so that image *b* reveals many otherwise invisible stars. Image *b* shows the image we would see if our eyes were sensitive to the infrared; the longer the wavelength, the redder the color used in this reproduction. Blue tones in image *b* correspond to light at 1.2 μm wavelength; reddest tones 2.2 μm. (National Optical Astronomy Observatories.)

to provide a "missing link" between the ragged cocoon nebula stage and the later dusty disk required to form planets. Among the stars studied have been HL Tauri, Beta Pictoris, and other main-sequence stars. The edge-on disk around Beta Pictoris can be seen in Figure 18-8.

The Mystery of Mass Ejection from New Stars

Studies of gas velocities (from Doppler shifts) around T Tauri stars indicate chaotic motions of gas around these new stars, sometimes infalling but more often blowing outward as fast as 50–200 km/s. The evidence suggested that T Tauri stars blow off as much as 0.4 $M_\odot$ of gas and dust before they reach the main sequence. Astronomers visualized very strong "solar winds" blowing out from T Tauri stars, but no one was sure how the mass was ejected. (See the nontechnical review by Lada, 1982.)

Observations in the 1980s indicate that the outflow involves a mysterious phenomenon we will encounter again and again: **bipolar jetting.** Gas flows out in two narrow streams in opposite directions, apparently along the polar axes of a disk, as seen in Figure 18-9. Gas clouds moving in opposite directions have actually been observed near new infrared stars (Bailly and Lada, 1983). The theoretical picture of these objects is that a cocoon nebula forms a disk around the central star, and its gas spirals turbulently inward (as if along the grooves of a phonograph record). Somewhere near the central star, it gets accelerated by an unknown means in diffuse jets "upward" and "downward," away from the disk, as seen in the figure.

In 1986 a team of astronomers from Arizona and Missouri announced another "missing link" observation. They studied an infrared star named IRAS 1692A, first detected by the IRAS satellite. Bipolar jets had been found soon after the discovery of the star, but the 1986 observations revealed a gaseous disk lying per-

Figure 18-8 The best pictorial evidence yet available of planetary dust around a nearby star, Beta Pictoris, 16 pc away. Thermal infrared radiation from the dust had been detected in 1983 by the IRAS infrared satellite telescope in space, leading astronomers to make this 1984 visual-light image from a ground-based telescope in Chile. The black central disk and crosshairs represent a system used to block the light of the star itself. With the star glare blocked, sensitive imaging equipment recorded bright material extending on either side of the star. This is an edge-on view of a dust disk extending to about the distance of the inner Oort swarm of comets in our own planetary system. The Beta Pictoris dust nebula is thicker and brighter than the zodiacal dust of our own system. Researchers suggest that planets may have recently formed in the Beta Pictoris system, though none have yet been detected. (Photo by S. Larson and S. Tapia, 1987, at Cerro Tololo International Observatory, Chile.)

pendicular to the jets and reaching 800 AU from the star. The outer parts of the disk were found to be orbiting the star, but the inner parts were collapsing inward and feeding the jets. This observation strongly supports Figure 18-9's theoretical model of young stars' disks and bipolar jets. An imaginary view of a disk–jet system such as IRAS 1692A is shown in Figure 18-10.

In summary, gas shedding by infrared stars and T Tauri stars, and the associated creation of bipolar jets, is a challenging new area of astronomy unknown just a few years ago (Schwartz, 1983; Hartmann and Raymond, 1984). It may be teaching us not only about formation of stars per se but also about what happens during formation of planetary materials and planet systems (Welch and others, 1985).

Young Clusters and Associated Young Stars

Because stars form in groups, T Tauri and infrared stars are often found in groupings associated with star-forming regions. In the beginning of this chapter, we described clusters of stars proven to be very young by their dynamical properties and H–R diagram position. Because low-mass stars take the longest to evolve to the main sequence, low-mass stars in these clusters ought to be found in the T Tauri stage. This has been abundantly confirmed by observations.

One of the best-known examples is the young star cluster NGC 2264, shown in Figure 18-11. Most of the low-mass stars (spectral classes A to K) in this cluster

Figure 18-9 Recent evidence suggests that as a new star turns on in the center of its turbulent disk-shaped cocoon nebula, gas is blown outward along the disk's axes by an unknown mechanism. Outflowing clouds move "upward" and "downward," at speeds up to 200 km/s, ramming into surrounding interstellar gas.

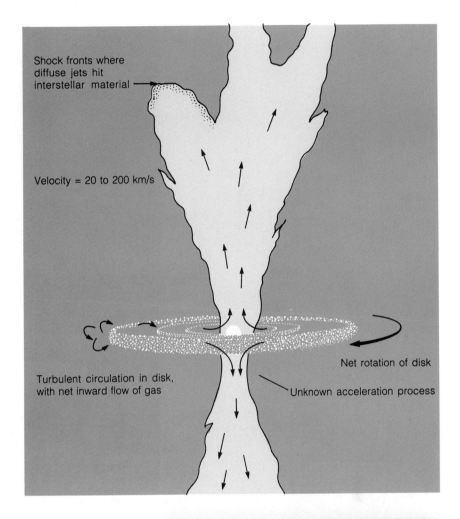

Shock fronts where diffuse jets hit interstellar material

Velocity = 20 to 200 km/s

Turbulent circulation in disk, with net inward flow of gas

Net rotation of disk

Unknown acceleration process

lie distinctly to the right of the main sequence, as shown in Figure 18-12, and many of these are identified as T Tauri stars. The calculated lines of constant age (dashed lines in Figure 18-12) show that the T Tauri stars match the positions predicted for an age of 3 to 30 million years. *This demonstrates how the H–R diagram can be used, together with theoretical data, to determine the* **ages of star clusters.**

Recall our rule of thumb from Chapter 17 that massive stars evolve fastest. Brilliant, massive, bluish, O-type supergiants last only a few million years. For this reason they are found only in star-forming regions a few million years old. By the time stars have dispersed from such regions, the O supergiants have already burnt out. Thus they are rare among representative field stars but dominate pictures of star-forming areas such as Figure 18-11.

Figure 18-10 Probable appearance of a disk-shaped cocoon nebula of dust and gas around a newly formed star. Dust grains may be aggregating into asteroids or planetary bodies within the disk. High speed jets of gas are being expelled along the polar axes of the disk. In the background is a star-forming region and a dense cloud of dark dust. (Painting by the author.)

Figure 18-11 This swirling mass of glowing gas, young stars, and dark dust clouds, called NGC 2264, is a star-forming region. The brightest stars are all newly formed, massive, hot, blue stars. Molecules of formaldehyde (H_2CO) and other gases have been found in the region. Blue star at bottom left marks top of Cone Nebula seen on page 309. The region is mapped in Figure 18-5b. (Copyright Anglo-Australian Telescope Board.)

THE MAIN-SEQUENCE STAGE

A star reaches the **main-sequence stage** when it begins to generate energy by consuming hydrogen in nuclear reactions deep in the star's central regions. Prior to that time, the star generates energy primarily by its Helmholtz contraction (see page 216). As a star enters the T Tauri pre-main-sequence stage, it still radiates by this means and still contracts slowly, as seen by comparing the evolutionary tracks and lines of constant radius in Figure 18-2. As the pressure and temperature near the star's center increases, individual nuclei eventually hit each other so hard that they begin to fuse together. These nuclear fusion reactions generate new heat and

Figure 18-12 This H–R diagram of stars in the region around Figure 18-11 reveals stars lying to the right of the main sequence along age lines (dashed), suggesting ages of only a few million years. (Data from M. Walker, 1972, with age lines from I. Iben, 1965.)

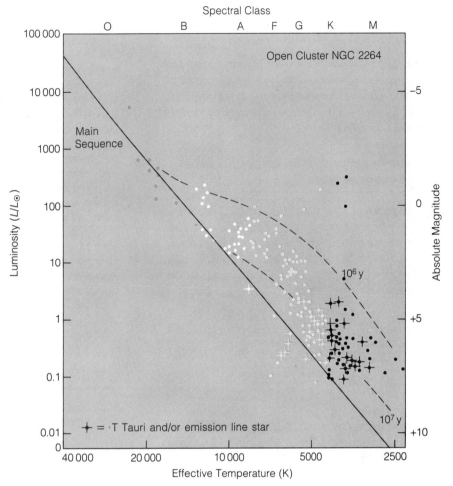

pressure that stop the contraction, causing the long-term stability of the star on the main sequence.

In the H–R diagram, an imaginary array of stars with different masses, which have all just reached the main sequence and just begun to consume hydrogen, is called the **zero-age main sequence.** Calculations show that it would be a very narrow band of stars on the H–R diagram. In reality, the main sequence on the H–R diagram is a bit broader because it contains stars of different ages, which have converted different fractions of their hydrogen to helium. Thus they have slightly different structures and slightly different positions on the H–R diagram.

Once the star reaches the zero-age main sequence, it begins its life as a true star and commences a long period of hydrogen consumption by means of various nuclear reactions. Because there is so much hydrogen, stars spend most of their lives on the main sequence.

During the nuclear reactions, tiny amounts of matter are converted to energy, providing the heat and light of the star. We can't overemphasize the basic cause of further stellar evolution: *The nuclear reactions convert certain elements into other elements, changing the star's composition and thus the energy generation rates; this in turn causes the star to change to new structures.* To clarify these changes, we will now review the nuclear reactions in stars in more detail.

Nuclear Reactions in the Smaller Main-Sequence Stars

As described in Chapter 15, studies of our own main-sequence star, the Sun, reveal that its energy comes from a series of nuclear reactions called the **proton–proton chain.** We list it again for emphasis; you can review its details on page 232.

$$^1H + {}^1H \rightarrow {}^2H + e^+ + \text{neutrino}$$
$$^2H + {}^1H \rightarrow {}^3He + \text{photon}$$
$$^3He + {}^3He \rightarrow {}^4He + {}^1H + {}^1H + \text{photon}$$

The proton–proton chain is the primary energy-producing process not only inside the Sun but also inside all main-sequence stars smaller than F stars of about $1\frac{1}{2}$ $M_\odot$. It dominates if the central temperatures are less than about 15 million Kelvin.

Nuclear Reactions in the Larger Main-Sequence Stars

In main-sequence stars larger than about $1\frac{1}{2} M_\odot$ where interior temperatures are higher than about 15 million Kelvin, another reaction series dominates in producing energy. This is the **carbon cycle,** sometimes called the *CNO cycle* to reflect the involvement of carbon, nitrogen, and oxygen. The reactions are as follows:

$$^{12}C + {}^1H \rightarrow {}^{13}N + \text{photon}$$
$$^{13}N \rightarrow {}^{13}C + e^+ + \text{neutrino}$$
$$^{13}C + {}^1H \rightarrow {}^{14}N + \text{photon}$$
$$^{14}N + {}^1H \rightarrow {}^{15}O + \text{photon}$$
$$^{15}O \rightarrow {}^{15}N + e^+ + \text{neutrino}$$
$$^{15}N + {}^1H \rightarrow {}^{12}C + {}^4He$$

Again the net result is that four hydrogen atoms are used up to produce helium-4 atoms with an associated release of energy. In a sense, carbon acts as a catalyst (a stimulant of change), because carbon-12 reappears at the end of the cycle, to be used again in the first reaction of a subsequent cycle.

In main-sequence stars, one or the other of these reactions consumes hydrogen in the core. Finally, major structural changes happen as the hydrogen is used up. In the next chapter we will see what dramatic events ensue.

SUMMARY

Stars are forming even today within a few hundred parsecs of the Sun and in more distant regions of space. The starry sky is not a static scene but the site of continual births of new stars out of interstellar dust and gas. Many stars and star systems are less than a few million years old—much less than 1% as old as our galaxy. Some have become visible since humanity evolved, though prominent newly formed stars have probably not appeared in our sky during recorded history.

Star formation begins with protostars, which are clouds of dust and gas that begin to contract due to their own gravity. They collapse fairly rapidly to stellar dimensions and become pre-main-sequence starlike objects. Theoretical calculations in the last two decades have revealed what features these objects probably have. Many of these features have been confirmed by observation, especially by infrared equipment, which detects the radiation from relatively low-temperature dust in nebulae around the newly formed stars.

Among the objects revealed in this way are groups of stars evolving toward the main sequence. Many groups of stars and individual stars are surrounded and obscured by cocoon nebulae consisting of dust particles (probably silicates and ices similar to those that formed the first planetary material in our own solar system). More evolved objects, such as the T Tauri stars, appear to be shedding their cocoon nebulae and have almost reached the main sequence.

Once a star reaches the main sequence, energy is generated as hydrogen converts into helium by means of the proton–proton chain for smaller stars and the carbon cycle for larger stars.

CONCEPTS

evidence for "present-day" star formation

theory of star formation

gravitational contraction

subfragmentation

protostar

collapse

pre-main-sequence star

high-luminosity phase

cocoon nebula

infrared star

brown dwarf

Bok globule

Orion star-forming region

T Tauri star

bipolar jetting

age of star clusters

main-sequence stage

zero-age main sequence

proton–proton chain

carbon cycle

PROBLEMS

1. Did the Sun and solar system form in the first half or last half of our galaxy's history?

2. Suppose that you magically smoothed out all the clots in the interstellar gas so that it was all uniform.
    a. Would this help or hinder star formation?
    b. What processes would keep the gas from staying uniform indefinitely?
    c. Would star-forming conditions tend to return to normal?

3. Why do stars form in groups instead of alone?

4. How do theories of solar system formation and theories of star formation support each other? Contrast the sources of information on these two subjects.

5. Suppose a cluster of stars formed 3 million years ago. Why would you expect the H–R diagram of the cluster to show no stars on either the very high mass end of the main sequence or its very low mass end?

6. Suppose a new 1-$M_\odot$ star began forming about 10 pc from the Sun. What would Earth-based observers see during the next few million years? Consider infrared observers as well as naked-eye observers.

7. Why doesn't gravity immediately cause the collapse of all interstellar clouds?

8. Why does the structure of a star stabilize when it reaches the main sequence?

PROJECT

1. On a clear night (an early evening in February or a late evening in December is ideal) scan the region of Orion with your naked eye and compare it to other regions of the sky. Note the concentration of bright blue O-, B-, and A-type stars (such as Sirius and Rigel) and star clusters (such as the Pleiades, Hyades, and the Orion Belt region) in this broad area. How do these features indicate that star formation has been going on in this general direction from the Sun in the last few percent of cosmic time?

CHAPTER 19

Stellar Evolution II: Death and Transfiguration

Stars are born and stars die. By this we mean that the nuclear reactions converting mass to energy in stars have a beginning and an end. The Sun, for example, spent a brief 10^7 y in its formative stages; it will spend about 10^{10} y on the main sequence, roughly 10^9 y in the giant state, and a shorter time in later unstable states consuming heavier elements until energy generation stops. This evolutionary sequence might be likened to the periods of human life: 9 months in the womb, 65 years of normal life, 6 years of rapid aging, and perhaps a year of terminal illness. As we will soon see, stars do evolve to pathological terminal states, involving incredible forms of dense matter and processes unimagined until a few years ago.

Writer Ben Bova (1973) summed up star deaths by quoting one of Ernest Hemingway's characters, who was asked how he went bankrupt. "Two ways," he says. "Gradually and then suddenly." The gradual part is the slow expenditure of a star's hydrogen, which is scarcely noticeable to an observer. The sudden part—the subject of this chapter—comes as the star runs out of hydrogen. It goes into fits of activity, searching for new sources of fuel until, as Bova says, "gravity forecloses all the loans."

Gravity forecloses the loans by causing a continual tendency toward contraction. When the core of a main-sequence star runs out of hydrogen, several things happen. At first, the core may burn its way into outer layers that still have hydrogen. This can cause an expansion of the outer layers, so that the star temporarily balloons to a giant size. But eventually this process of hydrogen burning spreads outward into a layer that is too cool to sustain the reactions, and they wind down. This means there is less outward pressure generated by reactions. Thus the core must contract due to gravitational forces pulling inward. Although some of the upper layers may be blown outward, most of the star's mass contracts to a small size. This contraction is what we mean by "gravity forecloses the loans."

What happens after the hydrogen is exhausted and the core starts to shrink? Remember from the discussion of Helmholtz contraction (page 216) that as a mass of gas contracts, it must grow hotter. Thus the atoms in the core move faster and collide even harder than before. Soon they reach a condition where the helium atoms in the core collide hard enough to start fusion reactions. We say that the core has stopped burning hydrogen and is now burning helium. After the helium has burned, more contraction occurs and still other elements may become fuels. Eventually, there are no more fuels to react and the star contracts to a very dense state.

This chapter discusses the four major post-main-sequence stages of stellar old age: the giant stage, variable stages, explosive stages of several types, and terminal high-density stages. Table 19-1 lists examples of stars in each of these stages, which are described throughout the chapter.

Note that the most common stars—those about one solar mass—go through the giant phase and then contract rather smoothly to a small, high-density stage. But the rare, high-mass stars beyond about $8\,M_\odot$ go through explosive instabilities. Eventually they reach even higher-density stages than Sun-like stars, because they have more mass and gravity pulling inward. They produce some of the most exotic objects yet discovered by astronomers, such as pulsars and black holes.

THE GIANT STAGE

As described briefly in Chapter 17, a star evolves off the main sequence as hydrogen fuels run out. At this time a thin shell, in which hydrogen is still burning, surrounds the core. Outside the shell, temperatures have never risen high enough to "ignite" the hydrogen nuclei (that is, to fuse them into helium) so there is still plenty of hydrogen fuel. This allows the hydrogen-burning shell to migrate further toward the surface. The increased temperatures cause great expansion of the outer layers, making the star evolve rapidly toward the giant state. The outermost atmosphere becomes huge, thin, and cool, even though the inner core is smaller, hotter, and denser than ever. From the outside, the outer atmospheric layers are seen to glow with a dull

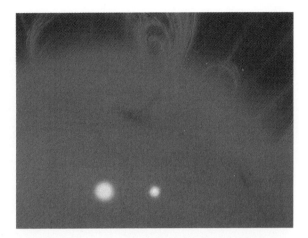

Figure 19-1 A schematic representation of the size of a typical red giant in comparison with more familiar stars. The larger of the two stars at the bottom is Sirius; the smaller is the Sun. (Painting by Tom Miller.)

red color, and the star is perceived as an enormous **red giant,** such as was shown in Figure 17-4.

The largest giants are truly huge—approaching 1000 times the size of the Sun. If the Sun were replaced by one of these stars, its thin, red-glowing outer atmosphere would reach beyond the orbit of Mars! A typical red giant's scale is shown by Figure 19-1.

As charted on the H–R diagram (Figure 19-2), stars from all points along the main sequence display a **funneling effect:** As they move off the main sequence, their evolutionary tracks funnel into the red giant region. They resemble patients from all walks of life crowding into the same hospital because they are victims of the same malady—hydrogen exhaustion.

Meanwhile, the star cores contract until they reach temperatures near 200 million K. This is hot enough to begin to fuse helium nuclei in the central regions of most stars, primarily by the **triple-alpha process** (named after the alpha particle, another name for the helium-4 nucleus):

$$^4\text{He} + {}^4\text{He} \rightleftharpoons {}^8\text{Be} + \text{photon}$$

$$^8\text{Be} + {}^4\text{He} \rightarrow {}^{12}\text{C} + \text{photon}$$

In this process, three helium-4 nuclei combine to produce a carbon-12 nucleus. Notice that we are now burning helium to make an even heavier element, carbon.

The triple-alpha process produces prodigious energy, which at once heats the rest of the core, creating a

TABLE 19·1

Examples of Stars in Middle and Late Evolution

Stage of Evolution[a]	Distance from Earth (pc)	Mass $(M/M_\odot)$	Radius $(R/R_\odot)$	Luminosity $(L/L_\odot)$	Spectral Type	Average Density[b] (kg/m^3)
Main Sequence						
Sun	<<1	1	1	1	G2	1420
α Centauri B	1.3	0.85	1.2	0.36	K4	700
Procyon A	3.5	1.7	2.3	6	F5	200
Algol A	31	4	3.0	100	B8	210
Giant						
Arcturus	11	~4	25	115	K1	0.36
Aldebaran A	21	~4	45	125	K5	0.07
β Pegasi	64	~5	140	400	M2	0.003
Supergiant						
Antares A	160	~12	700	9000	M1	0.00005
Betelgeuse	170	~18	~700	11,000	M2	0.00007
VV Cephei A	200–1200?	20–80?	400–1600?	200–10,000?	M2	0.00003?

Source: Data from Burnham (1978), Liebert (1980), Baize (1980), and Bonneau and others (1982).
[a]Listed in order of increasing age.
[b]For comparison, the following are densities of familiar materials: atmosphere at sea level, 1.2 kg/m³; water, 1000 kg/m³; lead, 11,300 kg/m³.

sudden burst of helium burning often called the **helium flash.** In some stars, the helium flash may consume the central core's helium in only a few seconds. Because the energy from the flash diffuses through the star slowly, the heating effects seen at the surface may last thousands of years. The flash occurs as the star enters the giant region. In a Sun-sized star, it may occur 300 million years after evolution off the main sequence.

After the helium flash, "squiggles" may develop in the evolutionary track as shown in Figures 19-2 and 19-3. Reduction of pressure in the outer atmosphere may allow the star to contract, temporarily moving it back toward the main sequence.

Note that the core's evolution begins to be independent of the evolution of the outer atmosphere. The char-

acteristics of the star as perceived by an astronomer on Earth are those of the outer atmosphere. These characteristics determine the star's position on the H–R diagram. When you go outside at night and look at Betelgeuse or Antares, you are seeing the cool, red, outer atmosphere of a swollen star. Hidden inside is a very hot, dense core. Thus although we may say "the star" is cooling and getting redder, it is really its atmosphere that is cooling. Instabilities after the helium flash may cause temporary heatings and coolings of the outer layers and temporary expansions to greater size; these appear as squiggles in the evolutionary track on the H–R diagram. But all this time, the core is shrinking and growing hotter.

As the core contracts and gets hotter, it initiates

Examples of Stars in Middle and Late Evolution, *continued*

Stage of Evolution[a]	Distance from Earth (pc)	Mass ($M/M_\odot$)	Radius ($R/R_\odot$)	Luminosity ($L/L_\odot$)	Spectral Type	Average Density[b] (kg/m^3)
Variable and Explosive Stars						
Mira A (long-period variable)	77	2	230	Up to 1100	M6	~0.0002
Polaris A (Cepheid variable)	120?	~6	~25	830	F8	~0.5
HD 193576B (Wolf–Rayet)	?	12	~7		"WR"	~50
DQ Herculis B (nova)	?	0.2	~0.1	<0.1	M?	300,000
White Dwarf						
Sirius B	2.7	1.0	0.022	0.002	A5	130,000,000
Procyon B	3.5	0.65	0.02	0.0005	F	120,000,000
Pulsar (neutron star)						
Crab Nebula pulsar	1100	~2?	<0.00002	High in UV, X ray	—	~10^{17}?
Possible Black Hole						
Cygnus X-1	≥2500	6–15?	<0.000004	0	—	~10^{22}?

reactions involving even heavier elements. Carbon burning may occur. Some of the reactions create floods of neutrons that strike nearby atoms and are added to them, building up very heavy elements, including metals. Without such reactions, the universe would still consist of almost pure hydrogen and helium and would lack sufficient heavy elements to make planets such as Earth. Table 17-3 reviewed this sequence of element evolution in stars.

THE VARIABLE STAGE

A **variable star** is a star that varies in brightness on a time scale of hours to years. Most variable stars apparently represent *postgiant stages* of evolution (although, as we saw in the last chapter, some variables are pre-main-sequence stars). Some pulse with a constant period; others flare up sporadically, often brightening by many magnitudes and then fading again. Such stars were recorded as long ago as 134 B.C. when Hipparchus recorded the flare-up of a "new star," proving that the heavens were not eternally constant. The star Omicron Ceti has long been known as Mira (Wonderful) for its brightening from invisibility (magnitude 8–10) to the second magnitude every 331 d.

Shakespeare used the simile "constant as the northern star." He might have been interested to know that Polaris is constant neither in brightness nor in marking the north celestial pole. Polaris is a variable star that

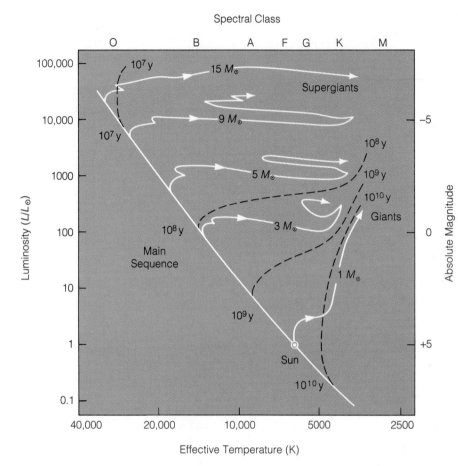

Spectral Class

Figure 19-2 Evolutionary tracks off the main sequence toward the giant region, plotted on the H–R diagram for stars of different mass. Dashed lines give the length of time since star formation; massive stars evolve fastest. All the stars funnel toward the upper right corner of the H–R diagram, the region of giant stars. (After calculations by Iben.)

changes from magnitude 2.5 to magnitude 2.6 in just less than 4 d. Because of precession (Chapters 1 and 2), it is the North Star for only a few centuries every 26,000 y.

Some 22,650 variable stars have been cataloged and divided into as many as 28 types. The type most important to astronomers is the **Cepheid variable,** which has regular variations in brightness, usually with periods from 5 to 30 d. In 1784, the 19-year-old English astronomer John Goodricke[1] discovered the first Cepheid variable, Delta Cephei. It varies from magnitude 4.4 to 3.7 with a period of 5.4 d; Cepheids were named after that star.

Cepheids are important for two reasons. First, because their variations are regular, they are somewhat better understood than stars whose brightness changes are unpredictable. Second, and more important, the period of variation of each Cepheid directly correlates with its average luminosity. This relation was discovered in 1912 by a famous astronomer, Henrietta S. Leavitt, who measured hundreds of photos of Cepheid variables in the first years of this century at Harvard Observatory. Two types of Cepheids and a related type of variable called an RR Lyrae star were eventually found to have distinct period–luminosity relationships. These relationships allow astronomers to determine the luminosity of any Cepheid at any distance simply by measuring its period. This, in turn, leads to a new way to measure the distances of stars:[2] Find a Cepheid; mea-

---

[1]Goodricke offers an inspirational example of accomplishment during a short, difficult life. Deaf from infancy, he lived during the first European generation to recognize that deafness was not the same as idiocy. He attended the first English school for deaf children, took up astronomy, and made numerous interesting observations, but died at age 21 from pneumonia, possibly contracted from nighttime exposure during prolonged observing efforts.

[2]For discussion of this very important technique in mapping distances of star clusters and similar galactic features, see Chapter 22.

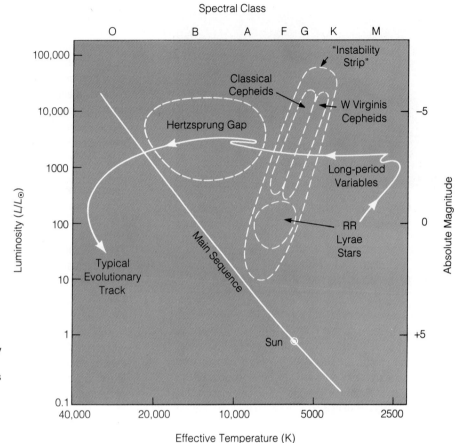

Figure 19-3 An H–R diagram showing the later stages of evolution, indicated by a typical evolutionary track (in white). The star evolves across an "instability strip" that contains Cepheid variable stars and RR Lyrae stars and then evolves rapidly across the Hertzsprung gap.

sure its period and hence its luminosity, or absolute magnitude; then measure its apparent magnitude and derive its distance from the relations in Figure 16-7.

Cause of Brightness Variations

Most variables change brightness because of changes in their interior nuclear reactions and in the flow of radiation from their centers. Variables vary in physical size and other properties as well as in brightness. Delta Cephei itself varies in diameter (from roughly 38 million to 41 million kilometers) and surface temperature (from 5600 to 6500 K) during its cycle.

Astrophysicists have long sought the precise cause of the internal physical disturbances that make Cepheid variable stars pulsate. Various researchers, especially the Soviet astronomer S. Zhevakin, have established that the H–R diagram contains a nearly vertical *instability strip,* shown in Figure 19-3. Stars become unstable where their evolutionary tracks cross this strip. Inside

such stars ionized helium absorbs outgoing radiation and thus becomes doubly ionized in the outer atmosphere, not far below the star's visible surface. This "dams up" the radiation, leading to expansion of the outer atmosphere. Once the atmosphere gets thin enough for the radiation to escape, the star shrinks again, reblocking the radiation. Hence, we get periodic variability.

The Hertzsprung Gap

Just to the left of the instability strip in the H–R diagram (Figure 19-3) is the *Hertzsprung gap,* a vertical band nearly empty of stars. Why do no stars have the combination of luminosity and temperature that would put them in this gap? Apparently the gap represents an evolutionary stage so unstable that stars evolve very quickly across it. This rapid evolution may involve a final blowoff of any remaining cool, outer atmospheric layers of gas, as well as exposure and final collapse of the dense inner core.

Figure 19-4 A star undergoing mass loss has produced a shell of diffuse gas around itself. The star is associated with a star grouping cataloged as IC 2220. (Copyright Anglo-Australian Telescope Board, courtesy D. Malin.)

MASS LOSS AMONG EVOLVED STARS

As the outer atmospheres of evolved stars expand, some of the gas may be blown entirely free of the star, as shown in Figure 19-4. After expansion of a red giant, for example, some of the gas of the outer atmosphere may continue to move outward and be blown out of the star altogether. Many red giants are losing mass. This process can be observed from blue-shifted absorption lines arising in gas layers moving out of the star toward the observer. Probably this process grades from a smooth shedding of gas by means of a strong stellar wind in stars of about a solar mass, to explosive shedding in stars of a few solar masses, which are less stable.

Planetary Nebulae

The gas being blown out of mass-shedding stars expands into space. It may cool enough for grains of dust, such as carbon grains, to condense in it. It may collide with other nearby gas clouds at high speed, creating glowing shock waves. Often, however, it is shed in relatively spherical bubbles of gas that surround the central star. Ultraviolet light from the star excites and ionizes the gas atoms, causing them to glow. Clouds of gas in space are called nebulae, and decades ago these particular nebulae came to be called **planetary nebulae** because the palely glowing bubbles looked like disks of planets in small telescopes. This term is a misnomer, however, because they have nothing to do with planets. As shown in Figure 19-5, they are among the most beautiful celestial features, with wispy symmetry and delicate colors (Kaler, 1986).

Supergiants

The more massive a star, the more violent and unstable its evolution. Stars of more than a few solar masses are very hot and evolve rapidly. When their atmospheres expand, these stars leave the main sequence at a luminosity already brighter than most giants. They are thus called **supergiants**—not because they are larger than giants (some are and some aren't) but because they are brighter. They are often brighter than absolute magnitude −6 and are spread across the top of the H–R diagram, and some are hot stars classified as spectral type O and B.

THE DEMISE OF SUN-LIKE STARS: WHITE DWARFS

Let us now consider the further evolution of Sun-like stars, by which we mean stars with initial mass similar to that of the Sun, say from about 0.1 to a few solar masses. The outer atmosphere is very expanded and may be shedding its outer layers, but most of the mass is still in a dense hot core. Eventually the core runs out of fuel and gravity causes a final collapse of the whole star, including the parts of the atmosphere that have not blown off. This final collapse creates a very small, very hot, very dense star about the size of Earth. Such a star is known as a **white dwarf**. About 200 are cataloged.

The Discovery and Nature of White Dwarfs

In 1844 the German astronomer Friedrich Bessel studied the motions of the brightest star in the sky, Sirius, and found that it is being perturbed back and forth by a faint, unseen star orbiting around it. This star was not glimpsed until 1862, when American telescope maker Alvan Clark detected it. It is almost lost in the glare of Sirius, as shown in Figure 19-6. In 1915, Mt. Wilson observer W. S. Adams discovered that it was a strange, hitherto-unknown type. It is hot, bluish-white, and lies below the main sequence on the H–R diagram. It has about the mass of the Sun, but it is so faint that its total radiating surface cannot be much more than that of Earth.

A star the size of Earth? If a solar mass were packed into an Earth-sized ball, it would be astonishingly dense. A cubic inch would weigh a ton! A thimbleful would collapse a table!

Figure 19-5 The Helix Nebula (NGC 7293) in Aquarius is a glowing spheroidal shell of gas blown off the surface layers of an evolved star by strong stellar winds. The remainder of the star is visible at the center—a very hot, blue, exposed stellar core. Strong ultraviolet radiation from this star core excites atoms of different elements at different distances in the gas around the star, causing the atoms to emit colorful glows. Dominant color is the red glow of hydrogen alpha emission from ionized hydrogen atoms. Additional blue-green comes from doubly ionized oxygen atoms. Delicate radial filaments may be related to outrushing motions of the gas. The distance and diameter are probably around 100 pc and 0.4 pc, respectively. (Copyright Anglo-Australian Telescope Board, courtesy D. Malin.)

Dutch–American astronomer W. J. Luyten, who discovered a number of these stars, described what happened as astronomers first recognized Adams' amazing results (quoted by Shapley, 1960):

The figures were too staggering to be believed without further evidence. Could there be an error somewhere? What had happened? But while we pondered these things, two more stars of the same kind were found: small, extremely faint, and white, and all three were named white dwarfs.

Theoretical astrophysicists such as Sir Arthur Eddington and the Indian–American theoretician S. Chandrasekhar explained the evolution of white dwarfs. Once no more energy is available to generate outward pressure, a star collapses until all its atoms are jammed together to make a very dense material. But what does "jammed together" really mean? At these densities, some electrons move at almost the speed of light and matter loses its familiar properties. It is no longer either gas, liquid, or solid, but a new form known as **degenerate matter.** By a physical law called the

Figure 19-6 Two exposures of Sirius with the Lick 3-m telescope. Short exposure (left) reveals the faint white dwarf (below Sirius) that is the binary companion roughly 25 AU from Sirius; in the long exposure (right) the white dwarf is lost in the overexposed image. "Rays of light" are artifacts of the telescope optics. (Lick Observatory.)

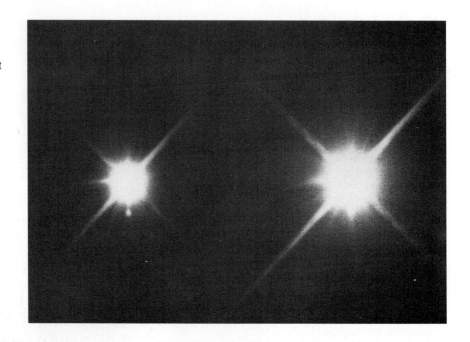

Pauli exclusion principle, only a certain number of electrons can be jammed into a given small volume of space. Just as there is empty space between the atoms of a quartz crystal, there is still some empty space between the electrons and other atomic particles of a white dwarf—but repulsive forces hold the electrons apart and thus they resist denser packing. The atomic nuclei no longer react, but are held apart by a sea of electrons—a so-called degenerate electron gas—at densities of about 10^8 to 10^{11} kg/m³. White dwarfs are sometimes called *degenerate stars.*

Because white dwarfs have low luminosities and large amounts of stored internal energy, they take a long time to radiate enough energy to cool significantly. White dwarfs thus last a long time. The coolest white dwarfs have temperatures around 5000 K, and may have cooled for at least 7 billion years to reach these temperatures (Helfand, 1983). Indeed, some white dwarfs may be among the oldest stars we can observe.

Although the interiors of white dwarfs may have densities of billions of kilograms per cubic meter, the outer layers of some may consist of ordinary matter—possibly hot gas at the surface and crystalline rocklike or glasslike solids in a crust 20 to 75 km deep. At the base of these crusts, densities may be as high as 3 million kilograms per cubic meter!

The Most Massive Possible White Dwarf: The Chandrasekhar Limit

White dwarfs cannot have more mass than 1.4 $M_\odot$ because the white dwarf structure becomes unstable at this point. If you tried to dump more mass on the surface of a 1.4-$M_\odot$ white dwarf, its gravity would become so strong that it would overcome the resistance of the electrons to denser packing. A still denser state of matter would arise.

The critical mass, 1.4 $M_\odot$, is called the **Chandrasekhar limit** after its discoverer. Stars with masses from about 0.08 to 1.4 $M_\odot$ apparently evolve smoothly to the white dwarf state. Current data suggest that stars from 1.4 to about 8 $M_\odot$ also evolve into white dwarfs by developing strong stellar winds (like the solar wind) or eruptive explosions that blow off mass until they are below the Chandrasekhar limit (Liebert, 1980). According to calculations, the Sun will probably blow off about 40% of its mass when it goes through its red giant stage about 5 billion years from now. Thus it will collapse into a white dwarf of about 0.6 $M_\odot$ (Burrows, 1987).

THE DEMISE OF VERY MASSIVE STARS: SUPERNOVAE

Now let us consider the terminal evolution of the rarer, more massive stars. In stars of around 10–12 $M_\odot$, a

Figure 19-7 Schematic diagram of onion-skin layering of different elements as calculated for the precollapse core of a 15-$M_\odot$ star. Dominant element in each layer is given at top; each layer's mean temperature is given at bottom. Figures at left give mass of material out to designated layer. Nuclear reactions in successively deeper layers fuse nuclei into successively heavier elements. (After diagram by Burrows, 1987.)

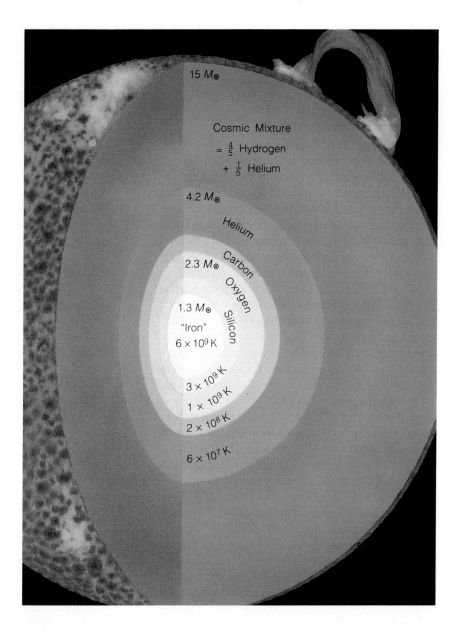

long series of reactions will fuse nuclei into elements as heavy as iron. But nuclear structure is such that no more energy can be produced by burning iron. Energy is consumed, not produced, as iron atoms fuse into heavier elements. Thus the iron cores of stars do not continue to ignite if they contract and get hotter. As astrophysicist Frank Shu (1982) comments: "Iron . . . is the ultimate slag heap of the universe."

Thus if the star is massive enough, the core-building process leads to a core consisting of shells of different elements (like an onion skin), surrounding an inner core of iron, as illustrated in Figure 19-7. The shell structure can be so complex that there may be different fusion reactions happening at the surfaces of different shells, all at the same time. Hydrogen might be burning at the base of the hydrogen-rich atmosphere, while helium might be fusing into carbon at the hot base of the helium layer. Eventually, however, this situation becomes unstable, because one or more layers of the onion skin is running out of fuel, and the fusion reactions start to decline.

This is when gravity gets the upper hand, and the collapse of the core begins.

The Nature of a Supernova Explosion

At the moment of collapse, the core of a massive red giant has consumed as much fuel as possible, burning its way out into the outer parts of the star. The core collapse starts because there is no point in the core where there is potential fuel *and* the possibility to create a hot enough temperature to ignite it. In a star larger than about 8 $M_\odot$, this core encompasses more than 1.4 $M_\odot$, exceeding the Chandrasekhar limit. What will happen when it tries to collapse into a white dwarf?

Chandrasekhar was right: It cannot make a white dwarf. Its increased mass causes stronger gravity and compresses the star to a much smaller, denser state. Such states include the strangest objects known in the universe: neutron stars and black holes. These are much smaller than white dwarfs and involve an extraordinary collapse. The collapse of the core to tiny size is aided by the fact that the iron "slag-heap" core will not ignite, even though it gets extremely hot and dense. The supercollapse releases a tremendous outburst of energy inside the already barely stable star. This outburst of energy blows off all the outer layers of the star in a titanic explosion called a **supernova** (plural: *supernovae*).

What happens inside the star during the collapse and explosion? Although there are several types of supernovae,[3] which may have somewhat different processes, the main process appears to be as follows. When the incredible temperature of several billion degrees Kelvin is reached, the iron nuclei are hitting each other so hard that although some fuse into still heavier nuclei, many simply break apart into a shower of protons, neutrons, neutrinos, and other subatomic particles. Calculations suggest that once it starts, the collapse takes only a few seconds! The core material moves so fast toward the center that it overshoots,

finds itself too compressed to be stable, and rebounds. This starts an outward-moving shock wave that blows away the outer layers in what astrophysicist A. Burrows calls "the fabulous explosion that we call a supernova" (Shu, 1982; Reddy, 1983; Wheeler and Nomoto, 1985; Burrows, 1987).

A supernova explosion takes the star to the amazing luminosity of 10 billion times that of the Sun! For a few days the exploding star blazes with the light of an entire *galaxy*. Then, after months, the star itself fades. During the explosion an expanding cloud of gas is launched. Initially it is too close to the star to be seen from Earth, but years later the site will be marked by a colossal, expanding nebula called a supernova remnant.

Examples of Supernovae

Many supernovae have been close enough to the solar system to produce temporarily prominent "new stars" that were recorded by ancient people. The ancient Chinese called them "guest stars." Some astronomers have suggested such an explanation for the Star of Bethlehem.

The most famous supernova was the explosion that produced the Crab Nebula (Figure 19-8). It was visible in broad daylight for 23 d in July 1054 and at night for the subsequent 6 mo. It was recorded in Chinese, Japanese, and Islamic documents, and perhaps in American Indian rock art, as seen in Figure 19-9. The nebula is the expanding, colorful gas shot out of this supernova. Other supernova remnants are scattered throughout our galaxy. Records indicate about 14 supernovae in our galaxy during the last 2000 y, an average of one every 140 y. None has been very bright in the last two centuries, but any of us might live to see one bright enough to be visible in daylight.

Supernovae occur in other galaxies, too. In 1885, a supernova in the nearby Andromeda galaxy briefly doubled that galaxy's brightness. More than 300 have been recorded in other galaxies, and their remnant clouds can sometimes be identified (Figure 19-10).

The most recent important supernova blazed forth on February 23, 1987,[4] in a neighboring galaxy called

---

[3]One type of supernova (called Type I) occurs in double-star pairs when a white dwarf is orbiting a star that is shedding mass. The mass crashes onto the dwarf, pushing it over the Chandrasekhar limit, making it unstable, and causing it to collapse and then explode. We will discuss this process of mass transfer further in Chapter 21. The type of supernova described above, caused by a star's own internal evolution, is called Type II.

---

[4]This is the date when the light corresponding to the event reached Earth. Because the star's location was nearly 170,000 ly away, the light took some 170,000 y to reach us. Thus the explosion actually occurred around 168,000 B.C.!

Figure 19-8 The Crab Nebula is the remnant of a supernova explosion recorded in A.D. 1054. Red outer filaments are splatters of excited hydrogen glowing with Hα emission. Milky-green glow comes from high-speed electrons moving in the nebula's magnetic field. The nebula is roughly 2000 pc away, 3 pc across, and expanding at a speed of about 1000 km/s. At the center is a pulsar that flashes 60 times per second, as shown in Figure 19-12. (Copyright California Institute of Technology and Carnegie Institution of Washington; by permission of Hale Observatories.)

the Large Magellanic Cloud. Within hours, the newly brightening star was sighted by astronomers, as seen in Figure 19-11, and word was flashed to observatories around the world. For many days the distant supernova was bright enough to see with the naked eye, but was visible only near the equator and in the Southern Hemisphere. Telescopes on Earth and in space were pointed toward the supernova in a coordinated observing effort (Helfand, 1987). Space telescopes used to study the supernova included a new X-ray telescope module attached by the Soviet Union to their Mir space station only a few weeks after the supernova was first sighted.

One of the most exciting findings was the detection of a burst of neutrinos in the first few seconds of the explosion. Detectors in Japan and America confirmed what, until then, had been only theoretical predictions of neutrino production. It was a good example of how scientific theory can predict features of the universe (Spergel and others, 1987).

In the following months the supernova confirmed other basic aspects of stellar theory. Astronomers detected gamma rays emitted by the heavy element

a

b

c

Figure 19-9 American Indians may have recorded the appearance of the Crab Nebula supernova on the morning of July 5, 1054. **a** Painting on overhanging rock ledge. **b** Closeup of pictograph as seen by artist facing eastern horizon. **c** Sketch based on computer reconstruction of appearance of rising crescent moon and supernova as seen to east across the canyon from the site shortly before dawn on July 5, 1054. (Photo by author.)

Figure 19-10 This glowing bubble of gas is believed to be an expanded remnant cloud from a supernova explosion. This nebula lies on the outskirts of a neighbor galaxy known as the Large Magellanic Cloud. It is about 110 pc in diameter. The star that exploded to form it may have been a massive member of the central cluster of stars. (Copyright Anglo-Australian Telescope Board, courtesy D. Malin.)

cobalt, for example, confirming that the massive parent star did synthesize heavy elements in its heart.

As the first readily observable supernova in three centuries, the 1987 supernova gave astronomers their first look at a stellar explosion with modern equipment. A good wrap-up about this supernova and its ramifications is in a nontechnical *Scientifc American* review by California physicists S. Woosley and T. Weaver (1989).

NEUTRON STARS (PULSARS): NEW LIGHT ON OLD STARS

What is left after a supernova explosion? As early as 1934, American astronomers W. Baade and F. Zwicky made a correct guess:

With all reserve we advance the view that a supernova represents the transition of an ordinary star into a neutron star, consisting mainly of neutrons. Such a star

Figure 19-11 The 1987 supernova. **a** The region of the supernova before the explosion. Red wisps are gas excited by ultraviolet light of massive young stars in the region and glowing by the light of hydrogen alpha emission. **b** A few days after the explosion, the same region (shown at same scale) is dominated by the brilliance of the supernova. (Copyright Anglo-Australian Telescope Board.)

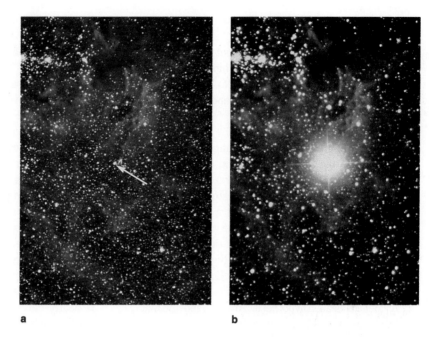

a   b

may possess a very small radius and an extremely high density.

Nuclear physicists were already working on the problem of very dense forms of matter, and they quickly took up the challenge of dense star forms. By 1938, J. Robert Oppenheimer and R. Serber showed how stars, after exhausting their nuclear fuel, could collapse to states much denser than white dwarfs.[5]

The Nature of a Neutron Star

The concept of a **neutron star** was merely an extension of the concept of a white dwarf. In an ordinary gas, atoms are held apart by their motions, and when they collide, their nuclei are held apart by their surrounding electron swarms. In an ordinary star's core, the electrons are stripped off and nuclei begin to hit each other, reacting and releasing nuclear energy. In a white dwarf, the nuclei that are left can no longer react and are crowded randomly among a dense sea of electrons. A white dwarf,

therefore, might be called an electron star because electrons control the spacings of particles.

But if a burnt-out star core is still more massive than a white dwarf, the gravity is so strong that even the repulsion between electrons is overcome. Theory indicates that this could happen in objects between 1.4 $M_\odot$ and about 3 to 5 $M_\odot$. In such objects, the nuclei themselves begin to be jammed into contact. Because whatever reactions might occur have already run their course, there are no new energy-generating reactions. Instead, the nuclei are broken into their constituent neutrons and protons. The positively charged protons coalesce with negatively charged electrons to make more of the neutrally charged neutrons. The important particles are now the neutrons, and the star is a neutron star, smaller than the white dwarf. Its density approaches the incredible densities of atomic nuclei themselves: roughly 10^{17} kg/m³. A thimbleful would weigh 100 million tons! A skyscraper-full could contain all the mass of the Moon! A whole neutron star could contain twice the mass of the Sun but be no larger than a small asteroid—perhaps 20 km across!

Because a neutron star is a mass of nuclear matter comparable in density to the nucleus of an atom, some astronomers have pictured a neutron star as a giant atomic nucleus with mass around 10^{57} times as much as a hydrogen nucleus!

[5]Oppenheimer is most famous for his work as leader of the Los Alamos team that developed the atomic bomb in 1945. His physics research was halted when he lost his security clearance in the anticommunist political wrangles of the 1950s.

The Discovery of Neutron Stars

For decades, astrophysicists talked about neutron stars, but, like the weather, nobody did anything about them, because nobody *could* do anything. No known observational technique could detect them and no one could prove they existed.

But in November 1967, a 4.5-acre array of radio telescopes in England detected a strange new type of radio source in the sky. Analyzing the surveys (each equaling a 120-m roll of a paper chart), a sharp-eyed graduate student, Jocelyn Bell, was astonished to find that one celestial radio source (about a centimeter of data on the chart) emitted "beeps" every 1.33733 s. At first, project scientists speculated that they might have actually discovered an artificial radio beacon placed in space by some alien civilization! But by January, another source was found, pulsing at a different frequency, arguing against the beacon hypothesis. Analysis showed that the first source was less than 4800 km across, much smaller than ordinary stars. These pulsing radio sources came to be called **pulsars.**

Within a year, astrophysicists realized the mysterious pulsars were the long-sought neutron stars. In an exciting burst of research, the number of scientific papers on pulsars jumped from zero in 1967 to 140 in 1968. By 1973, about 100 pulsars had been discovered. The codirectors of the original discovery project, Anthony Hewish and Martin Ryle, shared the 1974 Nobel Prize in physics.

Why do neutron stars pulse? After collapsing to a small size, supernova remnant stars have very strong magnetic fields and very fast spins, rotating once every second or so. Recall from Chapter 14 that any spinning object—even a figure skater—spins faster as it contracts. Ions trapped in the magnetic fields spin around with velocities near the speed of light. In 1968, Cornell researcher Thomas Gold showed how ions trapped in magnetic fields of spinning neutron stars produce strongly beamed radio radiation, so that the pulsar acts like a lighthouse with a beam sweeping around. Most pulsars have rotation periods of ¼ to 13 seconds. The fastest-spinning pulsar found thus far was discovered in 1982 and flashes with a 0.0016-s cycle, meaning that it spins 642 times every second (Waldrop, 1983)!

The discovery of a pulsar in the center of the Crab Nebula (Figure 19-12) and in other supernova remnants proves that pulsars are related to supernovae. Careful studies of such pulsars show that they pulse not only in radio waves but also in X rays and visible light.

REMNANTS OF EXTREMELY MASSIVE STARS: BLACK HOLES

As soon as pulsars were discovered, the search for still denser objects began. From this search came evidence for the strangest of all astrophysical concepts—**black holes.** Black holes are bodies so dense that their gravitational fields can keep most light (or other forms of energy and matter) from escaping.

Black holes can be visualized through an analogy with Newtonian physics made as early as 1798 by the French astronomer–mathematician Pierre Laplace. He reasoned that some bodies might be dense enough to have an escape velocity (at their surfaces) faster than the speed of light. Laplace thus assumed light could never escape from such bodies and that they would be permanently black and opaque. Although this basic idea is nearly correct, we now know that the situation is more complex because of Einstein's theory of relativity, which more correctly predicts phenomena involving speeds near that of light. In any case, theorists using Einstein's theory have concluded that black holes probably do exist.

Physical Nature of Black Holes

Theorists believe that a black hole would be a very dense mass surrounded by a so-called **event horizon,** or imaginary surface from which no radiation or matter could escape.

The event horizon concept is illustrated in Figure 19-13, which shows the escape velocity at various distances from a hypothetical black hole of 1 $M_\odot$. A rocket passing at a great distance would experience the same gravity field and motions as a rocket at a great distance from the Sun. At 1 AU from the object, for example, the velocity needed to escape into interstellar space would be 42 km/s, the same as the speed needed to leave Earth's orbit. But the black hole itself might be only a kilometer or less across! As we get within a few thousand kilometers of it, the escape velocity would be thousands of kilometers per second. At a distance of around 3 km, the speed needed to escape would be the speed of light. From *within* the event horizon, an object on a ballistic trajectory (such as an atomic particle or a meteoritic dust grain) would have to move outward faster than light in order to keep from falling back—hence its escape would be impossible. This imaginary boundary region where the escape velocity equals the velocity of

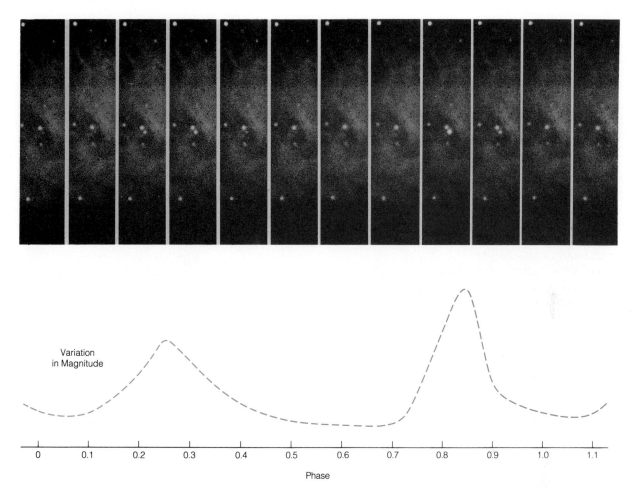

Figure 19-12 A complete sequence of flashes from the pulsar NP 0532 in the center of the Crab Nebula. Inner portions of the surrounding nebula can be seen. The entire cycle, including two flashes, lasts about 1/30 s, equaling one rotation of the pulsar. The pulsar and nebula are about 2000 pc away. Figure 19-8 shows the full nebula. (Kitt Peak National Observatory.)

light is the event horizon. Calculations indicate that black holes would form from the collapse of only the most massive stars (25 $M_\odot$ or more?) or the most massive exhausted post-supernova remnants (5 $M_\odot$ or more).

Two theoretical results on black holes came in the mid-1970s, especially from the English physicist Stephen W. Hawking. First, he pointed out that because matter was more densely packed in the earliest days of the universe, objects much smaller than stars could have gravitationally contracted to form black holes as small as subatomic particles, like protons and neutrons. Some of these tiny, primitive black holes might still exist in space.

Second, Hawking (1977) applied the theory of quan-

tum mechanics and showed that *black holes need not be entirely black*, as was once thought. Quantum mechanics shows that subatomic particles often act in ways unpredicted by older theories. Thus proton-sized black holes (containing up to 10^{12} kg—microscopic motes with the mass of mountains) could radiate energy, often as an explosion emitting gamma rays. Large, kilometer-scale black holes (containing as much mass as a star, or about 10^{30} kg) would radiate virtually nothing, just as in the original black hole theory.

If black holes exist, they must have truly unfamiliar properties. If an instrumented probe were dropped toward it, we would stop receiving signals after the probe fell through the event horizon, beyond which vir-

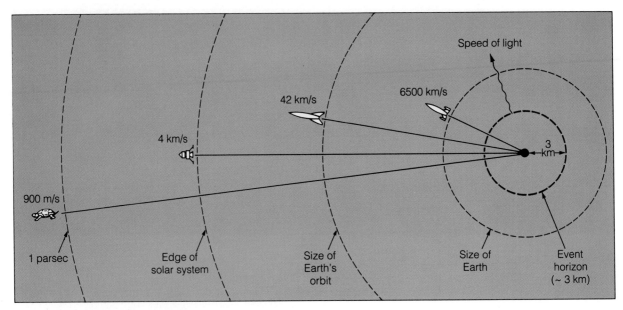

Figure 19-13 Exploring the gravitational field of a black hole of one solar mass (right). The schematic diagram (not to scale) shows the velocity needed to escape from the black hole at different distances from it. At a parsec away (left) a mere 900 m/s would suffice, but this speed increases as we get closer. At a distance of about 3 km, a body would have to be moving at the speed of light to escape; this distance is called the event horizon. No object on a ballistic trajectory could escape from inside this distance.

tually no radiation or matter can escape. The probe, then, would seem to disappear from the observable universe. Where would it go?

Pondering this strange question, some scientists speculate that black holes amount to separate universes. Others feel that when mass "pops out of existence" by collapsing to superdensity, new mass or energy emerges somewhere else in the universe. But perhaps the probe would merely "squish" to very high density and collide with the black hole, adding a bit to its mass. Could our own universe have begun as a black hole in some other "universe"? Do black holes make fact out of the "space warps" invented decades ago by science fiction writers to allow their spaceships to wink out of sight in one place and reemerge instantaneously at some distant point? No one knows. As the British geneticist J. B. S. Haldane put it, "My suspicion is that the universe is not only queerer than we suppose, but queerer than we *can* suppose."

Detecting a Black Hole

If a black hole itself cannot emit radiation or mass, can we ever hope to detect one? Yes. Outside their event horizons, black holes have gravity fields indistinguishable from those of ordinary stars of the same mass. Thus they can orbit around stars just like planets or binary star companions. If we observed such a star from a distance, we would not see the black hole, but we could see the star's orbital motion and calculate the mass of the unseen companion, just as astronomers routinely do in the case of ordinary faint companions. The result would indicate an unusually high-mass companion for an unseen star—maybe 5 or 10 $M_\odot$—which should tip us off that we are dealing with a black hole candidate. Furthermore, such strong gravity exists close to black holes that any matter falling into them undergoes terrific acceleration.

Suppose the black hole is orbiting around an evolved star that has expanded into the giant state and is shedding mass.[6] Some of the expanding gas would fall toward the black hole at terrific speed. Because this gas would,

---

[6]Many pairs of coorbiting stars of this type are known. They are discussed in more detail in Chapter 21.

on the average, have some angular momentum around the star, rather than falling directly toward it, it would form a disk of gas spiraling inward toward the black hole. This disk is called an **accretion disk.** Its gas would be extremely hot, because it would be constantly hit by new gas streaming in from the other star. Because of the high temperature, the disk would radiate very short-wave radiation, such as UV, X-ray, and gamma-ray radiation. Therefore X-ray and gamma-ray telescopes launched into orbit around Earth have played an important role in searching for black holes. The possible appearance of the inner accretion disk and the black hole is indicated in Figure 19-14.

The best evidence for a black hole, then, would be a massive, high-temperature X-ray or gamma-ray source orbiting around another normal star. Two such candidates have been discovered. Cygnus X-1 (so named because it is the first X-ray source discovered in the constellation Cygnus) is a binary system about 2500 pc away. The visible star is a supergiant O or B star with a mass of 15 to 30 $M_\odot$. Orbiting around it is a small, hot, X-ray source with a mass of 10 to 15 $M_\odot$. The X-ray emissions suggest that the gas temperatures reach a billion Kelvin in the X-ray source (Nolan and Matteson, 1983). The second candidate is much more distant—55,000 pc away in a nearby galaxy called the Large Magellanic Cloud. It is called LMC X-3 and was first detected by the Uhuru X-ray satellite in the early 1970s. The visible star is a type B main-sequence star, which orbits in 1.7 d around a small, hot, X-ray source with a mass of 6 to 14 $M_\odot$. Many astronomers believe that the hot X-ray source in each of these systems is the accretion disk surrounding a black hole. Continuing observations will document these and other black hole candidates in more detail.

PRACTICAL APPLICATIONS?

People often ask, "What good is astronomy? How can it be of use to know about places too far away to see, to touch, or almost to imagine?" We've argued that we attain firmer philosophical footing if we know about our cosmic surroundings and that exploration of the planets helps us understand the climate and resources of the Earth. This chapter provides another reason. Astronomers' understanding of dense star forms like white dwarfs, neutron stars, and black holes has developed hand in hand with physicists' understanding of dense

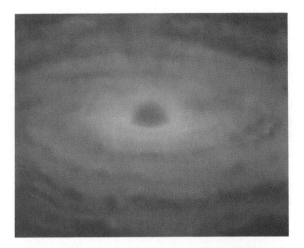

Figure 19-14 Possible appearance of the heart of an accretion disk around a black hole. The gas spiraling around the center is extremely hot and radiating blue and ultraviolet light, X rays, and gamma rays. The event horizon of the black hole appears as a reddish disk because we see red-shifted light from material falling into it but no radiation from the hot material inside it. (Painting by author.)

forms of matter in the lab. But there are limits to the experiments that can be conducted with such weird forms of matter. The discovery of actual examples of such matter in space, together with the measurement of masses, temperatures, magnetic fields, and so on, has provided physicists with examples impossible to create in their labs. Physicist D. Pines (1980) comments that such discoveries "may constitute an almost unique probe of . . . high-temperature plasma in super-strong magnetic fields, . . . neutron solids," and other superdense materials.

If research on such material still sounds esoteric to you, recall that a promising long-range solution to the energy shortage on Earth is to control nuclear fusion in the lab by jamming nuclei together in dense plasmas constrained by strong magnetic fields. Fuel for hydrogen fusion could be the cheapest and most accessible material on Earth: seawater! Experimental work in this direction is already under way worldwide. Controversy erupted in 1989 when fusion at low temperatures was claimed by some researchers but not confirmed by others. This controversy is unresolved. In any case, a direct line stretches from studies of neutron stars to our manipulation of matter to create new forms of usable energy.

SUMMARY

The search for the forms of aging stars has yielded some of the most fascinating objects now being studied both by physicists and by astronomers. Stellar old age leads to two basic phenomena: high-energy nuclear reactions, as ever-more-massive elements interact, and inexorable contraction, as energy sources are eventually exhausted.

During the first stages of old age, as stars evolve off the main sequence, high energy production causes expansion of their outer atmospheres, producing giants and supergiants. As heavier elements go through quick reaction sequences, various kinds of instability may produce variable stars and slow mass loss. After energy generation declines to a rate too low to resist contraction, low-mass stars contract into dense white dwarfs, with final mass less than $1.4\ M_{\odot}$.

Rarer massive stars, which start out with $8\ M_{\odot}$ or more, undergo supernova explosions and blow off much of their initial material. If the remnant cores of stars end up between $1.4\ M_{\odot}$ and about $5\ M_{\odot}$, they form dense, rapidly rotating neutron stars known as pulsars. If the remnant cores have more than about $5\ M_{\odot}$, they may form black holes. Although virtually no radiation escapes from these strange objects, they may be detectable by orbital motions of their companion stars and by high-energy radiation from material falling into them. The detailed physics of these dense, small star forms is an area of intense current research.

CONCEPTS

red giant	Pauli exclusion principle
funneling effect	Chandrasekhar limit
triple-alpha process	supernova
helium flash	neutron star
variable star	pulsar
Cepheid variable	black hole
planetary nebula	event horizon
supergiant	accretion disk
white dwarf	
degenerate matter	

PROBLEMS

1. Why do stars just moving off the main sequence expand to become giants instead of starting to contract at once? Why does contraction ultimately win out?

2. Many red giants are visible in the sky, even though the red giant phase of stellar evolution is relatively short-lived. Why are so many red giants visible?

3. List examples of evidence that certain stars can lose mass.

4. Which stars will eventually become:
 a. White dwarfs?
 b. Neutron stars?
 c. Black holes?

What will be the ultimate fate of the Sun?

5. A main-sequence B3 star has about 10 times the mass of the Sun and therefore has about 10 times as much potential nuclear fuel. Why then does it have a main-sequence lifetime only 1/200 as long as that of the Sun?

6. According to the law of conservation of angular momentum, a figure skater spins faster as she pulls in her arms. How does this principle help explain why neutron stars spin much faster than main-sequence stars?

7. Comment on the roles and relations of theorists and observers in the three decades of work on white dwarfs, pulsars, and black holes. Are black holes fully understood today?

PROJECTS

1. Locate the star Mira (R.A. = 2^h14^m; Dec. = $-3°.4$) with a small telescope and determine whether it is in its faint or bright stage. If it is bright enough to see with the naked eye, record its brightness from night to night by comparing it with other nearby stars of similar brightness. By checking brightnesses of these stars with star maps showing magnitudes, plot a curve of Mira's brightness over time.

2. With binoculars or a small telescope locate the star Delta Cephei (R.A. = 22^h26^m; Dec. = $+58°.1$) and compare it from night to night with other nearby stars of similar brightness. Can you detect its variations from about 4.4 to 3.7 magnitude in a period of 5.4 d?

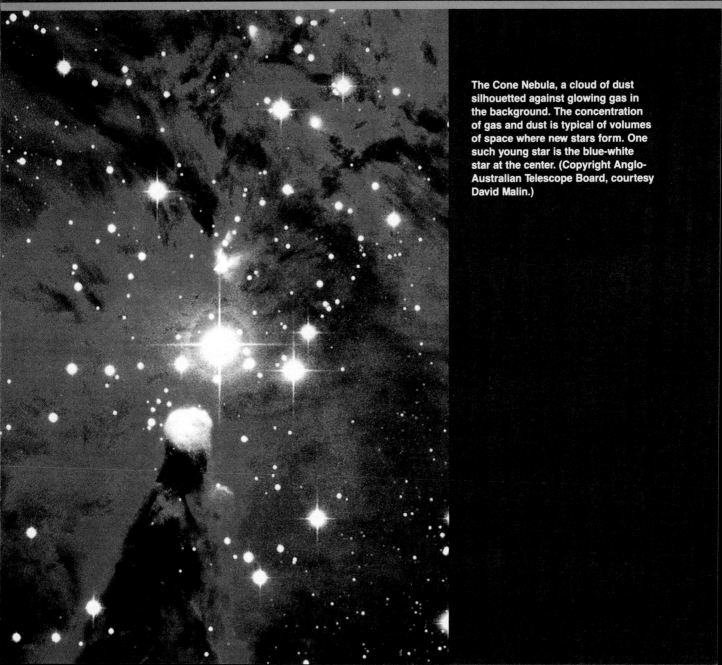

The Cone Nebula, a cloud of dust silhouetted against glowing gas in the background. The concentration of gas and dust is typical of volumes of space where new stars form. One such young star is the blue-white star at the center. (Copyright Anglo-Australian Telescope Board, courtesy David Malin.)

CHAPTER 20

Interstellar Atoms, Dust, and Nebulae

The preceding chapters discussed stars as if they were isolated individual objects. But they are not isolated. They are engulfed in a thin but chaotic medium of gas, dust, and radiation. They form from this thin material, interact with it, recycle it, and expel it to form new interstellar material.

Among the individual clouds of gas and dusts, particularly vivid nebulae were cataloged as early as 1781 by the French astronomer Charles Messier. They are thus known by their **Messier numbers,** or M numbers. The well-known Orion nebula, for example, is M 42. Others are known by **NGC numbers** or **IC numbers,** based on the more recent New General Catalog and Index Catalog, respectively. Tradition also names most bright nebulae according to their appearance in small telescopes; examples include the Crab, Dumbbell, and Ring nebulae. Some examples are listed in Table 20-1.

We sometimes casually say "space is a vacuum," but this is not quite true. Although space is a better vacuum than can be achieved in labs (Table 20-2), its material cannot be neglected. What significance can this thin material have for us? For one thing, it dots our sky with **nebulae,** or vast clouds of dust and gas, some twisted into beautiful wispy forms, some dark, and some glowing with different colors. For another thing, as Joni Mitchell said in her song "Woodstock," "We are stardust." As Chapter 14 pointed out, our solar system, our Earth, and we ourselves are formed from atoms that were once part of the interstellar gas and dust. More provocatively, recent discoveries have demonstrated that interstellar material does contain complex organic molecules, and some scientists have speculated that primitive biochemical processes, perhaps related to the origin of life, may have occurred in nebulae.

TABLE 20·1

Characteristics of Selected Nebulae

Name	Constellation	Approx. Distance from Earth (pc)	Approx. Diameter (pc)	Estimated Atoms per m³	Mass ($M_\odot$)	Spectral Type of Associated Star
Nebulae Probably Associated with Young Objects						
NGC 2261 Hubble's (R Mon)	Monoceros	700	.00001	10^{18}	0.1	F
Kleinmann–Low IR	Orion	500	0.1	10^{12}	100	Protostar?
Dark Nebulae						
Coal Sack	Crux	170	8	2×10^6	15	None
IC 434 Horsehead	Orion	350	3	2×10^7	0.6	B
Emission Nebulae						
M 42 Orion (central)	Orion	460	5	6×10^8	300	O
Eta Carinae	Carina	2400	80	2×10^8	1000	Peculiar
M 8 Lagoon	Sagittarius	1200	9	8×10^7	1000	O
M 20 Trifid	Sagittarius	1000	4	10^8	1000	O
Reflection Nebulae						
M 45 Pleiades	Taurus	126	1.5	?	?	B
Cocoon	Cygnus	1600	2	7×10^7	7	B
Planetary Nebulae						
M 57 Ring	Lyra	700	0.2	10^9	0.2	White dwarf?
M 27 Dumbbell	Vulpecula	220	0.3	2×10^8	0.2	White dwarf?
NGC 7293 Helix	Aquarius	1400	0.5	4×10^9	0.2	White dwarf?
Supernova Remnants						
M 1 Crab	Taurus	2200	3	10^9	0.1	Pulsar
NGC 6960/2 Veil (Loop)	Cygnus	500	22	?	?	?
Gum	Puppis–Vela	460	360	10^5	100,000	Pulsar?

Source: Data from Allen (1973); Maran, Brandt, and Stecher (1973); and other sources.

OBSERVED TYPES OF INTERSTELLAR MATERIAL

Interstellar material includes gas (that is, atoms and molecules), microscopic dust grains, and possibly larger objects.

Interstellar Atoms

Atoms of interstellar gas were discovered in 1904, when German astronomer Johannes Hartmann accidentally discovered absorption lines caused by interstellar calcium atoms while studying the spectra of a binary star.

TABLE 20·2

Gas Densities in Different Environments

Locale	Density[a] (kg/m³)	Particles per m³	Typical Distance Between Particles
Air at sea level	1.2	10^{25}	1 nm
Circumstellar cocoon nebula	10^{-5}	10^{22}	50 nm
"Hard vacuum" in terrestrial lab	10^{-9}	10^{18}	1 μm
Orion Nebula	10^{-18}	10^{9}	0.1 cm
Typical interplanetary space	10^{-20}	10^{7}	0.5 cm
Typical interstellar space	10^{-21}	10^{6}	1 cm
Interstellar space near edge of a galaxy	10^{-25}	10^{2}	20 cm
Typical intergalactic space	10^{-28}	10^{-1}	2 m

[a]Average density of observable matter in the whole universe is estimated to be about 3×10^{-28} kg/m³ (Shu, 1982), but many astronomers believe the average density may be somewhat higher (3×10^{-27} kg/m³?) due to nonluminous unseen matter.

Certain other **interstellar atoms,** such as sodium, were soon found to produce additional prominent interstellar absorption lines. These lines are identified as interstellar by the fact that they have different Doppler shifts than the stars in whose spectra they appear.

Further studies have convinced astronomers that, although atoms such as calcium and sodium have prominent absorption lines, the most common interstellar gas is the ubiquitous hydrogen. Like the Sun and stars, the total mass of interstellar gas is nearly three-fourths hydrogen and one-fourth helium.

Radio Radiation from Interstellar Atoms

In 1944, the Dutch astronomer H. C. van de Hulst predicted that the most important type of interstellar radiation would be an emission line with wavelength 21 cm, caused by a change in the spin of hydrogen atoms' electrons. These electrons can have only certain spin rates and directions, and they emit the 21-cm radiation when they change from one state to another. Such a long wavelength is not visible light, but radio radiation. The predicted emission was confirmed in 1951 when Harvard astronomers, using radio equipment, detected this **21-cm emission line** of atomic hydrogen.

This discovery not only confirms the importance of hydrogen as a main constituent of interstellar gas, but gives radio astronomers a tool to detect where clouds of interstellar gas are concentrated. Because of its long wavelength, this radiation can penetrate much greater distances through the interstellar gas and dust than ordinary light. The 21-cm line of interstellar hydrogen is thus the most important emission line in radio astronomy.

Interstellar Molecules

By 1940, astronomers at Mt. Wilson Observatory in California had built spectrographs that detected absorptions due not only to interstellar atoms but also to **interstellar molecules,** such as CH and CN. Figure 20-1 shows spectral absorption lines caused by interstellar CH.

Because of molecular structure, a great many of the molecular absorptions lie in the infrared or radio parts of the spectrum. Not until after World War II did astronomers have available the technology of infrared and radio detectors to search for interstellar molecules. Putting some of the new detectors in satellites and large telescopes sparked an explosion of interstellar discovery in the late 1960s and 1970s. The hydroxyl (OH) molecule, water (H_2O), ammonia (NH_3), and formaldehyde (HCHO) were found by 1969. More and more complex forms have been discovered, such as the nine-atom molecule ethyl alcohol (C_2H_5OH), found in 1974. By 1983, some 56 varieties were cataloged.

Interstellar molecules may shed light on the chemical processes that led to life. The atoms recurring again and again in these large molecules are carbon, hydro-

Figure 20-1 Spectral absorption lines caused by interstellar molecules of CH gas. Background light is part of the blue portion of the spectrum of the star Zeta Ophiuchi. Wavelengths are marked in nanometers. (G. Herbig, Lick Observatory.)

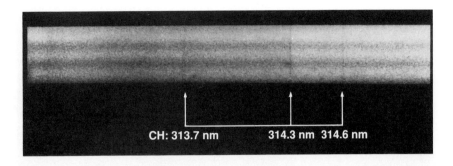

CH: 313.7 nm   314.3 nm 314.6 nm

gen, oxygen, and nitrogen—the "building blocks of life"! Still more provocative is the fact that two of the detected molecules, methylamine (CH_3NH_2) and formic acid (HCOOH), can react to form glycine (NH_2CH_2COOH), one of the amino acids. These large molecules can join to form the huge protein molecules that occur in living cells.

Interstellar Grains

Still larger than interstellar molecules are **interstellar grains.** They cause two observational effects: the reddening already explained and a general dimming of starlight at all wavelengths, called **interstellar obscuration.** Some grains are clumped in distinct clouds (as in Figure 20-2), but others are widely distributed throughout the interstellar gas, producing a general haze or "interstellar smog." All distant stars are harder to observe because of this haze.

If we look along the Milky Way plane, dust grains dim stars by an average of about 1.9 magnitudes for every 1000 pc traversed by the beam of light. Of this amount, about 1.6 magnitudes of dimming are caused by grains concentrated in clouds and about 0.3 magnitude by grains dispersed between clouds. Over a distance of 10,000 pc, stars are dimmed by 19 magnitudes! Thus it is hard for us to study regions far away across our own galaxy.

Astronomical studies reveal various properties of the interstellar grains: (1) They range in size from one-sixth to twice the wavelength of visible light; (2) they are concentrated in clouds; (3) compositions may vary somewhat from cloud to cloud; (4) many grains are elongated; and (5) many elongated grains are aligned parallel to other nearby grains, possibly because they are iron-rich and aligned with magnetic fields in space.

Grain composition has long been debated. Carbon compounds, silicates, and ices have been identified spectroscopically in dust grains near other stars, especially in star-forming regions. Reactions initiated by ultraviolet light in these materials apparently create complex organic molecules on many grains' surfaces (Greenberg, 1984). Interstellar grains of carbon and spinel (Al_2MgO_4), recognized as interstellar through their isotopes, have been identified in some meteorites. Magnetite grains similar to those in meteorites are thought to contain as much as 16% of all the iron atoms in the galaxy and may account for the magnetically controlled alignment. Only a few years ago, astronomers would have judged it implausible to study interstellar grains at firsthand. Now, according to the meteorite studies, we have them in our labs!

Interstellar Snowballs?

Astronomers are now looking for **interstellar snowballs**—hypothetical bodies much larger than grains. These could be BB-sized, baseball-sized, or even kilometer-scale bodies in interstellar space. If light interacted with such particles, it would be blocked equally at all wavelengths much less than the particle size. Although microscopic grains are revealed by reddening, large particles would be difficult to detect because virtually no color effects would occur. Nonetheless, if single atoms join into molecules, and molecules into dust grains, why not expect still larger particles?

There are some indications that interstellar snowballs exist (Greenberg, 1974; Herbig, 1974). For example, the interstellar gas is strangely lacking in certain elements. Calcium, titanium, iron, and magnesium are all much rarer than we would predict from their abundance elsewhere in the universe. Aluminum has not been found at all. Because the interstellar grains apparently do not contain enough of these materials to make up for all the atoms missing in the gas, another class of interstellar material—snowballs—may contain the

Figure 20-2 An example of obscuration by interstellar dust. A black dust cloud is silhouetted against a crowded field of stars in the constellation Sagittarius. Close examination of the dust cloud reveals that dimmed stars seen through it are strongly reddened. Near the cloud is the star cluster NGC 6520. (Copyright Anglo-Australian Telescope Board, courtesy D. Malin.)

missing atoms. They might look like our present conception of comet nuclei—icy bodies with silicates and other "dirt" mixed in—justifying the term *snowball* coined by Greenberg.

EXPLAINING THE COLORS OF INTERSTELLAR CLOUDS

Clouds of interstellar gas and dust are some of the most beautifully colored objects in space. To understand why, we need to consider how dispersed particles, whether floating in space or in the atmosphere, interact with radiation. When light from a distant star passes through clouds of interstellar atoms, molecules, and dust grains, the interaction changes the light's properties, such as intensity and color. Several complex physical laws describe these changes in some detail, but the changes can be grouped under two main principles:

> **1. When radiation (ultraviolet light, visible light, infrared, radio waves, or any other type) interacts with particles, the type of interaction depends on the types of particles and their sizes relative to the wavelength of the light.**
>
> **2. The appearance of the light and the particles may depend on the direction from which the observer looks.**

Three important types of interaction between radiation and matter involve atoms, molecules, and dust grains.

In reality, the interstellar material is always a mixture of gas and dust, but it is easier to understand the effects if we imagine separate interactions of light with atoms, molecules, and dust grains.

Interaction of Light with Interstellar Atoms

Several things can happen if starlight passes through a cloud of interstellar gas atoms. Suppose a star radiates light of all wavelengths, and the photons of light enter a cloud of gas. Photons corresponding to certain wavelengths will have just enough energy to **excite** this gas, that is, knock electrons from lower to higher energy levels. They may even **ionize** the gas—knock electrons clear out of the atoms. Each time a photon excites or ionizes an atom, that photon is consumed and disappears from the light beam. Thus an observer looking at the star through the cloud would see absorption lines created by the interstellar material, as shown in Figure 20-3.

But the energy absorbed by the cloud must be reradiated, assuming that the cloud is in equilibrium. This reradiation occurs as the electrons cascade back down through the energy levels of the atoms, creating emission lines. The photons in these emission lines leave the cloud in all directions, as shown in Figure 20-3, so that an observer off to one side would see the cloud glowing in the various colors corresponding to the emissions. Colors of nebulae are hard to see with the

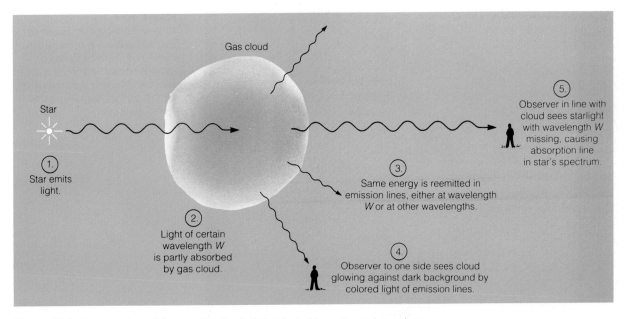

Figure 20-3 A cloud of gas (atoms and molecules) illuminated by a star and seen by an observer to one side (item 4) and an observer in line with the cloud and the star (item 5).

eye, even with large telescopes, because the light's intensity is low and the eye's color sensitivity is poor at low light levels (the reason why a moonlit scene looks less colorful than in daylight). Sensitive films and other detectors can record the extraordinary colors quite accurately, however. If a red emission line is especially strong, the cloud looks red. In another cloud, struck by photons of different wavelengths or containing atoms at different levels of excitation, a green emission line might be strongest, and the cloud would glow with green light. Photos through filters of different colors reveal different patterns (see Figure 20-4). Because hydrogen is the

Figure 20-4 Four views of the Crab Nebula (Messier catalog number M 1), photographed in light of different colors. Red and yellow views show the outer filamentary structure corresponding to clouds of excited gas giving off spectral emission lines (especially red lines from hydrogen). Infrared view emphasizes inner amorphous clouds. Many of the color figures in this book are made by combining such images. (Hale Observatory.)

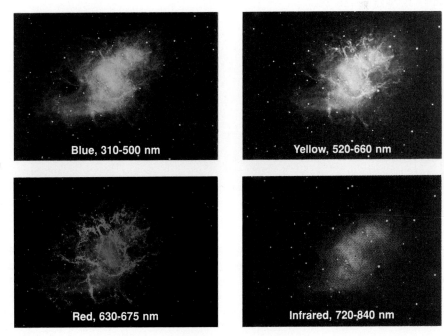

Figure 20-5 The nebula NGC 6559 and surrounding clouds of dust and gas in the constellation of Sagittarius display a variety of delicate colors. Red glows come from the hydrogen alpha emission line. (Copyright Anglo-Australian Telescope Board.)

most abundant gas, and because the red Hα emission line is one of its strongest emissions, many clouds of excited gas glow with a beautiful deep red color, as seen in Figure 20-5. Other colors also appear in nebulae, as seen in the same figure.

Interaction of Light with Interstellar Molecules

Molecules have interactions with starlight similar to those of atoms except that each absorption or emission affects

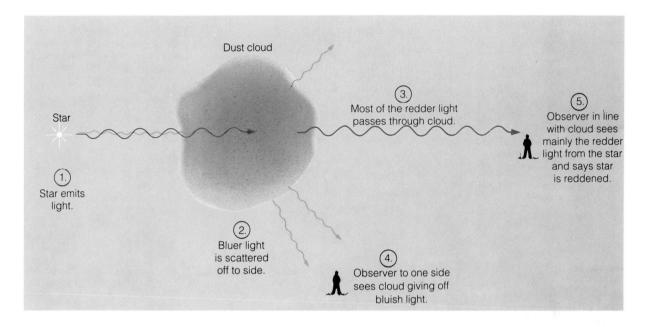

Figure 20-6 A cloud of dust grains illuminated by a star and seen by an observer to one side (item 4) and an observer in line with the cloud and the star (item 5). Compare with Figure 20-3 (p. 315), showing effects for a gas cloud.

a range of wavelengths, creating an absorption or emission *band* instead of a narrow line. Thus a gas cloud containing atoms and molecules involving different elements produces a variety of absorption lines and bands and might emit light with emission lines and bands of different wavelengths. Many molecular absorption and emission features are concentrated in the infrared part of the spectrum.

Interaction of Light with Interstellar Dust Grains

Atoms are as small as 0.001 times the wavelength of visible light, and molecules are a few percent of this wavelength, but many interstellar dust grains are much bigger than molecules, but smaller than light waves, and because of this, their interactions are quite different. They absorb some starlight, dimming distant stars. They also affect colors over a much broader range of wavelengths than individual spectral lines or bands. The most important effect is that redder light (longer wavelengths) passes through clouds of dust, whereas bluer light (shorter wavelengths) is scattered out to the side of the beam (see Figure 20-6).[1]

The scattering of blue light results from a property of all particles smaller than the wavelength of light: They scatter more blue light than red light out of the beam. This preferential scattering of blue light is called **Rayleigh scattering** after its discoverer. It occurs in interstellar dust and gas because many grains and all atoms are smaller than the wavelengths of visible light.

Thus the observer who looks through the dust cloud at a distant star sees most of its red light, but not much of its blue light. In this way, interstellar dust makes distant stars look redder than they really are—an effect called **interstellar reddening**. An observer who looks at the dust cloud from the side will see the blue light scattered out of the beam, however, so that a nebula illuminated in this way will have a bluish color. A prominent example of a blue nebula is seen in Figure 20-7.

[1]Scattering of light is different from absorption, mentioned a few lines earlier, but we will not emphasize the difference. Think of absorption as a disappearance of a photon into an atom (or ion or molecule), making the atom more energetic. Think of scattering as a bouncing of a photon off an atom out of the light beam into a new direction. Either process reduces the amount of light in the beam.

Figure 20-7 The beautiful nebula NGC 1977, in the constellation Orion, shines mostly by reflected light of nearby stars. Scattering of light in the nebula's gas makes it appear blue by the same principle that makes our sky blue. (Copyright Anglo-Australian Telescope Board, courtesy D. Malin.)

Why Is the Sunset Red and the Sky Blue?

These same rules apply to material in our own atmosphere. The lower few kilometers of the atmosphere are full of dust and large molecules, including many that are slightly smaller than the wavelength of light. When we look at the Sun through these particles, the same effects can be seen. If the Sun is high in the sky, we look through the minimum amount of dust, as shown in Figure 20-8a; thus the reddening is minimal, and the Sun is perceived as white.

At sunset, as shown in Figure 20-8b, the sunlight passes through much more gas and dust, and much of the blue light is lost from the beam by Rayleigh scattering. This strongly reddens the Sun and adjacent parts of the sky. At any time of day, if we look at some other part of the sky, as in Figure 20-8c, the light we see is the blue light scattered out of the beam of sunlight and then scattered by air molecules and dust back toward our eyes. In contrast, as seen in Figure 10-7, the sky on Mars is reddish because many of the reddish Martian dust particles are larger than the wavelength of light; hence there is no Rayleigh scattering of blue light, but simply the reflection of their own red color.

FOUR TYPES OF INTERSTELLAR REGIONS

Based on the principles discussed so far in this chapter, astronomers have classified interstellar space into four types of regions, based on their temperature. Each type has a different appearance.

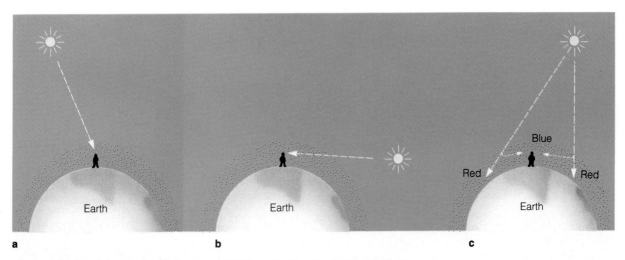

Figure 20-8 Explaining colors in the sky. **a** Light from the Sun high in the sky passes through minimal dust and is minimally reddened. **b** Light from the Sun at sunset passes through the maximum amount of dust and is strongly reddened. **c** Light from the sky at any time of day is the blue light scattered out of the sunlight beam.

Cold, dense dust clouds at about 10 K are called **molecular clouds.** They contain many molecules because the atoms frequently collide and merge. They are common in star-forming regions, and often appear as dark clouds silhouetted against nearby brighter nebulae. Figure 20-2 shows such clouds.

Probably the most common interstellar regions have temperatures around 100 K. At this temperature the hydrogen atoms are neutral, and these are called **HI regions** ("H one regions") after the symbol for neutral hydrogen (HI). These regions are generally transparent.

Near the hottest stars, O and B stars, the hydrogen gas is strongly heated to about 10,000 K, and the star's ultraviolet photons excite the hydrogen atoms, or even knock electrons clear out of them, ionizing them. These are therefore called **HII regions,** after the symbol for ionized hydrogen (HII). Here, the electrons in the H atoms lead an up-and-down life, constantly being excited to higher energy levels and then dropping back to lower levels. The latter process gives off emission lines of certain colors, as we have seen in Figure 5-4. Therefore, HII regions typically give off colorful glows, and are among the most beautiful nebulae. They are also called **emission nebulae.** As we have seen, the bright-red H alpha line is often the strongest, so that emission nebulae usually are predominantly red. However, under certain conditions in the gas, other atoms may give off other strong colors. Figure 20-9 shows a beautiful

example. Because massive O and B stars last only a short time, and burn out before their vast, surrounding star-forming region can dissipate, they are usually found only in the midst of regions where stars are still forming.

A special subclass of emission nebula, called a bubble, occurs when the star itself has blown out its surrounding hot gas. Often this forms a spherical cloud that looks something like a planetary disk in a small telescope. They are thus sometimes called planetary nebulae, although they have nothing to do with planets. Figure 20-10 shows a colorful example.

The fourth type of region is still hotter. Temperatures around 1 million Kelvin can exist where gas has been superheated by expanding blasts from supernovae. The hot, supernova-heated gas is usually expanding into the cooler surrounding gas, creating a huge bubble of superhot gas. These are called **superbubbles.** The supernovae that form them are produced only when short-lived massive stars explode; thus superbubbles, like HII regions, are usually associated with star-forming regions.

LIFE CYCLE OF A STAR-FORMING REGION

As shown in Figure 20-11, the endless cycle of star birth and death is bound to produce superbubbles that

Figure 20-9 *(Opposite)* The Trifid Nebula, about 1600 pc away in the constellation Sagittarius, beautifully combines red and blue. The hot gas within the cloud glows with the red light of Hα emission. It is roughly 5 pc across. Dark lanes of colder dust superimposed in front create a flowerlike pattern. Associated cool clouds have a soft blue coloration from scattering of starlight. The bright star in its center is a very luminous, hot O star with a blue-white light; it may be the star that excites the nebula to glow. (Copyright Association of Universities for Research in Astronomy; NOAO.)

play a key role in redistributing the interstellar material. Star formation produces a few very massive stars that rapidly explode; this heats and expands local gas, which pushes against adjacent gas. The resulting compression starts new star formation on the outskirts of the expanding cloud. This leads to new explosions and expansions until the whole star-forming complex blows itself apart.

Example: The Orion Nebula, a Nearby Star-Forming Region

In the direction of the constellation Orion lies a region about 400 pc away dominated by nebulae and star-forming activity. Orion itself is the constellation of the hunter—a great figure raising a club over his head. As can be seen in the sketch map of Figure 18-5b, three bright stars mark his belt and three below it mark his sword. Orion's shoulder on the left side is the bright-red giant star Betelgeuse, 180 pc away. His opposite knee is the blue B-type star Rigel, 270 pc away.

As shown in Figure 18-5, if you look toward Orion on a February evening, swinging your head back and forth past this part of the sky to compare it with other regions, you will see a great concentration of bright stars. Orion contains 7 of the 100 brightest stars in the sky, and except for Betelgeuse, all of them are massive, hot O- and B-type stars 140 to 500 pc away. Because

Figure 20-10 The Dumbbell Nebula, named for its appearance in small telescopes, is an example of a moderate-sized bubble (also classified as a planetary nebula). It is about 220 pc away and 0.3 pc across. It was blown off the central star, whose radiation excites the atoms to glow. Red Hα radiation dominates the outer part; bluish light from ionized oxygen atoms, excited in the space closer to the star, dominates the interior. (Copyright California Institute of Technology and Carnegie Institution of Washington; by permission from Hale Observatories.)

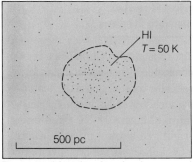

Dense, cold molecular cloud, contracting

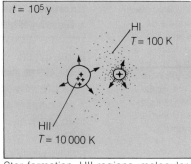

Star formation, HII regions, molecular clouds

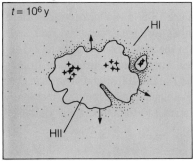

More star formation, HII expansion by heating

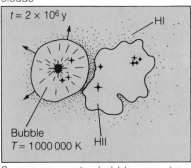

Supernova creates bubble, new star formation

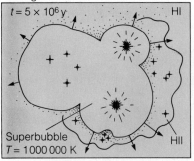

More supernovae, superbubbles, and star formation

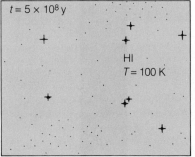

Clouds and star clusters dispersing, star formation and supernovae over

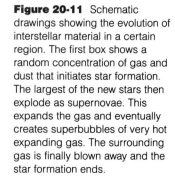

Figure 20-11 Schematic drawings showing the evolution of interstellar material in a certain region. The first box shows a random concentration of gas and dust that initiates star formation. The largest of the new stars then explode as supernovae. This expands the gas and eventually creates superbubbles of very hot expanding gas. The surrounding gas is finally blown away and the star formation ends.

massive stars are short-lived, their mere existence shows that they and other stars must have formed recently in the nebulosity around Orion. These are well shown in Figure 20-12, an overview of Orion.

Consistent with its fame as a star-spawning region, Orion is full of vast clouds of gas and dust. Because the dust cloud temperatures are around 30 K, the dust radiates in the far infrared, where it has been imaged by satellites as shown in Figure 20-12b. Here we see the most intense dust clouds centered on the Orion Nebula in Orion's "sword" and at the left end of the "belt." By coincidence, a shell of dust has been blown outward from stars in the head region, making a ring-shaped infrared halo (invisible to the naked eye) around Orion's head.

Because O and B stars are extremely hot, they radiate prodigious amounts of ultraviolet radiation. For this reason, UV photos of Orion, such as Figure 20-12c, reveal it to be one of the most dazzling regions in the "UV sky." The massive stars have ionized much hydrogen, and the hot gas is being blown outward from the bright stars in central Orion. Some filaments of this gas show up in the outer parts of the overexposed region of the belt and sword in Figure 20-12c.

a

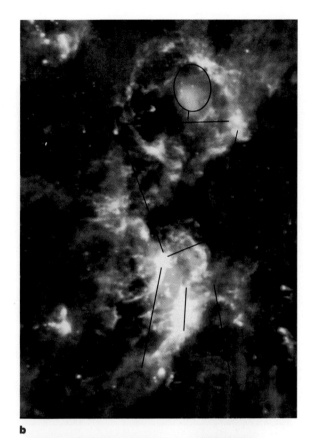

b

c

Figure 20-12 Three faces of the constellation Orion (see overview of the region in Figure 18-5). Stick figure of constellation gives orientation of each aspect; dotted box shows location of Figure 20-13. **a** Visible-light photo shows dominance of blue-white, hot, young stars (and the prominent exception—red giant Betelgeuse in the upper left). Arrow shows pink glow of Orion Nebula in the "sword." Very faint red Hα glows can be seen left of the left star in the "belt." (Photo by author; guided 35-mm camera, 55-mm lens, f2.8, 20-min exposure on 3M ASA 1000 slide film.) **b** Same region in false color view by the IRAS satellite at infrared wavelengths of 12 to 100 μm. Redder colors are cooler; blue, hotter. Orange regions are clouds of interstellar dust. Brightest sources are Orion Nebula (bottom) and region at left end of belt. Note that most individual stars are too hot to radiate much light at these wavelengths; Betelgeuse appears as a blue spot, hotter than most dust but cooler than most stars. Blue region at top is part of background Milky Way. (NASA photo.) **c** Black and white rendition of a view at ultraviolet wavelengths (125–160 nm) emphasizing the hottest O stars, with temperatures around 20,000 K. Betelgeuse is too cool to show up at these wavelengths. (Naval Research Lab photo from rocket; courtesy George Carruthers.)

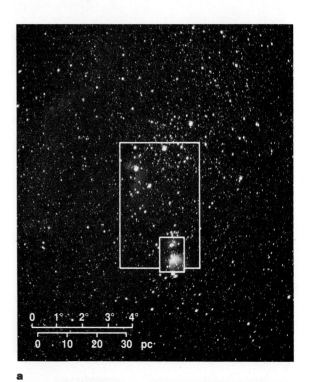

a

b

Figure 20-13 The "belt" and "sword" of Orion: heart of a star-forming region. **a** Photo shows angular scale of region in the sky and linear scale in parsecs calculated at 460 pc average distance. Boxes show region of *b* and Figure 20-15. (Photo by author; 12-min guided exposure on Ektachrome ISO 1600 film, 100-mm telephoto lens, f2.8.) **b** Color photo shows the region of the Horsehead Nebula (upper left) and overexposed Orion Nebula (bottom). (Copyright Anglo-Australian Telescope Board.)

Figure 20-13 zeroes in on the star-forming area of the belt and sword. At the top we see a colorful region where the massive stars of the area, with their high temperatures and blue colors, contrast with a remarkable sheet of red-glowing gas. Silhouetted against this red curtain is the most striking dark nebula: the Horsehead Nebula, a dust cloud resembling an ominous cosmic chess piece (Figure 20-14). At the bottom of Figure 20-13b is the still brighter Orion Nebula, which is somewhat overexposed and therefore whitish in this picture.

Figures 20-15 and 20-16 highlight the **Orion Nebula,** perhaps the most famous of nebulae. It is about 460 pc away, in the heart of the Orion star-forming area. To the naked eye it appears as the middle star in the sword, but even a small telescope or good pair of binoculars

reveals it as a misty luminous haze about 5 pc across. As early as two centuries ago, English astronomer William Herschel examined this region with his pioneering large reflecting telescope and prophetically described it as "an unformed fiery mist, the chaotic material of future suns." And the details are spectacular: intermingled wisps of red-glowing hydrogen, dark dust, and filaments dominated by pale greens of ionized oxygen and colors of other excited atoms. Here new stars are being born.

Astronomers observing at infrared wavelengths have been able to map individual clouds of dust and newly forming stars inside the nebula. Various clouds move at about 8 to 10 km/s relative to each other, and backward tracing of their motions suggests some may have formed within the last 100,000 y.

Figure 20-14 The Horsehead Nebula. Not readily visible to the eye in telescopes, the horsehead shape was first described from photographs around 1900 as a gap in the nebulosity. Some years later, it was recognized as an unusually dense dust cloud silhouetted in front of the Hα-glowing red cloud. (Copyright Anglo-Australian Telescope Board.)

Figure 20-15 The Orion Nebula is the core of a star-forming complex about 460 pc away and 5 pc across. This image is especially processed to show faint outer details without overexposing the much brighter central region. A separate star cluster and nebula lie to the north. The brightest part of the nebula is centered on a tight cluster of four young stars called the Trapezium, located in the white region just to the right of the projecting dark cloud. (Copyright Anglo-Australian Telescope Board, courtesy D. Malin.)

Radio astronomers have estimated that some 110,000 $M_\odot$ of HI and HII are involved in the expanding gas shell around the extended Orion star-forming region. Backward tracing of the motions indicates that the shell started to expand about 6 million years ago, when very massive, hot stars must have formed, heated nearby gas, and created a giant HII region that has been expanding ever since. Certain O and B stars are racing out from the nebula like sparks from a blast.

Orion has changed dramatically in the last few million years, since humanlike creatures emerged on the plains of Africa. New stars have blazed up, clouds of hydrogen have been expelled, and star formation is apparently continuing there today.

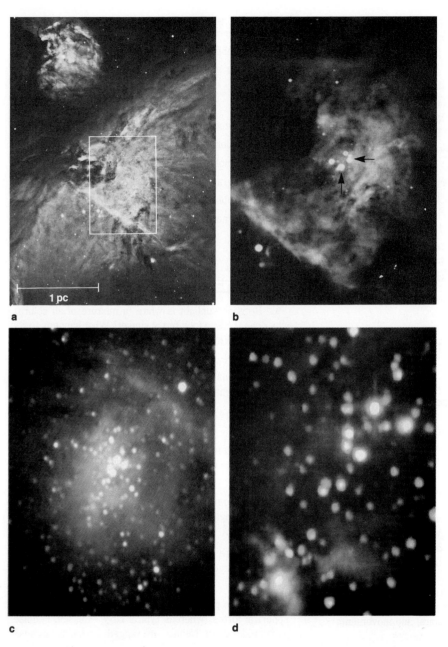

Figure 20-16 In the midst of the Orion Nebula lie many secrets.

a The core of the nebula, with scale. Box shows size of next image.

b The central region, showing the quartet of newly formed stars called the Trapezium (arrows).

c Same region as *b* but imaged at infrared wavelengths and rendered in false color. The infrared images are made at wavelengths 1 to 5 μm and reveal a compact cloud of infrared-radiating dust (orange), undetected in visible light. Also seen are many stars that are too cool and faint to be prominent in *b*, but warmer (hence rendered bluer) than the dust cloud.

d Enlarged view shows the Trapezium (upper right) and "hotspots" in the infrared nebula. The infrared nebula is probably a mass of dust and gas that is contracting and fragmenting into a group of protostars. (Photos *a*, *c*, and *d* copyright Anglo-Australian Telescope Board, courtesy D. Malin; photo *b*, Allegheny Observatory, courtesy W. A. Feibelman.)

Example: The Gum Nebula, an Ancient Spectacular

In the 1950s, a graduate student named Colin Gum used photographs sensitive to hydrogen alpha radiation and discovered a ring of nebulosity with the amazingly large angular diameter of 60°, which has come to be called the **Gum Nebula**. It has been identified as an expanding bubblelike shell of gas from an ancient supernova explosion in the Southern Hemisphere constellation Vela.

Studies of the expansion rate suggest the explosion might have been witnessed around 9000 B.C. Because the explosion was so close (about one-third the distance of the Crab supernova) and had such high energy, it was probably extremely brilliant, reaching an estimated apparent magnitude of −10 (about as bright as the first quarter moon). This raises the intriguing prospect that late Stone Age humans (south of latitude +47°) may have seen a brilliant celestial beacon blaze forth for a year or two, rivaling the Moon. What effect might this have had on their concepts of the heavens and gods in the sky?

The effect of the Vela supernova may have been more than a dramatic skyshow. Theorists have concluded that the gamma-ray and X-ray radiation from such a close supernova would deplete Earth's ozone layer, increase the nitrogen oxide production, and cool the atmosphere. The fallout of the NO in rain would have a fertilizing effect, increasing photosynthesis and deposition of organic sediments. G. R. Brackenridge (1981) describes 11 dramatic cases of black, organic sediments deposited around 9000 B.C., and suggests that they were caused when the Vela supernova produced a worldwide but brief climatic change. Earth may be affected more than we know by "winds" that blow from interstellar space.

SUMMARY

Space is not empty but thinly filled with atoms and molecules of gas, grains of dust, and possibly larger debris, which partly obscure distant stars and redden their light. Concentrations of these materials are nebulae. Some are excited to fluorescence by nearby stars, forming emission nebulae; some merely reflect the light of nearby stars; and some are dark silhouettes against distant backgrounds. Various processes of emission, Rayleigh scattering, and absorption give nebulae different colors. Various processes of expansion and turbulence give nebulae different shapes.

Nebulae are the material from which stars are born and into which the larger stars blow some of their material when their fuel runs out. Their existence shows that matter in our galaxy has not dispersed uniformly and stably, but rather is continually stirred, formed into clouds, dispersed, and disturbed by influences such as the formation of new stars, the explosions of old stars, and movements of all local material around our galaxy's center.

From nebulae and the general interstellar complex we know that all material visible from Earth has participated in a vast cosmic recycling. Interstellar matter forms clouds; clouds may contract to form stars; young and old stars blow out their material, creating a new interstellar medium. Nebulae also reveal clear evidence of cosmic events that have markedly changed our celestial, and perhaps even terrestrial, environment within the last million years—a period less than 0.01% of cosmic time.

CONCEPTS

Messier number / ionization
NGC number / Rayleigh scattering
IC number / interstellar reddening
nebula / molecular cloud
interstellar atom / HI region
21-cm emission line / HII region
interstellar molecule / emission nebula
interstellar grains / superbubble
interstellar obscuration / Orion Nebula
interstellar snowball / Gum Nebula
excitation

PROBLEMS

1. Because sunlight is white (a mixture of all colors), why does the Sun look red at sunset? What happens to the blue light? Why does the part of the sky away from the Sun look blue?

2. Why was the sky red as photographed on Mars by Viking cameras?

3. How do massive stars help keep interstellar gas stirred up?

4. How do interstellar molecules illustrate the fact that complex organic chemistry is likely elsewhere in the universe?

5. How do masses of prominent nebulae compare with masses of single stars? Are stars more likely to form singly or in groups?

6. Why are O-type supergiant stars likely to be associated with large emission nebulae, whereas solar-type stars are not?

7. Why do planetary nebulae often have simple, nearly spherical forms, whereas typical large emission nebulae, such as the Orion Nebula, are ragged, irregular masses?

PROJECTS

1. Observe a cloud of cigarette or match smoke illuminated by a single light source, preferably a shaft of sunlight in a darkened room or a strong reading lamp. Compare the color of light transmitted through the smoke (by looking into the beam) with the color of light scattered out of the smoke (by looking at right angles across the light beam).

Are there any differences? What can you conclude about the size of particles in the smoke cloud, assuming that the light wavelength is mostly 400 to 800 nm? Compare forms in the drifting smoke cloud with forms of nebulae illustrated in this book.

2. In a dark area away from city lights, by naked eye, observe and sketch the Milky Way in the region of Cygnus. Can you observe the dark "rift" that divides the Milky Way into two bright lanes in this region? The rift is caused by clouds of obscuring interstellar dust close to the galactic plane that are between us and the more distant parts of the Milky Way galaxy.

3. Observe the Orion Nebula with a telescope. Sketch its appearance. Locate the Trapezium (four stars near the center). The dark wedge radiating from the Trapezium is a dense mass of opaque dust. If a large telescope (50–100 cm) is available, look carefully for color characteristics. Generally, the eye is unresponsive to colors of very faint light, but large telescopes gather enough light so that colors can sometimes be perceived, especially with fairly low magnifications, giving a compact, bright image.

4. Observe the Ring Nebula in Lyra or other nebulae such as the Crab (in Taurus) or the Trifid (in Sagittarius). Comment on differences in form and origin.

CHAPTER 21

Companions to Stars: Binaries, Multiples, and Possible Planetary Systems

We have been soft-pedaling a fundamental fact: Most stars are not solitary wanderers in space. This is beautifully revealed in Figure 21-1, where a sequence of photos shows the motion of one star around its companion. As seen in Table 17-1, four of the first six stars we encounter beyond the solar system have known companion stars. Surveys of this kind suggest that around three-fourths of all "stars" are really coorbiting star systems.

Each pair of coorbiting stars is called a **binary star system,** or sometimes just a binary star. Each system with more than two stars is called a **multiple star system.** How do these systems affect our understanding of the universe? For one thing, we would like to know if our own multiple system, the Sun and its planets, is related to other multiple star systems, or whether it is a different kind of phenomenon. Second, we must be sure that our theories of star formation and evolution account for binaries.

OPTICAL DOUBLES VS. PHYSICAL BINARIES

Among the star pairs that appear to be close together in the sky, some are at different distances and merely aligned by chance (as seen from Earth)—they are not actually coorbiting. Such star pairs are called **optical double stars.** They can be identified by the absence of any orbital motion around each other (revealed by photos or Doppler shifts) and are of little consequence in astronomy.

Pairs that are close enough to orbit around each other are called **physical binary stars.** They are the ones that clarify our knowledge of star properties and are the ones we will discuss here.

Early observers thought that *all* close star pairs

were merely optical doubles, but in 1767 John Michell pointed out that so many chance alignments were unlikely. He proposed a physical association. This was confirmed in 1804 when William Herschel discovered that Castor (brightest star in the constellation Gemini) has a companion orbiting around it. In 1827 F. Savary showed that the orbit of the physical binary Xi Ursae Majoris is an ellipse fitting Kepler's laws. These results marked the *first discovery of gravitational orbital motion beyond the solar system,* an important confirmation that gravitational relations are universal.[1]

Among physical binaries, the brighter star is usually designated A and the fainter star B—for instance, Castor A and Castor B. Analysis of orbits reveals which is the more massive star, and it is usually called the *primary.* The less massive one is called the *secondary.* Normally, the primary is also the brighter, or star A.

TYPES OF PHYSICAL BINARIES

Astronomers have developed two ways of classifying physical binaries. The traditional method relies on observing technique, and classifies the binaries according to how they are detected. Further work led to a second method—involving the separation distance and evolutionary state—which we will consider later in the chapter. But the first set of classifications is important, because each class teaches us distinct information about stars, such as mass or size.

A **visual binary** is a binary pair in which both members can be detected individually in a large telescope. In a **spectroscopic binary,** orbital motion is revealed by periodic Doppler shifts in the spectral lines as one star moves around the other. The individual stars cannot be resolved, and the only clue about their binary nature comes from the spectrum.

An **eclipsing binary** is a binary pair (also generally unresolved) whose orbit is seen nearly edge-on, so that one star moves in front of the other. As this happens, the back star is hidden, and the observed brightness of

---

[1]Herschel realized that this discovery was much more important than his original goal of simply measuring distances, because the orbital motions reveal many properties of the stars. This illustrates how important, unexpected scientific results often derive from mundane research on another topic. Herschel reportedly likened himself to the biblical character Saul, who went out to find his father's mules and discovered a new kingdom.

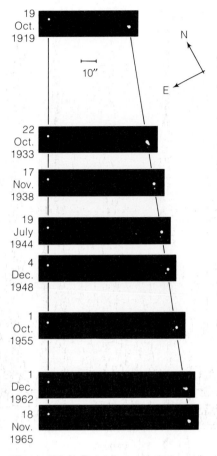

Figure 21-1 Forty-six years' photography of the visual binary Krüger 60. Binary pair Krüger 60, on the right, is moving past a background star, on the left. The photos show the orbital motion of one component of the binary around the other with a period of 45 y and a separation of 1.4 to 3.4 seconds of arc. The two stars have magnitudes about 9 and 11 and masses of 0.3 and 0.2 $M_\odot$. (Leander McCormick Observatory and Sproul Observatory, after Wanner, 1967.)

the system dips, as in Figure 21-2. Analysis of the shape and timing of these brightness dips reveals information about the sizes, orbits, and other properties of the system, as shown in the figure.

Most eclipsing binaries are really **eclipsing–spectroscopic binaries,** in which both Doppler shifts and eclipses can be detected. This is the most informative type of binary, permitting very detailed analysis of the motion, mass, and size of the stars, through application of Kepler's laws and other principles. To take a simple example, suppose the Doppler shifts reveal

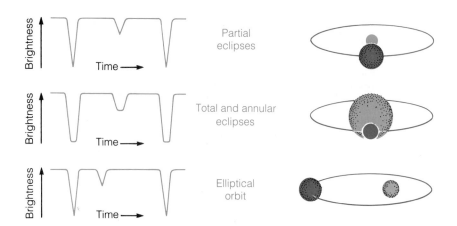

Figure 21-2 Different types of light curves (left) from eclipsing binaries reveal different geometric properties of the eclipsing stars and their orbits (right), even though the stars themselves cannot be resolved with the telescope. The brightness dips when one star is hidden behind the other.

TABLE 21·1

Selected Binary and Multiple Stars

System Name	Component A		Component B			Component C		
	Distance (pc)	Mass ($M_\odot$)	Separation from A (AU)	Mass ($M_\odot$)	Eccentricity	Separation from A (AU)	Mass ($M_\odot$)	Eccentricity
Visual Binaries								
α Centauri	1.3	1.1	24	0.9	0.52	10,000	0.1	
Sirius	2.6	2.2	20	0.9	0.59			
Procyon	3.5	1.8	16	0.6	0.31			
Eclipsing Binaries								
Algol (β Persei)	27	3.7	0.73	0.8	0.04	2?	1.7	0.13
ε Aurigae	1350	15?	35	15?				
Eclipsing–Spectroscopic Binaries								
β Scorpii	118	13	0.19	8.3	0.27			
η Orionis	175	11	0.6	11	0.02			
Astrometric Binaries								
Krüger 60	4.0	0.28	9.5	0.16	0.41			
Barnard's star	1.8	0.14	2.7?	0.001?				
Visual and Astrometric Binaries								
61 Cygni	3.5	0.6	83	0.06			0.0008?	
BD +66°34	10	0.4	41	0.13	0.05	1.2	0.12	0.00
Solar System								
Sun, Jupiter, Saturn	0.0	1.0	5.2	0.001	0.05	9.5	0.0003	0.06

Source: Heintz (1978); Popper (1980).
Note: Some properties of these systems are not yet known.

that one star moves in a circular orbit at 100 km/s, and timing of the eclipse reveals that it takes 10,000 s (about 3 h) to pass in front of the other star. Then the diameter of the star must be about 1 million kilometers (100 km/s × 10,000 s). This type of observation is one of the most accurate checks on theories of stellar structure.

Imagine a binary consisting of a bright star and a companion too faint to see from Earth. The bright star is moving around the unseen companion. Extremely careful measurements of its position, relative to background stars, can reveal this motion, in turn revealing that it is a binary and has an unseen companion. Such a binary is called an **astrometric binary.** Interestingly, information can be derived in such systems about the *unseen* companion, as well as the visible star.

Table 21-1 lists examples of the most important types of binary and multiple stars. The famous star Mizar, in the middle of the Big Dipper's handle, illustrates several types of doubles. Mizar forms an optical double with a fainter star, Alcor, less than a degree away. The Arabs called this pair the "horse and rider" and regarded it as a test of good eyesight. In 1650, the Italian observer Jean Riccioli discovered that Mizar itself is a double with a visible 4th-magnitude companion, Mizar B, just 14 seconds of arc away. This pair was later found to be a physical binary. In 1889 Mizar A was found to have double spectral lines with periodic Doppler shifts, revealing that it is a spectroscopic binary. In 1908, the same was found true of Mizar B. Thus Mizar is really a quadruple star system, with one close coorbiting pair revolving around another close coorbiting pair.

One interesting facet of binary stars is that very different types of stars can be paired: massive and not so massive, giant and dwarf, red and blue. An imaginary example is seen in Figure 21-3. The variety offers challenging opportunities for both measurement and imagination.

HOW MANY STARS ARE BINARY OR MULTIPLE?

This question is a challenge to observers. The nearest stars are easiest to observe but give too small a statistical sample to be reliable. At greater distances there are more stars, but faint companions might not be detected. Spectroscopic binary statistics are biased toward pairs with small separation distances, because according to Kepler's laws these have the fastest velocities and greatest Doppler shifts, thus being the most

Figure 21-3 Binary stars show a wide variety of pairings. In this imaginary view, a white dwarf orbits around a red giant that is shedding mass into a disk of gas and dust around itself. The white dwarf is a pinpoint of light relative to the swollen giant in the background. (Painting by author.)

likely to be discovered. Visual binary statistics are biased toward wide separation distances, which make the two stars easier to resolve. All these biases, which tend to make the data nonrepresentative of the whole population, are called **selection effects.**

Table 21-2 lists three estimates of the **incidence of multiplicity** for systems ranging from single to sextuple. By the time we reach six-member systems, definitions of multiple systems become hazy. It is unclear whether close groupings like the Trapezium in the Orion Nebula should be counted as multiple systems. The 1982 Yale Catalog of 9096 prominent stars lists multiple systems ranging as high as one system with 17 members. Such systems would present an interesting spectacle —if they could be seen from nearby (Figure 21-4). We don't know if there is a physical relation between such

Figure 21-4 A hypothetical view within the sextuple star system Castor—brightest star in the constellation Gemini. What appears to the naked eye as a single star is in fact three pairs of close binaries, all orbiting around each other. Here we are on an imaginary planet orbiting the faintest pair, two low-mass red stars of about 0.6 $M_\odot$ each. More than 1000 AU away are the two other pairs (left). All four of the distant stars are whitish A stars about 50% more massive than the Sun and averaging 10 times as luminous. Although astronomers are searching near other stars for planets like that depicted in the foreground, none has been positively identified. (Painting by Ron Miller.)

TABLE 21·2

Incidence of Multiplicity Among Stars (Estimated Percentage of Systems Containing *n* Members)

n	25 Systems Within 4 pc	Average of 7 Estimates by Various Authors[a]	Estimate by Batten (1973)[b]
1	48%	41%	30%
2	36	41	53
3	12	14	13
4	04[c]	03	03
5	—	01	008
6	—	—	002

[a]Data from Batten (1973); Abt (1983, p. 345).
[b]Batten's estimates attempt to average over all stars, using data from various sources. Differences between the estimates are measures of our uncertainty about multiple systems.
[c]This entry represents our own multiple system, considered as a central star with numerous smaller companions.

large multiple systems and small clusters or if they are different phenomena. Even though Table 21-2 illustrates our uncertainty about the frequency of binaries, it does show that single stars are a minority.

Kitt Peak astronomers Helmut Abt and Saul Levy (1976) found that seemingly single stars probably *all* have companions too small to detect! Some companions might be brown dwarfs with only a few percent of a solar mass; still smaller ones may be planets. According to this estimate, virtually all stars have at least one companion.

Thus although many people mistakenly assume that most stars in the night sky are single, *binary and multiple systems are more common than single stars*. Obviously, then, we must understand the origin of systems of two, three, four, and more stars in order to claim any understanding of stars in general. We will return to the problem of origin after reviewing a second classification system based on the evolution and separation distance of the system.

EVOLUTION OF BINARY SYSTEMS: MASS TRANSFER

The evolution and spacing of individual systems provides a natural way for astronomers to classify them. The theory of dynamics of binary stars indicates that a system of two coorbiting stars can be pictured as con-

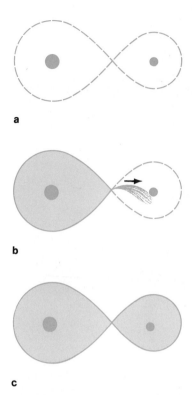

a

b

c

Figure 21-5 Evolution of a binary pair is related to the configuration of Lagrangian lobes. **a** In the first class of systems, neither star fills its lobe (dashed line). **b** After the larger star expands to become a giant, it may fill its lobe, taking on a teardrop shape and perhaps ejecting some mass through the tip, which interacts with the second star. **c** In the third class of systems, both stars fill their lobes.

taining an imaginary surface called the **Lagrangian surface**, which is of special importance; it is often called the Roche surface—both names come from French theorists who worked on the dynamics of such systems. Its cross section is a figure 8, with one lobe around each star, as shown in Figure 21-5a. Slow-moving material inside either lobe orbits around the star in that lobe. Material that moves out of either lobe is not gravitationally bound to either star individually. It may transfer from one to the other or, with enough velocity, eventually leave the system entirely. When either star evolves toward a red giant state, it expands until it fills its lobe, as in Figure 21-5b. Its outer layers then assume the peculiar teardrop shape of the lobe, and the Lagrangian surface becomes the real surface of the star. Using these ideas, astronomers realized that **three structural classes of binaries** exist, as shown in Figure 21-5:

1. Systems in which neither star fills its Lagrangian lobe

2. Systems in which one star fills its Lagrangian lobe

3. Systems in which both stars fill their Lagrangian lobes; that is, stars in contact with each other (**contact binaries**)

Evolution can take a single binary through all three types. A pair might start as class 1. The more massive star swells into a giant as it evolves. This giant may fill its Lagrangian lobe, creating class 2. Any further tendency to expand causes matter to be shed, mostly through the point common to the two lobes, as shown in Figure 21-5b. Like sand in an hourglass, matter lost through this point enters the lobe of the smaller star. Now the first (larger) star loses mass as the second (smaller) one gains mass. Obviously, the evolution of both stars is altered from what it would have been if they had been isolated. At some later time, the second star evolves toward its own giant stage and expands, either because of mass gain or because of its own normal evolution. Both stars may now simultaneously fill their lobes, producing class 3, a contact binary. The two stars actually touch as they orbit around each other as a single unit! Such a configuration would make a strange "sun" in the sky of some imaginary world (Figure 21-6). If two stars start very far apart, the giant expansion may not fill either Lagrangian lobe, and the binary may never evolve out of class 1.

The grouping of stars in binary and multiple systems with different masses and different states of evolution in classes 1, 2, and 3 ensures that binary stars provide a rich variety of types.[2] Even backyard telescopes can reveal pairs and multiples with beautiful color combinations.

NOVAE: EXPLODING MEMBERS OF BINARY PAIRS

Consider the case of a class 2 binary when the smaller star is a white dwarf of nearly $1.4\ M_\odot$. If the red giant's atmosphere dumps unburned hydrogen onto the surface of the white dwarf, the hydrogen may ignite in nuclear reactions that blow excess gas outward. This type of

---

[2]The colorful nomenclature of star types encourages shameless wags to concoct such descriptions as "degenerate dwarf in the company of a young starlet."

explosion is called a **nova** (plural: *novae*). It is generally smaller than a supernova explosion, discussed earlier.

Some novae are believed to occur in cycles. Hydrogen may accumulate on a dwarf until nova explosions occur, perhaps 10,000 y apart. Because the individual explosions blow off only a small fraction of the star's mass, leaving the rest intact, the process can start over. Among closer novae, the cloud of expanding debris can sometimes be seen in telescopes a few years after the explosion.

CONTACT BINARIES AND OTHER UNUSUAL PHENOMENA ASSOCIATED WITH BINARIES

Among the most unusual stars are the contact binaries, stars of class 3 in the evolution sequence described above (Figure 21-5c). The most famous of these are the **W Ursae Majoris stars,** named after the prototype in the constellation Ursa Major (better known as the Big Dipper). These consist of stars of rather similar mass, both filling their Lagrangian lobes. They have total masses ranging from 0.8 to 5 $M_\odot$, and because they are so close together, their common revolution periods are very short, less than 1.5 d.

How did such a pair form? Some theorists believe they formed from protostars rotating so rapidly that they split, or *fissioned*, into two components. In contrast to the widely separated types of binaries, which have mass ratios expected from random pairings of field stars, close binaries show a tendency toward mass ratios of 1:1 (Abt, 1979), consistent with the fission theory of origin. Theorists have also proposed mechanisms that might cause widely spaced pairs to evolve into contact binaries.

Possibly related are stars with very close orbits and extremely short orbital periods. The shortest known period for any binary, only 11 min, was discovered in 1986 by the European orbiting X-ray observatory, Exosat (Thomsen, 1986). The binary is apparently a neutron star orbiting a white dwarf, cataloged as 4U1820-30 and about 6000 pc from the solar system. Theoretically, mass loss could produce contact binaries with periods as short as 2 min! Somewhere in our galaxy there might be a planet in whose sky is a giant glowing figure 8 doing cartwheels like some bizarre advertising gimmick, as is illustrated in Figure 21-6.

Observations suggest that binary evolution accounts

Figure 21-6 Hypothetical view of a contact binary system as seen from a distant planet. The planet is imagined to circle a third star in the system, another red giant offstage to the right. (Painting by author.)

not only for novae and short-period contact binaries but also for other unusual stars. Gamma-ray bursts, X-ray bursts, and irregular brightness fluctuations are probably associated with mass transfer from an evolving member of a binary pair to a companion, for example (Ventura and others, 1983). Many irregular variables are binary systems in which matter has streamed from a giant into a disk surrounding a smaller companion. Several stars once thought to be short-period variables have been recognized as probable short-period binaries in which matter is streaming from one star to another, causing irregular variations in brightness. As we learn more, we may find that many peculiar types of stars are explained by strange processes of evolution in binary or multiple systems.

Figure 21-7 Hypothetical view of the strange binary SS-433. Interpretation of spectra suggests that gas is flowing out of an evolved star (left) into an accretion disk of hot gas surrounding a neutron star (right). Gas spiraling into the center of the disk is intensely heated and somehow ejected into 78,000-km/s jets moving outward in two opposing directions. Hydrogen alpha emission from the glowing jets is Doppler red-shifted to a deep red color in the receding jet (bottom) and blue-shifted to a more yellowish color in the approaching jet (top). (Painting by author.)

EXAMPLES OF BINARY AND MULTIPLE SYSTEMS

Algol and Similar Systems: Altered Masses

Algol, a variable star with a regular period of 2.9 d, is an eclipsing–spectroscopic binary. The primary is a hot, blue B8 star of about $5\,M_\odot$. The secondary is a K0 giant of about $1.0\,M_\odot$. Here lies a paradox: The more massive B or A star should be the one to expand first; yet the less massive star is the more evolved giant. Why? Is there a fundamental mistake in our ideas of stellar evolution?

Theoretical studies have resolved the paradox: What is now the smaller star was *originally* the more massive star. Reaching the giant stage first, it transferred so much mass to the other star that it became the less massive of the pair. In one theoretical study of a similar system, the more massive star starts out with $9\,M_\odot$ and is reduced to a mass of about 2 or $3\,M_\odot$ in only about 50,000 y! During this very rapid mass exchange, stars reached more nearly similar masses.

SS-433: A Bizarre X-Ray Binary

As noted in Chapter 19, observations from space have revealed binary systems that strongly emit X rays and

gamma rays, probably from gas heated to extremely high temperatures as it transfers from one evolved star onto the surface of a dense companion, such as a neutron star. One of the most exciting is SS-433, a binary in which mass transfer not only heats the gas but also somehow accelerates the atoms into two jets that shoot out of the system in opposing directions at 26% of the speed of light—an extraordinarily high speed unprecedented in most stellar phenomena.

Figure 21-7 shows a hypothetical view of the system as interpreted by some astronomers. Gas is being transferred from one normal, evolved star into an accretion disk surrounding a dense stellar core, probably a neutron star (Margon, 1980, 1982).

Astronomers continue to study SS-433 with considerable excitement. Some models call for a central object as small as a $1\text{-}M_\odot$ neutron star. However, Harvard astronomer J. E. Grindlay and colleagues (1984) studied X rays from SS-433 with the Einstein Orbiting Observatory and concluded the central object is a black hole of $10\,M_\odot$, surrounded by a gaseous accretion disk some 10 to 100 AU across. Time will tell who is right!

The most important features of SS-433 are its two 100-AU-long, brilliant, narrow, high-speed jets shooting out of the accretion disk. As mentioned in Chapter 18 (Figures 18-9 and 18-10), diffuse bipolar jets have been found emanating from various accretion disks, but nothing like the tightly beamed, ultra-high-speed jets of SS-433. What mechanism creates the jets? In 1984, using gamma-ray observations from a satellite telescope, Iowa astronomer Richard Lamb discovered that thermonuclear reactions of the carbon cycle (see page 288) may occur in the jets. Could the jet source regions contain dense matter in which these reactions occur, as in stellar interiors? How does nature accelerate material to a sizable fraction of the speed of light without producing a diffuse explosion of million-degree gas? The answers aren't known yet.

THE SEARCH FOR PLANETS ORBITING OTHER STARS

One of the most provocative searches in current astronomy is the search for possible planets around other stars. None has yet been positively identified or confirmed, because a Jupiter-sized object near another star is just at the threshold of sensitivity of current instruments. However, a number of candidate objects close to the boundary line between brown dwarfs and planets have been suggested in recent observations.

The question of whether true extrasolar planets exist has important implications. It represents a possible extension of the Copernican revolution in which we learned that Earth is not located at the center of the universe. Perhaps we are not the center, but are we unique? One philosopher might speculate that the solar system is a special creation—the only place in the universe where a rare accident allowed formation of planets that could support life. Another might counter that the process of planet formation, as far as we can tell from evidence in meteorites and planetary samples, was a normal part of the Sun's birth and that other "single" stars probably had similar histories that yielded many planets—and at least a few habitable ones resembling Earth. But without data there is no way to settle this intriguing argument. And we have no data one way or the other because we haven't been able to look at stars with sensitive enough equipment to detect Jupiters, or Saturns, or Earths near them.

Techniques now being developed, however, should make it possible within a decade to detect such planets, if they exist. First is improvement in *astrometry,* the technique for making precise measures of a star's position in the sky. If a star has an unseen Jupiter-like planet moving around it, the star itself "wiggles" back and forth due to gravitational pulls of the planet first on one side of its orbit and then on the other. Thus the star is

revealed as an astrometric binary (see page 333), and in some cases the mass of the unseen companion can be estimated. A mass smaller than $2\,M_{\text{Jupiter}}$ could be considered a planet, according to our discussion on page 278. A slightly larger mass, from 2 to $85\,M_{\text{Jupiter}}$, would be classified as a brown dwarf.

A second technique is extremely precise measurement of Doppler shifts. If the same "wiggle" motions are detected in this way, the star is a spectroscopic binary. Again, an estimate of the companion's mass might be possible, especially if this technique is combined with one of the other techniques. And the mass estimate might reveal the companion to be a planet.

A third technique is to detect the infrared radiation coming from the planet. You might think direct photography with giant space telescopes could reveal a planet, but generally the glare of the star is probably too bright for that technique to work. In the infrared part of the spectrum, however, the hot star is putting out less radiation than at visible wavelengths, whereas the planet is putting out its peak radiation. (Review Wien's law, pages 67–68, if this is not clear.) Thus astronomers can look for excessive infrared emission that might be coming from a planet at a few hundred degrees Kelvin. Already this technique has led to discovery of systems of dust particles, possibly from asteroids or comets, near other Sun-like stars, as shown in Figure 18-8—a promising harbinger of possible planets.

Planet Formation vs. Star Formation

By contrasting planet formation (Chapter 14) with star formation (Chapter 18), we can see the problem we face in interpreting possible future discoveries of low-mass companions around stars. Planets apparently formed primarily by a condensation of dust and ice particles followed by accumulation during their collisions; the surrounding gas medium eventually blew away. This happened inside a cocoon nebula near a star. Most stars formed when the entire contents of an interstellar region of gas and dust collapsed gravitationally, trapping all the gas, dust, and ice particles in the region. But suppose we find an intermediate-sized object of, say, $3\,M_{\text{Jupiter}}$ orbiting a star. Did it form by dust-particle aggregation or by gravitational collapse? Could there be some unusual planets and brown dwarfs that formed by collapse? Could there be some unusual companion stars in binary systems that formed by particle aggregation within the cocoon nebula? Is its orbit nearly circular, like our planets' orbits, or highly elliptical, like many binary star orbits, and does this give us a clue to its origin? The answers are uncertain to today's astronomers, but the search for planetary systems and brown dwarf companions may clarify the situation.

Examples of Extrasolar Planetary Systems?

If we want to confirm that planetary systems can form elsewhere in the universe, and if we want to understand how they differ from multistar systems, we will have to do more than simply detect objects with mass less than about 2 $M_{Jupiter}$. We will also need to build up statistics on brown dwarfs to find out if there is a smooth transition from planets to brown dwarfs or whether most brown dwarfs form by the same process as stars. And we will need statistics on orbits, especially in systems with more than two bodies, to see if the orbits lie near the same plane like the planets in our solar system.

The star BD +66°34, for example, which is 10 pc from Earth, reportedly consists of two main-sequence M-type stars of mass 0.4 $M_{\odot}$ and 0.13 $M_{\odot}$, with a third unseen object of mass 0.12 $M_{\odot}$, or 120 $M_{Jupiter}$. The third object would be classed as a small star, but the three bodies may have orbits in the same plane, like planets. Thus this system might have formed by solar-system-like processes in a cocoon disk, rather than the normal multistar-forming process that yields inclined orbits.

With each year, the size of detected objects seems to get nearer the intriguing planet/brown dwarf boundary. The famous newly forming star T Tauri has been known since the 1970s to have a surrounding cocoon nebula containing silicate dust grains from which a planet could form. In 1982, Hawaiian observers announced discovery of a faint infrared companion interpreted as an accreting brown dwarf of mass 5 to 80 $M_{Jupiter}$ and temperature 800 K (Dyck and others, 1982; Hansen and others, 1983). It is reportedly 80 to 150 AU from T Tauri, about two to four times as far as Neptune and Pluto from our Sun.

In 1987, Canadian observers announced still more exciting results after monitoring 16 nearby stars by the Doppler shift technique. They reported possible unseen companions of only 1 to 10 $M_{Jupiter}$ around seven stars! Two cases were especially provocative. The nearby solar-type star Epsilon Eridani reportedly has a companion of only 2 to 5 $M_{Jupiter}$. A more distant solar-type star,

Gamma Cephei, seems to have a tiny companion of about 1.7 $M_{Jupiter}$. If confirmed by others, this might be listed as the first true extrasolar planet. An interesting feature of these two detections is that both stars resemble the Sun, being of spectral class K2 and K1, respectively, and having around 80 or 90% of the mass of the Sun. Perhaps many "single" solar-type stars will turn out to have systems of planets!

THE ORIGIN OF BINARY AND MULTIPLE STARS

Clearly there are many questions about the origin of multistar systems. What determines whether a star forms singly or forms with companions? What determines the distribution into systems of two, three, four, or more members? (This may be a question of the angular momentum and its distribution in the original collapsing protostellar cloud.) And what determines whether a companion to a star is a planet, a brown dwarf, or another star? Most astronomers believe there may be several processes at work to form binary and multiple systems. Different processes—fission, capture, subfragmentation—may produce different types of systems.

Fission

Theories that picture a fast-spinning protostar as splitting in two are called **fission theories.** Many researchers believe that very close pairs such as the W Ursae Majoris contact binaries are formed in this way. However, as discussed below, Kitt Peak astronomer Helmut Abt and his co-workers (1990) believe that fission may be less common than once thought, and that contact binaries may form as orbits evolve closer together by other mechanisms.

Capture

According to **capture theories,** systems of widely separated stars that share weak gravitational attraction are chance configurations arising when one star approaches another. The chance of encounters among random field stars is far too slight to explain the observed numbers of binaries. Furthermore, randomly paired stars would be of widely different ages, but this is not observed among binaries. Where could stars of similar ages interact in a closely packed group? In a newly formed star cluster. Massachusetts astronomers T. Arny and P. Weissman (1973) showed that fully half the protostars in a cluster probably undergo collisions or close encounters. Therefore astronomers believe at least some binaries and multiples formed inside young open clusters when protostars made close approaches to each other. Many widely binary pairs probably formed in this way; Abt and others (1990) found that the statistics of spectral types among wide pairs matched that expected for random captures, supporting this theory.

Subfragmentation During Gravitational Collapse

In a third category of theory, the **subfragmentation theory,** the prestellar cloud breaks up into several clouds which independently contract into stars. Such a subfragmentation would happen before a star forms. It might produce widely spaced or close double, triple, or even multiple systems. In contrast, in the fission theory, the split occurs during or after star formation and produces only a close binary system. An example of a system that might be produced by subfragmentation is the so-called Trapezium of four stars at the heart of the Orion Nebula (Figure 20-16, p. 327).

Early Evolution of Binary and Multiple Systems Inside Star Clusters

As we will see in the next chapter, stars form in groups of hundreds of members, called star clusters. Soon after such a cluster forms, it breaks up, and the stars—single, binary, or multiple—disperse among the surrounding field stars. Recent work on binary and multiple systems proves that the types of systems we see among field stars today are strongly influenced by the crowded conditions these stars experienced during their early days within these clusters. As mentioned above, researchers as early as 1973 showed that many close encounters occur in the clusters and produce capture of one star into orbit around another star or pair of stars. But then the orbits can evolve through various influences, including mass transfer in close pairs and gravitational interactions with more distant stars. Some binaries are split into single stars by gravitational forces during close encounters between binary pairs within the cluster (McMillan and others, 1990). On the other hand, if a pair of moderately close stars captures a third star,

gravitational interactions among the three stars, and among nearby passing stars, often result in the third star being flung out at high speed, while the original two are brought closer together in tighter orbits. In one theoretical study, of some 800 imaginary triple-star systems, about 97% are found to be gravitationally unstable, eventually kicking out one star while the other two become even closer binary pairs. Similarly, as the whole cluster disperses, individual widely spaced pairs can evolve into more closely spaced pairs. This process creates more tight binary pairs than would be seen otherwise.

Such processes accentuate the distinction between close binaries and more widely separated binary or multiple pairs. For example, many close systems are nearly touching, as little as 0.01 AU apart or less, while Trapezium-like multiple systems can have members 2000 to 50,000 AU apart (Fekel, 1987). The closest systems probably form by fission or evolution of an initially close subfragmented pair into a tight system during encounters; the looser systems probably form by capture or by subfragmentation.

SUMMARY

At least half of all the seemingly single stars in the sky are binaries or multiple systems. Many of these may have formed by interactions of stars in crowded new clusters, but some may have formed by other means such as fission or subfragmentation.

Binary and multiple systems can be detected in different ways. Binaries were once classified by these different methods of detection, which yield different types of knowledge. Examples include spectroscopic binaries, eclipsing binaries, and astrometric binaries. Certain types of binaries give the best available data on certain properties of stars, such as mass and diameter (see pages 331–333).

A classification based on evolution includes binaries that have filled neither Lagrangian lobe, one Lagrangian lobe, or both Lagrangian lobes. Once one lobe has been filled by expansion of a star to the red giant stage, gas may flow from one star to the other, causing flare-ups, X-ray emission, and nova explosions.

Companions to stars include not only other stars but also brown dwarfs and probably planetary systems. Astronomers are actively searching for conclusive examples of planetary systems, which may extend the Copernican revolution by proving that the solar system is not unique. Whether companions as small as planets form in a contracting protostar may depend on the protostar's rotational properties.

CONCEPTS

binary star system

multiple star system

optical double star

physical binary star

visual binary

spectroscopic binary

eclipsing binary

eclipsing–spectroscopic binary

astrometric binary

selection effect

incidence of multiplicity

Lagrangian surface

three structural classes of binaries

contact binary

nova

W Ursae Majoris stars

fission theories

capture theories

subfragmentation theories

PROBLEMS

1. Describe verifications of Kepler's laws other than the planets' motions around the Sun. What was the first verification outside the solar system?

2. How do binary and multiple star systems generally differ from planetary systems?

3. Give evidence that at least a subclass of binary and multiple star systems might be generically related to planetary systems.

4. How are novae related to binaries? Are supernovae related to binaries?

5. How will the evolution of a 1-$M_\odot$ star in orbit close to a 3-$M_\odot$ star differ from the evolution of a 1-$M_\odot$ star by itself?

6. Why are binaries more likely to have formed in star clusters than as isolated field stars?

PROJECTS

1. Observe Mizar and Alcor with the naked eye. They are the middle "star" (actually a close pair) in the handle of the Big Dipper. Can you see the faint star Alcor? Sketch its position. (Inability to see Alcor may be due to insufficiently keen eyesight, a hazy sky, or a sky illuminated by city light.)

2. Observe the eclipsing binary Algol with a telescope or binoculars each evening for 10 to 20 d in a row. (This can be done as a class project with rotating observers.) Using neighboring stars as brightness reference standards, estimate the brightness of Algol. Can you detect the eclipses, which occur at 2.9-d intervals?

3. The star Epsilon Lyrae is famous as the "double double." It consists of a binary pair 208 seconds of arc apart, easily seen in a small telescope. But each of these is a binary only 2 to 3 seconds apart. These pairs are a test of good optics and good atmospheric observing conditions. Does your telescope reveal the two close pairs? Sketch them.

CHAPTER 22

Star Clusters and Associations

If you could roam through space to ever greater distances, you would eventually lose track of individual stars and see the galactic disk defined primarily by clusters of stars. Writing about clusters in 1930, Harvard astronomer Harlow Shapley pointed out that "their problems are intimately interwoven with the most significant questions of stellar organization and galactic evolution."

Yet at the same time Shapley noted that scientific study of clusters had hardly begun. Nobody knew how to measure their distances or plot their distribution in space until the 1920s. Although some clusters, such as the Pleiades (or Seven Sisters), are easy to recognize with the unaided eye, others are so far away that they require large telescopes to detect. Still others are so close that they cover much of our sky and were not even recognized until recent years.

In other words, our cosmic voyage has brought us to clusters so far-flung that they were recognized as a class only in this century. Clusters have been of major importance because they have revealed to us the shape and age of our own galaxy, as will be clarified in this chapter.

THREE TYPES OF STAR GROUPINGS

The three basic types of clusters are *open clusters, associations,* and *globular clusters.* Examples are listed in Table 22-1.

Open Star Clusters

Open star clusters are moderately close-knit, irregularly shaped groupings of stars. They usually contain 100 to 1000 members and are usually about 4 to 20 pc in diameter. Our Sun is possibly inside or on the edge of a loose open cluster centered about 21 pc away toward

TABLE 22·1

Selected Star Clusters and Associations

Name	Distance (pc)	Z^a (pc)	Diameter (pc)	Estimated Mass ($M_\odot$)	Estimated Age (y)
Open Clusters					
Ursa Major	21	18	7	300	2×10^8
Hyades	42	18	5	300	5×10^8
Pleiades	127	54	3	350	5×10^7
Praesepe	159	84	4	300	4×10^8
M 67	830	450	4	150	4×10^9
M 11	1900	99	6	250	2×10^8
h Persei	2250	156	16	1000	1×10^7
χ Persei	2400	167	14	900	1×10^7
O Associations					
I Orionis	470	150	?	3000	?
I Persei	1900	164	?	180	?
T Associations					
Ori T2	400	132	28	800	?
Tau T1	180	52	?	50?	?
Globular Clusters					
M 4	2000	550	9	150,000	$\sim 1.4 \times 10^{10}$
M 22	3500	460	9	530,000	$\sim 1.4 \times 10^{10}$
47 Tucanae	4400	3100	5	1,600,000	$\sim 1.4 \times 10^{10}$
M 13	6600	4300	11	660,000	$\sim 1.4 \times 10^{10}$
M 5	8000	5500	12	850,000	$\sim 1.4 \times 10^{10}$
M 3	10,000	9900	13	1,100,000	$\sim 1.4 \times 10^{10}$

Source: Globular cluster data from C. J. Peterson (1986, private communication).

$^a Z$ = perpendicular distance north or south of Milky Way plane. Note that only globulars are far "above" or "below" the plane of our galaxy.

the constellation Ursa Major, many of whose stars belong to this cluster, as shown by Figure 22-1. The best-known *clusters*, the Hyades and the Pleiades (see Figure 18-5), lie 12° apart in our winter evening sky, about 42 and 127 pc away, respectively. Figure 22-2 shows the relation of the Hyades and Pleiades in the sky, whereas Figure 22-3 shows the Pleiades in more detail. About 900 known open clusters are concentrated along the Milky Way band, indicating that they lie in the plane of our galaxy. (Open star clusters are sometimes called *galactic clusters* for this reason.)

As described in Chapter 18, stars form in open clusters, and most open clusters have prominent young stars or associated clouds of star-spawning gas. Then why aren't all stars in clusters? The reason is that most open clusters break apart into individual stars within only a few hundred million years because of dynamic forces acting on them. In comparison with most cosmic lifetimes, open clusters are short-lived.

Associations

Associations are cousins of open clusters. They often have fewer stars, but are larger in size and have a looser structure. Some large associations include an open star cluster within them. They may have 10 to a few hundred members and diameters of about 10 to 100 pc. They are rich in very young stars, such as O and B stars

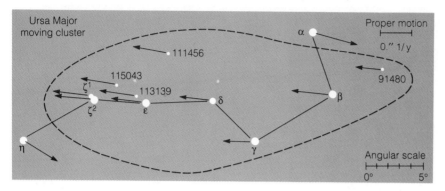

Figure 22-1 Some members of the closest open cluster, the Ursa Major cluster, form part of the familiar figure of the Big Dipper (solid lines). Arrows show proper motions of the stars, that is, their rates of angular motion relative to distant background stars. All but two of the stars in the figure are moving together at almost exactly the same rate, thus defining the cluster (dashed line). Stars at each end of the dipper are moving in different directions and are not true cluster members. Compare with Figure 1-4.

(which burn their fuel too fast to last long), or T Tauri stars (which evolve toward the main sequence too fast to last long). Associations are classified as **O associations** or **T associations,** depending on whether the prominent stars are O and B blue stars or T Tauri variables. Some 70 O associations were listed in a 1970 catalog. The general region of Figure 22-2 includes a T association.

The smallest associations grade into small, multiple-star-like groups, such as the Trapezium in the Orion Nebula. Sometimes called *trapezium systems,* these might be a link between multiple stars and small clusters. Like open clusters, associations are involved with regions of recent star formation and are short-lived.

Globular Star Clusters

Globular star clusters are quite different from the other two types. They are much more massive, more tightly packed, more symmetrical, and very old. Figure 22-4 emphasizes the remarkable symmetry and compactness of globulars. They typically contain 20,000 to

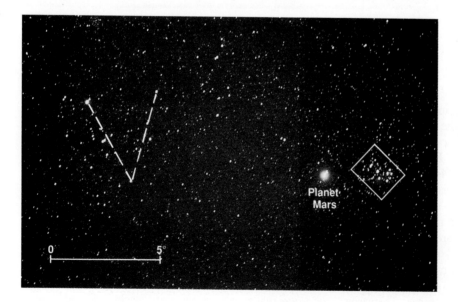

Figure 22-2 Two open clusters are prominent in this portion of the January evening sky. The Pleiades is the tight group in the box, about 127 pc away. Closer to us is the V-shaped group, the Hyades, about 42 pc away. Its brightest star (upper left) is the 1st magnitude red giant Aldebaran, whose color shows here. Even more prominent is the plant Mars, which happened to be passing through this part of the sky when this 1990 photo was made. Location of this photo can be compared with wider-angle Figure 18-5 on pp. 280–281. (Photo by author; 12 min, f2.8, 100-mm lens, on Ektachrome ISO 1600 film.)

Figure 22-3 The Pleiades, also known from mythology as the "Seven Sisters," is an open cluster of young stars about 50 million years old. It is about 127 pc away and about 3 pc in diameter. The brightest stars are massive, hot, blue stars. Dust and gas clouds around them scatter blue light, creating wispy blue reflection nebulae. (Copyright Anglo-Australian Telescope Board, courtesy D. Malin.)

several million stars, although many of these stars crowd too close to be resolved by Earth-based telescopes, especially in the central regions.

Typical diameters of the central concentrations range from only 5 to 25 pc. To imagine conditions inside a globular cluster, picture 10,000 stars placed around the Sun at distances no farther than Alpha Centauri, our nearest star!

Even the nearest globular clusters are thousands of parsecs away from us. It is only because they have so many and such very bright stars that we see them at all. Yet a modest backyard telescope can reveal many prominent examples. The total of known globulars around our galaxy is about 150, and the total around our sister galaxy, M 31 in Andromeda, is at least 200. In three-dimensional space they are not confined to the galactic

plane but are distributed in a spherical *halo* surrounding our galaxy. Similar distributions are found around other galaxies. They are among the oldest objects in our Milky Way galaxy, with ages estimated around 14 billion years.

DISCOVERIES AND CATALOGS OF CLUSTERS

Recognition of clusters is an enterprise spread out over many centuries. When the French astronomer Charles Messier tabulated nebulae in 1781, he included many star clusters that are still known by their "M numbers." Between 1864 and 1908 many clusters were cataloged in the New General Catalog (NGC) and the Index Catalog (IC) and became known by their NGC and IC num-

1 pc

Figure 22-4 Globular star cluster M 5 (NGC 5904). The cluster is estimated to be about 12 pc across and 8000 pc away. It contains roughly 60,000 stars. (Kitt Peak National Observatory.)

bers. Not until the 1920s was the difference between various types of clusters and the still more remote galaxies understood, and only then could reasonable research begin. Shapley's 1930 book was the first substantial work on star clusters.

MEASURING DISTANCES OF CLUSTERS

Because of clusters' enormous range of distances, astronomers have devised varied and sometimes inge-

nious methods to determine their distances. Details of all the methods are beyond the scope of this book, but a few examples can be given.

Parallax

The most basic method of measuring star distances, the measurement of parallax (described in Chapter 16), is almost useless on clusters because it is not very accurate for distances beyond about 20 pc. Parallaxes give checks on distances of some stars in the Ursa Major cluster, but they mostly provide a foundation on which other distance-measuring methods are built.

Star Luminosity

If observers can determine the luminosity of any star in a cluster, the distance from Earth to the cluster can be estimated if there are no complications from obscuring interstellar matter. (This method is similar to the example in Chapter 16 of estimating the distance of a certain light if we know that it is a candle and that there is no intervening fog.) One way to determine luminosity is simply to use spectra to identify spectral types and to estimate luminosities from H–R diagrams that plot luminosity versus spectral type, as derived from nearby stars.

Cluster Diameter

Once the linear diameters (in parsecs) of a few nearby clusters were known and found to be roughly comparable, remote clusters could be assumed to have the same linear diameter. Then the distances of remote clusters were estimated from their angular diameters, by assuming that all clusters of similar appearance have similar linear size. If a globular cluster 10 pc across subtends $1/10°$, for example, it must be about 6000 pc away.

The Complication of Interstellar Obscuration

Before 1930, these methods were carried out under the assumption that all intervening interstellar space is transparent. However, early estimates of distances, brightnesses, and sizes of clusters gave inconsistent results. In 1930 Robert Trumpler used the cluster results to show that diffuse interstellar dust dims stars and clusters that are more than a few dozen parsecs away. This throws off the estimates of stars' luminosities and distances. Fortunately, the total amount of dimming can be estimated by measuring the amount of interstellar reddening, or color change, caused by the dust. The more reddening is observed, the greater the degree of dimming. Once the obscuration is measured and taken into account, distances can be accurately measured if the luminosity of any star or class of stars in the cluster is known.

Cepheid Variables as Distance Indicators

Luminosity is the key to distance measurement, and this brings us to the most useful technique for measuring distances of remote objects. As described in Chapter 19, Henrietta S. Leavitt discovered in 1912 that the periods and luminosities of **Cepheid variable stars** are correlated. The Cepheid variables have periods—the duration of their brightness variation cycle—of 1 to 50 d. Once you measure the period of one of them, say 5 d, that tells you the luminosity of the star.

By the 1950s a small complication was discovered and overcome. The Cepheids found in open clusters form one group, called Type I Cepheids, and have a special period/luminosity relation. Those in globular clusters and in our galaxy's central bulge form another group, called Type II Cepheids, and have a somewhat different period/luminosity relation.

Thus measuring the distance of a cluster requires several steps. A Cepheid in the cluster must be located, its period and type measured, and its luminosity read from the appropriate period/luminosity diagram and combined with the apparent magnitude (corrected for interstellar obscuration) to calculate the distance. Few astronomers are equipped to carry out all the necessary observations, such as measuring light variations and periods, determining spectral properties, measuring absolute magnitudes, and measuring interstellar reddening and obscuration. Thus they have specialized. Some study periods of variable stars; some make photometric measures of absolute brightness; some study interstellar reddening. The simple statement that cluster A is x parsecs from Earth may represent years of work by many astronomers.

DISTRIBUTION OF CLUSTERS AND ASSOCIATIONS

Once the distances of clusters and associations could be measured, their positions in three-dimensional space could be mapped simply by plotting their distances and directions. This process began in the late 1920s and 1930s, and striking results began rolling in at once. The distribution of clusters is not random, but defines the shape of our galaxy. The open clusters and associations lie in a disk, and the globulars form a spherical cloud around the disk. Most globular clusters are concen-

Figure 22-5 A globular cluster peeks through the "clouds." Much of this picture is dominated by multicolored emission and reflection nebulae around stars in the constellation Ophiuchus. These are foreground clouds, lying between the solar system and the Milky Way's center. Through a gap in these clouds, we see a globular cluster (top edge just left of center). It is one of the swarm of clusters that forms a halo around the Milky Way's center. (Copyright Anglo-Australian Telescope Board.)

trated toward the center of our galaxy and thus lie in the part of our sky toward the galactic center. Some are seen in Figure 22-5. Shapley, a pioneer in these measurements, said that the clusters reveal "the bony frame of our galaxy."

OPEN CLUSTERS: AGES AND ULTIMATE DISRUPTION

It is relatively straightforward but time consuming to construct the **H–R diagram of a cluster.** The appar-

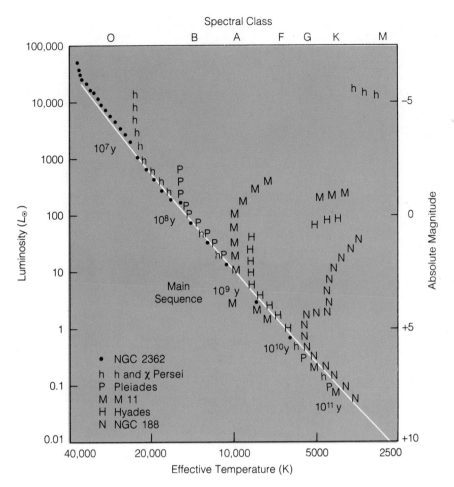

Figure 22-6 An H–R diagram for several open clusters. Numbers along the main sequence give ages in years for groups of stars turning off the main sequence. Clusters can be dated by measuring these turnoff points. (Based on the work of A. R. Sandage, O. J. Eggen, and L. Aller.)

ent magnitudes and color temperatures of hundreds of stars must be measured. If the distance to the cluster is known, the apparent magnitudes can be converted to absolute magnitudes, or luminosities. Astronomers and their graduate students have plotted the H–R diagrams of many clusters.

The importance of such work, of course, lies in the evolutionary information it yields. Figure 22-6 brings together data on several open clusters. Marked along the main sequence is the age at which various stars evolve off the main sequence. (Compare with Figure 19-2, where the dashed lines show predicted positions of stars of various ages.) As discussed in Chapter 19, the point at which stars in a cluster have left the main sequence is a measure of the cluster's age. Note in Figure 22-6 that some clusters, such as NGC 2362, are so young that hardly any stars have evolved off the main sequence. **Ages of open clusters** range from about 1

or 2 million years for NGC 2362 to 5 billion years for most clusters. Among published ages for 27 open clusters, about half (55%) are less than 100 million years old. The solar system is nearly 50 times as old as this.

These results confirm that in cosmic terms, *open clusters are mostly young,* and that (as stated in Chapter 18) star formation is continuing in open clusters to this day. Many open clusters, as in Figure 22-7, are associated with dense nebulae where star formation is still proceeding. Often their brightest stars are the hot, blue O and B stars, which formed recently, burn their fuel fast, and cannot last long. In slightly older clusters some of the more massive stars may have evolved into red giants, adding interesting color effects. The cluster in Figure 22-8, for example, has two bright orangish giants among a field of young blue stars.

Velocities of cluster stars, measured by Doppler shifts, indicate that some clusters are expanding or los-

Figure 22-7 Because stars form in clusters, many open clusters are still surrounded by the nebulosity that gave them birth. Here the impressive Lagoon Nebula (M 8 in the constellation Sagittarius) is a 25-pc-wide cloud punctuated by the bluish-white brightest stars of the open cluster to which it is giving birth. The complex is about 1200 pc away. (Copyright Association of Universities for Research in Astronomy; NOAO.)

ing members, or both. Some high-velocity stars exceed the escape velocity of their parent clusters. Thus these clusters are breaking apart as we watch. Clusters tend to disperse after a few hundred million or a billion years. This **disruption of open clusters** occurs by several simultaneous mechanisms. Fast-moving stars, or stars that are accelerated by interaction with others, may escape. This reduces the mass and gravitational self-attraction of the cluster, so that other stars can escape. According to Kepler's laws, stars on the side of the cluster closer to the galactic center orbit around the galactic center faster than stars on the other side, thus stretching and shearing the cluster. The Hyades, for example, are barely stable now, and the outer parts of

the Pleiades are already dissipating, though the central, tighter grouping may be stable.

ASSOCIATIONS: AGES AND ULTIMATE DISRUPTION

Like open clusters, associations are young. They are gravitationally bound together, but because of their loose structure, they may break up even faster than ordinary open clusters. For example, a T association of eight T Tauri variables, about 100 pc away and 25 pc across, has been estimated to be only about 10 million years old; it may soon break apart because of so-called "tidal"

Figure 22-8 Open cluster NGC 3293 in the constellation Carina. The cluster is about 8 minutes of arc across and contains about 50 stars brighter than the 13th magnitude. At least two of the brightest are red giants, indicating that enough time has passed since the cluster's formation for a few of the more massive stars to evolve into giant status. (Copyright Anglo-Australian Telescope Board.)

gravitational forces of the galaxy acting on the cluster. Most associations cannot last more than a few tens of millions of years, because of these disruptive forces and the tendency for their stars to follow individual orbits around the center of the galaxy.

GLOBULAR CLUSTERS: AGES AND INTERIOR CONDITIONS

The technique of H–R diagram analysis is especially well suited to determining the ages of globular clusters because they have well-defined turnoff points along the main sequence. It turns out that globulars are extraordinarily old. Star formation has ceased in them. In all globulars of our galaxy, the O, B, and A stars have already evolved off the main sequence and have become red giants, as seen in Figure 22-9. For this reason, *the bright stars in almost all globulars have a reddish color and there are no bright blue stars.*

Assigning numerical ages to globulars must be done with some care, because spectra show that their stars contain fewer heavy elements than the more familiar stars of the solar neighborhood. This means that the details of their inner processes of energy generation and transport are different. Because the evolutionary time scales may thus be different, different calculations are used to determine the ages corresponding to various main-sequence turnoff points.

The best techniques set the **age of globular clusters** at about 14 ± 4 billion years (Jones and Demarque, 1983; King, 1985; C. Peterson, 1987). Most of

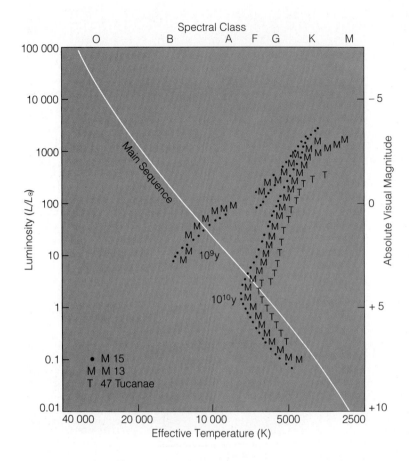

Figure 22-9 An H–R diagram for three selected globular clusters. Estimated main-sequence turnoff positions are indicated for ages of 10^9 and 10^{10} y. Globular cluster stars have fewer heavy elements than stars in open clusters and near the Sun (see next chapter), causing their main sequences to be shifted leftward from the reference line, which is derived from open clusters and nearby stars. (Based on data of A. R. Sandage.)

them seem to date from a single, brief period of formation, believed to be the period when our galaxy formed. We will return to this exceptionally important relation between globulars and Milky Way formation in the next chapter.

By 1975, orbiting astronomical satellites discovered that the cores of 10 globular clusters emit unexpectedly strong X radiation. In cluster NGC 6440, the rate of energy radiation in X rays alone reaches as much as 17,000 times the Sun's total luminosity! The **globular cluster X radiation** is believed to be produced as material crashes at high speed into accretion disks surrounding neutron stars or black holes, as discussed in Chapter 19 on pp. 306–307.

The discovery of globular cluster X radiation led astronomers to realize that the extremely crowded inner parts of globular clusters are really unusual stellar environments. Hubble Space Telescope data in 1991 showed that the central parsec of globular cluster M15 has stars spaced typically only 0.02 pc apart (only 4000 AU). Many old stars would exist—red giants, neutron stars, and

perhaps occasional black holes. Near misses would be common as stars move among each other. Studies suggest that during these near misses, some of the abundant neutron stars could capture other stars into orbit around themselves, creating many binaries. As discussed in the last chapter (Figure 21-5), class 2 binaries would evolve, and material shed from one star would crash into an accretion disk around the neutron stars, explaining the X rays. Furthermore, the massive binary systems would serve to accelerate some other stars passing them in near misses, thus stirring up the velocities in the globular cluster.

Other theorists estimate that the centers of some globular clusters contain massive black holes that swallow passing stars, growing in mass and explaining the X radiation.

How do observations check with these theories? In particular, do the observations favor black holes or just neutron stars? From 1988 to 1991, at least 40 radio-emitting neutron star pulsars were discovered in globular clusters, proving that the neutron stars do exist

Figure 22-10 Imaginary view from a planet about 60 pc from a globular cluster. In such a case, the globular cluster would dominate the sky in a blaze of pinkish stars. Such a planet is unlikely to circle a star within the cluster, because globular clusters are deficient in the heavy elements that form solid planets; however, such a scene might occur as a globular cluster passed through the galactic disk near a star that had planets. (Painting by Chesley Bonestell.)

there. Orbiting telescopes in space give additional data favoring neutron stars as the source of most X rays. In the 1980s, the Einstein X-Ray Observatory found that X rays in M15 originate not from a massive object at the center, but rather from many sources scattered around the center. In 1991, images from the Hubble Space Telescope also indicated that M15's core does not have a central bright, massive object, but a smoother

dispersion with only 1% as many stars/pc^3 as predicted by the black hole model.

Globular clusters are so old that magnetic fields around neutron stars might have been expected to slow down the neutron stars' spin rates, ending their careers as pulsars. The discovery of active pulsars indicates that some have been rejuvenated by the processes described above. After an old neutron star captures a binary companion, and as the companion begins to dump material into an accretion disk around the neutron stars, the material spirals inward and increases the mass and spin rate of the neutron star. It becomes more energetic and it becomes a "born-again" pulsar, in the words of Caltech astrophysicist Shrinivas Kulkarni.

ORIGIN OF CLUSTERS AND ASSOCIATIONS

The theory of gravitational collapse in Chapter 18 gives an excellent introduction to understanding how different types of clusters formed. The gas in halos around newly formed galaxies was extremely thin, with densities perhaps 10^{-22} to 10^{-26} kg/m^3. According to this theory, in such a thin gas, large masses around 10^{35} kg became gravitationally unstable and began to contract into discrete entities. These masses—around $10^5 \, M_\odot$—became globular clusters on the fringes of galaxies, probably around 14 billion years ago. In cool dust clouds in the disks of galaxies, gas densities were higher, around 10^{-17} kg/m^3. This led to the contraction of clouds containing around 10^{32} kg—a few hundred solar masses. Those clouds became open clusters and associations strewn through galactic disks. Although most open clusters are young due to their rapid breakup, the oldest are around 8 billion years old (Jones and Demarque, 1983); thus we can be sure open clusters began forming at least that long ago. Contracting clouds of these masses ultimately broke up into individual stars rather than a single short-lived superstar of, say, $10^4 \, M_\odot$.

SUMMARY

Although three types of star groups have been defined, they can be grouped in two main categories. The open clusters and associations are young groups of about 10 to 1000 newly formed stars located in the galactic disk. The globular clusters, which are old groups of 20,000 to a few million stars, are located within a spherical volume extending above and below the galactic plane, which is centered on the galaxy's center. They formed about 14 ± 4 billion years ago.

Star clusters are important for four major reasons. First, they have played a vital role in mapping our galaxy and clarifying its history. Second, they clarify stellar evolution by presenting groups of stars formed at about the same time, so that their H–R diagrams clarify evolutionary tracks. Third, their stars reveal two different stellar populations in our galaxy. Stars in the disk and in open clusters and associations have solar-type compositions, with a few percent heavy elements, a certain type of Cepheid variable, and other distinctive properties. Stars in globular clusters have very few heavy elements, a different type of Cepheid, and other distinctive properties. Fourth, the crowded conditions in globular clusters provide intriguing natural laboratories for studying evolution of neutron stars and other massive objects under extreme conditions. The importance of these properties of clusters will be clarified in the next chapter as we turn our attention to our galaxy as a whole.

CONCEPTS

open star cluster	H–R diagram of a cluster
association	age of open clusters
O association	disruption of open clusters
T association	age of globular clusters
globular star cluster	globular cluster X radiation
Cepheid variable star	

PROBLEMS

1. Why are O and B stars the brightest stars in open clusters? Why are red giants the brightest stars in globulars?

2. If you saw the galaxy from a great distance, which would be brighter, open or globular clusters? Which redder? Which farther from the galactic disk?

3. Sketch the H–R diagrams of open and globular clusters and associations.

4. Describe a view of the sky near the center of a globular cluster.

PROJECTS

1. Observe the Pleiades with your naked eye and make a sketch. How many stars can you count in the group? Can you see all "Seven Sisters"? (The number of stars seen depends on keenness of vision, darkness of the observing site, and the clarity of the atmosphere.)

2. Observe the Pleiades and Hyades or open clusters h and χ Persei in a telescope. Move the telescope and compare star fields in and out of the cluster. Estimate how many times more stars are in the cluster than in the background region.

3. Locate a globular cluster with the telescope. Make a sketch. Can you resolve individual stars? Compare the view in the telescope with photos, where the central region is often overexposed and "burned out."

NGC 2997, a normal spiral galaxy in the southern constellation of Antlia. Uncounted thousands of such galaxies make up the basic building blocks of the universe. (Copyright Anglo-Australian Telescope Board.)

CHAPTER 23

The Milky Way Galaxy

Our exploration of space has taken us out to distances of a few thousand parsecs. By looking at the distribution of stars and clusters throughout volumes of this size, we begin to perceive the **Milky Way galaxy.** To a remote observer the Milky Way would be a disk with a central bulge. The disk is about 30,000 pc across, 400 pc thick, and packed with open clusters, individual stars, dust, and gas, mostly arranged in ragged spiral arms. Globular clusters surround the disk in a spherical swarm concentrated toward the center of the disk.

These distances are difficult to comprehend. In a model of the Milky Way galaxy the size of North America, stars like the Sun would be microscopic specks less than a thousandth of a centimeter across and scattered a block apart. The solar system would fit in a saucer.

The view from Earth is not from the outside, of course, but from inside the galaxy. From our position partway out in the disk, we see a band of unresolved, faint, distant stars when we look out along the plane of the disk. This is the Milky Way, shown in a wide-angle photographic view in Figure 23-1.

DISCOVERING AND MAPPING THE GALACTIC DISK

Even before the invention of the telescope, people could plainly see a band of light arching across midnight skies at certain seasons.[1] Democritus (c. 400 B.C.) correctly

---

[1]The naked-eye prominence of the Milky Way is unknown to most modern urbanites. It can't be overemphasized that faint celestial displays must be viewed away from urban lights. My own astonishment at the Milky Way's clarity was greatest when I was living in a deserted region at about 3 km altitude on Mauna Kea volcano in Hawaii, prior to the establishment of the observatory at that site. One night when my eyes were already dark-adapted and I stepped outside, I thought the sky was partly cloudy, because I

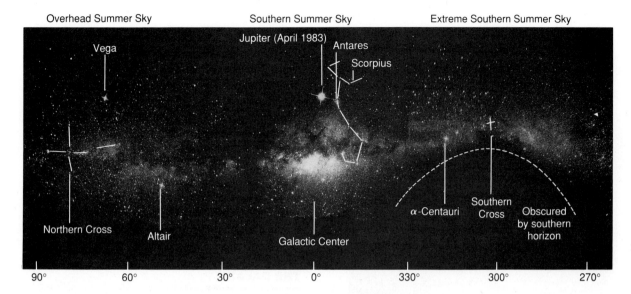

Figure 23-1 A panorama of the inner Milky Way galaxy from our position near the outer edge. This 180° view clearly shows the bright central bulge and the ragged dust clouds that lie along the plane of the galaxy between us and the center. The galaxy's nucleus is hidden behind dark dust clouds. Galactic longitude scale, at bottom, measures angular distance in degrees along the Milky Way. Seasons of visibility for the different regions, as seen in U.S. evening skies, are at the top. On the date of the photo, Jupiter happened to lie north of the galactic center. (Mosaic of photos from Hawaii by author, exposures 32 to 77 min with 24-mm wide-angle lens, f2.8, 35-mm camera with commercially available 2475 recording film.)

attributed this glow to a mass of unresolved stars, which came to be called the *Via Lactea,* or Milky Way. In 1610 Galileo turned his telescope on the Milky Way and confirmed Democritus' idea. The proof is seen in Figure 23-2. Galileo wrote (quoted by Shapley and Howarth, 1929):

The galaxy is nothing else but a mass of innumerable stars planted together in clusters. Upon whatever part of it you direct the telescope, straightaway a vast crowd of stars presents itself to view.

Many astronomers following Galileo observed the Milky Way. By 1918 Harvard astronomer Harlow Shapley showed that a spheroidal swarm of globular clusters

---

could see dark patches blotting out parts of the softly glowing Milky Way. Then I realized the air was crystal-clear. The dark patches were clouds, all right, but instead of being clouds of water droplets 1 km away, they were clouds of dust grains 100 million billion km away! Such a view makes one believe we really do live in an immense, disk-shaped system of dust, luminous nebulae, and distant stars!

extends above and below the disk (Figure 23-3). This swarm of clusters, which also includes some sparsely scattered individual stars and gas, is called the **galactic halo.** Shapley also showed that the halo is centered not on the Sun but on a distant point in the disk in the direction of the constellation Sagittarius. He correctly hypothesized that this point is the center of the galaxy. This gave the first proof that the solar system is far from the galaxy's center, at one side of the disk, and it gave astronomers their first chance to measure the distance from our solar system to the center.

According to current estimates of the **dimensions of our galaxy,** the disk is about 30,000 pc across with the Sun about 9000 pc from the center.

An important step in establishing the nature of our galaxy was recognizing that other galaxies like our own exist. Until 1924, many astronomers thought that certain disk- or spiral-shaped glowing patches were nebulae of gas, like the Orion Nebula. Many were incorrectly called "spiral nebulae." In 1924, however, photos made with the new 2.5-m telescope on Mt. Wilson in California showed that at least one of these objects, *a spiral-shaped object in the constellation Andromeda, consisted*

Figure 23-2 Star clouds of the Milky Way. This region of the constellation Sagittarius, looking toward the center of our galaxy, testifies to the fact that the hazy light of the Milky Way is really composed of innumerable distant stars. The central brightest star cloud, cataloged as M 24, is an estimated 5000 pc away; in other words, it is part of a spiral arm about halfway between us and the galactic center. At top is the red-glowing nebula M 17, estimated to be about 1800 pc away. The vertical height of the picture is 5°. (Copyright Anglo-Australian Observatory.)

of very faint, distant stars, not gas. Most importantly, some of the stars were very faint Cepheids, which could be used to estimate the Andromeda galaxy's distance. The results revealed that the Andromeda object was far beyond any stars in our own galaxy. The Andromeda object and others were soon recognized to be complete galaxies like our own. This recognition allows us to perceive galactic shapes all at once, instead of probing through the murk from inside one such system. As shown in Figure 23-4, photos of other galaxies seen edge-on reveal striking similarity to our own view of our galaxy, seen edge-on from the inside.

THE ROTATION OF THE GALAXY

We have noted that cosmic systems of particles tend to become flattened if they are rotating. The galaxy's flattened shape suggests that it, too, is rotating. All stars, including the Sun, are in fact orbiting around the massive central bulge. But which way is the galaxy turning?

To answer, we need some frame of reference outside the galaxy itself. The distant galaxies provide such a frame, and their Doppler shifts reveal a systematic motion of the Sun and nearby stars toward the constellation Cygnus and away from Canis Major. The velocity of the Sun in this direction is about 220 to 250 km/s. This is the orbital velocity of the Sun and nearby stars around the galactic center.

If the Sun travels at about 235 km/s, then the **Sun's revolution period,** or time required to travel all the way around our circular orbit, which has a radius of 9000 pc, is about 230 million years.

Nearby stars are at nearly the same distance from

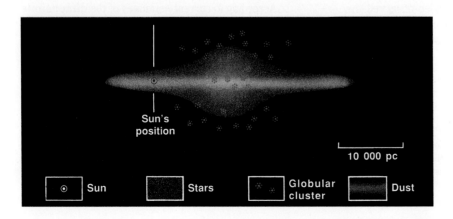

Figure 23-3 A schematic edge-on view of the galaxy, showing the position of the Sun.

Sun's position

10 000 pc

Sun | Stars | Globular cluster | Dust

Figure 23-4 Comparison of a wide-angle view of the Milky Way extending from the horizon (*a*) and a telescopic view of a distant edge-on galaxy (*b*), showing similarity in form as an observer looks edgewise through a galactic disk. **a** About 50° of the Milky Way near the central region in Sagittarius; compare with lower left of Figure 23-1. (35-mm camera; 24-mm lens at f2.8; 16-min exposure on 2475 Recording film; photo by author.) **b** A portion of the galaxy NGC 55. (Cerro Tololo Inter-American Observatory.)

a

b

the center as we are, and so they are moving around the center at almost the same speed as our Sun. However, orbital speeds around the center are different at different distances from the center,[2] which means that no galaxy rotates as a solid disk. No spiral galaxy, for instance, rotates like a pinwheel painted on a phonograph record; rather, the inner parts turn faster than the outer parts. This difference in speed at different distances is called **differential rotation** of the galaxy. Differential rotation, together with random motions of stars (typically about 20 km/s), means that the stars of the galaxy are ceaselessly changing their positions relative to each other as they move around the center.

THE AGE OF THE GALAXY

In the last chapter we saw that globular clusters have an average age of around 14 ± 4 billion years. Because globulars are believed to have been among the first objects formed as the galaxy itself took shape, this age of roughly 14 billion years is believed to be the approximate **age of the galaxy.**

---

[2]The relation of orbital speed to distance departs from Kepler's third law, because the galaxy's mass is spread throughout many stars in the large central bulge, rather than being concentrated in one central object at the center, as is the case in the solar system.

MAPPING THE SPIRAL ARMS

Once astronomers realized we live in an independent galaxy, they faced the challenge of mapping its structure to find out what type of galaxy we live in. Many galaxies, but not all, have their stars arranged in **spiral arms,** which are spiral-shaped patterns formed by the brightest hot stars and their associated emission nebulae.

Our own galaxy has prominent spiral arms. But how did astronomers discover **evidence of spiral structure,** because we cannot see our galaxy from outside? First, they mapped the positions and directions of bright, young stars and emission nebulae in three-dimensional space. That is, they plotted their directions and distances. These young stars and nebulae are the best markers of spiral arms, and the maps revealed they are not distributed at random, but lie in arms, coiling out at an angle from our galaxy's center.

These maps, however, reveal only parts of nearby spiral arms—the regions not far enough away to be obscured by interstellar dust. What about the rest of the galaxy?

To map these more distant regions, astronomers turned to radio telescopes. Unlike visual light, which is blocked by the dust, radio waves pass through most of the dust. The **21-cm radio waves** produced by neutral hydrogen, or HI, are especially useful for galactic mapping, because they allow us to detect HI clouds, which are concentrated in the spiral arms.

HI clouds have been mapped over a large part of

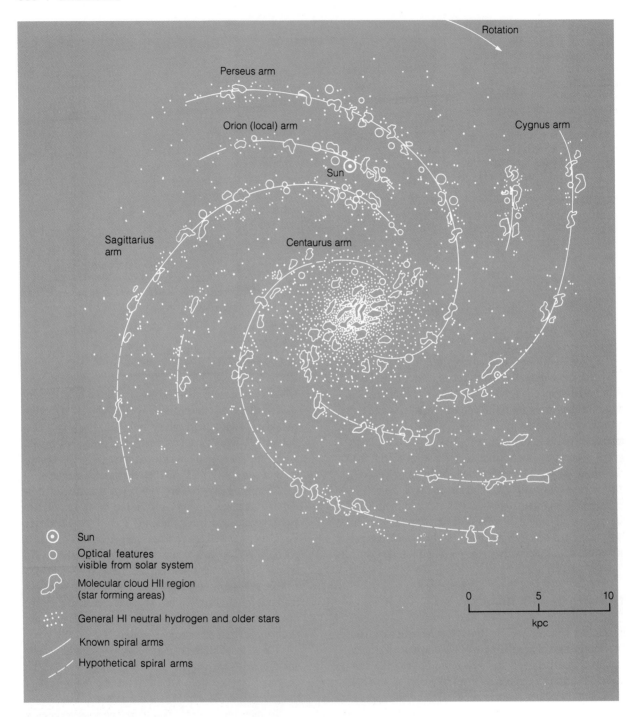

Figure 23-5 Currently known and estimated features of the Milky Way galaxy viewed from the north. Features nearest the Sun are the most certain; features on the far side are unmapped. The spiral arms are named for the constellations in the directions of prominent features in each arm.

the galaxy. Their positions roughly define the spiral arms and offer our first view of our galaxy's spiral shape, as seen in Figure 23-5. Most of the galaxy's material is concentrated in a disk only about 200 to 400 pc thick.

Radio astronomy roughly triples the distances to which we can probe across this galactic disk. We can map structures that lie some 15,000 pc away, even farther than the galactic center. However, there are regions directly on the far side of the center that we cannot map. Therefore, we cannot tell exactly how tightly wound the arms are or where the Milky Way falls in the range of forms of other galaxies. Nonetheless, using Figure 23-5 as a guide, we can say that if our galaxy could be seen from "above," it would probably look something like the view in Figure 23-6. Many other galaxies have a similar appearance.

The View from a Spiral Arm

The maps show us to be located on the inner edge of a spiral arm. On an evening with a clear sky, rural observers have a commanding view of the galaxy from our home in the Orion arm. If you look at the brightest parts

of the Milky Way in the southern summer sky,[3] as in the lower left of Figure 23-1, you are looking toward the center of the galaxy between the constellations Sagittarius and Scorpio. The true center itself is hidden behind the intervening 9000 pc of dust and gas.

Higher in the sky, stretching toward Cygnus, the Milky Way is divided by the Great Rift, a band of nearby dark dust clouds that lie in the plane and obscure background stars. These can be seen in Figure 23-1. Overhead on a summer evening, the bright star clouds around Cygnus (sometimes called the Northern Cross) mark our view down our own spiral arm.

Because we live on the inner edge of a spiral arm, our view away from the galactic center directly crosses our arm. This view is spectacular on a winter evening, as seen in Figure 23-7. In this direction are open clusters of the Pleiades and Hyades, regions of recent star formation within our Orion arm. The concentration of bright, bluish stars in the region from the Pleiades through

---

[3]The seasonal and directional references are for Northern Hemisphere observers.

Figure 23-6 Our home, the Milky Way galaxy, as it might look from "above" the plane of the disk. The orientation is the same as in the preceding figure. The solar system would be a microscopic dot among the stars of the Orion arm in the upper left center. The spiral arms are composed of the youngest, hottest stars and clumps of star-forming nebulae. The central region and a foreground globular cluster (upper right) are marked by reddish Population II stars. Brightest object is the nucleus. (Painting by author.)

Figure 23-7 A 120° panorama of the winter Milky Way, looking away from the galactic center. Sirius and the Orion star-forming region lie in the direction looking down our local spiral arm. The Hyades and Pleiades clusters are in our arm in a direction opposite from the center. The *W*-shaped constellation Cassiopeia lies to the right. Compare with narrower-angle view in Figure 18-5. (35-mm camera; 15-mm fish-eye lens at f2.8; 25-min exposure on 2475 Recording film; photo by Floyd Herbert and author.)

Orion to Sirius helps convince us that we are looking into a dense star swarm—our spiral arm—in this direction. The central Orion Nebula and neighboring star associations are located about 500 pc away down the arm.

Why Does the Galaxy Have Spiral Arms?

This question has puzzled astronomers for many years. Some plausible "common sense" explanations do not fit well with observations. For example, a common sense analogy might be a rotating garden sprinkler, where the water jets spray out in spiral arm forms. The trouble with this model is that most material in the galactic spiral arms is not moving radially outward from the central region, as the water droplets do.

A better analogy is a stirred cup of coffee with a few drops of cream. Because the coffee surface rotates faster at the center than at the rim, which slows it, the cream is sheared into long spiral streamers that look like galactic arms. The trouble with this model is that if arms are primordial features of the galaxy, twisted by rotation, they should be very old. Because the galaxy is about 14 billion years old and its rotation time is only about 230 million years, there has been time for spiral arms to be twisted into 60 complete windings. Yet, observations show that spiral arms of various galaxies (including ours) consist of the youngest stars and clusters. They rarely show more than one complete winding.

A modification of the coffee cup model, which we will call the **chain reaction theory,**[4] addresses these points. It is based on the fact that star formation has not happened smoothly and continuously since the galaxy formed, but rather in chain reaction bursts. As discussed in the last chapter, star formation occurs in open clusters, and the expanding gas from massive supernovae in one cluster compresses neighboring clouds of gas and dust, initiating formation of new, adjacent clusters. Therefore, during a period of up to 100 million years, a large region of new clusters containing brilliant, hot, massive, short-lived stars may be produced in one part of a galaxy. During the galaxy's rotation time of 200 million years or so, the inner edge of this region pulls ahead of the outer edge because of differential rotation, and the region of bright, new stars is sheared into a spiral-trending segment. After a few hundred million years, this arm segment runs out of new, young stars and fades, which explains why the spiral pattern rarely achieves more than one winding. This theory also explains why spiral arms frequently appear to be made up of bright segments, instead of a continuous winding from the center outward to the edges of the galaxy.

This concept resembles the coffee cup model, but with new droplets of cream being added intermittently instead of all the cream being added at the beginning. This theory also explains certain correlations of spiral shapes and rotation speeds observed in other galaxies.

MEASURING THE GALAXY'S MASS

Knowing the Sun's velocity around the galactic center (about 235 km/s) and its distance from the center (about 9000 pc, or 2.7×10^{17} km), we can calculate the amount of mass in the central bulge around which we are orbiting. This is essentially an application of Kepler's laws, giving a result of 4×10^{41} kg, or 200 billion solar masses in the central region, within our orbit.

Until the 1980s, this was thought to represent most of the mass of our galaxy. In the early 1980s, however, increasing evidence indicated that much more mass lies in the galactic halo. Recent estimates of the **mass of the galaxy** amount to 1000 billion (10^{12}) $M_\odot$ or even more. Discovery of a distant star falling at about 465 km/s into the galaxy from 45,000 pc above the plane and 59,000 pc from the center gives an estimated total galactic mass of some 1400 billion $M_\odot$ (Hawkins, 1983). Much of this unseen halo mass may be gas and dust, with only 200–400 billion $M_\odot$ being luminous stars. In any case, we live in an enormous system!

COMPREHENDING GALACTIC DISTANCES

We have been casually discussing distances of thousands of parsecs, but it is good to pause and consider what these distances mean. Table 23-1 lists some distances to various objects in our galaxy. To put them in more meaningful terms, the right column lists the time their light takes to reach us (or, conversely, the time it would take us to reach these objects if we could travel at the speed of light). Although we could reach the nearby stars in a few years even at less than the speed of light, we would have to build spacecraft accommodating several generations of travelers to reach well-known open clusters. To reach distant spiral arms, globular clusters, or the galactic center would require more than the entire recorded history of humanity so far![5] If twentieth-century physics is correct in saying that no matter or energy can ever travel faster than the speed of light, then prospects for astronautical exploration of the rest of our galaxy seem dim.[6]

When we consider objects in distant parts of our galaxy, we are dealing with distances of thousands of parsecs. For this reason, astronomers often use a larger unit of distance called a **kiloparsec (kpc)**—note that the same prefix, kilo-, is used throughout the metric system to indicate 1000:

---

[4]It was cumbersomely called the "stochastic, self-propagating, star-formation model" or "SSPSF model," by its original authors.

[5]As the spaceship approached the speed of light, the elapsed time experienced by the astronauts would decrease according to effects predicted by the theory of relativity. In any case, the sponsoring civilization back on Earth would have to wait centuries or millennia for the spaceship or its messages to return from such distant sites.

[6]But maybe twentieth-century physics is not the last word. After all, anybody in any other century who claimed to know the ultimate truth about physics and astronomy would later have been proved wrong. Historically, it has been risky to predict that some deed is impossible.

1 kpc = 1000 pc

Galactic astronomers say, for example, that we are 9 kiloparsecs from the galactic center, and that our galaxy is roughly 30 kiloparsecs in diameter.

THE TWO POPULATIONS OF STARS

One of the most startling discoveries about our galaxy is that it contains a range of star types of different composition, age, distribution, and orbital geometry. These are commonly divided into two major groups called *Populations I and II,* depending on the composition of the gas in their surface layers. **Population I stars** have compositions similar to the Sun's, are relatively young, and are distributed in nearly circular orbits in the gal-

actic disk. A few percent of the mass of Population I stars consist of "heavy elements"—the elements heavier than helium, including carbon, oxygen, silicon, and iron. **Population II stars** are nearly pure hydrogen and helium with only a fraction of a percent of heavy elements.[7] Old stars, they are associated with the bulge of stars in the center of our galaxy and with globular clusters that have orbits taking them far above and below the galactic plane. You can recall the two types more easily by remembering that Population I stars were the *first* group astronomers became familiar with, because

[7]It is important to note that when speaking of the compositions of either population, we are referring to the composition of the visible gases in the surface layers, not the interior composition. Of course, even Population II stars have synthesized heavy elements in their interiors, by normal nuclear reactions. But the unburned surface layers reveal the *initial* composition of the star.

TABLE 23·1

Distances to Selected Destinations in the Milky Way Galaxy

Destination	Distance (pc)	Travel Time of Light from Object to Earth (y)[a]
Nearest star beyond Sun	1.3	4.2
Sirius	2.7	8.8
Vega	8.1	26
Hyades cluster	41	134
Pleiades cluster	125	411
Central part of our spiral arm (Orion arm)	400	1300
Orion Nebula	460	1500
Edge of galactic disk in Z direction (perpendicular to plane)	1000	3300
Next-nearest spiral arm (Sagittarius arm)	1200	3900
47 Tucanae globular cluster	4600	15,000
Center of galaxy	9000	29,000
M 13 globular cluster[b]	11,000	36,000
Far edge of galaxy	24,000?	78,000?

[a]These numbers, by definition, also equal the distance as expressed in light-years.
[b]Target of first beamed radio transmission from Earth directed to hypothetical extraterrestrial civilizations, November 1974. Signal will reach the cluster in A.D. 38,000.

Stellar Populations and Their Properties

Property	Extreme Population I	Intermediate Populations[a]	Halo Population II
Orbits	Circular	Elongated, perturbed by galaxy	Elliptical
Distribution	Patchy, spiral arms	Somewhat patchy	Smooth
Concentration toward galactic center	None	Slight	Strong
Typical distance above or below galactic plane (pc)	120	400	2000
Heavy elements (%)[b]	2–4	0.4–2	0.1
Total mass ($M_\odot$)	2×10^9	5×10^{10}	2×10^{10}
Typical ages (y)	10^8	10^9	10^{10}
Typical relative velocities of stars (km/s)	10–20	20–100	120–200
Typical objects	Open clusters, associations, gas and dust, HII regions, O and B stars	Sun, RR Lyrae stars ($P < 0.4$ d),[c] A stars, planetary nebulae, giant stars, novae, long-period variables	Globular clusters, RR Lyrae stars ($P > 0.4$ d),[c] Population II Cepheids

Source: Data based on tabulations by D. O'Connell, A. Blaauw, J. Oort, C. Allen, and others.
[a]Includes older Population I and intermediate Population II stars.
[b]Elements heavier than helium, sometimes loosely referred to as metals.
[c]P = period of light variation.

they are the type located near the Sun; Population II stars were discovered *second*. Some aspects of the two populations are shown in Table 23-2.

The evidence for chemical difference between the two populations is seen in Figure 23-8, which compares spectra of a Population I G-type star (the Sun) and a Population II G-type star. They have hydrogen lines of equal strength, but the latter has very weak lines of metals and other heavy elements, showing that Population II stars have very low abundances of heavy elements.

Population II stars include a useful type of variable related to Cepheids but not found in Population I. Called **RR Lyrae stars,** they all have a distinctive period around 0.3 to 1 d and a luminosity nearly 100 times that of the Sun (100 $L_\odot$). This means that they are easy to recognize, and their apparent brightness can be used to measure the distance of the cluster or other system in which they lie.

Ages of Populations I and II

Ages of populations are especially intriguing. Population I stars, like the Sun and other stars in the galactic disk, have varied ages, from billions of years to zero age in regions where stars are still forming. But H–R diagrams show that Population II stars are all around 14 ± 4 billion years old—an indication that the globular clusters and other objects in the halo formed during one era roughly 14 billion years ago.

Motions and Orbits of Populations I and II

As shown in Figure 23-9, Populations I and II have different kinds of motions. Population I stars move in circular orbits around the galactic center, in the plane of the galaxy. Population II stars do not share these motions.

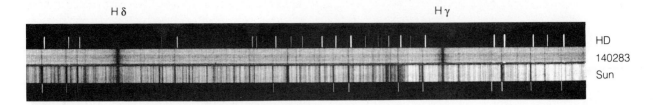

Hδ   Hγ

HD
140283
Sun

Figure 23-8 Comparison of spectra of a Population II star (upper, HD 140283) and a Population I star (lower, the Sun) of similar spectral type. (Bright lines at top and bottom are matching comparison spectra produced in the laboratory.) Both stars have prominent absorption lines of hydrogen (Hδ and Hγ), but the Sun has many additional absorption lines caused by various heavy elements. In the Population II star these lines are very weak or absent, indicating that the heavy elements are virtually absent from its gases.

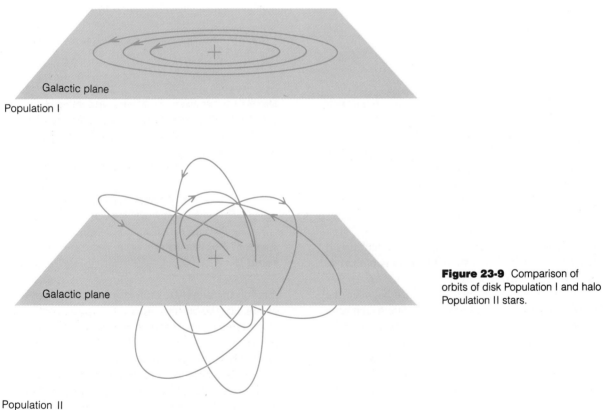

Galactic plane

Population I

Galactic plane

Population II

Figure 23-9 Comparison of orbits of disk Population I and halo Population II stars.

They have highly inclined, elongated orbits. As a result, when Population II stars move through the solar neighborhood they appear to be moving unusually fast. Visualize an analogy with a race track. Suppose you are a driver of one car racing around the track at 120 mph.

A neighboring car seems to have low speed relative to you. Similarly, a Population I star moving around the center of the galaxy near us moves at about the same speed, and seems to have a low velocity relative to the Sun. But a car moving the wrong way around the track would seem to have enormous speed, just as a Population II star on a retrograde or inclined orbit seems to zip by at high speed.

Origin of Populations I and II

As noted earlier, the gas from which old, Population II stars were formed contained mostly hydrogen, and almost no visible elements heavier than helium. But the newer, Population I stars formed from gas that included around 2% heavy elements. We also noted that Population II objects, like globular clusters, form a spheroidal swarm surrounding the galaxy, whereas Population I stars lie in a flat disk. These facts tell us that when Population II originated some 14 billion years ago, the gas cloud from which it formed consisted of about three-fourths hydrogen (by mass) and about one-fourth helium, and less than 1% of heavy elements, and it had not yet flattened into a disk shape. This primordial spherical cloud, from which the galaxy itself was forming, is called the **protogalaxy.** The volume of the protogalaxy at that time matched the volume now occupied by the globular clusters, which were just forming then.

Now we can understand a dramatic connection between the populations and formation of elements. Heavy elements are therefore *created* inside stars and then blasted by disruptive events (such as supernova explosions) from old stars into interstellar space. Later generations of stars form from this heavy-element-enriched interstellar gas and create still more heavy atoms. As the protogalaxy contracted and the earliest stars evolved, more heavy elements were added to the gas. When the galaxy reached its present disk shape, nebulae and later generations of stars in the disk thus slowly grew richer and richer in heavy elements These recent stars, formed in the disk, are Population I.

The two sets of data fit beautifully together. Observations show that the recently formed Population I contains many more heavy elements than the older Population II, and theories of stellar interior processes show why this is so. Observation of star populations thus confirms theories of stellar evolution.

PROBING THE GALACTIC CENTER

In many galaxies, most of the light comes from a brilliant central core, or **galactic nucleus,** buried in the heart of a large central bulge. The concentrated nucleus of the nearby Andromeda galaxy, for example, is as small as 5 pc, but it is much brighter than any of the rest of the 40,000-pc-diameter disk. What mysteries are hidden in a galactic nucleus? Does our galaxy have such a nucleus? A large telescope alone does not help us to find out, because from our vantage point in a spiral arm, our galaxy's center is hidden behind kiloparsecs of dust, as seen in Figure 23-10.

In the 1930s Bell Telephone Laboratories put the young researcher Karl Jansky to work on sources of radio static interfering with long-distance radio signals. Jansky built, in effect, the first radio telescope and discovered that one major source was a steady hiss from the Milky Way galaxy (Jansky, 1935). The technique was next taken up not by traditional optical astronomers, but by an amateur astronomer and radio buff, Grote Reber, who built a radio telescope in his backyard and made the first radio maps of the Milky Way. He established that the strongest radio emission comes from the galactic center (Reber, 1944). This type of mapping succeeds in revealing the nucleus because radio waves (and some infrared wavelengths) can penetrate all the way from the galactic center, even though visible light is blocked by dust.

Subsequent radio, infrared, and gamma-ray studies identified a unique radio source at the center of our galaxy. It is called Sagittarius A—the brightest radio source in the constellation Sagittarius. A region about 150 by 300 pc on the east side of Sagittarius A emits nonthermal radiation believed to result from the interaction of ionized gas and strong magnetic fields. This region may be a supernova remnant. A magnetic field about two to five times stronger than the interstellar field in our region has been found. A region about 35 by 80 pc centered on the west side of Sagittarius A is a concentration of dust clouds, HII regions, and molecular clouds rich in hydroxyl, ammonia, and other molecules. The whole region is also crowded with red giants.

The western part of Sagittarius A (often called Sgr A West) itself is about 10 pc across with a still brighter core. This complex resembles the nucleus of the Andromeda galaxy. It contains some 60 million stars, together with HII regions and dust. Its gas clouds are ionized by radiation like that of stars as hot as 35,000 K (Lacy and others, 1980). Its central 1-pc core contains an amazing 2 to 3 million stars along with gas and dust (Lacy and others, 1980; Genzel and others, 1984)! This compares with 10,000 stars in the central parsec of a globular cluster and only one star within about a parsec of the Sun! In the galactic center stars may average only some 1600 AU apart; the sky would be ablaze with stars as bright as our Moon (Figure 23-11).

ENERGETIC EVENTS IN THE GALACTIC NUCLEUS

The nucleus is unique not only because of its high density of stars and dust, but also because it releases prodigious amounts of energy. For example, a mass of hydrogen that resembles a spiral arm is expanding from the nucleus at a distance about 3000 pc from the center. This cloud, called the **3-kiloparsec arm** (or 3-kpc arm), contains about 10 million solar masses of neutral hydrogen. It revolves around the center at about 210 km/s, but is also rushing outward toward the Sun at about 53 km/s. To set this mass of gas in motion required about 10^{46} joules (J) of energy—over 1000 times the energy production of the Sun during its lifetime! Yet the 3-kpc arm is only about 10 to 100 million years old. Apparently a large energy release occurred "recently."

A second line of evidence for energetic events in the nucleus is simply the high luminosity of the nucleus,

a

Figure 23-10 a A 30° segment of the Milky Way in the direction of the galactic center, showing the reddened color of the bright central bulge, protruding from behind ragged black clouds of intervening dust. Box shows region of *b*. (Copyright Anglo-Australian Telescope Board.) **b** The same region photographed at infrared wavelengths of 12 to 100 μm, which penetrate through the foreground clouds of *a*. The false color image shows a band of 20–100 K dust tightly confined within about 60 pc of the galactic plane. Knots of warmer dust (yellow) and a 200-pc-wide bulge of warmer material at the galactic center can be seen. The box shows the approximate outline of Figure 23.12. (NASA photo from IRAS infrared satellite.)

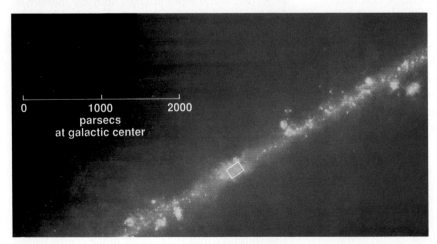

b

Figure 23-11 Hypothetical panorama from a position only a few hundred parsecs from the Milky Way's nucleus. Here we are surrounded by thousands of brilliant red giants and other Population II stars and by dense clouds of dark dust and glowing nebulosity lying along the Milky Way's plane. In the right distance is the actual nucleus—possibly a large black hole surrounded by a brilliantly glowing accretion disk and partly obscured by clouds. (Painting by author.)

much of it in the form of infrared radiation from hot dust clouds. Some energy source must be available to heat these clouds.

A third line of evidence comes from other galaxies. The Andromeda nucleus is bright but relatively stable; yet certain other galaxies emit strong radio radiation and jets of gas, indicating occasional explosions with energies comparable to the 10^{46} J needed to produce our own 3-kpc arm.

What **conditions in the galactic center** could produce the strange environment and energetic events just described? This area of research is very active and exciting, partly because of the increasingly detailed maps of the galactic nucleus being made by radio astronomers.

One theory to explain the energetic events of the nucleus is that enormous, unstable "stars" containing millions of solar masses may form in the dense clouds there. As we saw in Chapter 19, such huge bodies would soon explode, creating powerful supernovae. This could explain the energies needed to create the motions and luminosity observed in and near the nucleus—for example, in the 3-kpc arm around 10 million to 100 million years ago. Analysts have proposed explosions of short-lived objects of as much as $4 \times 10^8 \, M_\odot$ (Sanders and Prendergast, 1974; Lo and Claussen, 1983). The wisps in Figure 23-12, only 40 pc from the nucleus, might be

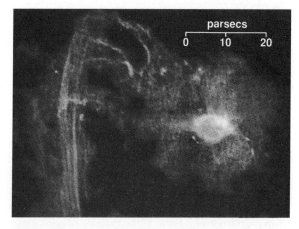

Figure 23-12 Gas wisps (left) and shell surrounding the Milky Way's nucleus (blob at right). Wisps up to 50 pc long and only 1 pc wide are believed to be organized in this pattern by magnetic fields surrounding, and generated in, the nucleus. The process generating the fields, however, is uncertain. The radiation used to make this image is radio radiation at 20-cm wavelength. (National Radio Astronomy Observatory, operated by Associated Universities, Inc., under contract with the National Science Foundation. Observers: F. Yusef-Zadeh, M. Morris, D. Chance.)

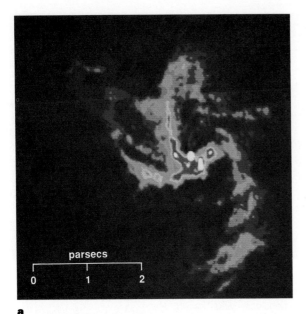

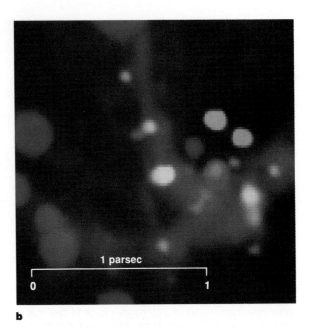

a   **b**

Figure 23-13 The central few parsecs of our galaxy. Radio astronomers discovered this small spiral, believed to be associated with matter spiraling into the accretion disk around the galaxy's massive nucleus (black hole?). The spiral appears tipped up toward us, out of the galactic plane, possibly indicating that the gas was chaotically disturbed (by an explosion?) before the accretion began and is now rotating in a new plane. **a** Radio image at 6-cm wavelength. Dark central features are due to image processing that contoured the image according to brightness levels; central region is actually brightest. (California Institute of Technology, courtesy K. Y. Lo.) **b** This curious false color image is a composite of radio data (redder tones) and shorter-wavelength infrared data (bluer tones). It reveals the radio-emitting spiral structure and several other sources that are probably infrared-emitting dust clouds. Such images allow us to map the features of the galactic nucleus at a wide range of wavelengths. (Copyright Anglo-Australian Telescope Board.)

gas originally thrown out of such explosions and forced into arclike shapes by magnetic fields. Radial and spiraling motions observed only 2 pc from the nucleus (Figure 23-13) suggest some of the gas is falling back toward the nucleus. Some of the motions suggest explosive disturbances as recently as 10,000 y ago, and luminosities of 10 to 30 million $L_\odot$ in the central parsec were needed to heat the gas and dust in the central region (Waldrop, 1985)!

Additional mechanisms are needed to explain the object at the heart of the central star cluster of our galaxy. Observations in the 1980s have shown that a few million solar masses must be crammed into the central parsec (Genzel and others, 1984). This strongly suggests that the object at the center is a huge black hole of several million solar masses (Lacy and others, 1980)! More and more observations are supporting this

theory in recent years. It is remarkable to think that a vast black hole could be the defining object identified with the nucleus and its powerful energy output.

Radio and infrared data indicate that the gas in the central region is not as highly ionized as might be expected if numerous hot stellar cores were the main heat source. Rather, the heat source might be a huge accretion disk formed by matter slowly spiraling into the black hole at an estimated rate of 10 $M_\odot$ every million years. The 4-pc-wide spiral structure of some 60 $M_\odot$ of (infalling?) ionized gas, discovered by radio mapping (Lo and Claussen, 1983) and shown in Figure 23-13, may be our first view of accretion onto a black hole.

Since 1977, Levanthal and colleagues (1982) have monitored gamma-ray emission from this area. The wavelength is exactly that created when electrons collide with positrons (positive particles of electron mass).

gas and dust from staying in a perpetual orbit like Mercury's around the Sun; rather the gas and dust spiral inward into the accretion disk and keep it hot. Ultimately the material spirals into the black hole at the center, adding to its mass. Infall of large objects into the disk might cause explosive outbursts.

Clearly, astronomers are still groping for explanations of what is happening in the center of the Milky Way, but new techniques of observing the center with balloon and satellite gamma-ray telescopes, ground-based radio telescopes, and infrared telescopes promise exciting results in the next few years. Even from what we know so far, we can imagine that the scene near the galaxy's nucleus must be extraordinary (Figure 23-11). It provokes the imagination to glance across thousands of parsecs toward the star clouds of Sagittarius on a summer evening away from the city lights, and wonder what lies beyond those softly glowing star fields (Figure 23-14).

SUMMARY

Our discussion in the last few chapters can now be fitted into a compelling theory of the origin and evolution of the galaxy. Globular clusters give us the best estimate of the age of the galaxy—about 14 billion years.

What was happening 14 billion years ago? Evidently the galactic mass was a single cloud of hydrogen-rich gas, with one-fourth helium (by mass) and virtually no heavier elements. The mass of this protogalaxy equaled the present galactic mass, roughly $10^{12} M_\odot$. The protogalaxy must have been rotating and contracting. When it reached a rather spherical shape perhaps 30 to 40 kpc across, it became dense enough that separate, large, self-gravitating masses formed within it. These became globular clusters with certain characteristics we observe today:

Age: *about 14 billion years*

Composition: *few heavy elements*

Distribution: *spheroidal halo 30–40 kpc across; elongated orbits*

Mass: 10^5–$10^7 M_\odot$ *each*

Because of rotation and collisions among the atoms and clouds of gas, the inner part of the protogalaxy, still containing some $10^{11} M_\odot$, flattened into a disk shape. The globular clusters, too far apart to interact, were left behind and did not form a disk. Thus two populations (with some intermediate objects) arose from the earlier- and later-

Figure 23-14 The Milky Way rising over telescope domes at Mauna Kea Observatory, Hawaii. This is the Milky Way in the constellation Sagittarius, the direction toward the nucleus of our galaxy, which lies beyond the brightest star clouds in the photo. The camera tracked on the stars for 10 min, making the domes and horizon slightly blurred. (Photo by author; 24-mm wide-angle lens, f2.8, Ektachrome ISO 1600 film.)

Their interpretation favors the idea that the electrons and positrons boil off the hot gas of the central accretion disk surrounding a supermassive black hole.

Radio data indicate a unique object in the densest core of Sagittarius A. Special techniques using combined signals from several radio telescopes indicate this object is no larger than 20 AU across. It may be the outer part of an accretion disk surrounding the massive black hole (Geballe, 1979; Waldrop, 1985). In this view, turbulence and drag effects in the gas and dust surrounding the nucleus dissipate energy. This keeps the formed systems. The stars and gas in the disk took on the properties we see today in nearby space:

Age: *youthful; a few billion years for stars; only tens of millions of years for nebulae*

Composition: *few percent heavy elements*

Distribution: *flat disk; 30 kpc across, 400 pc thick*

Mass: *stars formed in clusters of a few hundred $M_\odot$*

In the center, an extraordinary, massive black hole may have formed, and violent explosions may have taken place. In the disk, nebulae traveling in elliptical orbits quickly collided with other nebulae until the gas, dust, and nebulae all moved together in relatively circular orbits. The spiral arm pattern probably emerged after a number of galactic rotations, perhaps within a billion years. In the spiral arms, the densest clouds contracted and spawned associations and open star clusters. Each group broke apart into scattered stars a few hundred million years after its formation, but new star groups continued to form, so that the galaxy kept its present general appearance. Supernovae blew out gas laced with heavy elements created inside stars, so that later generations of stars had more heavy elements than the earlier stars. Perhaps 7 or 8 billion years after the galaxy's formation, in one of the spiral arms, an obscure star formed—our Sun—and in its surrounding dusty nebula, Earth was born.

CONCEPTS

Milky Way galaxy	mass of the galaxy
galactic halo	kiloparsec (kpc)
dimensions of the galaxy	Population I stars
Sun's revolution period	Population II stars
differential rotation	RR Lyrae star
age of the galaxy	protogalaxy
spiral arms	galactic nucleus
evidence of spiral structure	3-kiloparsec arm
21-cm radio waves	conditions in the galactic center
chain reaction theory	

PROBLEMS

1. During what percent of human recorded history (define as you think appropriate) have people *not* known that we live in an isolated galaxy of stars similar to other remote galaxies?

2. From the appearance of the Milky Way, how do we know that the solar system is in the galactic disk and not far above or below it? How might the appearance of the central region differ in the latter case? (*Hint:* See Figures 23-1 and 23-10.)

3. Compare the shapes of the volumes occupied by the swarm of open clusters and by globular clusters. Relate the difference to differences in stellar populations.

4. How do we know the size of the galaxy?

5. How do we know the location and distance of the galactic center?

6. Describe evidence for spiral structure in the Milky Way. Which of the following types of objects reveal spiral structure when their positions are mapped on the Milky Way plane?

a.	O stars	**d.**	Open clusters
b.	M stars	**e.**	Globular clusters
c.	HII clouds	**f.**	Star-forming regions

7. Will the present constellations be recognizable in the Earth's sky 100 million years from now? Why or why not?

8. Why are no O- or B-type stars found in the galactic halo?

9. Summarize evidence for violent, energetic activity in our galaxy's central region.

PROJECTS

1. Compare views of the Milky Way with the naked eye, binoculars, and a small telescope. Scan along the Milky Way with each instrument and record the number of stars per square degree in different constellations (or at different galactic longitudes). Relate these densities to the actual structure of the galaxy. Note that if observations of the summer evening Milky Way can be obtained in the regions of Scorpio and Sagittarius, the direction toward the center can be studied. Compare star counts with each instrument in the Milky Way plane with counts near a point 90° from the plane, where we are looking directly "up" out of the disk. Can you account for the differences?

2. Compare the Milky Way as seen with the naked eye or binoculars:

a. In the heart of your city

b. On the edge of town

c. As far from lights as you can possibly get

City lights have little effect on the telescopic views of individual bright stars. But when it comes to broad areas of faint nebulosity or unresolved star clouds, even a single street light can illuminate local smog or fog and cause the iris of the eye to contract, thus destroying faint contrast and the ability to see faint glows.

CHAPTER 24

The Local Galaxies

What lies in the vastness of space beyond our own galactic disk and its surrounding swarm of globular star clusters? Our first problem is to see out of our own galaxy. Thick clouds of interstellar galactic dust keep us from seeing out along the Milky Way's plane, just as they keep us from seeing our galaxy's nucleus. However, when we look "up" or "down" out of the Milky Way disk (at high angles to the Milky Way plane), our line of sight passes through little dust. The view in these directions reveals galaxy upon galaxy far outside our own Milky Way galaxy, as far as we can see. Some are like the Milky Way; some are not.

To understand the spacings of galaxies, consider a scale model where we let a 12-in. dinner plate represent the disk of the Milky Way. The closest galaxies would be irregular objects like crumpled balls of cotton 2 or 3 in. across only about 2 ft above the plate. A dozen other galaxies of various shapes, $\frac{1}{2}$ to 4 in. across, would be scattered around the room. In a hallway, about 20 ft away, is the nearest galaxy resembling the Milky Way— a 16-in. platter representing the great spiral galaxy in Andromeda. A three-dimensional sketch of the situation is seen in Figure 24-1, and properties of these and some other galaxies are listed in Table 24-1. These objects are so remote that their true nature was recognized only in the 1920s.

THE NEARBY GALAXIES

In small telescopes (up to about 30-cm aperture), most galaxies look like mere blurs of light. When the Messier, New General, and Index catalogs of celestial objects were compiled, galaxies were not distinguished from nebulae, and so they received M, NGC, and IC numbers in sequence with nebulae. The Andromeda galaxy, for example, is M 31.

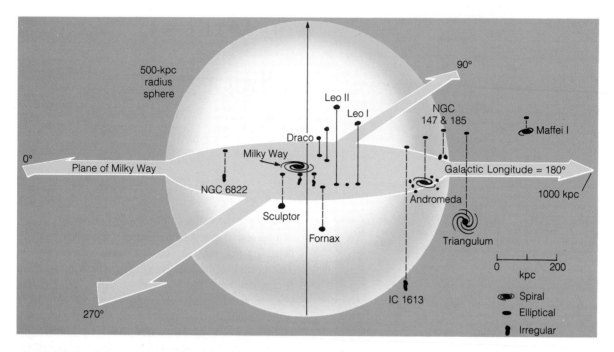

Figure 24-1 A three-dimensional plot of most of the Local Group of galaxies. Solid lines extend north of the galactic plane; dashed lines extend south. Table 24-1 lists details of these systems. Clustering of dwarf satellite galaxies around the great Milky Way and Andromeda spirals can be seen.

Later photographs made with large telescopes show some galaxies as pinwheels with beautiful spiral arms—like the Milky Way—but many others as ellipsoidal groups of stars like giant, flattened globular clusters or irregular masses of stars and nebulae. These objects were proved to be galaxies and not amorphous clouds of gas by identifying individual stars, Cepheid variables, and supernovae in them.

Once these objects were identified as galaxies, their distances could be estimated. For example, Henrietta Leavitt's 1912 discovery of the Cepheid's period/luminosity relation (page 294) could be used in the following way: Periods of Cepheids in other galaxies could be observed; then their apparent brightness and known luminosity combine to give a distance. Another method for measuring a galaxy's distance uses the galaxy's apparent angular size. For instance, a galaxy with about the size and shape of ours that subtends an angle of $2°.6$ in the sky would have to be around 670 kpc away. This is the situation with the Andromeda galaxy.[1]

Doppler shifts help us learn about galaxies' rotations. By measuring the Doppler shift on each side of a galaxy's center, we can see stars approaching on one side and receding on the other due to the stars' orbital motions around the galaxy's center. Combining the rotation speeds with the stars' distances from the galactic center, we can apply Kepler's laws and determine the mass of the galaxy, as was done for the Milky Way in the last chapter.

The Magellanic Clouds: Satellite Galaxies of the Milky Way

Medieval Arab astronomers recorded a glowing patch far down in the southern sky, visible from southern Saudi Arabia. Amerigo Vespucci, who sailed farther south, reported that there were really "two clouds of reasonable bigness moving about the place of the (south celestial) pole" (quoted by Shapley, 1957). Because there is no bright south polar star, the clouds helped navigators

---

[1]It is surprising to realize that one of the objects subtending the largest angle in our sky is not our own nearby Moon ($\frac{1}{2}°$), our Sun ($\frac{1}{2}°$), a planet (Jupiter's whole satellite system subtends $2°.1$, whereas the disks of Jupiter and Venus reach only about $1'$), or a cluster like the Pleiades ($1\frac{1}{2}°$), but the enormously remote Andromeda galaxy ($2°.6$ to $4°.5$, depending on the detector).

Known Galaxies Out to 1000 kpc

Name	Catalog Number	Distance (kpc)	Diameter (kpc)	Mass ($M_\odot$)	Absolute Magnitude (M_v)	Type[a]
Milky Way Subgroup						
Milky Way	—	9	30	2×10^{11}	−20.6	Sb
Large Magellanic Cloud	—	52	8	10^{10}	−18.4	Irr
Small Magellanic Cloud	—	63	5	2×10^9	−17.0	Irr
Sculptor system	—	71	2	3×10^6	−10.6	Dwarf E
Sextans system	—	86	1	?	?	Dwarf E
Draco system	DDO 208	90	1	10^5	− 8.0	Dwarf E
Ursa Minor system	DDO 199	90	2	10^5	− 8.2	Dwarf E
Carina	—	90	2?	?	?	Dwarf E
Fornax system	—	150	6	2×10^7	−12.9	Dwarf E
Leo I system	DDO 74	180	1	4×10^6	− 9.6	Dwarf E4
Leo II system	DDO 93	180	2	10^6	− 9.2	Dwarf E1
Barnard's galaxy	NGC 6822	550	2	3×10^8	−15.1	Irr
WLM	DDO 221	610	?	?	−15.0	Irr
—	IC 5152	610	?	?	−14.6	Irr
Andromeda Subgroup						
Andromeda companion	NGC 205	640	3	8×10^9	−15.7	Dwarf E5
—	NGC 147	670	2	10^9	−14.4	Dwarf E
—	NGC 185	670	2	10^9	−14.6	Dwarf E
Andromeda companion	M 32 (NGC 221)	670	2	3×10^9	−15.5	Dwarf E2
Andromeda galaxy	M 31 (NGC 224)	670	40[b]	3×10^{11}	−21.6	Sb
Andromeda I	—	670	0.7	?	−10.6	Dwarf E
Andromeda II	—	670	0.7	?	−10.6	Dwarf E
Andromeda III	—	670	0.9	?	−10.6	Dwarf E
Triangulum galaxy	M 33 (NGC 598)	770	18	1×10^{10}	−19.1	Sc
—	IC 1613	770	4	3×10^9	−14.5	Irr
Aquarius	DDO 210	920	?	?	−11.0	Irr
Pisces	LGS 3	920	?	?	− 9.7	Irr
Maffei 1	—	1000	?	2×10^{11}	−20	S0

Source: Data from Allen (1973), van den Bergh (1972), Hirschfeld (1980), Hodge (1987), and Irwin and others (1990).
[a]S = spiral (subtypes 0, a, b, c); Irr = irregular; E = elliptical (subtypes 0 through 7).
[b]Diameter of Andromeda based on angular diameter 3°.1.

Figure 24-2 The Small Magellanic Cloud, about 63 kpc away and 5 kpc across, is an irregular galaxy exhibiting a barlike configuration of bluish stars and red HII regions. (Copyright R. J. Dufour, Rice University.)

Figure 24-3 The Large Magellanic Cloud, about 52 kpc away and 8 kpc across, is the closest galaxy to us. Color view shows a barlike distribution of bluish stars and a scattering of red Hα-glowing HII regions, including the massive Tarantula Nebula (upper left). (Copyright Anglo-Australian Telescope Board, courtesy D. Malin.)

to mark the pole. Well described during Magellan's around-the-world expedition of 1518 through 1520, they came to be called the **Magellanic clouds.** They are shown in Figures 24-2 and 24-3.

The clouds turned out to be small galaxies, probably

moving in orbits around the Milky Way. The Large Magellanic Cloud is 52 kpc away, and the Small Magellanic Cloud 63 kpc, less than three times as far as the far edge of our own galaxy. The large cloud is only 8 kpc across, and the small cloud only 5 kpc, compared with about 30 kpc for the Milky Way. Star counts and radio measures of hydrogen indicate that these clouds are only a few percent as massive as the Milky Way. In addition, the clouds do not show the Milky Way's beautiful spiral structure. Because they are ragged, they are classified as **irregular galaxies.** Each contains a softly glowing, barlike structure composed of stars, as shown in Figures 24-2 and 24-3.

Can we identify a nucleus in either galaxy? Somewhat off the end of the bar in the Large Magellanic Cloud is the **Tarantula Nebula,** also known as **30 Doradus** or NGC 2070. It is shown in Figure 24-4. This nebula, although 52 kpc away, can be seen with the naked eye, so it must be extremely luminous. In fact, if it were moved to the position of the Orion Nebula, it would fill the whole constellation of Orion and be bright enough to cast shadows on Earth! In its center is a cluster about 60 pc in diameter, containing at least 100 massive, bluish supergiant stars (Figure 24-4b). This cluster is several hundred times brighter than ordinary globular clusters near the Milky Way. It is surrounded by intricate wisps of red-glowing hydrogen, similar to those blown out by the Crab supernova explosion (see Figure 19-8). In 1981, Wisconsin astronomers proposed, on the basis of observations from a UV space telescope, that this cluster may contain the most massive star known—a behemoth named Radcliffe 136a, with an estimated mass of $1000-2000 \, M_\odot$, temperature 56,000 K, diameter equaling that of Mercury's orbit, and a lifetime of only a million years before a supernova outburst (Humphreys and Davidson, 1984; Mathis and others, 1984). Other astronomers propose a central black hole or a cluster of massive O stars. Controversy abounds.

The perceptive reader will have noticed that we are really facing the mystery of the galactic nucleus all over again. Many observers regard the Tarantula Nebula and Radcliffe 136a as a partially formed nucleus of the large cloud. Reminiscent of our galaxy's nucleus, it contains a thermal radio source about 60 pc across, surrounded by a source of nonthermal radio radiation about 200 × 400 pc. The central object, in both cases, is mysterious.

Strange features of both Magellanic clouds are a few blue globular clusters, unlike the red-giant-dominated clusters of our galaxy. To contain hot, blue stars (which

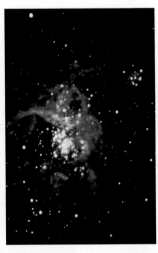

Figure 24-4 Zeroing in on the Tarantula Nebula—a possible nucleus of the Large Magellanic Cloud. Also known as 30 Doradus, this nebula is 30 times as big as the Orion Nebula. The orientation of Figures 24-3 and 24-4 is the same. **a** Long exposure showing outer wispy hydrogen clouds. **b** Short exposure revealing cluster of young, blue stars whose radiation excites the red Hα glow of the nebula. (*a*, copyright Anglo-Australian Telescope Board; *b*, copyright AURA, Inc.)

are massive and have short lifetimes), these globulars must have formed less than a billion years ago, and others may still be forming. An example is seen in Figure 24-5.

In 1983 astronomers announced that the Small Magellanic Cloud may actually be two galaxies, one in front of the other, about 9 kpc apart. This finding was based on their stars' velocities, which seem to divide into two groups.

The Magellanic clouds are important to modern astronomy for providing a collection of varied objects at essentially constant distances. In the words of South African astronomer A. D. Thackeray (1971):

The opportunity to compare the luminosities of supergiants, giants, main sequence stars of all types, Wolf–Rayet stars, cepheids, eclipsing variables, Mira-variables, RR Lyrae variables, globular clusters, novae, and so on, in an endless sequence all at the same distance simply never occurs in our galaxy.

Dwarf Elliptical Galaxies

As indicated in Table 24-1, the next seven objects beyond the Magellanic clouds are small systems of a type known as **dwarf elliptical galaxies.** These small elliptical-shaped or spheroidal galaxies resemble giant globular clusters. They are more symmetrical than the Magellanic clouds, but lack the disk shape and spiral arms of the Milky Way. One is shown in Figure 24-6.

Dwarf elliptical galaxies are apparently the most common type of galaxy. They are dominated by old Population II stars and have little gas or dust. In most respects they are less impressive than our own giant spiral disk, with its chaotic clouds of gas and dust and regions of continuing star formation. But interest in them revived in the 1980s when studies indicated that some of their stars must be fairly young, perhaps 2 to 5 billion years old. Thus they must be somewhat more active than globular clusters, whose stars are all around 14 billion years old. Furthermore, even though they look somewhat like overgrown globular clusters, at least one (the Sculptor system) has its own swarm of globular clus-

Figure 24-5 A globular cluster in the Large Magellanic Cloud exhibits bluer stars than the red-giant-dominated globular clusters of our own galaxy. The evolution of these blue globulars is not clear. (Copyright Anglo-Australian Telescope Board.)

Figure 24-6 One of the closest galaxies is this loose dwarf elliptical called Leo I, about 1 kpc in diameter. At a distance of 180 kpc it is close enough to be easily resolved as separate stars, and it looks somewhat like a very large globular cluster. (Copyright Anglo-Australian Telescope Board, courtesy D. Malin.)

ters, which are about the size of our own globulars or slightly larger. Because "dogs have fleas but fleas don't have fleas," this fact strengthens the classification of Sculptor-like dwarf ellipticals as galaxies in their own right (Hodge, 1987). Most dwarf elliptical galaxies may be satellite attendants to larger systems. Astronomer Paul Hodge calls the group near our galaxy "the Seven Dwarfs."

THE GREAT ANDROMEDA SPIRAL GALAXY—A SISTER OF THE MILKY WAY

Finally, at 670 kpc, we encounter the first **spiral galaxy** truly comparable to the Milky Way, along with several of its smaller satellite galaxies. This galaxy was first recorded in a star catalog by Arab astronomer al-Sufe in A.D. 964, but it must have been known much earlier, being easy to see with the naked eye as a hazy patch on a dark, clear night. The **Andromeda galaxy** is the one that settled the argument about the nature of so-called spiral nebulae. Edwin Hubble identified Cepheid variables there and showed that it was far outside the disk of our own galaxy.[2] His announcement of this finding at the American Astronomical Society in 1924 caused

[2]Even until the 1960s, some books maintained the misnomer "Andromeda nebula," a name carried over from the 1920s when it was still thought to be only a nebula inside our own galaxy.

a sensation and marked the first real understanding of galaxies.

The Andromeda galaxy can hardly be distinguished from the Milky Way in shape or stellar content, though it may be a little larger. The naked eye sees it as a faint patch (Figures 24-7 and 24-8), but this is really only the brightest, innermost region, a few kiloparsecs across. Telescopic photographs show that the spiral arms form

Figure 24-7 How to find the Andromeda galaxy with your naked eye. This color view shows the Milky Way extending across the region of the constellation Cassiopeia (M-shaped figure). The angular distance from the Andromeda galaxy to Cassiopeia's middle star is about 20°. Although the Andromeda galaxy is about 670 kpc away, on a clear, dark night it can be seen by the unaided eye as a faint smudge of light. This view records details somewhat fainter than the eye can see and reveals the reddish color of the galaxy's nucleus. Box shows the area of Figure 24-8. (35-mm guided camera; 24-mm lens at f2.8; 13-min exposure on 3M ASA 1000 film; photo by author.)

Figure 24-8 The region of the Andromeda galaxy, showing an angular extent approaching 3°. Vertical height of the photo is about 8°. The bright star, upper right, is Beta Andromedae. The box shows the outline of views in Figure 24-9. (35-mm camera; 135-mm lens at f2.8; 10-min exposure on 2475 Recording film.)

a disk at least 20 kpc across (Figure 24-9), whereas outlying associations of B stars suggest a diameter as great as 50 kpc. As with the Milky Way, there are globular clusters and a halo of HI gas perhaps 100 kpc in diameter.

Spiral Arms

The spiral arm pattern of Andromeda resembles that of the Milky Way, though it is much easier to observe because we see it from "above" the disk. *Inner arms* have conspicuous dust; *outer arms* have less dust but more HII regions and young stars. Most star formation now occurs in the intermediate and outer arms.

Stellar Populations

In 1943, Walter Baade, working at Mt. Wilson observatory near Los Angeles, took advantage of wartime blackouts of the city lights. With photographic time exposures as long as nine hours, Baade discovered that the central bulge of Andromeda and its two nearby elliptical galaxies are dominated not by blue stars (like the young, brightest stars of the spiral arms) but by, as he put it, "thousands and tens of thousands" of Population II red giant stars. This discovery showed that the Andromeda pattern of populations matches that of our own galaxy. It also clarified the role of star populations in galaxies. *In spiral galaxies, Population I star-forming regions are concentrated in the spiral arms, which are marked by open clusters, giant nebulae, and massive blue stars. The central bulges of spirals are redder than the arms, because star formation has ceased and there are no bright, young blue stars; the brightest stars are old red giants.* This effect can be seen in several other galaxies in color figures in this and the next chapter. The same

a

b

Figure 24-9 Images of the Andromeda galaxy with different types of photographic and telescopic equipment. **a** This picture was printed to show faint details in the spiral arms. (Steward Observatory, University of Arizona.) **b** Mosaic of photos with an amateur's 32-cm telescope, printed to show maximum detail in both nucleus and spiral arms. (Clarence P. Custer.) Neither view captures the extreme outer faint regions.

pattern applies to elliptical galaxies, which resemble central bulges of spirals but without the spiral arms. *Thus Population II stars predominate in elliptical galaxies and central bulges of spiral galaxies.*

Dynamics and Mass

As in our galaxy, the density of stars is greatest near the center. A region of low rotational velocity and low star density about 2 kpc from the center may be a feature related to our own 3-kpc expanding arm. The total mass of the main disk of the galaxy is roughly 3×10^{11} $M_\odot$, but there may be more unseen halo mass, as in the Milky Way (Rubin, 1983).

The Nucleus

The Andromeda galaxy gives us a chance to look directly at the light of the nucleus of a galaxy like ours, rather than trying to study it through 9 kpc of dust. The results are intriguing.

Figure 24-10 shows the similarity between our moderate-angle view into the Andromeda galaxy's center and our lower-angle view of our own galaxy's dust-obscured center. Figure 24-11 shows a shorter color exposure of the central region, which reveals its reddish tones—due partly to the Population II red giants that dominate the region. This figure also hints at the fact that the brightness increases the closer we get to the

a

b

Figure 24-10 A comparison between the ragged, dust-cloud-strewn heart of our galaxy and the central part of the Andromeda galaxy shows striking similarities between the two systems. **a** From our position perhaps slightly above the plane of the Milky Way, we peer over dark dust clouds in intervening spiral arms toward the glowing central bulge. **b** From a slightly higher tilt, we look at the central bulge of the Andromeda galaxy. Box shows approximate size of Figure 24-11, oval shows position of nucleus (Figure 24-12) overexposed here. High-contrast print brings out dust clouds similar to those in Milky Way. (Image a, 70° panorama from two wide-angle photos by author; 24-mm lens, f2.8, 40- and 46-min exposures on 2475 Recording film; image b, Kitt Peak National Observatory.)

Figure 24-11 This color photo of the central bulge of the Andromeda galaxy emphasizes the great brightness of the central region relative to the faint spiral arms. The photo covers about the area of the box in Figure 24-10; only the inner edges of the spiral dust lanes are visible. Population II stars with reddish color dominate the central bulge. (Copyright Association of Universities for Research in Astronomy.)

center. Most photos overexpose this region in an effort to record the fainter, outer spiral arms. But short exposures reveal a brilliant star-like object at the very center, as seen in Figure 24-12. This is the nucleus—the brightest object in the whole galaxy.

About a dozen X-ray sources lie within a few hundred parsecs of this nucleus (Figure 24-13), possibly marking accretion disks of neutron stars. Sharp photos show that the bright nucleus is elliptical, about 3×5 pc across (Figure 24-12). It contains about 10^8 stars crowded only 0.01 pc (2000 AU) apart on average, similar to the results we found for our galaxy's nucleus (Chapter 23). Special techniques used by Wyoming infrared astronomer E. J. Spillar and colleagues (1990) reveal a dense bluish core, not more than about 0.35 pc across, possibly the accretion disk of a black hole.

SURVEYING AND CLASSIFYING GALAXIES

As seen in Table 24-1, an extension of our survey of galaxies all the way out to 1000 kpc from the Milky Way nets 26 objects, including 14 dwarf ellipticals, 8 irregulars, and 4 spirals. What do the statistics of these galaxies tell us about galaxies in general?

First, galaxies are not randomly distributed through space. They tend to be clumped in different-sized groups

called **clusters of galaxies.** As seen in Figure 24-1 and Table 24-1, for instance, most of the galaxies out to 1000 kpc are clumped in two subgroups: those around the Milky Way and those around the Andromeda spiral. The whole population of 26 galaxies out to 1000 kpc is a concentration called the **Local Group.** Just as stars form in clusters, something about the formation process of galaxies also produces clusters.

Nonluminous Mass in Galaxies

As mentioned earlier, only a fraction of the total material in the Milky Way and Andromeda galaxies is luminous, visible stars. Surveys also confirm this result for many other galaxies (Rubin, 1983). The rest is invisible. Where is this mass? This question is sometimes called the **problem of the missing mass.** Perhaps the missing mass is in the form of faint stars, planetary matter and dust, black holes, gas, or subatomic particles like neutrinos, but the answer is uncertain. The invisible mass has been detected by using Doppler shifts to study orbital velocities of stars around galactic centers; the velocities prove that the stars are attracted to the galaxies by more than the visible mass. As astronomer Vera Rubin (1983) concluded from her study of such velocities, "as much as 90 percent of the mass of the universe is evidently not radiating at any wavelength with enough

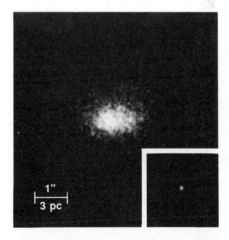

Figure 24-12 Highest-resolution photo of the bright nucleus of the Andromeda galaxy, a dense cluster of stars about 3 by 5 pc in diameter. The photo was made by a 36-in. telescope carried to 84,000 ft by a balloon. The inset at right shows the image of a star made by the same telescope. (Princeton University, Project Stratoscope, supported by NSF, ONR, NASA.)

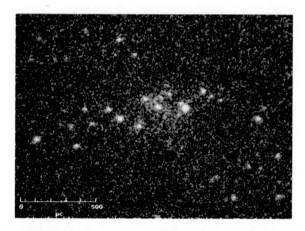

Figure 24-13 An X-ray image from the orbiting Einstein Observatory shows a cluster of X-ray sources in the central kiloparsec of the Andromeda galaxy. One of the sources (in upper center group) corresponds to the nucleus, shown in Figure 24-12. It varied in X-ray output from 10^{30} to 10^{31} watts during the 6-mo period of observation. Other sources include globular clusters and may include accretion disks around black holes. (Harvard/Smithsonian Center for Astrophysics, courtesy Leon Van Speybroeck.)

intensity to be detected on the Earth." Solving the problem of why the missing mass is invisible is important in understanding the interactions of galaxies. Future observations from above the atmosphere may help; astronomers are especially looking forward to results from the orbiting Hubble Space Telescope.

A Classification System: Making Sense of Galaxies

Most of the known types of galaxies appear within the Local Group (Table 24-1), which takes us out to 1000 kpc. Even if we expand our horizon to, say, 15,000 kpc, as in Table 24-2, we discover only a few new types. M 87, shown in Figure 24-14, is a **giant elliptical galaxy** much larger than the ones mentioned so far; it is about 30 kpc across and more massive than the Milky Way. It contains about $4 \times 10^{12} M_{\odot}$, or perhaps 4000 billion stars! These *giant* ellipticals give a good reason for distinguishing the smaller, globular-cluster-like ellipticals as "dwarf ellipticals." Another class is illus-

TABLE 24-2

Selected Galaxies Out to 15,000 kpc

Catalog Number	Distance (kpc)	Diameter (kpc)	Mass ($M_{\odot}$)	Absolute Magnitude (M_v)	Type[a]	Radial Velocity (km/s)
NGC 55	2300	12	3×10^{10}	−20	Sc	+ 190
NGC 253	2400	13	10^{11}	−20	Sc	− 70
M 82 (NGC 3034)	3000	7	3×10^{10}	−20	Irr	+ 400
M 81 (NGC 3031)	3200	16	2×10^{11}	−21	Sb	+ 80
M 83 (NGC 5236)	3200	12	10^{11}	−21	SBc	+ 320
M 51 "Whirlpool" (NGC 5194)	3800	9	8×10^{10}	−20	Sc	+ 550
NGC 5128 (Centaurus A)	4400	15	2×10^{11}	−20	E0p[b]	+ 260
M 101 "Pinwheel" (NGC 5457)	7200[c]	40	2×10^{11}	−21	Sc	+ 402
M 104 "Sombrero" (NGC 4594)	12,000	8	5×10^{11}	−22	Sa	+1050
M 87 (NGC 4486)	30,000	13	4×10^{12}	−22	E1	+1220

Source: Data from Allen (1973).
[a]S = spiral (subtypes 0, a, b, c); Irr = irregular; SB = barred spiral (subtypes 0, a, b, c); E = elliptical (subtypes 0 through 7); p = peculiar.
[b]Centaurus A is a strong radio source, appearing as an elliptical galaxy with a peculiar dense dust lane across its face.
[c]M 101's distance was revised by A. Sandage and G. Tammann (1974) from 3800 to 7200 ± 1000 kpc.

Figure 24-14 The giant elliptical galaxy M 87 (NGC 4486) is about 13,000 kpc away and 30 kpc across (including its halo of globular clusters). It is some 5 or 10 times larger than typical dwarf elliptical galaxies. The faint, fuzzy starlike images around M 87 are thousands of globular clusters swarming in a halo like bees around a hive. M 87 is about 40 times as massive as our galaxy. Two background spiral galaxies appear in the image. Note the reddish Population II stars. (Copyright Anglo-Australian Telescope Board.)

Figure 24-15 Barred spiral galaxy NGC 4650, in the constellation Centaurus, displays a broad central bar instead of the normal spheroidal central bulge. The photo shows yellowish-red colors in the central bar due to Population II stars, and bluish colors in the spiral arms due to hot, young Population I stars forming there. (National Optical Astronomy Observatories.)

trated by M 83, a spiral galaxy whose central region is a bright, barlike object instead of an ellipsoidal bright area; such galaxies are called **barred spiral galaxies.** An example is seen in Figure 24-15.

How can we make sense of these different shapes of galaxies and their different types of stellar populations? The standard scientific approach is to try to classify the different types of objects and then arrange them in a system that shows smooth transitions from one type to another. This system can then be studied for possible causes of the transitions, such as evolution, mass differences, or rotational differences.

The best-known classification scheme was developed by Edwin P. Hubble in the 1920s and 1930s and extended by his colleagues after his death. The system is based purely on the apparent shapes of the galaxies, not on measured properties such as spectra or mass. The scheme, shown in Figure 24-16, distinguishes three main types of galaxies: *ellipticals,* both dwarf and giant, (designated E), *spirals* (S), and *irregulars* (Irr). Each type is subdivided according to form. For example, as shown in Figure 24-16, ellipticals are subdivided according to apparent flattening, which may be due either to our angle of view or to true flattening of the system.

The general appearance of spirals is strongly dependent on the angle of their tilt toward us, because the disk can appear either face on or edge on. This is shown by the pair of galaxies in Figure 24-17. Edge-on views of spiral galaxies show that many of them are very thin. The subclassification of spirals, however, is based on the relative sizes of the disk and central bulge and on the tightness of winding of the arms. Forming a transition between ellipticals and spirals is a special class, S0, which has a spiral's disk shape but lacks clear spiral arms. Sa spirals have the tightest arms and Sc spirals the loosest. Spirals are subdivided into two parallel sequences—barred and normal spirals—giving Figure 24-16 its common designation as the **"tuning fork" diagram.** Thus a tight spiral with a central bar is designated SBa, instead of Sa.

Irregular galaxies, as their name implies, have a ragged shape and may or may not display prominent nebulae.

This classification system correlates well with certain physical properties of galaxies, as summarized in Table 24-3. For example, from ellipticals through irreg-

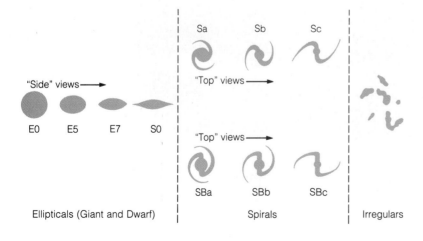

Figure 24-16 The simplified classification scheme for galaxies known as Hubble's "tuning fork" diagram after the originator of the system.

a

b

Figure 24-17 Spiral galaxies (type Sc) face on and edge on. **a** NGC 2997 is seen nearly face on, emphasizing the spiral arm pattern. **b** NGC 253 is seen nearly edge on, emphasizing the thick lanes of dust in the spiral arms that cross in front of the nucleus. (Both photos copyright Anglo-Australian Telescope Board.)

TABLE 24·3

General Characteristics of Galaxies

Characteristic	Ellipticals	S0	Sa (Spirals)	Sb	Sc	Irregulars
Dust	None	Little	Some			Some
Percent neutral hydrogen (mass HI/total mass)	0	0	1	3	9	20
Most prominent populations	II	II	II (halo, center); I (arms)			I
Typical rotation periods (million years)	?	59	63	140	200	300?
Dominant color	Red	Red	Red (halo, center); blue (arms)			Blue
Spectrum of central regions	K	K	K–G	G	F	F–A
Luminosity $(L/L_\odot)$	Giants $\leqslant 10^{11}$ Dwarfs $\geqslant 10^5$		10^8–10^{10}			10^7–10^9
Diameter (kpc)	Giants $\leqslant 200$ Dwarfs $\geqslant 1$		5–50			1–10
Mass $(M_\odot)$	Giants 10^9–10^{12} Dwarfs 10^6–10^9	?	5×10^{11}?	3×10^{11}	10^{11}	10^9

Source: Data from Allen (1973) and de Vaucouleurs (1974).

ulars there is a progression from older to younger populations of stars and from less to more dust and gas. The implication is that star formation is still occurring in spiral arms and some irregulars, but not in ellipticals.

Many people tend to think of all galaxies as beautiful spirals. Yet, just as giant and supergiant stars are overrepresented if we scan the prominent stars in the sky, spirals are also overrepresented because they are bright and spectacular. For example, an early survey of prominent galaxies by Hubble netted 80% spirals, and a sample of 100 photos in three recent astronomy texts included 67% spirals. But the actual distribution of galaxies is more like the following:

Dwarf ellipticals	50%
Giant ellipticals	5%
Spirals	20%
Irregulars	25%

That is, although perhaps four-fifths of the *big* galaxies are spirals, only about one-fifth of *all* galaxies are spirals. Although spirals are not very common, they do contain much of the most luminous material in the universe.

COLLIDING GALAXIES

As can be calculated from Table 24-1, the Milky Way and Andromeda galaxies are separated by a distance amounting to only 15 or 20 of their own diameters. Other galaxies, drifting through space, are separated by similar relative spacings. If you tried to keep a roomful of frisbees aloft—enough that they were only 15 or 20 diameters apart on the average—many of them would collide. Galaxies collide, too.

This is an extraordinary situation, very different from that of stars. Stars are separated typically by 20 million of their own diameters; therefore, in spite of their random motions, few if any stars in a galaxy have collided during the whole history of the universe. If you made a model of our galaxy the size of North America, stars would be microscopic dust grains a city block apart. You can imagine that dust grains a block apart could drift around a long time without hitting each other. But many galaxies have collided with other galaxies during their lifetimes. Indeed, we can actually see this process occurring among numerous sets of close galaxies, as shown in Figure 24-18.

Each collision lasts a few hundred million years, because galaxies drift with respect to each other at speeds averaging tens of kilometers per second, and at this collision speed it takes that long for two galaxies to pass through each other. This is a relatively brief encounter in the history of a galaxy—two vast ships passing in the night. But it has important consequences.

The idea that galaxies collide was first proposed in 1940 by a Swedish astronomer, Erik Holmberg. In the 1980s astronomers began to realize that collisions between galaxies may be fundamental to the evolution and shapes of many galaxies.

Results of Galaxy Collisions

Think of a galaxy as a giant swarm of gas with stars embedded in it. When the random motions of galaxies cause them to penetrate each other, individual stars are unlikely to collide. Even if you squashed the Milky Way and Andromeda together, the number of stars per cubic parsec would increase only by a factor of 2, and the individual stars would still be many millions of diameters apart—like the dust grains in the example discussed above.

Even though the individual stars don't hit each other, there are four important consequences of collisions between galaxies. The first is that the gas components of the two galaxies *do* interact. The dusty gas in each galaxy is more like a continuous medium, so that when two galaxies collide at many tens of kilometers per second, there is a dramatic interaction of the two gas swarms. You might think of it as if the contents of two buckets of water were sloshed toward each other—the two masses of water would hit and make a splash. In fact, two galactic masses of gas do make a splash. And when the gas atoms collide, the gas and dust get heated. The gas atoms get excited (electrons kicked to higher energy levels) or even ionized (electrons knocked off). Hydrogen is the most abundant gas, of course. Thus colliding galaxies are often *intense sources of radio radiation from excited hydrogen* and infrared radiation from the hot dust. Any galaxies that emit strong radio noise are called **radio galaxies.** Colliding galaxies are one class of radio galaxies. Some astronomers believe that in addition to simply exciting the galactic gas, galaxy collisions also somehow trigger formation of energetic nuclei in galaxies, perhaps extraluminous accretion disks around massive black holes (Gunn, 1979; Simkin and others, 1987). We will return to this possibility in the next chapter.

The second important consequence of colliding galaxies is distortion of shapes and creation of certain rare **peculiar galaxies** that don't fit in the "tuning fork" scheme of classification. A galaxy's shape is determined by its gravitational field acting in conjunction with its spin rate and total mass. If another galaxy with comparable or larger mass approaches, that galaxy's gravitational forces will disturb the equilibrium shape of the first. The closer the two galaxies approach, the greater the distortions. If two comparable-sized galaxies draw close, the shape of each may be wildly distorted from its initial disk shape. Streams of gas may be drawn out of the pair, forming a connecting bridge, as seen in Figure 24-18b. Streamers of gas may be ejected, too,

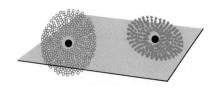
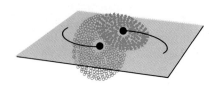

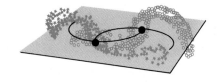
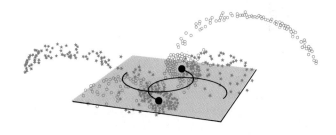
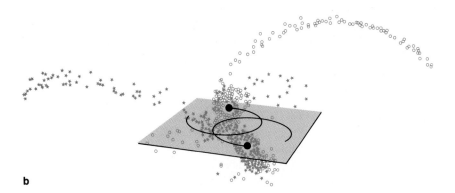

Figure 24-18 a Radio galaxies NGC 4038 and 4039 have prominent curved streamers and may be an interacting close pair about 15,000 kpc away. The picture width is about 9 minutes of arc, or 40 kpc at the galaxies' distance. (Hale Observatories.) **b** This computer analysis of a near collision between two galaxies is designed to simulate the appearance of NGC 4038 and 4039. Circles are material from one galaxy; stars are material from the other. Gravitational forces eject a streamer from each galaxy. Successive images are spaced about 200 million years apart. (Toomre and Toomre, 1973.)

a

b

forming galaxies with long tails. Dynamicists have been able to calculate results of such encounters with amazing success. As shown in Figure 24-18, computer simulations beautifully predict some shapes actually observed among pairs of galaxies.

A third result of galaxy collision is **starbursts**— intense bursts of star formation and supernova explosions (resulting from blowup of the most massive, most short-lived stars created during the star-forming activity). Evidence for this phenomenon comes from radio telescopes and orbiting infrared telescopes. Often 99% of the radiated energy from these galaxies comes in the infrared part of the spectrum from warm dust heated by the activity.

Galactic Cannibalism: Formation of Giant Elliptical Galaxies

The fourth and most important result of galaxy collision is that it may explain the formation of giant elliptical galaxies (Silk, 1987; Schweizer, 1986). When two galaxies collide, depending on the circumstances, one galaxy may merge with the other instead of passing on by. The larger galaxy gobbles up the smaller, so to speak; the phenomenon is called **galactic cannibalism.**

Dynamical studies in the 1980s show how galactic cannibalism can explain giant elliptical galaxies. As shown in Figure 24-19, two disk-shaped rotating protogalaxies may collide. According to recent views, if the conditions of collision are right, the disk shapes may be destroyed, producing a huge ellipsoidal mass of stars. Starbursting may result in much of the gas and dust being consumed to form stars. After this burst of star-forming activity, there will be little remaining gas and dust to form later generations of stars. Star formation stops. An elliptical galaxy forms. An observer who compares this elliptical galaxy with a spiral galaxy, a few billion years later, sees a galaxy with less gas and dust and a population of old stars with few young stars. In other words, the observer sees Population II, the typical content of a giant elliptical.

EVOLUTION IN GALAXIES

Colliding galaxies thus provide an important clue about the evolution of at least one class of galaxies: giant ellipticals. But what about the larger question? How do galaxies in general form and evolve into different types?

Current evidence suggests that most galaxies formed

some 10 to 18 (probably closest to 14) billion years ago during a restricted interval. This agrees with conclusions based on studies of the Milky Way galaxy (preceding chapter). Galaxies formed by the gravitational collapse (described in Chapter 18) of huge clouds of hydrogen and helium called **protogalaxies.** This explains why the oldest star groups in all galaxies, such as the 14-billion-year-old globular clusters, consist of hydrogen-rich Population II stars with very few heavy elements. The collapse of protogalaxy clouds was relatively fast, taking about 10^8 y for a Milky Way–size galaxy (Strom and Strom, 1982).

The various galactic types observed today are believed to have evolved due to differences in initial conditions such as mass, density, and rotation rates as well as due to the cannibalistic collisions that may have yielded the giant ellipticals. Many galaxies during collapse may have resembled an extended elliptical galaxy. According to recent theories, the smallest protogalaxies (around $10^6 M_\odot$) produced large globular clusters or dwarf ellipticals. Faster-than-normal rotation of the protogalaxies favored formation of a bar-shaped region during collapse of the protogalaxy, possibly explaining barred spiral systems. And, as we've noted, giant ellipticals probably formed by collision.

The subsequent evolution of a galaxy depended partly on the number of massive stars that formed from the initial gas of the protogalaxy. This initial gas was apparently all hydrogen and helium (judging from the absence of heavy elements in the oldest stars), hence it produced Population II stars. In the central Population II bulges of spiral galaxies, apparently the gas was used up in the first generation of stars. The evidence for this is seen in Figure 24-20, where a comparison of optical and radio images shows most gas in the arms and little in the center. If the initial generation of stars in the arms contained many massive stars of $10 M_\odot$, $20 M_\odot$, or even more, these would quickly explode as supernovae (see Chapter 19), spewing out gas containing heavy elements manufactured by nuclear reactions in these stars.

The further evolution of the galaxy now depended on the fate of this gas (Ostriker, 1981). The heavy elements would condense into dust grains. In intermediate-mass disk-shaped galaxies, the dust would accumulate in the plane of the disk. The gas and dust would form nebular clouds and produce later generations of stars with heavy elements (Population I stars). Computer studies, as shown in Figure 24-21, have demon-

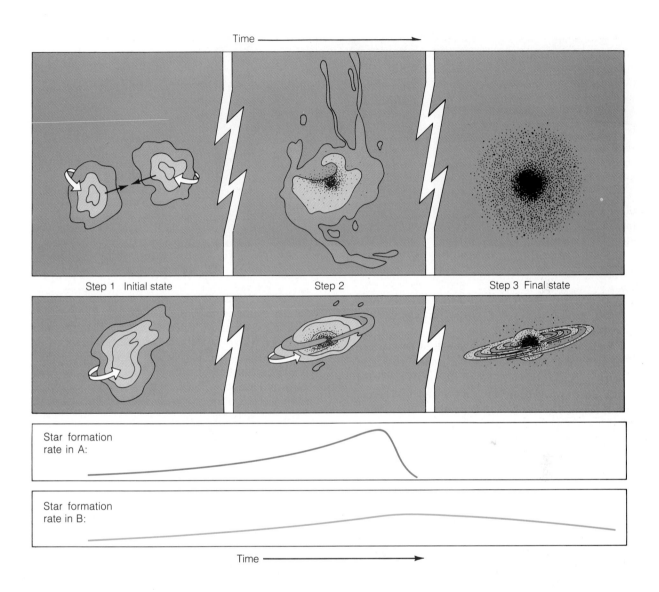

Time ⟶

Step 1   Initial state   Step 2   Step 3   Final state

Star formation
rate in A:

Star formation
rate in B:

Time ⟶

Figure 24-19 Possible formation processes leading to creation of a giant elliptical
galaxy (A) and a spiral (B). In case A, two protogalaxy clouds have collided and
coalesced. This excites and concentrates the gas, leading to a starburst of star formation
(shown in bottom graphs). In case B, a rotating protogalaxy cloud contracts to form a disk
with more leisurely star formation. In the final stages, cloud A has formed a full-fledged
giant elliptical with little remaining star formations but cloud B has formed a spiral. Each
type of galaxy is surrounded by a halo of globular clusters formed from initially outlying
material. (Loosely based on a diagram by Silk, 1987.)

a

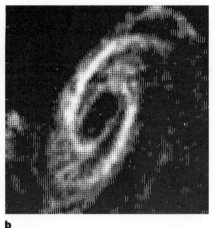

b

Figure 24-20 Optical and radio images of spiral galaxy M 81 (NGC 3031), about 3200 kpc away. Scale is about the same in both images. **a** The optical image shows stars and star clusters. Scale bar applies in unforeshortened direction. **b** The radio image shows radiation from neutral hydrogen gas. Important results are that (1) neutral hydrogen gas coincides with the position of Population I stars in the spiral arms, (2) there is virtually no neutral hydrogen in the center, and (3) the hydrogen spiral pattern is much larger than the visible star pattern. (Leiden Observatory.)

strated how disk-shaped dusty galaxies of this type could produce spiral arms of young, Population I star clusters in only half a billion years or so.

Similarly, if certain galaxies produced no massive first-generation stars, supernovae would not have occurred, and there would be no dust or Population I stars. There would be no dusty disk. We might have a dwarf elliptical galaxy lacking Population I stars. Similarly, if a few giant supernovae blew gas and dust clear out of the system, no disk or spiral arms could form.

Study the galaxy in Figure 24-22. Nuclei of spirals have the pale yellowish-orange color of the old, Population II giants. But the spiral arms are composed of bluish clusters of young, Population I O and B stars, along with red-glowing HII regions scattered like rubies in a jewelled brooch. Thus we can actually *see* evidence of evolution among star populations in galaxies simply by looking at color pictures of galaxies!

evolved and used up their gas and dust, thus producing different stellar populations in galaxies.

The most massive galaxies are giant ellipticals with Population II stars and little gas or dust. Spirals appear to be intermediate-sized galaxies with "skeletal frameworks" resembling elliptical galaxy systems—central, flattened bulges of Population II stars surrounded by a spheroidal halo of globular clusters; however, gas and dust blown out of evolved stars have also formed disks of Population I stars. Other galaxies of intermediate and small size have elliptical and irregular shapes. Peculiar galaxies' shapes may have been affected by collisions.

As we look at the familiar rocks, mountains, buildings, and trees of Earth, it is strange to realize that our silicate-, carbon-, and iron-rich materials may have become concentrated by certain processes of galactic evolution in which debris from the insides of ancient stars aggregated into star-spawning clouds in the dusty disk of a spiral galaxy. As Harlow Shapley commented, "We are brothers of the boulders, cousins of the clouds."

SUMMARY

The Local Group of galaxies, out to about 1000 kpc, contains a variety of types and sizes. Combining observations of these galaxies with data from more distant galaxies, we find that most galaxies can be classified as elliptical, spiral, or irregular. A more elaborate classification scheme, the "tuning fork" diagram, subdivides them, including barred as well as unbarred spirals. Most galaxies lie in groups called clusters.

The observations discussed here combine with the theories of stellar evolution, discussed in Chapters 17 to 19, to explain how some galaxies or parts of galaxies have

CONCEPTS

Magellanic clouds

irregular galaxy

Tarantula Nebula (30 Doradus)

dwarf elliptical galaxy

spiral galaxy

Andromeda galaxy

clusters of galaxies

Local Group

problem of the missing mass

giant elliptical galaxy

barred spiral galaxy

"tuning fork" diagram

radio galaxy

peculiar galaxies

starburst

galactic cannibalism

protogalaxy

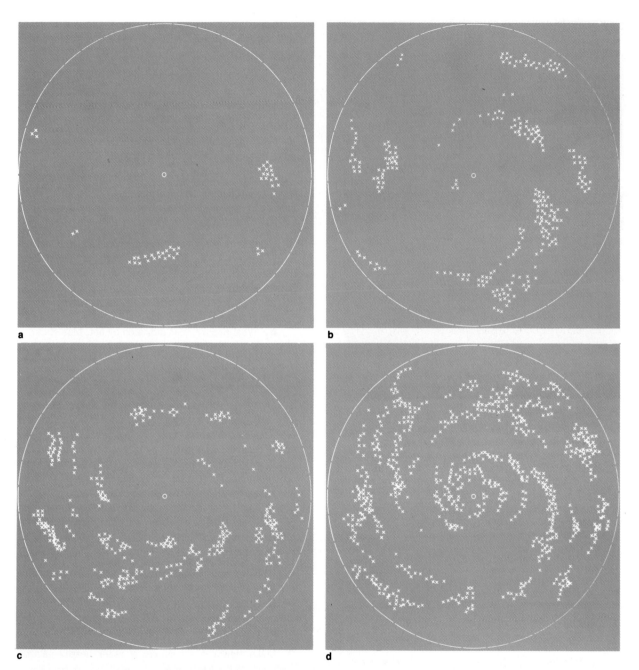

a
b
c
d

Figure 24-21 Theoretical model of spiral galaxy evolution: *x*'s represent regions of recent star formation (open clusters and concentration of bright, bluish stars). This model starts with a centrally concentrated disk of gas. Probability of star formation is related to the local gas density and the number of nearby young stars, which trigger adjacent star formation by blowing out material that compresses adjacent gas. Regions of star formation rapidly shear into spiral arms due to orbital motions of the material around the center at different rates. The model allows gas to be consumed in formation of young stars, but also returned to the interstellar medium as older stars expel gas. This example shows how a rotating, galaxy-sized gas disk develops familiar features of a spiral galaxy in only about half a billion years. (After computer models by W. Freedman and B. Madore, 1983.)

Figure 24-22 The great barred spiral galaxy M 83 (NGC 5236) beautifully displays the consequences of stellar and galactic evolution. Here the yellowish Population II central regions are surrounded by spiral arms dominated by bluish Population I stars, massive and short-lived, forming in clusters. Scattered through the arms are red-glowing HII regions associated with local areas of star formation. M 83 is a type SBc barred spiral about 3200 kpc away from us and about 12 kpc in diameter. (Copyright Anglo-Australian Telescope Board.)

PROBLEMS

1. Which type of galaxy tends to be largest? Brightest? To contain the fewest young stars? (See Tables 24-1 and 24-2.)

2. Of giant ellipticals, ordinary ellipticals, spirals, barred spirals, and irregulars, which type is most common? (See Tables 24-1 and 24-2.)

3. How many years would a radio signal take to reach the Andromeda galaxy?

4. Two very faint galaxies were detected on photos but they were too distant to allow resolution of spiral arms and other identifying features. If spectra showed that one was reddish with a spectrum of K-type stars, whereas the other was bluish with a spectrum of B and A stars, what types would you expect the two galaxies to be?

5. If you were to represent the Milky Way and Andromeda galaxies in a model by two cardboard disks, how many disk diameters apart should they be to represent the true spacing of the two galaxies?

6. Why are small ellipticals not found in catalogs of the more distant galaxies, such as Table 24-2?

7. Irregular galaxies are dominated by stellar associations, open clusters, and gas and dust clouds, all of which indicate stellar youthfulness. Does this prove that the galaxies themselves have only recently formed?

8. Using principles of star formation and stellar evolution from Chapters 17 and 18, explain why the prominent light from star-forming regions in galaxies comes from massive, hot, blue stars.
   a. Why are these stars not seen in regions where star formation has ended?
   b. What population is indicated by hot, blue stars?

PROJECTS

1. Locate the Andromeda galaxy by naked eye. Compare its visual appearance with its appearance in binoculars and telescopes of different sizes. Across what diameter can you detect the galaxy? (1° = 12 kpc at the Andromeda galaxy's distance.) Why do most photographs show a large central region of constant brightness, whereas visual inspection through a telescope reveals a sharp concentration of light in the center? Can dust lanes or spiral arms be observed? (Check especially with low magnification on telescopes with apertures of 0.5 to 1 m [about 20 to 40 in.], if available.)

2. If photographic equipment is available, take a series of exposures of the Andromeda galaxy with different times, such as 1 min, 10 min, and 100 min. Describe some of the differences in appearance. What physical relations are revealed among different parts of the galaxy?

3. Make similar observations of other nearby galactic neighbors of the Milky Way.

CHAPTER 25

The Expanding Universe of Distant Galaxies

The galaxies we have seen up to now vary in size, shape, and distance, but they are basically understandable. They have the same two populations of stars found in our galaxy, and we can at least partially explain their varied forms by imagining different conditions of mass and rotation in various regions of the early universe, as matter broke into protogalactic clouds.

But as we go still farther away from Earth, we discover strange objects. We find that all the clusters of galaxies are rushing away from each other. Remember that we are now looking at objects billions of parsecs away; thus the light from them has been traveling for billions of years. This means that we are seeing them as they were soon after their formative era. Thus, study of the most distant galaxies brings us face to face with questions about the origin and large-scale structure of the universe itself.

THE LARGEST DISTANCE UNIT IN ASTRONOMY

In dealing with stars, we used the parsec as a distance unit. In dealing with the Milky Way and nearby galaxies, we used the kiloparsec, equal to 1000 pc. The distances to remote galaxies are so great that we must use the **megaparsec,** which is a million parsecs, or 1000 kpc. A megaparsec (Mpc) is about 3×10^{19} km, or 30,000,000,000,000,000,000 km!

CLUSTERS AND SUPERCLUSTERS OF GALAXIES

Among distant galaxies, as among our own Local Group, clustering is the rule. As shown in Figure 25-1, galaxies often appear in groups ranging from pairs to collections

of thousands, with ellipticals, spirals, and irregulars mixed together in a volume about 5 Mpc across. These groups are the **clusters of galaxies** mentioned in the last chapter. Mapping of galaxies, both on the plane of the sky and in terms of distance in three-dimensional space, shows that even the clusters of galaxies are clumped together. These clumps of clusters are called **super-clusters of galaxies.** Although clusters and superclusters are among the most distant known entities in the universe, they may cover an angle of many degrees in our sky! For example, the Virgo cluster (Figure 25-1), located about 180 Mpc away, is the center of a supercluster as much as 50° (170 Mpc) across (Chincarini and Rood, 1980). As we will see in the next chapter, mapping of superclusters indicates that they themselves may be part of an even larger-scale clumpy or filamentary distribution.

A cluster of galaxies is typically held together by the gravitational attraction of its member galaxies and is thus likely to be stable throughout the history of the universe. The stability of superclusters is less certain. Some of them may be expanding, and the member clusters may be insufficient in number or closeness to hold the superclusters together indefinitely.

DISCOVERING THE RED SHIFT

As you will remember from Chapter 16, spectral emission or absorption lines emitted by objects moving away from us are shifted toward a redder color, and lines emitted by objects approaching us are shifted toward the blue. Cataloging of the Doppler shifts of galaxies led to an unexpected and amazing finding. By 1914, when Doppler shifts had been published for 13 of the brightest galaxies, most were red shifts, rather than a random mixture of red and blue shifts.

Around 1920, Edwin Hubble began further Doppler shift observations with the 2.5-m reflector at Mt. Wilson, and he soon made new findings. By 1925, about 45 galactic Doppler shifts had been published, and Hubble (1936) noted that, again, red shifts completely dominated the list. This phenomenon, called the **red shift of galaxies,** indicates that most galaxies are moving away from us. Red shifts corresponding to recession speeds as high as 1800 km/s appeared on the 1914 list. As Table 25-1 shows, dramatically higher speeds have since been measured.

In 1928, American physicist H. P. Robertson discovered another curious fact: The more distant the galaxy, the greater the red shift. This result is demonstrated in Figure 25-2. And the relation is linear; the red shift is directly proportional to the distance. This finding applies at distances out to at least 1000 Mpc.

INTERPRETING THE RED SHIFT: AN EXPANDING UNIVERSE?

According to everything we have learned so far, a red Doppler shift means that the source is moving away from us. Therefore, astronomers have concluded that all distant galaxies are moving away from us, and the farther away they are, the faster they are receding. In 1933, English astronomer Arthur Eddington titled his book on this subject *The Expanding Universe,* the

TABLE 25·1					
Characteristics of Selected Clusters of Galaxies					
Cluster Name	Estimated Distance (Mpc)	Diameter (kpc)	No. of Galaxies Counted	No. of Galaxies/Mpc^3	Recession Velocity (km/s)
Local Group	0.6	1000	~ 27	50	0
Virgo	19	4000	2500	500	+ 1180
Pegasus I	60	1000	100	1100	+ 3700
Cancer	70	4000	150	500	+ 4800
Perseus	80	7000	500	300	+ 5400
Coma	100	8000	800	40	+ 7000
Hercules	130	300	300	20,000	+ 10,300
Ursa Major I	190	3000	300	200	+ 15,400
Leo	240	3000	300	200	+ 19,500
Gemini	290	3000	200	100	+ 23,300
Boötes	450	3000	150	100	+ 39,400
Ursa Major II	460	3000	200	400	+ 41,000
Hydra	700	1000	10	40	+ 60,600
3C 123 and cluster	1500	?	?	?	+ 135,000?

name that has been given to this concept ever since. However, the phenomenon might better be called the **mutual recession of galaxies.**

Hubble's Relation and Recession Velocities

Hubble and Robertson emphasized the regular relationship (now known as **Hubble's relation**) between the distance of galaxies and their recession velocities: As distance increases, recession speed increases. Hubble measured the constant proportion between velocity and distance, known as **Hubble's constant** H. Measurements suggest H is about 77 ± 14 km/s per megaparsec (Hodge, 1984; Bartel and others, 1985). This means that galaxies 1 Mpc away recede from us at 77 km/s on the average; galaxies twice as far away recede twice as fast; galaxies at 10 Mpc recede at about 770 km/s, etc.

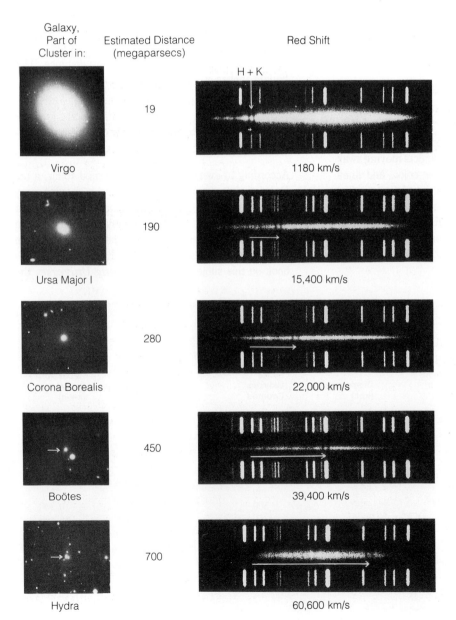

Galaxy, Part of Cluster in:	Estimated Distance (megaparsecs)	Red Shift
Virgo	19	1180 km/s
Ursa Major I	190	15,400 km/s
Corona Borealis	280	22,000 km/s
Boötes	450	39,400 km/s
Hydra	700	60,600 km/s

Figure 25-2 Photographic proof of red shifts of remote galaxies. Left column shows galaxies (note decreasing angular size, interpreted as caused by increasing distance). Right column shows spectra: White lines at top and bottom of each spectrum are emission lines produced in the instrument for comparison, being similar in each spectrum. A pair of dark absorption lines (the *H* and *K* lines of gaseous calcium) can be detected in each galaxy's spectrum, above the head of the white arrow. This pair is farther right (red) in each succeeding galaxy. The center column shows the inferred distance. (Hale Observatories.)

Conversely, now that this relation has been measured, we can use it to estimate distances. A galaxy with a red shift corresponding to recession at 7700 km/s is estimated to be about 100 Mpc away.

Hubble's constant H is one of the most important parameters in astronomy—not only because it helps us estimate galaxy distances, but also because it helps us measure the age of the universe (as will be clearer in Chapter 27). Modern astronomers are expending much effort with very large telescopes to observe faint galaxies and pin down a more precise value of H, about which there is still some controversy.

Are We at the Center of the Universe?

If all distant galaxies are moving away from us, does this mean that we are located at the *center* of a vast explosion—that is, at the *center* of the universe? That concept would mean that we have come full circle since the Copernican revolution removed us from a special position at the center of our solar system. *If* we are at the center of the universe, we would occupy a special position. The answer is that *we are not necessarily at the center.* Observers in any other galaxy would see galaxies moving away from them, too, because *all galaxies are moving away from each other.*

In other words, we may very well be riding out the aftermath of a vast explosion, but the view from one flying spark is like the view from any other. There is no outside reference point to define who is closest to the center, unless we found we were close to the edge of the swarm of sparks. But no astronomer has yet found an edge to the swarm of galaxies. They seem to go on and on.

Red Shift Theory I: Cosmological Red Shifts

Everything we've said so far assumes that the observed red shifts should truly be interpreted as **Doppler shifts.** The Doppler effect has been confirmed in Earth laboratories, with planets, and with stars. But the idea that it explains *all* or nearly all red shifts of galaxies is only an assumption, called the **theory of cosmological red shifts.** Most astronomers believe this assumption, because red shifts correlate so well with other evidence of distances, such as diminishing angular size (Figure

25-2). This theory, incidentally, underlies the last column of Table 25-1. But there are other theories.

Red Shift Theory II: Noncosmological Red Shifts

Physical effects besides motion away from us could also cause red shifts of some galaxies. Then the shifts would be ambiguous: Part might be caused by recession and part by something else. In fact, we know at least one other possible cause of red shifts. The theory of relativity shows that light emitted from regions of very high gravity, such as the environs of a black hole, will be red-shifted. Thus some of the shifts we observe may be at least partly **gravitational red shifts.**

The minority of astronomers who argue that red shifts of some galaxies are noncosmological point to certain examples of strongly red-shifted galaxies that seem to be closer than some distant, less red-shifted galaxies. This theory—that some red shifts of galaxies are not Doppler shifts—is called the **theory of noncosmological red shifts.**

For the moment, we will side with the majority who believe the evidence for theory I, that most red shifts correlate with distance.

INTENSELY RADIATING GALAXIES

As we move further and further away in intergalactic space, and to greater red shifts, we encounter seemingly endless numbers of ordinary galaxies. At very great distances, they are almost too faint to detect. Nonetheless, there are a few superluminous, intensely radiating galaxies that can be spotted at very great distances. This is reminiscent of the situation among stars (page 265). Most stars are similar to, or fainter than, the Sun; we can see them only in our part of the galaxy. But at larger distances across the galaxy, we spot the "whales among the fishes"—superluminous stars such as red giants. Here the whales are the brightest galaxies.

These intensely radiating galaxies give off stupendous amounts of energy, so they can be detected at extremely large distances. The extra energy, compared to normal galaxies, may appear as a combination of excessive visible light, infrared emission, radio emission, X-ray emission, and other forms of radiation. The big question in research on intensely radiating galaxies is: What processes cause such tremendous amounts of radiation from certain galaxies?

Colliding Galaxies

One process is collision. Thus, one subclass of intensely radiating galaxy is colliding galaxies, described in the previous chapter. Their energy comes from gas that is excited as the galaxies crash into each other. Many are radio galaxies, as mentioned in the previous chapter. In this chapter the focus will be on the other class of intensely radiating galaxies, whose radiation comes from mysterious galactic nuclei with extraordinary luminosities.

Galaxies with Active Nuclei

In this second subclass of intensely radiating galaxies, called **galaxies with active nuclei,** or sometimes just "active galaxies," most of the excess luminosity (beyond that of normal galaxies) comes from the nucleus. The nucleus may be visibly much brighter than the nucleus of an ordinary galaxy. At least 5% of all galaxies have active nuclei, according to a 1974 study by French-American astronomer Marie-Helene Ulrich.

In these galaxies, the intense radiation is associated with some extraordinarily strange process occurring in the nuclei. This process is not well understood but is the subject of challenging research. It seems to work in varying degrees, producing a range of galaxy types. Some galaxies merely have an unusually bright nucleus. Some are strong radio galaxies, even though they are not colliding; their radio radiation is of a type called **synchrotron radiation** produced by electrons accelerated to nearly the speed of light by strong magnetic fields. Still other active galaxies have nuclei that produce enormous explosions and blow out clouds of gas or strange jets of material that speed out of the galaxy in opposite directions. Many are strange in appearance (for example, Figure 25-3). The most extreme active galaxies are the weird objects known as quasars.

Before describing each type, we note that collisions between galaxies may play a role in "turning on" an active nucleus, perhaps triggering formation of giant black holes with their surrounding, high-temperature accretion disks (Gunn, 1979). This theory has been difficult to prove, but it continues to gain evidence. A 1987 study revealed an example of a galaxy with a modestly bright nucleus that has been distorted in shape by a close encounter with another galaxy (Simkin and others, 1987). Another 1987 theoretical study concluded that a crowded cluster of stars at the center of a galaxy could collapse to produce a gigantic black hole with 10,000 to many millions of solar masses—enough to power a typical active galaxy (Kochanek and others, 1987). Perhaps a collision between two galaxies could trigger such a collapse. But some galaxies may evolve to this state on their own when supernovae or other events in the nucleus trigger the formation of a giant black hole or periods of extreme luminosity. In any case, the various active-nucleus galaxies that we will now explore are among the most interesting and mysterious objects in the universe.

Seyfert Galaxies—Active Nuclei of the Weakest Kind

In 1943, astronomer C. K. Seyfert studied some unusual galaxies with small, intensely bright nuclei, bright spectral emission lines, and brightness fluctuations. Seyfert concluded that the nuclei of these galaxies had hot gas in violent motion, often with velocities of several thousand kilometers per second. About 2% of all galaxies have these properties. They came to be called **Seyfert galaxies.** In the 1960s Seyferts were found to have radio and infrared emission similar to that of radio galaxies. They seem to be a "missing link" between ordinary and radio galaxies.

The unusually large and varying amounts of energy of Seyfert galaxies imply very energetic, explosive bursts in their nuclei, usually within a core region about 100 pc across. Of course, even ordinary galaxies have strikingly energetic events occurring in their centers—synchrotron radiation and expanding hydrogen clouds occur in our own galaxy as well as others. But Seyferts are even more energetic, especially in emissions at radio and infrared wavelengths.

Explosions and Jets in Galactic Nuclei

Certain galaxies that emit still more radiation than Seyfert galaxies have ragged explosive filaments or long, straight jets of material speeding out of their centers. Figure 25-4 shows an explosive galaxy only 3 Mpc away. This is radio galaxy M 82 (NGC 3034), which looks like a curious edge-on disk about 7 kpc wide and partly obscured by dust clouds. Splattering out from the radio-

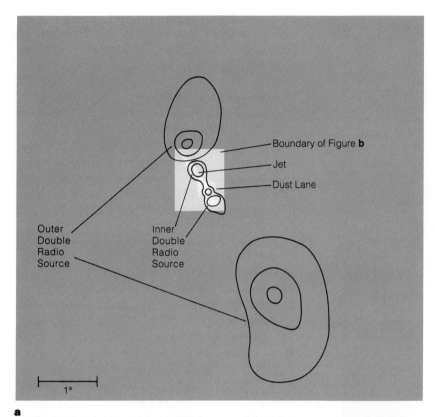

Outer Double Radio Source

Inner Double Radio Source

- Boundary of Figure **b**
- Jet
- Dust Lane

1°

a

Figure 25-3 The unusual galaxy NGC 5128, coincident with intense radio source Centaurus A, about 4.4 Mpc away. The galaxy appears to be a giant elliptical with an unusual dust lane across its center. **a** Sketch of radio and optical features. Contoured "lobes" are clouds of radio-emitting gas on either side of the galaxy, which also emits. Inner rectangle represents area of *b*. **b** Long exposure shows the overexposed galaxy and a faintly glowing jet of gas extending along the direction of the radio lobes. (Cerro Tololo Interamerican Observatory, courtesy NOAO.) **c** Moderate exposure shows dust lane and features of the elliptical galaxy with its creamy Population II stars. **d** Shorter exposure of core region shows details of thick, dark dust. (Images *c* and *d*, copyright Anglo-Australian Telescope Board.)

b

c

d

emitting nucleus are filaments of gas glowing red with the light of Hα emission. These filaments testify to some extraordinary event! They are shooting out of the nucleus at 1000 km/s and contain an estimated 5 million solar masses of gas—5 million solar systems worth of exploded material!

Somewhat similar are the explosive hydrogen filaments of NGC 1275, which is also an X-ray source, Seyfert galaxy, and radio source Perseus A. Its fila-

ments are expanding at 2400 km/s! It is shown in Figure 25-5.

In both of these galaxies, the red-glowing hydrogen filaments resemble the debris from the Crab supernova explosion. (Compare with the Crab's appearance in Figure 19-8.) This suggests that explosions in galactic nuclei may resemble superviolent supernova explosions.

Figure 25-6 shows the related phenomenon of **galactic jets**—narrow, straight beams of material shoot-

a

b

Figure 25-4 Peculiar galaxy NGC 3034 (M 82) coincides with intense radio source 3C 231. This galaxy, 3 Mpc away, appears to have undergone explosive activity. **a** The galaxy displays thick, chaotic dust clouds silhouetted against a probable edge-on disk. Red-glowing excited hydrogen forms a projecting cloud near the center. (National Optical Astronomy Observatories.) **b** Semi-false color image is made by printing images made from blue light, white light, and the red Hα emission wavelength in blue, green, and red, respectively. This process reveals a bluish diffuse disk and a dramatic splatter of red-glowing, excited hydrogen gas, apparently blown out of the central regions. Hydrogen expansion speeds of 1000 km/s confirm the explosive violence and suggest major explosions as recently as 1.5 million years ago. (Original image, Hale Observatories; color image processing by Jean Lorre, Jet Propulsion Laboratory.)

ing out of galaxies. Here we have a 1300-pc-long jet, glowing by synchrotron radiation, shooting out of the nucleus of the giant elliptical galaxy M 87. (Compare with the full view of this galaxy's outer regions in Figure 24-14.)

There are many other examples. Of special interest are galaxies with bipolar jets extending in both directions. This feature is seen, for instance, in spiral galaxy NGC 1097, where jets extend in opposite directions from the galaxy's center (Figure 25-7). These bipolar jets are reminiscent of those we encountered in star-forming disks (page 283) and in binary star SS-433 (page 337). Assuming that the disk is about 25 kpc across, the jets extend out to 60 kpc but are only 2 kpc wide.

Recently radio astronomers have been able to use groups of radio telescopes linked in tandem to produce extremely detailed images of the radio radiation from such galaxies. Just as a photograph taken with a red filter records material emitting light at red wavelengths, these radio images record material emitting radio waves. These images show that in addition to having diffuse blobs of hydrogen blown outward from active nuclei (as was shown in Figure 25-3), many radio galaxies have remarkable radio-emitting jets streaking out in narrow beams reaching from within a few parsecs of the nucleus out to thousands of parsecs. Three examples are shown in Figure 25-8. In some of these examples, the jets are bent back—perhaps by motion of the galaxy through thin intergalactic gas, like the plume of a steam locomotive moving forward through the air.

What is the physical nature of these jets? How are they produced? We don't have very good answers. They are apparently narrow columns of plasma glowing by synchrotron radiation. They are very hot and have somehow been accelerated to high speeds. Some observations suggest that the material in some jets is moving outward at nearly the speed of light! These jet-emitting and exploding galaxies are among the most energetic known objects in the universe! Blobs of radio-emitting gas around certain radio galaxies have been estimated to contain 10^{54} joules of energy—as much as 10 billion supernova explosions—a "majestic phenomenon" indeed, in the words of astrophysicist Frank Shu (1982).

The Mystery of the Galactic Nucleus

An emerging theory that explains intense radiation, explosions, and jets in galactic nuclei involves gradual accumulation of dust and gas in the center, after it spi-

Figure 25-5 Evidence for explosive activity in the nucleus of Seyfert and radio galaxy NGC 1275. In the light of Hα emission, the central region is resolved into radiating streamers. Compare with similar patterns of eruptive hydrogen streamers in explosive objects of various sizes, such as Crab supernova cloud (Figure 19-8), Tarantula Nebula (Figure 24-4), and explosive radio galaxy M 82 (Figure 25-4b). Several apparent elliptical galaxies (fuzzy ovals) are nearby. (Sharp circular images are foreground stars in our galaxy.) The galaxy lies about 50 Mpc away; filaments extend about 14 kpc from the galaxy. (Kitt Peak National Observatory, courtesy Roger Lynds.)

a (1956)   **b (1978)**

Figure 25-6 The luminous jet extending from the nucleus (left blob) of the giant elliptical galaxy M 87 (NGC 4486; compare with Figure 24-14. M 87 coincides with the strong radio source Virgo A. The jet and radio emission suggest violent activity in the galaxy's nucleus. The jet is bluish in color and emits 10 times as much X-ray emission as it does visible and radio emission. Comparison of these views suggests possible changes in the relative brightness of knots along the jet between **a** 1956 and **b** 1978. The jet is about 20 seconds of arc long, corresponding to a length of about 1300 pc, with individual knots as small as 100 pc across. (Mt. Wilson and Palomar Observatories photo, computer processed by Jean Lorre, Jet Propulsion Laboratory.)

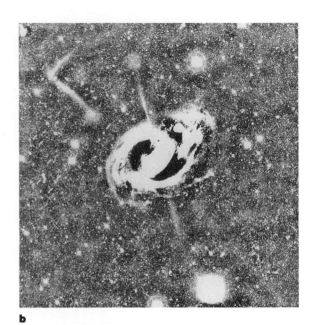

a

b

Figure 25-7 Two views of galaxy NGC 1097 and its system of jets. **a** A long-exposure view of this system, which appears to be a barred spiral galaxy roughly 10 Mpc away. Its angular size is about 9 minutes of arc. (Cerro Tololo Interamerican Observatory.) **b** A computer-enhanced version of *a* increases contrast and brings out faint jets and wisps extending in opposite directions out of the galaxy's disk. Origin of jets is uncertain but may involve extremely energetic events in the nucleus. (Courtesy Jean Lorre, Jet Propulsion Laboratory.)

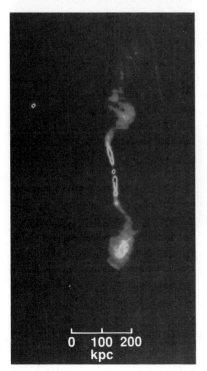

b

Figure 25-8 High-resolution radio images of jets being emitted from radio galaxies (bright central objects). The jets are hot, fast-moving gas giving off synchrotron radiation from electrons moving in magnetic fields. They expand into larger blobs of hydrogen at a distance from each galaxy. **a** Linear jets moving out from elliptical galaxy 3C 449, at an estimated distance of 66 Mpc. (Computer processing yielded dark cores in images of the galaxy and jets: they are actually bright.) **b** Linear jets and blobs streaming out of giant elliptical galaxy 3C 405, also known as Cygnus A. (National Radio Astronomy Observatory, operated by Associated Universities, Inc. under contract with the National Science Foundation; image *b,* courtesy R. Perley, J. Dreher, J. Cohan.)

0 100 200
kpc

a

rals in from outlying parts of the galaxy. A clue comes from our discussion of the jets emitted from double star SS-433, discussed on page 337 and illustrated in Figure 21-7. Recall that matter was believed to spiral from one star into a disk around the other star, finally to crash into that star, somehow ejecting jets "upward" and "downward" out of the disk.

A similar idea has been applied to galaxies. Dust and gas are slowed by drag during their orbital motions around the inner part of the galaxy and spiral in toward the center. Think of a galaxy's inner regions as resembling a giant phonograph record. Dust and gas spiral inward along the grooves and therefore accumulate in the region of the central label. The dense accumulation of gas in this region would promote the formation of extremely massive stars (100, 1000, or perhaps even 10,000 solar masses) that violently explode after brief lifetimes. This could explain some of the quieter active nuclei (such as Seyfert galaxies), but would not explain the most energetic ones, which have energies equivalent to a whole galaxy of supernovae!

But consider this model further. Massive black holes would be left as the remnants of supermassive supernovae. Some theorists therefore believe that in the very core of the galaxy's nucleus, corresponding to the central hole in our record, there may be clusters of black holes, or perhaps a single giant black hole, into which surrounding stars in the cluster would continue to crash. Detailed calculations (Kochanek and colleagues, 1987) show final black holes with up to 10 million solar masses, and additional calculations describe black holes forming with up to a billion solar masses! The journal *Science* aptly called such an object "the monster in the middle." Such monsters would be big enough to explain active galactic nuclei and quasars. Such a nucleus would be a region of extremely strong gravity and probably strong magnetic fields. Fresh gas would constantly spiral into this region from the outlying regions. It would be accelerated to speeds possibly approaching that of light because of the strong gravity.

Astronomers pursuing the monster in the middle suggest that the nuclei of various galaxies are objects with a few million to billion solar masses and note that their light does not match the light expected from this many stars. Hence these astronomers interpret these nuclei as black holes (Anderson, 1987). In particular, this work suggests that the great Andromeda galaxy has a black hole nucleus with 30 to 70 million $M_\odot$, whereas

its little elliptical galaxy satellite, M 32, has a black hole nucleus with 8 million $M_\odot$ (Dressler and Richstone, 1988).

As gas spirals into the accretion disks around such supermassive nuclei, it is heated and ionized and may also be whipped around by magnetic fields. Depending on physical conditions in the center, such as the mass of central black holes, the speed of sound in the central gas region, the magnetic field strength, and so on, the infalling gas may avoid crashing into the central massive black holes or stars. Instead it might be blown "upward" and "downward," out of the disk's plane through the twin exhausts of a magnetic nozzle, forming either a continuous stream of plasma blobs or a continuous narrow jet (Figure 25-9). If we lived in a galaxy with a luminous jet, the night sky might present an unfamiliar spectacle (Figure 25-10).

Before we hypothesize too much about energetic events in galactic centers, we should pause to remember that galactic nuclei are hard to study. Even the puzzling nucleus of our own galaxy—virtually on our doorstep only 0.009 Mpc away—is hidden from us by dust. Commenting on the mystery of galactic nuclei, astronomer W. C. Saslaw quotes the poet Robert Frost:

We dance around in a ring and suppose,
But the Secret sits in the middle and knows.

QUASARS: THE MOST ENERGETIC GALACTIC NUCLEI?

Quasars (an acronym for *quasi*-stell*ar* radio source)[1] are yet another strange type of object that may represent a still more energetic type of active galaxy. Quasars were discovered when optical astronomers began to photograph visible objects that could be identified with the sources of certain radio signals. By 1960, although many radio sources had been photographed and identified as galaxies, a few seemed to coincide only with faint, starlike objects. Were they really stars? If so, they were unusual, sometimes with unrecognizable emission lines in their spectra and sometimes accompanied by faint wisps of nebulosity, as shown in Figure 25-11. Hence they were designated *quasi-stellar.*

A review of old photographs showed that quasi-stellar object 3C 273 (number 273 in the Third Cambridge

---

[1]Many astronomers prefer QSS for *quasi-stellar* source.

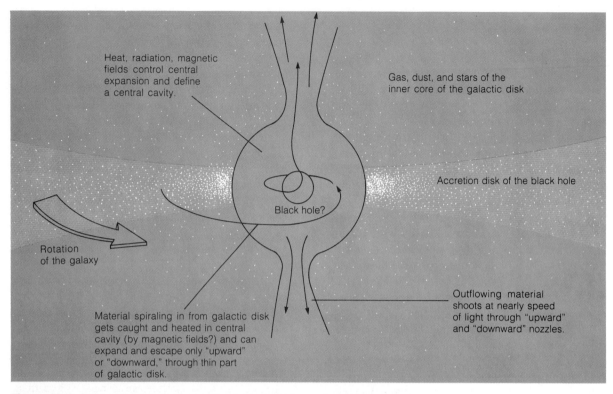

Figure 25-9 Schematic cross section well inside the central parsec of a galaxy with jets. According to some theories, a cavity and "nozzles" may form around a central black hole. Inward-spiraling material from the accretion disk is heated and shot out through the nozzles.

Figure 25-10 If we lived in the inner spiral arms of a galaxy with a jet, the "milky way" in our sky might present this curious appearance. Looking from a volcanic mountain at dusk, we see the "milky way" band at left, but the dim searchlightlike beam of the jet shoots across the sky at right angles to it. The picture assumes a planet at a somewhat higher distance above the galactic plane than Earth, so that we get a higher-angle view past the spiral arms into the bright region of the nucleus. (Painting by author.)

Figure 25-11 Examples of quasars, showing quasi-stellar images. These two quasars are of special interest because they show some detail around their quasi-stellar image. Quasar 3C 273 has a radial jet (compare with jetting galaxy in Figure 25-6), and 4C 37.43 has faint nebulosity adjacent to it (arrow). Distances may be around 1000 Mpc. (*a*: Hale Observatories; *b*: Mauna Kea Observatory, University of Hawaii, courtesy Alan Stockton.)

3C 273

a

4C 37.43

b

Catalog of radio sources) varied irregularly in brightness. In 1963 Palomar observer Martin Schmidt discovered that its red shift was unusually large for either stars or galaxies known at that time—0.16 of the normal line wavelength—implying that 3C 273 is moving away from us at roughly 16% the speed of light! This was the first proof of *large red shifts for quasars.*

Doppler shifts of many fainter quasars are even larger. The most distant identifiable *galaxies* known by 1989 have red shifts of 3.8 times the normal wavelength, but a few faint quasars have enormous red shifts exceeding 4.0, which means that the *shift* in wavelength exceeds four times the original wavelength, and that the quasar is moving away from us at more than 92% the speed of light!

In 1965 other objects were found that appeared to be faint bluish stars and had spectra similar to those of quasars, but they were not strong radio sources. These were identified as radio-quiet quasars, sometimes called **QSO**s, or *quasi-stellar objects.* Over 1500 quasars and QSOs are now cataloged. We will group both types under the term *quasars.*

What are quasars? The interpretation is based in part on spectra. The spectra of some quasars have absorption lines as well as emission lines. Star-formed elements such as carbon and oxygen are present, suggesting that quasars contain heavy elements produced inside evolved stars.

Quasars radiate at virtually all wavelengths—X ray, gamma ray, ultraviolet, visible, infrared, and radio. Although quasars were discovered in the 1960s through their radio and optical emissions, surveys with orbiting X-ray telescopes have revealed many more; indeed, only the most powerful quasars are strong radio sources.

Observations strongly suggest that quasars are somehow related to nuclei of active galaxies. Their spectral features, variability, and the tiny angular size of the radio-emitting region (often less than 0.001 second of arc) suggest a relationship with Seyferts. A breakthrough came in 1978 when South African and European observers identified a Seyfert galaxy that is also an X-ray source and has an intensely bright nucleus that is a quasar. This quasar galaxy, ESO 113-IG45, is shown in Figure 25-12. It is about 250 Mpc away and is probably a distorted spiral. Other links to active galactic nuclei are known. As seen in Figure 25-11, quasar 3C 273 has a radial jet resembling that in the brilliant radio galaxy M 87. A final breakthrough came in 1983 when Balick and Heckman showed that fuzzy regions *around* some quasars have the spectra, size, and brightness of large groups of stars; they concluded that "it seems fairly certain that quasars are the active nuclei of galaxies."

The main light-emitting parts of quasars are less than a few percent of a parsec across. These small regions are often surrounded by larger, expanding envelopes of diffuse gas and probably have intense magnetic fields

Figure 25-12 At least some quasars are bright nuclei of galaxies. This spiral galaxy, ESO 113-IG45, has a red shift corresponding to a recession at 13,630 km/s and a distance of roughly 150 Mpc. It was identified as a Seyfert galaxy in 1977. Its nucleus was found to have the spectral properties of a quasar in 1978. (European Southern Observatory, courtesy R. M. West.)

and high-energy particles, which account for synchrotron radiation and spectral features. They sound suspiciously like accretion disks around enormous stars, based on our discussion of SS-433 in Chapter 21 and Figure 25-9's description of galactic jets.

To gain a better understanding of quasars we need to know how much energy they radiate. To deduce this, we need to know their distances—which requires us to interpret their immense red shifts. This brings us back to the problem of our two theories of red shifts.

Interpretation I: Quasars as Very Remote Galactic Nuclei

Interpretation I, the more widely accepted interpretation of quasars, is based on the theory of *cosmological red shifts*. We apply the conventional conversion of red shift into distances and find that most quasars, moving away at 80 or 90% the speed of light, must be at enormous distances—thousands of megaparsecs, as shown in Figure 25-13. If they are that far away, they must have incredible luminosities to be visible to us at all. As shown in Table 25-2, radio and optical data yield luminosities of dozens to thousands of times those of normal galaxies.

For example, the nearby Seyfert and quasar ESO 113-IG45 has an estimated visual brightness about 15 times that of our galaxy, without even counting its X-ray and other nonvisual radiation. Another extremely luminous quasar, designated S5 0014 + 81, has an estimated visible light output of 1.2×10^{41} watts and an additional radio output bringing the total to 1.4×10^{41} watts, or 140 trillion trillion trillion kilowatts (Kühr and others, 1983)! This exceeds 10,000 times the luminosity of our galaxy. The red shift of 3.41 suggests recession of more than 90% the speed of light and a distance of more than 3000 Mpc! Other examples of interesting quasars and a related galaxy are shown in Table 25-3.

According to this interpretation, then, quasars have the most energetic of all known galactic nuclei. Galaxies would thus range from ordinary (nonactive) galaxies through the active Seyfert and radio galaxies to quasars.

According to this view, many quasars are so far away that their light takes billions of years to reach us. Thus we would be seeing *the oldest quasars as they were billions of years ago,* perhaps at about the time our own galaxy was forming (see reviews by Osmer, 1982 and Rees, 1990). Their radiation may represent conditions uncommon in galaxies today, but common during the first few billion years of galaxy formation. For example, as galaxies first formed, contraction of massive clouds near their centers may have produced many short-lived stars containing millions of solar masses in the galactic nucleus. These superstars exploded quickly, as the galaxy formed, leaving one or more massive black holes surrounded by an ultrahot accretion disk. Neighboring material falling into the accretion disk would keep it hot, causing it to radiate with the furious energy we see as quasar radiation.

A discovery by Caltech astronomers in 1990 suggested that collisions between forming galaxies might aid this process (Graham and others, 1990). The astronomers studied an obscure galaxy called Arp 220, which looks unimpressive in visible light, but radiates quasar-like energies, 1000 billion solar luminosities, in the infrared. Arp 220 shows telltale signs of a recent collision, including tails of debris (see Figure 24-18 for a depiction of a galaxy-collision process). Obscured by dust at the center of Arp 220 are two energetic nuclei,

TABLE 25·2

Energy Relations Between Quasars and Galaxies

Type of Galaxy	Total Luminosity[a] (W)
Normal galaxy	10^{36}–10^{37}
Strong radio galaxy or Seyfert galaxy	10^{36}–10^{38}
Quasar (if distant) (interpretation I)	10^{39}–10^{41}
Quasar (if near) (interpretation II)	10^{36}?

[a] From radio and optical sources. Note that $1\ W = 1\ J$ of energy radiated in each second, that is, $1\ J/s$.

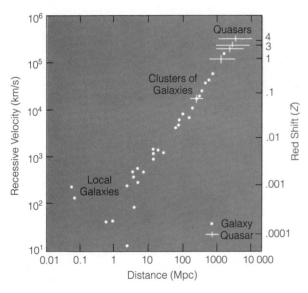

Figure 25-13 Estimated distances of galaxies and clusters of galaxies with various observed red shifts and recession velocities. Horizontal bars for quasars suggest uncertainties in their distances, based on interpretation I.

only 330 pc apart, which may be black holes. If the dust cleared, this system would probably appear as a quasar. The merger of two galaxies may have produced one disturbed galaxy with two nuclei, which will ultimately merge by tidal forces in about 20 million years. Then, gas will cease to spiral rapidly into the accretion disk, causing the quasar activity to shut off. The implication is that galaxy collisions and formation of temporary binary nuclei might be an important step in quasar formation. Such galaxy collisions were probably more common when the first galaxies were forming.

Most quasars are very red-shifted, and are believed to be very far away. Thus, they bring us light from the distant past. As astronomer Arthur Eddington said:

Cosmic radiation is a museum—a collection of relics of remote antiquity. These relics are stamped with an inscription indicating the [nature of the cosmos] in its earliest ages. Whoever ultimately identifies the sub-atomic process originating the rays will be able to read the inscription.

The bottom line of interpretation I is that quasars are a fairly common initial form of galaxies, during the first 20% of their existence, prior to 10 billion years ago. At a later state of evolution, quasars may turn into active galaxies or die out altogether, forming normal galaxies. Many neighboring galaxies' nuclei may contain "burnt-out quasars" in the form of black holes (Rees, 1990).

Gravitational Lensing: Evidence for Interpretation I

In 1936, Albert Einstein predicted that large enough masses of material could bend passing light rays enough for us to detect, so that distant astronomical objects might act as lenses to distort the light of even more remote objects. This type of distortion is sketched in Figure 25-14, and is called **gravitational lensing**. If a massive galaxy (perhaps too distant and faint to be seen at all by us) lies on the line of sight between us and a distant quasar, the quasar's light rays will pass very close to the galaxy, and will be bent to create distorted and displaced "images" of the quasar. As shown in Figures 25-15 and 25-16, this effect has actually been seen many times. It indicates that the quasars in question are very remote, far beyond the lensing galaxies, and this supports the general idea of interpretation I.

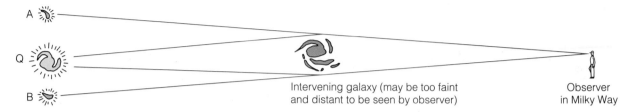

Figure 25-14 The gravitational lensing effect. If a quasar (Q) is very distant, there is some chance that a galaxy will lie along the line of sight from it to an observer. The gravity of the intervening galaxy deflects Q's light rays. The observer thus sees parts of Q's light coming from directions A and B, where the observer may see distorted images of Q. The quasar itself may be visible or it may be hidden by the galaxy.

Figure 25-15 One of the first gravitational lens images to be detected was the arc of light in the middle of this image of the remote cluster of galaxies Abell 370. As sketched in Figure 25-14, it is a distorted image of a distant quasar far beyond Abell 370. (National Optical Astronomy Observatories.)

Figure 25-16 Hubble Space Telescope photo shows the most spectacular case of gravitational lensing: G2237 + 0305, sometimes called Einstein's Cross. As sketched in Figure 25-14, multiple images of the distant quasar can be formed. Here, the nucleus of a galaxy 122 Mpc away (central blob) bends light rays and forms four distorted images of a quasar 2400 Mpc away. (NASA photo by European Space Agency Faint Object Camera on Hubble Space Telescope.)

Interpretation II:
Are a Few Quasars "Nearby"?

A minority of astronomers believe that the very high red shifts of quasars are caused not by recession velocity, but by some other mechanism, such as strong gravitational fields. They argue that most quasars are not very far away. This is a radical difference from interpretation I, because quasars would not be particularly old, and would not shed light on the early universe. Although interpretation II has not been proven, it is supported by two arguments.

First, a surprisingly large number of quasars (several hundred) lie within a few minutes of arc from galaxies with lower red shifts (Burbidge and others, 1990). Second, some of these quasars even appear to be physically connected to the nearby galaxy by bridges of faint gas. These results imply that the quasars are only as far away as these galaxies, and thus not at vast distances, in spite of their large red shifts.

Proponents of interpretation II often argue that most of the light from highly red-shifted galaxies and quasars is gravitationally red-shifted radiation from high-gravity regions, such as dense massive objects. These astron-

TABLE 25·3

Examples of Highly Red-Shifted Galaxies and Quasars

Object	Red Shift (Z)[a]	Probable Recession Velocity (% speed of light)	Comments
Galaxies			
PKS 1614 + 051	3.215	89%	Unusually distant galaxy.
4C 41.17	3.8	92%	Most distant known radio galaxy.
Quasars			
3C 273	0.158	15%	Quasar with greatest apparent brightness. Narrow jet 690 kpc long, <5 kpc wide.
PKS 1145 − 071	1.345	69%	Possible example of a binary pair of quasars orbiting around each other.
3C 275.1	—	—	First quasar found in a cluster of galaxies; supports identification of quasars with distant galaxies.
S5 0014 + 81	3.41	91%	Highest visible luminosity of any object known; absolute magnitude −33. (See text.)
PKS 2000 − 330	3.78	92%	Most distant quasar known by 1982.
0046-293	4.01	92%	First three quasars found with red shift exceeding 4; most
PC 0910 + 5625	4.04	92%	distant quasars known by 1987, when all three were first
Q0000-26	4.11	93%	announced. We see them as they were when the universe was only about 10% of its present age.

Sources: Chambers and others (1990); Djorgovsky and others (1987a, 1987b); Hayes and Sadun (1987); Peterson (1982); various press announcements.

[a] Numerically, these red-shift Z values are defined as the total shift in wavelength divided by the original wavelength.

omers suggest that some quasars are dense objects blown out of active galaxies, explaining the supposed faint, trailing gas bridges.

Critics of interpretation II note that the spectral emission lines of quasars are less broad than predicted if the red shifting were gravitational. Also, they argue that quasar–galaxy pairs are chance alignments of nearby galaxies and distant quasars, and that the supposed gas bridges are wisps that align only by chance with galaxies near the quasar's position on the sky.

Interpretation I continues to be the favored way to interpret most quasars and highly red-shifted galaxies. As shown in Table 25-3, current discoveries continue to add ever more interesting—and ever more highly red-shifted—quasars to the list of known objects. Astronomers stirred excitement in 1987 with the announcement that the first quasars with a red shift greater than 4 had been found. These objects are so far away that the light we see today probably left them when the universe was only 10% of its present age! Thus, as we will discover in the next chapter, quasars probably offer a window into the past, allowing us to examine the appearance of certain galaxies shortly after the primordial era of galaxy formation.

SUMMARY

Once we pass beyond the Local Group and other nearby galaxies, we encounter still more galaxies in large clusters at distances of some tens of megaparsecs. Such clusters, and perhaps clusters of clusters, extend as far as we can see, which is at least some thousands of megaparsecs. As we probe to farther distances, we encounter serious problems of interpretation, which might be summarized by the following list of statements:

Fact: *Galaxies at greater distances than a few Mpc have red shifts.*

Virtually certain assertion: *The greater the red shift, the farther the galaxy.*

Virtually certain assertion: *Among most galaxies, the red shifts indicate recession. These galaxies are moving away from us and from each other.*

Virtually certain assertion: *Some kinds of galaxies have active nuclei and emit much greater amounts of energy than normal galaxies.*

Virtually certain assertion: *Explosions or accelerations of gas, fed by enormous energy release, occur in the heart of the nuclei of some galaxies in regions less than a parsec across. They may involve super-supernovae, black hole accretion disks, or other energetic phenomena.*

Probable hypothesis: *Some faint objects with extremely high red shifts, called quasars, are intensely luminous galaxies (or the nuclei of such galaxies) thousands of megaparsecs away, sharing in the recession and moving away from us at appreciable fractions of the speed of light. They are now seen as they appeared billions of years ago, shortly after they formed. Explosive events in galactic nuclei may have been more common then than now.*

Nearly all astronomers accept the first three or four statements and agree that the universe is expanding, in the sense that the galaxies are rushing away from each other. A few astronomers are still debating whether some part of the largest red shifts might be caused by something other than recession. The most red-shifted objects are the puzzling quasars.

Most astronomers believe that quasars are remote galaxies with explosive events involving supermassive objects (possibly billion-$M_\odot$ black holes) in their nuclei. The formation of such supermassive objects may have been triggered by collisions among galaxies, or it may have been spontaneous. Formation of such objects and the resulting superluminous galactic nuclei may have been more common in primordial galaxies than in galaxies today. In any case, we would like to know more about these strange objects, which seem to offer unusual clues about the nature of the universe.

As we approach the frontiers—the farthest objects that astronomers can detect—we find many strange phenomena. However we interpret them, many galaxies apparently generate extraordinary energies. Explosions in their centers, possibly occurring millions of years apart, may cause strong radio and optical radiation and may expel filaments or condensations of matter. Observing distant galaxies, we see how these objects looked billions of years ago. Interpreting these observations has created some of the most exciting controversies in astronomy today. With research on these distant frontiers, we are probing the very nature of the universe itself.

CONCEPTS

megaparsec	gravitational red shift
cluster of galaxies	theory of noncosmological red shifts
supercluster of galaxies	
red shift of galaxies	galaxies with active nuclei
expanding universe	synchrotron radiation
mutual recession of galaxies	Seyfert galaxy
	galactic jets
Hubble's relation	quasar
Hubble's constant	QSO
Doppler shift	gravitational lensing
theory of cosmological red shifts	

PROBLEMS

1. How many miles is a megaparsec?

2. Suppose observers located in the Coma cluster of galaxies observe Doppler shifts in the spectra of our Local Group of galaxies, including the Milky Way.
   a. Would they see a red shift or a blue shift?
   b. What sizes of shift would they observe? (*Hint:* See Table 25-1.)
   c. What would they conclude about our Local Group's velocity if they believed the theory of cosmological red shift?

3. In what ways do Seyfert galaxies bridge the gap between ordinary galaxies and quasars?

4. Because our galaxy's nucleus is a radio source, why is the Milky Way not considered to be a typical radio galaxy?

5. Why is the study of the most distant galaxies we can see related to the study of conditions around the time that our galaxy (and perhaps others) was forming?

PROJECT

1. Examine the appearance of explosive radio galaxy M 82 (NGC 3034) in a telescope and compare with Figure 25-4. Note the elongated shape, which can be seen even with small telescope apertures of a few inches. Locate other galaxies and observe their general faintness. Study whether their central regions are brightest.

Frontiers

A special place in the universe. The evolution of life as we know it apparently depends on water and a changing but relatively stable planetary climate. Modern astronomy leads us to the question of whether planets exist near other stars and whether life has evolved on them. (Photo by author.)

CHAPTER 26

Cosmology: The Universe's Structure

We are approaching the limits of our ability to probe the skies. The previous chapter took us to galaxies so remote that they are not only almost too faint to see but also so red-shifted that much of their radiation is infrared or radio radiation. What can be said about still more remote regions? Are they like ours? Do galaxies exist there? These questions belong to the field of **cosmology**— the study of the structure or "geography" of the universe as a single, orderly system.

In the opening pages of this book we compared humanity to explorers on a strange island in an unknown sea. To extend the analogy, this chapter finds us at a point in our explorations where we know the island is made of grains of sand, we know how big it is, we know that the sea is large, and we know that there are many other islands out there in the distance. And now we ask: Do the islands go on forever? Is the world flat with islands dotted uniformly or in clusters? Does the world have an edge, or is it infinite? Or is the world round with no edges?

The same kinds of questions are asked in cosmology. Are galaxies dotted across space indefinitely, as photos such as Figure 26-1 seem to imply? Does space itself go on indefinitely, or does it come to some kind of end? Although these are exciting questions to debate, they are risky, because they tempt us to leapfrog beyond the limits of available observations and speak of such concepts as infinity. Cosmology tempts scientists to speculate about the **universe,** defined as all matter and energy in existence anywhere, observable or not. Yet scientists cannot be sure how concepts of infinity apply to the real universe. And, as J. D. North (1965) remarked in his history of cosmology, "It is easy to speak of the infinite . . . but it is difficult to speak of it meaningfully."

Why should normally cautious scientists attempt such speculations? There are several reasons. People have been asking cosmological questions and related religious questions for thousands of years. It is valid to do our

Figure 26-1 More galaxies. This image shows numerous elliptical and spiral galaxies in the Fornax cluster of galaxies, estimated to be 17 Mpc from the Milky Way. (Copyright Anglo-Australian Telescope Board.)

best to answer these questions. Innate in the character of curious, restless humanity is the desire to know our surroundings. Unfounded speculation should not be part of science, but scientists can legitimately take the principles we have learned about matter and energy in our part of the universe and then ask: "What would happen if these principles apply in all space and time?" or "What would happen if certain principles were different in early times or in distant space?" By such questions, cosmologists have been led to some startling ideas about the universe as a whole—ideas that lie far outside everyday experience.

THE MOST DISTANT GALAXIES

How do we really know that a certain galaxy is 1000 Mpc away? To say we know by the red shift is not a fundamental answer, because no one has been able to prove conclusively that the Hubble relation of red shift

and distance applies to all galaxies—although most astronomers think that it does. Furthermore, the Hubble constant rests on the interpretation of Cepheid variable stars and other distance indicators. And these, in turn, rest on other techniques, primarily trigonometric parallaxes of stars close to the solar system. In other words, we have built a whole ladder of distance indicators, each rung resting entirely on the security of the preceding rung, as shown in Figure 26-2. We have been tacking the ladder together as we climb it. We use this ladder because we think it is sound, but we use it only with healthy skepticism.

As noted in the last chapter, looking at distant galaxies involves looking far back in time, because their light takes a long time to reach us. Table 26-1 shows the estimated time required for light to reach us from different places in the universe. We see the Andromeda galaxy as it was over 2 million years ago. We see the Virgo cluster as it was when dinosaurs were dying. We see the Coma cluster as it was when early fishes were appearing in murky seas. We see the Hydra cluster as it was soon after the last lavas erupted in the Sea of Tranquillity on the Moon. We see some galaxies as they appeared before the solar system formed.

But now suppose we look for a still older galaxy or quasar. We run up against a very fundamental limit to our ability to observe for two reasons. First, the light of a sufficiently distant galaxy would have originated

Figure 26-2 The astronomical distance scale. The accuracy of distance measurement at each level depends on the accuracy of measurements at lower levels. Some methods of distance measurement are listed at left.

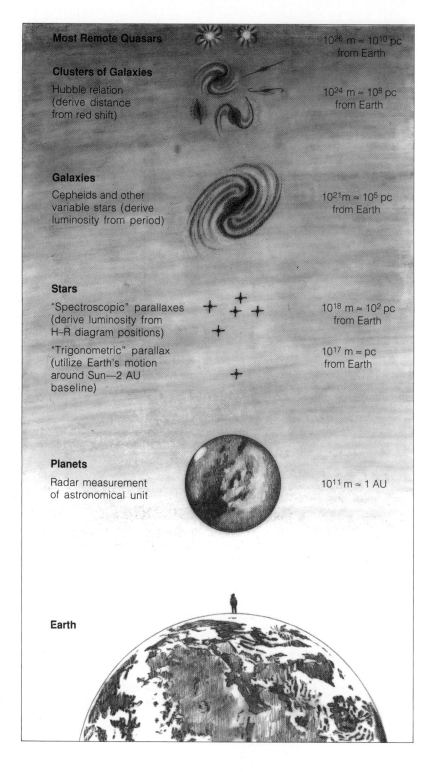

Most Remote Quasars — 10^{26} m $\approx 10^{10}$ pc from Earth

Clusters of Galaxies

Hubble relation (derive distance from red shift) — 10^{24} m $\approx 10^8$ pc from Earth

Galaxies

Cepheids and other variable stars (derive luminosity from period) — 10^{21} m $\approx 10^5$ pc from Earth

Stars

"Spectroscopic" parallaxes (derive luminosity from H–R diagram positions) — 10^{18} m $\approx 10^2$ pc from Earth

"Trigonometric" parallax (utilize Earth's motion around Sun—2 AU baseline) — 10^{17} m $\approx$ pc from Earth

Planets

Radar measurement of astronomical unit — 10^{11} m $\approx$ 1 AU

Earth

TABLE 26·1

Travel Times of Light from Distant Objects

Source of Light	Distance	Light Travel Time (y)
Typical visible stars	100 pc	326
Center of Milky Way	9 kpc	29,000
Magellanic clouds	60	200,000
Andromeda galaxy	670	2,200,000
Edge of Local Group	1000	3,300,000
M 51 "Whirlpool" spiral	3800	12,000,000
Centaurus A radio elliptical	4400	14,300,000
M 87 elliptical	16 Mpc	52,000,000
Virgo cluster	16	52,000,000
Coma cluster $(Z = 0.02)^a$	100	370,000,000
Hydra cluster $(Z = 0.2)^a$	800	2,300,000,000
Extremely distant galaxy 0902 + 34 $(Z = 3.4)$	4100?	7,700,000,000?
High-Z quasar OQ172 $(Z = 3.5)$	4100?	7,800,000,000?

Note: The calculated distances and light travel times for the highest-Z objects depend on the cosmological model assumed, and the light travel time is no longer simply the distance divided by the speed of light.

$^a Z$ = red shift expressed as fraction of original wavelength.

under conditions as they were about 14 billion years ago. We already saw in Chapter 23 that our own galaxy was just forming at this time, and astronomers believe that other galaxies formed at about this time, too (see Chapter 27). Therefore, if we try to look farther away in distance and further back in time, we may see no galaxies because none had yet formed. A second reason stems from the incredible recession velocities, which approach the speed of light at this distance. This means that the light from the galaxy would be extremely red-shifted. To look for objects much farther away, we would have to look for most of their light at infrared or radio wavelengths, where the photons have little energy and detection is difficult. This explains our difficulties in probing the most distant objects and earliest conditions in the universe.

EARLY COSMOLOGIES

The preceding discussion shows how cosmology, the study of the universe's structure, is related to the study of the universe's origin. One might think that we could study the present structure entirely separately from the ancient history of the system. In this chapter we will attempt to do this as much as we can, focusing mostly on the "geography" of space. But always in astronomy, the farther we probe in distance, the further back we

look in time. Ultimately, mapping the remotest regions means mapping the early history of the system. For this reason, the remote cosmic horizon contains important clues about the origin of our universe.

To understand how cosmologists have arrived at their present conceptions of the universe, we will review a series of cosmological theories developed at different times. These demonstrate best how various religious, philosophical, physical, mathematical, chemical, and astronomical studies have forced us toward our present conceptions. In each case, we will first present the model and then discuss some of its limitations, which led to new models.

Cosmology 1: Ancient Models (c. 3000 B.C.)

Even 5000 years ago philosophers speculated about the nature of the universe. Knowing nothing of the physical laws that govern matter, they tried to make sense of the universe by discussing nonmaterial attributes, or "essences," of things. They often gave these essences names, imaginary personalities, or abilities to control the affairs of the universe.

Early writers also speculated on the arrangement of the universe. According to a tradition in India, for example, the universe is a giant egg containing land, waters, animals, gods, and so forth, all created in primordial waters by Prajapati. A well-known book compiled around 500 B.C. considers both origin and structure:

In the beginning God created the heavens and the earth. The earth was without form and void, and darkness was on the face of the deep; and the spirit of God was moving over the face of the waters. And God said, "Let there be light!"

Thus many old traditions trace both the structure and the origin of the universe back to some underlying early spirit or principle. Sometimes this spirit was given personality and described as one or more gods. Sometimes it was described as a fundamental principle from which all else followed, as in the Book of John (c. A.D. 100): "In the beginning was the Word." *Word,* in this passage, was the Greek term *logos*—a principle of rational logic. So this famous cosmological thought might be translated, "In the beginning, underlying everything, was rational order." In many ancient cosmologies based on these ideas, the structure and events of the universe (including those involving living beings) were viewed as controlled by the characteristics of this initial creative essence.

Limitations These cosmologies have retained their appeal for thousands of years and provide us with an enduring link to our early ancestors. Historically, they helped set the stage for later investigations about initial conditions in the universe and about immaterial phenomena such as energy. They have the virtue of bringing us to the realization that there are forces in the universe greater than ourselves. The power of their poetry instills a beneficial respect for the sheer vastness and mystery of the universe.

But this kind of cosmology can produce the stultifying belief that all that *can* be said about the universe *has* been said. Such a belief can rob entire cultures of the motivation to question, explore, and see what they can learn for themselves. We humans have learned that we can deepen our understanding of the universe if we insist that assertions about nature be testable by observation, experiment, or calculation. But it is difficult to test the early cosmological ideas in this way or determine which is most accurate. Also, such cosmologies are of little value in predicting or interpreting the phenomena that we actually see in the distant universe. Ever-improving astronomical instruments reveal that much more can be said about the universe.

Cosmology 2: Newtonian–Euclidean Universe (c. 1700)

Around 1680, when Isaac Newton described how every particle in the universe gravitationally attracts every other particle, he realized that this principle might allow a simple description of the structure of the whole universe. First, he assumed that the principles of Euclid's geometry, such as the relations between angles, lines, and planes, would work just as well over the vast distances of universal space as they do in the farmyard or among the nearest stars. Euclidean geometry and Newton's gravitational law, then, allowed description of the separations and forces between particles. Galaxies and even clusters of galaxies can be considered as particles. Newton pictured the universe as infinite in extent and filled with these randomly moving "particles," a view called the **Newtonian–Euclidean static cosmology.**

In 1755, Immanuel Kant realized that although the Newtonian–Euclidean universe remained constant in the long term, many individual stars or galaxies might come and go. Kant thought that the *present* system of stars would burn out and pass away, but others might form from the debris of former systems, just as we have described stars forming from the debris of earlier stars. The universe was thus endlessly recycling—"a Phoenix of nature, which burns itself only in order to revive again . . . through all the infinity of times and spaces." The Newtonian–Euclidean universe was thus static (not expanding or contracting), but evolving.

Limitations Newton was right in predicting many suns far beyond our own, all obeying gravitational relations. And Kant was right in imagining worlds forming and reforming. But these ideas do not adequately describe the universe as we know it today. First, the recession of distant galaxies was neither predicted nor assumed in this theory. Second, why should matter in the universe remain dispersed? Gravitational attraction between particles might cause all the mass in a given region to collapse into a single star or galaxy. Mathematicians in the 1800s studied this problem extensively and even talked of a hypothetical repulsive force that might be important only over long distances, thus holding the particles apart and keeping the universe static. But the insurmountable objection is that the Newtonian–Euclidean cosmology fails to explain a problem known as Olbers' paradox.

OLBERS' PARADOX

Sometimes the simplest questions promote the most profound thoughts. The simple question, "Why is the sky dark at night?" leads to an astonishing paradox. For example, the night sky in the Newtonian–Euclidean universe ought to be ablaze with light! Astronomers are not sure who first realized this, but Edmond Halley indicated he had heard the idea as early as 1720. The idea was more carefully developed in 1823 by the German astronomer Wilhelm Olbers, whose name finally became attached to it. (See North, 1965, for a complete history.)

Olbers' paradox arises as follows. Assume that space extends indefinitely and is filled only with stars resembling the Sun. If we look at the Sun, we see that each unit of angular area of the Sun's surface (a square second of arc, for example) is intensely bright. If we now gaze in some other direction, our line of sight must ultimately intercept the surface of another star, because stars dot space to infinity. Each unit of angular area in that direction, therefore, would have about the same surface brightness as the Sun, because *surface brightness* (brightness per unit of angular area) does not depend on a star's distance. The whole day or night sky should look as bright as the surface of the Sun!

What is wrong with this argument? One early suggestion was that interstellar dust simply obscures the distant stars. But according to the Stefan–Boltzmann law (Chapter 16), the dust should absorb the stars' radiation and heat up until the amount of radiation it emits equals the amount absorbed and a constant equilibrium temperature is attained. Even if this radiation were not visible light, there would be infrared radiation that we would sense as intense heat. Because the dust is not radiating this intensely, this explanation fails.

The modern response to Olbers' paradox is more subtle, invoking the age and recession of distant galaxies. The most important factor is galaxies' ages. Astrophysicists have found that the total number of photons emitted by the galaxies in their finite lifetimes is too low to create the kind of pervasive bright glow described by Olbers (Wesson and others, 1987). To look at it another way, there is a "horizon" distance corresponding to a light-travel-time of 12 or 14 billion years, beyond which we see no galaxies because none had formed that long ago. The intense glow proposed by Olbers would apply only if the galaxies had been in existence for vastly longer times and our line of sight intercepted galaxies to virtually infinite distance. Additional

calculations have suggested that the time required for the radiation to permeate the available volume, heat the interstellar dust and gas, and achieve the equilibrium assumed by Olbers would be around 10^{23} y—more than a billion times the actual age of the universe! (See Harrison, 1974; Raychaudhuri, 1979.)

A secondary effect that helps explain Olbers' paradox is the recession of the galaxies. The red shifts of their light cause the apparent energies of the photons we receive from them to be reduced from high energies (short wavelengths) to low energies (long wavelengths) and cause them to be spread more thinly through space. Thus photons received from galaxies receding at almost the speed of light are strongly reduced in apparent energy.

In short, Olbers' paradox can be explained by modern discoveries concerning the age and recession of galaxies, but it can't be explained by the Newtonian–Euclidean model of the universe: extremely ancient, nonexpanding, static, infinite, and filled uniformly with stars.

MODERN MATHEMATICAL COSMOLOGIES

In most branches of astronomy, theoretical studies develop in response to, or at least hand in hand with, observations. For example, theories of stellar evolution were developed to explain observations of various star types and their positions on the H–R diagram. Cosmology is different: The outskirts of the universe are so far away and so hard to measure that only a few basic observations are available to distinguish correct models of the universe from incorrect ones. Fortunately, most cosmological theories *do* make predictions that may be testable in coming decades with space telescopes and other improved equipment. But for the moment, cosmological theorizing has certain aspects of a mathematical game. The rules are (1) invent some hypothetical natural laws and (2) show that they could, in theory, govern a complete universe, but (3) also show that they do not disagree with any of the few, available observations.

More than most theories, modern cosmologies involve many assumptions, because so few statistics are available about the universe out beyond about 1000 Mpc. Many modern cosmologies are essentially deductions from initial postulates, rather than experimental or observational tests of nature. As one researcher (Schatzman, 1965) has noted:

Once the basic ideas are understood, the deduction of their consequences is a simple exercise in geometry or algebra, which does not increase our knowledge of the properties of matter.

Non-Euclidean Geometries

In the late 1800s, mathematicians in Germany, Italy, and Russia became fascinated with geometries quite different from those of everyday experience. These so-called non-Euclidean geometries served as stepping stones to modern cosmologies.

Non-Euclidean geometries can be described by an analogy. Suppose we represent the three-dimensional volume of space by the two-dimensional surface of a chessboard. Instead of being able to move in any of the three dimensions of ordinary space (north–south, east–west, and up–down), light waves, sound waves, and inhabitants in the chessboard world could only travel along the two-dimensional (north–south and east–west) surface of the board. Imagine that the board is vast, covering many square kilometers. Inhabitants of this world might be visualized as tiny ants who can move and see only along the surface of the chessboard. To the ants, the chessboard represents all space.

Because the chessboard is flat, the ants would find that the principles of Euclid's plane geometry are satisfied. For example, parallel lines would never meet, and the sum of the angles in a triangle would be 180°.

But suppose instead that the chessboard covered the whole surface of the Earth. It can still be called a two-dimensional surface, because the ants (and light waves) can travel only along the *surface* in two dimensions (north–south and east–west). As long as the ants operate only in a small region, the surface would seem flat to them, and Euclidean geometry would seem true within the limits of the accuracy of their observations. But if the ants probed large enough regions, they would eventually discover that their surface, or "space," is curved. For example, as shown in Figure 26-3, they would discover that the triangle defined by the equator and two latitude lines 90° apart has three 90° angles whose sum is 270°!

If the chessboard covered the whole Earth, the ants might discover another interesting thing. They would find that their universe, instead of being infinite (as the ant-Newtons and the ant-Kants might have supposed),

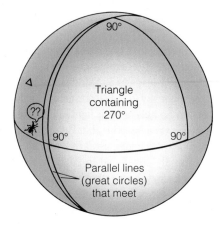

Figure 26-3 Ants living on a large spherical world would find that small figures, such as the small triangle, approximately obey Euclid's laws of geometry. But large figures, such as the large triangle, violate these laws. In the same way, studies of very large volumes of intergalactic space might reveal departures from Euclidean geometry.

was a finite amount of space,[1] even though it had no edge. The Earth has a finite area of about 500 million km², but no ant crawling around the surface could identify an edge or boundary.

Mathematically, these ideas can be applied to two-dimensional surfaces or to three-dimensional space. Following Euclid's laws of geometry, nineteenth-century mathematicians took these ideas one step further. Just as we can imagine a two-dimensional chessboard, which is curved to become three-dimensional, and adjust the geometrical math to that case, the nineteenth-century mathematicians imagined three-dimensional space to become curved into four dimensions, and they developed a new math to describe this **curved space.** Geometries of curved surfaces or curved spaces are called **non-Euclidean geometries.** After some controversy, cosmologists accepted the proposition that the real universe might actually be curved, or non-Euclidean.

Everyday experience gives us no clue as to whether or not space is curved. We normally interact with far too small a region to detect a slight curvature. The evidence for slight curvature of intergalactic space would come only from studying a very large part of space, including the remotest galaxies, just as the ants would need data from much of the Earth to test for curvature of their "space."

---

[1] The total area would be the area of Earth in this example.

Furthermore, different kinds of curvature are possible, each with its own properties. Examples of different types, shown in Figure 26-4, are based on analogies with surfaces. Case A shows a surface representing uncurved space, which could extend to infinity in all directions (dotted lines). Euclidean geometry would apply everywhere. Case B shows spaces with a slight curvature. These spaces could also extend to infinity. Case C shows an example of complex curvature, which could also extend to infinity. These types of space could have infinite volume.

Case D is an especially interesting case in which the curve has closed on itself, therefore having finite rather than infinite volume. If space is really curved in this way, the universe could have finite volume but still no boundaries, like the universe of the ants who lived on a globe. A traveler who went far enough in one direction would eventually come back to his starting point.

The **radius of curvature** is a distance that characterizes how tightly space is curved. It is analogous to the radius of Earth in the example of the ants. If space has only a slight curvature, the radius of curvature is very large. If space is infinite, the radius of curvature is infinite.

Space with infinite volume is called **open space**. Space that is curved and has finite volume is called **closed space**. Thus in Figure 26-4, A would be called flat and open with an infinite radius of curvature. B and C are curved and open, and D curved and closed.

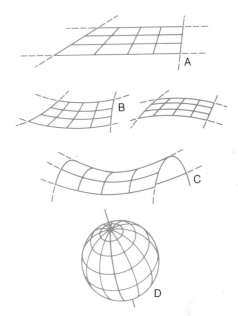

Figure 26-4 Examples of different curvatures of space, using surfaces as analogs to the volume of three-dimensional space. A is uncurved and infinite in extent; B and C are curved and infinite; D is curved and finite.

Cosmology 3: Static, Curved Universes (1917)

In 1917, Albert Einstein, who had just developed the theory of general relativity, tried to see how it applied to the universe as a whole. He made some basic assumptions that seemed reasonable at the time:

1. The universe is **homogeneous** (meaning that all its particles—stars, galaxies, clusters of galaxies, or whatever—are uniformly distributed on a moderately large scale) and **isotropic** (meaning that the view is similar in all directions from all galaxies). This assumption is sometimes called the **cosmological principle.**

2. Space is curved, as in non-Euclidean geometry, and the curvature is constant. (The surface analog would be a sphere.)

3. The universe is static, which means that galaxies stay at constant distances apart; mean density of matter is constant and the radius of curvature does not change.

But when he tried to solve the equations representing such a universe, Einstein found that it was impossible! *No static solution fit the principles of general relativity.* Gravitational attraction between galaxies would tend to make matter in such a universe collapse, instead of remaining in static dispersal.

Still assuming that the universe must be static, Einstein introduced a so-called cosmological constant—a hypothetical repulsive force between material particles, important only over long distances. The repulsion, similar to that suggested for Newtonian–Euclidean theories, could be chosen to balance the gravity of whatever mass was in the universe, so that a static state resulted, but only if there were no disturbances.

Limitations The original static model was scrapped in 1929 when Hubble proved that the galaxies were rushing away from each other. Moreover, there is no independent evidence for the repulsive force that the model requires.

Cosmology 4: Friedmann Evolving Universe (1922—1924)

From 1922 to 1924, the Russian mathematician Alexandre Friedmann studied a model that was evolving, not static. He simply assumed that the curvature of space varies with time. For example, space could increase in volume if the radius of curvature increased. In such geometries, the distances between all points continually increase, and galaxies would be seen to recede. This model was not widely discussed until Hubble published his discovery of actual galactic recession in 1929. What had seemed like an abstract theoretical model now seemed to be supported by astronomical observation—galaxies are moving apart! Einstein, de Sitter, Eddington, and others soon concluded that the Friedmann model might describe an expanding universe without invoking any repulsive force.

An analog of Friedmann's expansion model would be the surface of an expanding sphere, like a balloon. Galaxies could be imagined as dots on the surface of the balloon. The analogy is not perfect, because the size of the dots would expand with the rubber, whereas real galaxies could maintain their integrity as units of matter by their internal gravitational forces. Antlike observers on any dot would see other dots receding from them as the balloon expanded, just as we would see galaxies recede if the radius of curvature of space increased. In addition, no ant or any observer in any galaxy could say he or she lived at the center. Ants could move from dot to dot around the curved surface of the balloon, but no one dot (or galaxy) would be "centered" more than any other.

Limitations When Hubble discovered that galaxies are actually receding from each other, theorists were delighted that observations seemed to support the expanding model. Abstract mathematical theorizing seemed to be relevant to a subsequent direct observation of nature! In retrospect, however, astronomers do not say that the expansion of the universe was *predicted* in the normal scientific sense for two reasons. First, the observed expansion was only one of several possible types of Friedmann evolution dependent on the initial assumptions; second, the Friedmann model was developed almost completely outside the normal astronomical context. It was almost completely a mathematical model, with little to say about physical questions such as the distribution of galaxies or the density of matter in space. Thus the next step was to connect this exciting and provocative model more closely with astrophysical reality.

Cosmology 5: The Big Bang (1927—)

Around 1929, a Belgian priest trained in mathematics and astrophysics, Georges Lemaître, began to put more astronomy and physics into cosmology: He pointed out that if galaxies are now flying away from each other, there probably was a time in the past when all matter was closer together. He suggested that all matter exploded from this condition of high density. Lemaître thus became known as the father of what is popularly called the **big bang theory** of the universe's origin, which is the most widely accepted theory today.

The moment of maximum density is known as *the big bang*. Many people regard this as the actual creation of the universe. Modern evidence suggests that this moment occurred around 14 ± 4 billion years ago. The universe was supposed to have begun from an extraordinary state in which the radius of all space was nearly zero and all matter was concentrated in a virtual point with virtually infinite density.

Lemaître's picture of a primordial expanding space filled with matter and energy excited physicists and astrophysicists, who recognized that they could analyze the evolution of matter in such a model. This is the hallmark of a useful theory: It allows predictions to be made and tested. The Russian–American physicist George Gamow and his associates began this work in 1948 and found that the observed abundances of atoms of the different elements display trends expected for element formation in an expanding, hot, ancient fireball! This is an important example of how a successful cosmological theory ties into observations of nature. (Later studies showed that additional element formation inside stars was also important, especially in forming the heavier elements.)

Limitations The big bang theory says nothing about how (not to mention why) the "primeval atom" or initial fireball came into existence. Also, it has faced an increasing problem of explaining the distribution of distant clusters of galaxies, and their mode of formation from the primordial gas. Nonetheless, it has been very successful in saying that *if* there ever was such a fireball filling expanding space, then certain phenomena would result, and these phenomena have been *observed*.

Today, most astronomers accept the big bang theory as the best available cosmological picture of the universe.

In the 1980s the inflationary big bang model was modified, which we will take up in the next chapter, but first we need to consider some additional cosmological ideas to bring our story up to date.

Cosmology 6: Is the Whole Universe a Hole Universe? (1970—)

An interesting fact is that the total observable mass in the universe, divided by the observable volume, gives a density very close to the density that would be required by theory for a very large black hole. That is, the universe as a whole seems very close to the definition of a black hole! For example, a 1982 paper on various sizes of black holes listed the whole universe as a hypothetical black hole containing $2.8 \times 10^{23}\ M_\odot$, with a radius of 2.8×10^4 Mpc and a mean density of 2.4×10^{-28} kg/m^3. (A large enough black hole does not have to be extremely dense.) Perhaps the whole universe should be thought of as a black hole!

Limitations The physical meaning of this idea is not very clear, even though the density required over our several-thousand-megaparsec region is quite well defined. Could our "expanding universe" really be an expanding black hole filled with galaxies and expanding into some larger region from which we are forever detached? These ideas have led the famous British cosmologist, Stephen W. Hawking, to speculate that "our" big bang is just one detached event in a larger universe of sporadic big bangs, forever separate from each other. We don't know.

DIRECT COSMOLOGICAL OBSERVATIONS

Often it happens in science that out of a series of theories, such as we've just enumerated, several may contribute to our understanding. How can we pick and choose among them? We would like some observational tests to determine which theories make correct predictions. We've already mentioned that although few observational tests are currently possible, the big bang theory (cosmology 5) seems to have the most concepts matching observed nature. For instance, the galaxies do seem to be rushing apart, and the sky is dark at night instead of bright.

Our task in the rest of this chapter is to review some

direct observations that must be explained by any current theory of cosmology and to show how they fit the big bang picture. In the next chapter we will discuss the events around the time of the big bang.

Bubblelike Distribution of Galaxy Clusters

We have mentioned that galaxies are distributed in clusters, and the clusters in superclusters. Does this trend go on and on? Or do the superclusters on the largest scale "fuzz out" into a smooth distribution? Surveys of galaxy positions in three-dimensional space, done in the 1980s, reveal that galaxies and clusters of galaxies are *not* distributed uniformly. Rather, they are distributed as if on the surfaces of vast bubbles, enclosing galaxy-free voids. The large-scale structure of the universe might be likened to a froth of soapsuds, where galaxies are like water molecules on the surfaces of the touching bubbles, and clusters or superclusters are like the "nodes" at the junctions of several bubbles (Figure 26-5). Maps of galaxies in many regions of the sky reveal a filamentary pattern, like wisps of smoke.

To understand this *three-dimensional* bubblelike distribution of galaxies, we need to measure their distances (from their red shifts) and plot their positions in 3-D space. Early attempts at this, for selected parts of the sky, revealed large empty regions among the filaments (Gregory and Thompson, 1982). A 1986 survey of 1061 galaxies out to about 120 Mpc from our galaxy, in a narrow strip of the sky, revealed the bubblelike pattern, with bubbles about 20 to 50 Mpc across, containing voids in which very few galaxies or clusters exist.

Astrophysicists are now trying to understand how this bubblelike pattern, this froth of galaxies, resulted from the big bang. For example, Russian cosmologist Y. B. Zel'dovich identified conditions in which the initial, expanding gas would break into huge "wisps" with about the mass of a supercluster. These wisps then contract gravitationally and break up into clusters of galaxies. Today's galaxy froth may thus be a remnant pattern of the galaxies' parent gas (Silk, Szalay, and Zel'dovich, 1983; Burns, 1986).

WILL THE UNIVERSE KEEP EXPANDING FOREVER?

Is the expanding universe a finite closed space like an expanding balloon? Or is it infinite? Given that galaxies

Figure 26-5 Schematic conception of the large-scale structure of the universe. Galaxies are strung through three-dimensional space in a filamentary pattern, illustrated here in a two-dimensional representation. Giant "bubbles" of relatively unpopulated space appear throughout the wispy galaxy distribution. "Nodes" at intersections of bubbles may represent clusters and superclusters of galaxies. (Painting by Dennis Davidson.)

are rushing away from each other, we can ask another question: Can the universe recollapse? If there were enough material among the galaxies, their gravitational attraction for each other would be enough to overcome the recession; galaxies would slow down and eventually fall back together. In this case, the universe's history might be a cyclical series of big bangs! But if the density of material in the universe were low enough, gravity would be too weak to cause recollapse, and the galaxies would rush outward forever, like rockets launched from a planet at escape velocity. So the second question can be stated: Is the density of material in the universe great enough to cause recollapse?

Open Universe vs. Closed Universe

These two questions turn out to be related.[2] In general, cosmologists use the term **open universe** to refer to a universe composed of open space with infinite volume

and no boundaries, in which the expansion of galaxies continues forever. Cosmologists use the term **closed universe** to refer to a universe that is composed of closed space with finite volume and that has enough density to cause a reversal of galaxies' motions and eventual recollapse. If the average density of matter in the universe is less than about 6×10^{-27} kg/m^3 (about three hydrogen atoms per cubic meter), the universe is open; if more, it is closed. This critical density is called the **closure density,** or the density required for closure of the universe.

Thus much cosmological discussion boils down to the question: Is the universe open or closed? At least three observational tests are possible. To understand the first test, which deals with whether space is curved, let us return to our analogy of the ants. They ask themselves whether they live on an infinitely large flat chessboard (Figure 26-4, curve A) or on a surface more like the shape of Earth (Figure 26-4, curve D). Suppose they know of mileposts (galaxies) scattered every mile or so at random. If they gathered data on the number of mileposts within 10 km, 100 km, and 1000 km of their colony, they could infer curvature of their space. For instance, if their chessboard were flat, the number within distance d would increase indefinitely as d got larger. But if they lived on the curved Earth, the number would start out increasing as d increased until they began tabulating mileposts from far away on the other side of the Earth. Eventually, as d approached half the Earth's circumference, they would run out of new mileposts.

---

[2]There is some confusion in the literature over the terms *open* and *closed.* Some books define an open universe as being unbounded and others define it as expanding forever. Similarly, *closed universe* is sometimes defined as finite or bounded and sometimes as destined to recollapse. In most, but not all, cosmological theories, open universes are unbounded (as in Figure 26-4, cases B and C) *and* ever expanding, whereas closed universes are finite (as in Figure 26-4, case D) *and* will recollapse. We will thus use the terms in this latter sense.

Astronomers have tried to make this test by counting galaxies, but the test fails to give a clear-cut answer. At great distances, where the test is most sensitive, bright galaxies (quasars) are overabundant because we are seeing back to the era when many galaxies were forming and when many galaxies were brighter than they are today. Some astronomers have tried to correct for this effect, but the results are uncertain and controversial.

A second test is to plot red shifts (recession velocities) of galaxies versus their apparent brightness (distance). Different degrees of curvature give different predicted curves on such a diagram. Again we run into problems with interpreting the brightness of quasars, but the data seem to indicate that the universe is quite near the critical closure condition.

The third test is to tabulate all the known matter (galaxies, clusters and gas in galactic halos, possible intergalactic gas and stars, and so on) over a very large volume and use this to estimate the average density of the universe, to see if it is less than or greater than the critical closure value. As Figure 26-1 reminds us, the test is difficult. The tabulations of detectable material give less than 10% of the amount needed for a closed universe, which would imply strong evidence that the universe is open. But this is not the whole story. Further observations connect the question of endless expansion to one of the most important problems in modern astronomy: the problem of the missing mass.

The Problem of the Missing Mass

As early as 1933, Caltech astronomer Fritz Zwicky noted that in the Coma cluster of galaxies there does not appear to be enough mass to hold the galaxies gravitationally in the cluster. Many of the galaxies seem to be moving fast enough to escape from the gravity of the material we can see. Yet the cluster is still there. This suggests that an additional gravitational pull may come from mass that we cannot see—perhaps in the form of gas dispersed in the cluster, in the form of faint stars, or in some other form. This agrees with the evidence mentioned in Chapter 24 that individual galaxies have more mass than we can see.

This so-called **problem of the missing mass**—the fact that motions of stars and galaxies imply more mass than we can see in the universe—is one of the major research problems in modern cosmology. The reason it is important is that the density of visible material in the universe is about 3×10^{-28} kg/m^3, only a

few percent of the critical density that divides an open universe from a closed universe (van den Bergh, 1990). Recent observations have given stronger and stronger evidence that there *are* large amounts of unseen mass (e.g., Tyson and others, 1990). For example, several studies of motions of materials near other galaxies, or clusters of galaxies, suggest clouds of nonluminous mass, or galaxies with unaccountably high mass. So, the majestic glow of the thousands of stars and galaxies in our night sky probably represents only a small fraction of the material in the universe!

Astronomers are very troubled by the growing evidence that between 90 and 99% of the mass of the universe is in the form of material we have *not* detected yet. Many astronomers assume that the unseen material is in the form of so-called cold, dark matter, such as gas clouds or a haze of subatomic particles in intergalactic space. One idea is that the unseen material might be in the form of neutrinos. Usually, these subatomic particles are described as massless, but some physicists think they might have a tiny mass. If they have more than 3-millionths the mass of an electron, they would contain more mass than all other material in the universe; if more than 49-millionths, enough to make the universe reach closure density. Experiments indicate the mass is less than 52-millionths, and possibly zero (Thomsen, 1987). So this hypothesis is unresolved.

A similar idea is that the missing mass consists of hitherto undiscovered subatomic particles, called weakly interacting massive particles, or WIMPs. The WIMPs, some 10 times as massive as protons, have been predicted by some theories of the universe. Because they interact so weakly with other forms of matter, according to the theory, they pass through most detectors and have not yet been observed. Physicists are currently experimenting to see if WIMPs really exist and are sufficient to affect the universe's structure (Palca, 1991).

There are many reasons that astronomers are expending so much effort on the missing mass. One reason is simply that we can't make a complete cosmological picture until we know the form of the unseen material. A second reason is that surveys of distant galaxies show superclusters on a larger scale than explained by models of cold, dark matter. This leads to controversy. Some astronomers claim that there is not much cold, dark matter, while others say that there is a lot.

This leads to an even larger problem: Some astronomers think that if the big bang produced an expanding, uniform "fireball" of matter, then it would not lead to the clumpy distribution of superclusters of galaxies that we observe. For this reason, there is increasing talk that while the big bang theory does account for many gross features of the universe, it may require substantial revision. "Cosmology is in chaos" writes the respected Canadian astrophysicst Sidney van den Bergh (1990). "It seems likely that only a major paradigm shift can bring some kind of order into the plethora of puzzling observations."

Still another reason for interest is that the amount of missing mass determines whether the universe is closed or open. Thus, the future of the universe rides on this issue! Is the difference merely an abstraction, like medieval arguments over the number of angels that can dance on the head of a pin? Yes, in the sense that it will have no immediate bearing on Earth's evolution or human evolution within the next billion years. No, in the sense that it makes a real difference for the future of the universe we live in.

Some writers have whimsically claimed to prefer an open universe because then we (or our atoms sometime in the future) would not be gobbled up in a high-temperature recollapse and a new big bang. But the laws of thermodynamics predict that an open, ever-expanding universe would cool forever until the remaining matter consisted of black holes and burnt-out stellar debris. Our hypothetical descendants couldn't build a fire to warm up because no combustible fuel would be left.

Other writers have claimed to prefer a closed universe because the recollapse might initiate a new big bang, perhaps starting a new cycle of evolution of new galaxies and new life. The subject is really too speculative to be discussed seriously, because our twentieth-century physics may not be adequate to understand it. But one might like to recall Robert Frost's lines:

Some say the world will end in fire,
Some say in ice.
From what I've tasted of desire
I hold with those who favor fire.

SUMMARY

Cosmology is a risky business, and it is unwise to put too much faith in generalizations about the universe when there is still so much more observing to be done with new and larger instruments. The French–American astronomer G. de Vaucouleurs (1970) has said:

Less than 50 years after the birth of what we are pleased to call "modern cosmology," when so few empirical facts are

passably well established, when so many different over-simplified models of the universe are still competing for attention, is it, may we ask, really credible to claim, or even reasonable to hope, that we are presently close to a definitive solution of the cosmological problem?

De Vaucouleurs was healthily skeptical. What we think we know is that (1) galaxies are receding from each other through space that is either uncurved or only slightly curved, and (2) some singular event, called the big bang, initiated the expansion of all matter in the universe billions of years ago. We aren't sure if the expansion will continue indefinitely. New telescopes in space may help us complete several tests that will answer these questions more definitively.

CONCEPTS

cosmology	homogeneous
universe	isotropic
Newtonian–Euclidean static cosmology	cosmological principle
	big bang theory
Olbers' paradox	open universe
curved space	closed universe
non-Euclidean geometry	closure density
radius of curvature	problem of the missing mass
open space	
closed space	

PROBLEMS

1. Progress in many scientific fields, such as studies of stars, plants, or animals, has come by classification of different types of specimens, followed by comparisons of the different classes. How does a cosmologist suffer a disadvantage in this regard?

2. Telescopes much larger than present-day designs, perhaps located in space, would have much more resolving power and light-gathering ability and could reveal much fainter objects. Give examples of how this development would clarify current cosmological problems.

3. How is the real universe different from the Newtonian–Euclidean static model that dominated literary and cultural concepts in the 1700s and 1800s?

4. Why is the sky dark at night?

5. Many cosmologies, such as the big bang model, assume the cosmological principle that the universe is homogeneous at any given time, given large enough scale. Yet quasars seem to be more common per unit volume at very great distances than near our galaxy. Does this refute the big bang theory? Why or why not?

CHAPTER 27

Cosmogony: A Twentieth-Century Version of Creation

How did the universe begin? When did it begin? Or do these questions have any real meaning? **Cosmogony** is the attempt to decipher the origin of the universe and its major parts, such as galaxies. *Cosmology,* as seen in the last chapter, focuses on the universe's structure but often involves cosmogony.

The big bang theory is both a cosmogonical and cosmological theory, because it accounts for both the origin and present structure of the universe. It has come to be strongly favored among all such theories because it is based on supporting observations, some of which will become clearer in this chapter. These form an interlocking pattern. Therefore, instead of comparing different theories as we did in the last chapter, we now assume that the big bang theory is essentially true and use it to describe how the universe may have formed and evolved.

All cultures have their creation myths, and the big bang idea is ours. The word *myth* does not mean a falsehood, but a scenario widely told and widely believed. We must remember, as scientists and as humans, that we don't have final, dogmatic answers about cosmogony. The big bang theory certainly gives us an indication of events that actually occurred long ago, but the word *myth* should be kept in mind to remind us that this is only our best current solution to a grand mystery—what one satirist has called "the Vienna Philharmonic of scientific questions."

THE DATE OF CREATION

At least three kinds of observation indicate that the universe as we know it began 10 to 18 billion years ago.

Age of Globular Clusters

In Chapter 22 we saw that H–R diagrams allow estimates of the **ages of globular clusters.** Results indicate that the clusters formed about 14 ± 4 billion years ago. This period is believed to mark the formation of our galaxy out of hydrogen and helium gas that formed in the big bang. The big bang must have happened at least that long ago.

Hubble's Constant

Because **Hubble's constant,** H, measures how fast the galaxies are now rushing away from each other, we can calculate how long it has taken them to get this far apart if they have been receding at a constant speed. We can thus compute the age of the universe from H. If the density of matter in the universe is well below the critical closure density (page 422), then the universe will have expanded since the big bang at about a constant rate, and the age since the big bang will be estimated as the distance of a galaxy divided by its speed. This turns out to be simply $1/H$. If the density of the universe is greater than this minimum, the gravitational attraction of galaxies for each other will have caused the expansion to slow down over time. The universe would thus have been expanding faster in the past, so its age would be younger. If the density equals the critical closure density, it turns out that the age in most models would be estimated as two-thirds the age estimate given earlier, or $2/(3H)$. If we adopt $H = 77 \pm 14$ km/s per megaparsec (page 396), our age estimate will be about 11 to 16 billion years; this would be valid for a universe containing only the amount of mass we actually see. If we assume there is sufficient "missing mass" for closure (see page 423), the estimated age would be two-thirds as much, or about 7 to 10 billion years, which seems embarrassingly small, because this is less than the ages estimated for globular clusters.

Ages of the Elements

In Chapters 6, 7, and 13 we discussed how the radioactive elements can be analyzed to reveal the 4.6-billion-year age of the solar system. Chapters 15, 18, and 19 added evidence that the heavier elements were mostly synthesized by fusion reactions inside stars and distributed by supernova explosions during a period stretching back further than 4.6 billion years ago. The big bang theory allows physicists to estimate conditions inside the expanding environment of radiation and hot plasma that existed soon after the universe began. This environment is called the **primeval fireball.** Using the theory of the big bang, physicists can calculate how

many elements of various types were synthesized there. By comparing today's abundances of radioactive elements with their decay rates and with their calculated production rates inside stars and the primeval fireball, it is possible to calculate the total **time required to form the observed elements** in the universe.

For example, if all uranium formed in the big bang, the universe should be about 7 billion years old to account for uranium abundances today. But we know that such a heavy element did not form efficiently in the big bang. If it all formed slowly inside successive generations of stars, the universe should be about 18 billion years old (Peebles, 1971). By this calculation, the age of the universe must be between 7 and 18 billion years.

When researchers measured the radioactive thorium-232 detected spectroscopically in stars of different ages, they found that our galaxy is only about 10 billion years old, and the universe, 11 to 12 billion years old (Butcher, 1987). Astronomers are currently debating whether this result can be reconciled with the somewhat greater age estimated for globulars; one technique is presumably in error.

From the rough agreement among these three lines of evidence, most astronomers conclude that the big bang occurred 10 to 18 billion years ago. Perhaps the best estimate is 14 billion years ago, with the galaxies forming soon afterward. Some unique explosive event must have occurred at this time, apparently creating matter and sending it flying out on its expanding journey. Astronomers refer to the time interval since this event as the **age of the universe,** the age since the creation of the universe as we know it.

The term *explosive* may be misleading, because the big bang was not an ordinary explosion. No one could have floated in nearby space and watched the fireball expand, because, paradoxically, the fireball filled all space. *Space itself is viewed as contracted at the beginning.* An analog to these models would be an expanding balloon, whose *surface* represents space. An ant living on a spot on its surface could detect the expansion and other spots moving away, but the ant could not get off the surface to look at the whole balloon expanding. The ant could travel only *on* the surface and would perceive no "up and down" or any interior to the balloon—only the surface. At the beginning, the surface of the balloon would have been "small" and incandescent, but it still would have constituted the ant's whole universe. In this sense, the "small" balloon—the hot, high-density fireball—would fill all space perceived by the ant.

Instead of picturing an explosion, it may be more

helpful to think of the big bang just as a singular moment of high-temperature compression of the universe's matter, whose earlier history, if any, we can't trace.

THE FIRST MINUTES OF THE UNIVERSE

One of the astonishing things about the big bang theory is that it lets us describe **initial conditions** during the first moments in the history of the universe. The assumptions are simple, because if we say that all the observable mass was once as concentrated as possible, then at the initial moment the density should have been nearly infinite; it has been declining ever since. Packing all mass at high density would lead to extraordinary temperatures. A simple theoretical model developed around 1950 by George Gamow assumed infinite temperature and density at the zero instant! Because we know the behavior of expanding hot material, Gamow was able to calculate subsequent conditions just as an engineer might calculate conditions in a gas expanding in a piston. For example, he was able to predict a present mean temperature of gas in the universe to be about 20K, and its density about 10^{-27} kg/m^3, numbers that are close to the observed values.

QUANTUM MECHANICS AND THE INFLATIONARY BIG BANG THEORY

Although Gamow's early version of the big bang model had some successes, astrophysicists realized it left unanswered questions. In the first place, the universe at the instant of the big bang was viewed as a point (or as mathematicians say, a singularity). What made this point of mass explode? Moreover, observations of galaxies out to billions of parsecs indicate that the curvature of space is very small, and the universe is almost flat. But the big bang models suggest the curvature could have almost any value, so that the unique, Euclidean, flat geometry seemed merely a coincidence. Astrophysicists pondering these questions in the 1970s began to suspect that the original big bang model was too simple.

At this point in the story, we must face the problem that modern cosmogony involves as much basic physics as it does astronomy. Physicists are currently working on "grand unified theories"—models that attempt to combine an understanding of matter, gravitational fields,

magnetic fields, electric and other fields, on all scales. Modern quantum mechanics, which incorporates relativistic phenomena to describe the behavior of atomic particles, is a key part of this effort; it is currently the best mathematical description of nature. At tiny scales, such as the world of atoms and electrons, quantum mechanics gives the best description of the behavior of nature. At larger scales, such as billiard balls and planets, the laws of Newton, which are simpler, give virtually the same answers as quantum mechanics. For that reason, engineers don't bother using complex quantum-mechanical formulas to build a bridge; they use the simpler, more familiar Newtonian principles.

But astrophysicists, who can use Newtonian concepts (sometimes modified by relativity) to trace the motions of planets and the collapse of stars, have noted that if you trace the universe back to a small enough bit of space–time—the so-called point—it behaves as a quantum-mechanical system rather than as a familiar Newtonian system. This is equivalent to saying that although billiard balls follow Newton's laws of motion, electrons and other subatomic particles do not; they have to be described by quantum mechanics. Recent theories about the big bang have depended on quantum-mechanical descriptions of curved space or, more accurately, the combination of space and time called the space–time continuum.

Here, then, is a remarkable fact about our voyage through the universe. Although astronomy started out to describe the largest systems of mass in the universe—stars, galaxies, and clusters of galaxies—it ends up having to describe the smallest—the particles that constitute atoms.

Starting around 1980, MIT astrophysicist Alan Guth, British cosmologist Stephen Hawking, and others described new versions of the big bang scenario in which a phase of extremely rapid expansion occurs in the first fraction of a second; this inflates the universe to much larger scale than in the old models. In this **inflationary big bang theory,** space–time is described by certain mathematical variables of quantum mechanics. They describe "fields" in which instabilities can occur, which in turn trigger explosive expansion. The "fields" are not magnetic fields or gravity fields or descriptions of matter, but more of a description of the state of space–time itself. According to the new theories, in the incredibly short instant of 10^{-43} s after the big bang was triggered, space adopted a vacuumlike quality with no particles of mass in it. This unstable form of space did not expand at a constant rate, as space was imagined

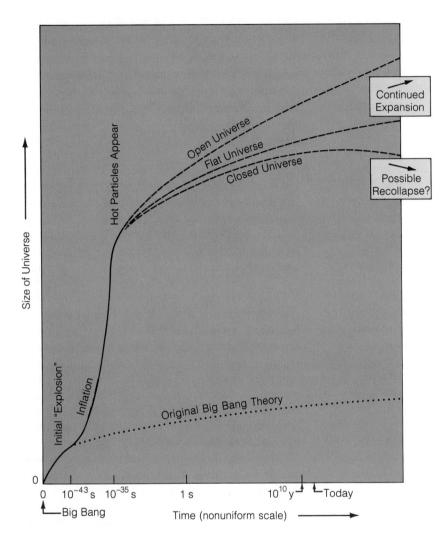

Figure 27-1 Schematic history of the size of the universe, according to modern big bang theory. "Size" refers to any property that measures the scale of space. Using Earth as an analogy, size could be Earth's radius or the distance required to traverse one degree of latitude along the curved surface. In the original big bang theory (bottom), expansion continues smoothly after the initial "explosion." In modern theories since 1980, the initial "explosion" is followed by an enormous inflation for about 10^{-35} s, after which the first energetic particles appear. The three dashed lines show possible subsequent histories. Note that it is hard in practice to distinguish among open, flat, and closed universes by present-day observations. (After diagram by Linde, 1987.)

to have done in the old theories, but at an ever-increasing, "exponential" rate. This process is shown in Figure 27-1. This rapid expansion, called **inflation**, continued for the first 10^{-35}s—still much less than a trillionth of a second. During this incredible inflation, the universe's volume made a sudden jump by as much as 10^{25} times! At the end of the inflation, this vacuumlike state changed and energetic subatomic particles began to appear in space. From that point on, the now-much-larger universe continued to expand at a smoother rate, as in the older big bang theories.

These new ideas inspired theorists in the 1980s. For the first time there were answers to various questions. The universe as we know it did not exist before the big bang, but one might say that the ghostly fields represented the precursor form of the universe. As shown in Figure 27-1, the universe in the new theories

is much bigger than imagined in the old theories. Thus it looks fairly flat, for the same reason that the vast Earth looks flat to an ant, because the scale of its curvature is so much bigger than the ant is. The new theory also helps explain the generally uniform appearance of the universe in all directions, a property that had begun to puzzle "old-big-bang" theorists.

PREDICTING ABUNDANCES OF ELEMENTS IN POPULATION II

The big bang theory in both its original and its inflationary form describes how the primordial particles joined together to form atoms of the various elements, and it predicts the abundances of those elements. In the high-temperature gas of the first seconds, matter was broken down into its simplest components—not grains or molecules or even atoms, but subatomic particles such as neutrons, protons, and electrons. Calculations based on known atomic physics show how these particles would interact under conditions after 1 s, 2 s, and so on.

The big bang seems to account for the formation of most of the material in the universe—the hydrogen, its heavy form deuterium, and helium—during the universe's first hour. One calculation (Reeves and others, 1972) gives these abundances at the end of the first hour:

Element	Percent Mass
Hydrogen (^{1}H)	75
Helium-4 (^{4}He)	25
Deuterium (^{2}H)	0.1?
Helium-3 (^{3}He)	0.001?
Lithium-6, lithium-7 (^{6}Li, ^{7}Li)	0.000001?
Heavier nuclei	Negligible

Perhaps it seems presumptuous to talk of events during the first minutes some 14 billion years ago, but the results can be confirmed by observation. For instance, the oldest stars we can find, the extreme Population II stars, appear to have formed from exactly the material shown in the table. This illustrates that when stars began to form, the heavy elements did not yet exist. Thus we can be pretty certain that shortly after the big bang, the universe consisted almost entirely of very hot hydrogen and helium.

THE FIRST MILLENNIA OF THE UNIVERSE: RADIATION VS. MATTER

In the big bang model, radiation was more important than matter for the first few thousand years. According to Einstein's famous equation $E = mc^2$, energy E is equivalent to mass m (c is the velocity of light). This means that each photon of radiation, carrying a certain energy E, will have an equivalent mass $m = E/c^2$. Thus any given amount of radiation corresponds to an equivalent mass of material. At the billion-Kelvin temperatures in the first moments after inflation, the universe amounted to a primeval fireball, full of intense radiation.

Slowly the radiation converted itself into hydrogen and helium mass. The radiation intensity declined. Today

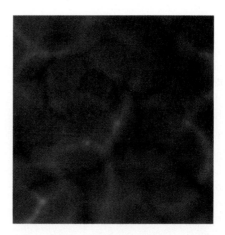

Figure 27-2 A schematic representation of a possible view in the primordial universe as matter starts to dominate over radiation. Seconds after the big bang, all of space was filled with a flash of ultraviolet light. As space expanded and the effective temperature of the radiation dropped, the light was eventually visible as a dull red glow in all directions. The first clouds of condensing matter are portrayed as luminous or nonluminous filaments silhouetted against this red glow. (Painting by Ron Miller.)

the radiation density is negligible. As Gamow pointed out, the mass of radiation in a cubic meter of air today is only about 10^{-26} kg. By comparison, the mass of the air in the cubic meter is about 1 kg. Even in the center of the Sun or an exploding atomic bomb, radiation amounts to only about 1 kg/m^3. But in the first tenth of a second it may have exceeded 1000 kg/m^3!

The importance of this result is twofold. When matter first appeared, the intense radiation agitated the particles of matter so violently that neither galaxies nor other clumps of matter could form. The gas was too hot. But after some thousands of years, radiation density dropped to below the density of matter. Within the first billion years, recognizable masses of material began to form.

The second consequence of the high radiation density is that in the far distant past (seen by looking toward distant, highly red-shifted regions), a different kind of radiation from that found near our own galaxy should be seen. It would be a trace of the dense radiation in the primeval fireball. The early universe was permeated by the kind of radiation that would come from high-temperature gas. As matter flew outward, the radiation was diluted and red-shifted (Figure 27-2). The red-shifting would lengthen the apparent wavelength of the radiation, and, according to Wien's law, the peak energy would thus come at a wavelength corresponding to quite a low temperature. Starting as early as 1948, this temperature has been variously predicted to be about 1 to 5 K. In other words, if the big bang theory is right, the sky should be uniformly filled with faint radio radiation resembling radiation from an object at about 1 to 5 K.

This line of thought was foreseen by the father of the big bang theory, Georges Lemaître, who wrote in 1931:

The evolution of the universe can be compared to a display of fireworks that has just ended: some few red wisps, ashes, and smoke. Standing on a cooled cinder, we see the slow fading of the suns, and we try to recall the vanished brilliance of the origin of the worlds.

But why just try to recall it? Why not point detectors at the sky to see if there is long-wavelength radiation left over from the primeval fireball?

Discovering the Primeval 3-K Radiation

The primeval radiation was eventually found, in 1965. It was discovered by Bell Laboratory researchers Arno Penzias and Robert Wilson, who were using radio telescopes in experiments on the first Telstar communications satellite. At first they thought the puzzling microwave radiation they detected was a problem in the instruments, but finally they established that it was coming from all over the sky. Further analysis by physicists at Princeton University showed that the radio waves corresponded to radiation from material at a temperature of 3 K, within the predicted range of radiation from the primeval fireball. Penzias and Wilson later received the Nobel Prize for this important discovery about the universe's beginnings.

This discovery is the strongest confirmation of the big bang theory, especially because the **3-K radiation** seems to be uniform in all directions. Just as predicted, observations at many radio wavelengths indicate no variations in intensity within an accuracy of a few percent.

Why Do We See the 3-K Radiation?

Radio radiation from all over the sky, emanating from material that appears to be at the specific but extremely low temperature of 3 K (three degrees above absolute zero) may seem like a strange concept at first. But it is not hard to understand, given the principles we have discussed. Consider the early history of the universe, shown in Figure 27-2. For the first few hundred thousand years, the temperature was so high that the hydrogen gas was ionized and opaque, like the gas in the Sun. When the temperature dropped to around 3000 K, the electrons could join with the protons to form neutral hydrogen atoms; by a million years after the big bang, the gas was no longer ionized. We know from studying the sharply defined surface of the Sun that ionized hydrogen can be quite opaque, even though interstellar cool hydrogen between the stars is quite transparent. So, for the first million years, the universe was filled with a brilliant glowing fog of plasma, and after a million years "the fog lifted" (Silk, 1980).

Now consider astronomers looking out through space. As they look out past nearby galaxies, they look through transparent space, very thinly populated by neutral hydrogen. But as they look farther away, they look back in time. This is shown schematically in Figure 27-3. At several thousand megaparsecs, their line of sight encounters the region where light left the brilliant fog about a million years after the big bang. This region is like a wall. It is opaque, and the astronomers can't see beyond it. But why doesn't it look like a 3000-K brilliant fog, resembling the surface of the Sun? The wavelength of the light they see from it is Doppler-shifted by about a factor of 1000. So the temperature they perceive (following Wien's law) is about 1000 times lower, or about 3 K. The 3-K radiation is really the light from the million-year flash that accompanied the big bang.

FORMATION OF GALAXIES

According to the big bang theory, including the inflationary versions, the densities of both radiation and matter decreased as material rushed apart after the big bang and the inflation. Probably within a few thousand years the density of matter became greater than that of radiation, allowing stronger gravitational forces between particles of matter. Clouds of gas organized themselves in the bubblelike pattern detected in galaxy cluster maps. After many millions of years, temperature and density were probably such that gravitational contraction began in the denser clouds of gas. Certain clouds or turbulent eddies had enough density and self-gravity to contract, perhaps to masses as large as entire superclusters of galaxies. Denser subregions eventually contracted to form individual galaxies, as seen in Figure 27-4. Thus

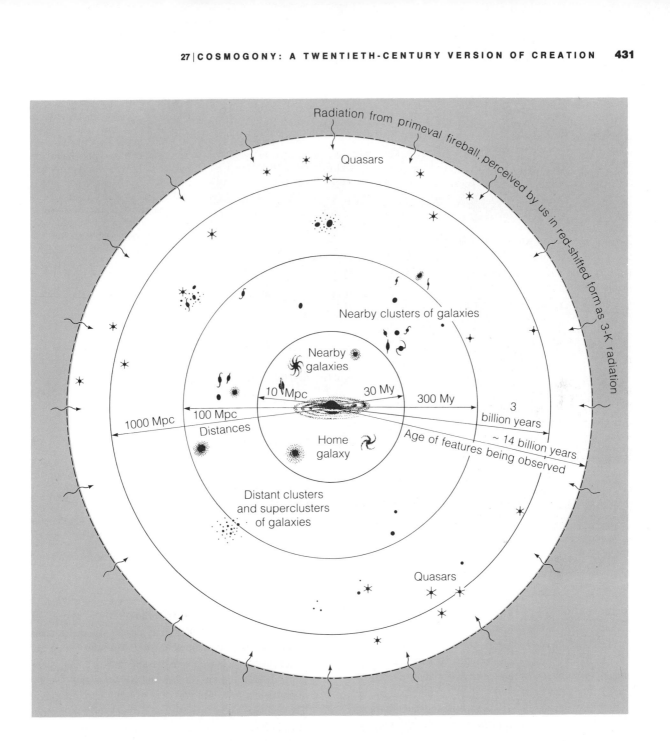

Figure 27-3 A schematic representation of the universe as seen from any galaxy, including ours. Note change of scale with increasing distance from the center. Clusters and superclusters populate space out to great distances. At very great distances, corresponding to a large red shift, we see light that was emitted billions of years ago, and the only visible objects are quasars. From an extremely red-shifted region beyond them comes radio radiation emitted during the era of the opaque fireball, about a million years after the big bang.

Figure 27-4 Galaxies without end. This very remote cluster of galaxies, the Coma cluster, is 100 Mpc away. All of the slightly elongated or fuzzy images are galaxies. The scene is dominated by two orangish giant ellipticals and one bluish foreground star (the image with diffraction spikes). Their lack of detail testifies to the difficulty of getting information about the most remote parts of the universe. (Courtesy L. Thompson, copyright N.O.A.O., Tucson, Arizona.)

formation of galaxies and galaxy clusters is a predictable result of the big bang.

Galaxies in large clusters show alignment of their rotation axes—supporting the idea that they are fragments of single, rotating, primitive clouds (Ozernoy, 1974). The era of galaxy formation can be dated by the statistics of quasars. A 1982 survey indicated a rapid drop-off in numbers of quasars as the red shift exceeds 3.53. This corresponds to an era about 1 to 4 billion years after the big bang. We see few earlier quasars. Therefore, astronomers infer that formation of most galaxies (some now seen as quasars) began about 1 to 4 billion years after the big bang, or perhaps 10 to 14 billion years ago (Schmidt, 1982). This is at least roughly consistent with a formation of our own galaxy roughly 14 ± 4 billion years ago.

Most galaxies finished forming by some 10 billion years ago. Therefore, researchers were taken by surprise when radio astronomers in 1989 discovered an interstellar hydrogen cloud that may be still contracting to form a new galaxy. It is about 20 Mpc away, near the Virgo cluster. It has enough mass to make a small galaxy but is $100 \times$ the size of the Milky Way galaxy. Does it imply that a few, late galaxies are still forming today? The answer is controversial, and the cloud is still under active study.

FORMATION OF HEAVY ELEMENTS

The first galaxies and stars were formed from the nearly pure hydrogen/helium mixture produced in the big bang. This is supported by the composition of the ancient Population II stars. As the theory of stellar evolution reveals (see Chapter 17), the **origin of heavy elements** came later, through nuclear reactions inside stars.

Heavy elements are still forming inside massive stars, and then being blasted into space when these stars explode as supernovae.

Later stars incorporate some of these ready-made heavy elements, as described in Chapter 23. Then, as they evolve, they cook up still more heavy elements in their "stellar pressure cookers."

BEFORE THE BIG BANG

What existed before the big bang? No one is sure that this question has meaning or that any observations could reveal an answer.

According to classic big bang theory, the mysterious explosive event at the beginning is *assumed* to have mixed matter and radiation in a primordial soup of nearly infinite density, erasing any possible evidence of earlier environments. If so, science would have no way of answering the question.

Conceivably, the big bang was not unique. This variant of big bang cosmogony is called the hypothesis of the **oscillating universe.** It can be visualized in the "closed universe" curve of Figure 27-1. In this view, after the primeval fireball explodes, gravity eventually pulls back the distant material and the universe collapses again, as seen in the turn-down at the right end of the curve. Another high-density period would follow, then another expansion, and so on.

In a variant of the oscillating universe hypothesis, the high-density era was not the infinite-density epoch of the classic big bang, but only a period of maximum, finite density, during which some structures from a preceding era might have survived.

In the non-oscillating big bang models, the big bang is the true beginning of the universe as we know it, a creation like those in the creation myths of many cultures. The oscillating universe model—that the big bang was not a true beginning, but only the beginning of the current cycle—resembles certain Eastern mythologies.

The new inflationary big bang theories add an interesting twist to these ideas. The physicists who developed these theories note that the properties of all the particles in the universe tend to cancel each other out. For example, there seem to be as many positively charged particles as negatively charged ones, so that the net charge in the universe is zero. Similarly, the net angular momentum, and other special properties of subatomic particles, seem to add up to zero. This has led some physicists to postulate that the big bang began from a special vacuumlike energy field in which all properties added up to zero and there were no material properties. It may have been, in cosmologist Alan Guth's words, a situation that "you can't distinguish from nothing." In other words, big bang theorists can now postulate, in a sense, that the universe did start from "nothing"— the ghostly creation field that exploded in a shower of positive and negative particles leading to the state we now see.

ARE THERE "OTHER PLACES"?

We have defined the universe as all matter and energy that exist anywhere. Until recent decades, most scientists had assumed that all parts of the universe had to be continuously connected, in the sense that light or a spaceship could go from one region to any other region and back again (even if the trip took a long time). But if there are black holes, there *might* be some regions permanently detached or unobservable from anywhere else. Some scientists speculate that there might be "other regions," such as regions "inside" black holes, in which events can occur but never be seen by us.

A still more intriguing speculation is that we might be in a black hole in someone else's universe. Contrary to earlier theories that black holes could never emit anything, recent work suggests that a black hole might be able to explode in a burst of gamma-ray radiation and subatomic particles. Thus the big bang itself might have been some kind of black hole explosion within some sort of larger universe.

Because black holes are thought to be permanently out of touch with each other and (so far) subject only to theoretical investigation, speculations on these problems are at the borders of what the scientific method can now deal with.

WHY?

Why did the universe come into being? If the preceding questions of "how?" seem difficult—if they seem at the borders of the scientific method—questions involving "why?" are clearly beyond the scientific method. Why was there a big bang? What started the bang? What was there before? What is the purpose behind it all? When thinking about these questions, we may find it helpful to recall a whimsical but deep comment by Mark Twain: "Why shouldn't truth be stranger than fiction? Fiction, after all, has to make sense. . . . " By this he meant that our cultural tradition tends to make us want to look for "sense" in a story—even the story of the universe. We like our stories to have a beginning, a middle, and an end—and perhaps even a nice moral that is consistent with what we believed before we heard the story. When we seek the story of the universe, we have no guarantee of this kind of sense. But we do have a virtual guarantee, based on experience, that if we keep making observations—for example, with improved telescopes in space—they will draw us on to new, fascinating discoveries about the universe.

Remember that the scientific method does not answer "why?" questions, except to show how events in nature follow from laws we have already discovered. The scientific method is basically a procedure for analyzing observations and predicting phenomena slightly beyond those we know. Thus people practicing the trade of science can only ask questions that can be answered by specific observations.

To questions of "why?" astronomers' answers may be little better than anyone else's. One can always make up answers, but it is more interesting to admit that there are many things we do not understand—and things we do not even understand *how* to understand.

SUMMARY

Three lines of astronomical evidence—ages of globular clusters, the expansion age calculated from Hubble's constant, and estimated ages of elements—indicate that the formative conditions existed in the universe roughly 10 to 18 billion years ago, probably close to 14 billion years ago. Most astronomers believe that an explosionlike event called the big bang occurred then. It takes some audacity to claim that we humans know something about this mysterious event, but there are several compelling observations:

1. Because galaxies are receding from each other, they must have once been concentrated in a much smaller volume; this is consistent with the big bang model of initial high density.

2. The big bang model explains the abundances of elements that we see both in Population II stars and, with the addition of later element-forming processes inside stars, in Population I stars as well.

3. The big bang model predicted a pervasive weak radio radiation, known as the 3-K radiation, *later* found by radio astronomers.

4. Theories of galaxy formation, counts of distant galaxies, and observations of the most ancient galaxies and

quasars, although not uniquely predicted by the big bang theory, are consistent with it.

The big bang theory takes us close to the limits of the ability of science to explain the properties of matter and energy in the universe. Questions such as why the universe came to exist are beyond the scope of science.

CONCEPTS

cosmogony
age of globular clusters
Hubble's constant
primeval fireball
time required to form observed elements
age of the universe
initial conditions

inflationary big bang theory
inflation
3-K radiation
formation of galaxies
origin of heavy elements
oscillating universe

PROBLEMS

1. Which phrase describes the big bang theory best?
a. An assumption with logical consequences that turn out to be verified by observation
b. A fact proved by repeated observation
c. A revelation

2. Why was the discovery of 3-K radiation from all over the sky heralded as strong evidence in favor of the big bang theory? Does comparing the radiation from different parts of the sky give any evidence of asymmetry or inhomogeneity in the universe?

3. Why would planets such as Earth be unlikely to exist in globular clusters? (*Hint:* Consider Earth's composition.)

4. How would spectroscopic observations help reveal that a cluster of very distant galaxies, which looks like a mere grouping of fuzzy stars on a photograph, is not a group of stars inside our own galaxy?

5. Why does the radiation from the primeval fireball as observed today have such long wavelength (radio waves) instead of the very short wavelengths (ultraviolet) that might be expected from the high temperatures theorized for the fireball?

6. Do you find any fundamental disagreement between the description of the universe's origin and structure, as described in this and the previous chapter, and any philosophical or religious beliefs you may hold? If so, do you believe such a disagreement might be clarified by further observations, or do you believe further observations are superfluous?

CHAPTER 28

Life in
the Universe

One of the most intriguing questions in astronomy is whether planets elsewhere in the universe harbor what we are pleased to call "intelligent life." Extraterrestrial life must either exist or not exist. Either case has striking consequences. The nineteenth-century Scottish writer Thomas Carlyle sardonically said that other worlds offer "a sad spectacle. If they be inhabited, what a scope for misery and folly. If they be not inhabited, what a waste of space."

If *intelligent* aliens exist, our society (if it survives) may someday be influenced by them, for better or worse. Anthropologist D. K. Stern (1975) notes that discovering such creatures "would irreversibly destroy man's self-image as the pinnacle of creation." If alien life does not exist, we are the only living creatures in the universe—a remarkable situation that implies a universe wide open for us to utilize and colonize.

What can astronomy say about these two possibilities? Some scientists argue that astronomy, lacking any definitive detections of extrasolar Earth-like planets or life, can say nothing. However, we certainly have data related to life's origins: **Organic molecules** (complex, carbon-based molecules) have been found in interstellar clouds; certain meteorites show that organic molecules have formed in our own solar system outside the Earth; and the paucity of organic molecules on Mars places intriguing limits on the biochemical possibilities of planets. Admittedly, discussions of the existence, intelligence, psychology, or appearance of higher alien life forms are almost entirely speculative. Nonetheless, a growing number of astronomers, biologists, chemists, anthropologists, and theologians are taking an interest in this new scientific adventure.

In this chapter, we consider the subject in several steps. First, we discuss what we mean by *life* and what conditions are necessary for life to exist on planets. Second, we review the long process that led to the evo-

435

lution of life on Earth. Third, we ask whether there are planets elsewhere in the universe that could support a similar evolution of life. Finally, we try to estimate the probability of intelligent life actually existing on other planets and whether we might contact it, or it, us.

THE NATURE OF LIFE

What do we mean by *life?* **Life** is not a status but a *process*—a series of chemical reactions, using carbon-based molecules, by which matter is taken into a system and used to assist the system's growth and reproduction, with waste products being expelled. The system in which the processes occur is the cell. All known living things are composed of one or more cells.[1] A **cell** is, in essence, a container filled with an intricate array of organic and inorganic molecules (protoplasm). Codes for cellular processes are contained in very complex molecules (such as the famous DNA) located in a central body called the nucleus. The elements most prominent in the organic molecules are carbon, hydrogen, oxygen, and nitrogen—all very common in regions populated by evolved (Population I) stars. Phosphorus, also important to life (in small amounts), is widely available. Carbon is especially critical because it can combine to make long chains of atoms—large molecules that encourage the complicated chemistry of genetics, reproduction, and so on. That is why the term **organic chemistry** (as well as *organic molecules*) refers not specifically to life forms but more generally to all chemistry (and molecules) involving carbon.

We often make the mistake of thinking of ourselves as static beings instead of dynamic systems. We casually assume that we are constant entities, as if our identity depended solely on the form of our bodies. But our bodies today are not the same ones we had seven years ago. Hardly a cell is still alive that was part of that body. Even our seemingly inert skeletons are living and changing, always replacing their cells. We *must* keep changing—the cells must keep processing new materials to stay alive. When the processing stops, we call it death.

The nature of living beings is illustrated in an analogy from the Russian biochemist A. I. Oparin (1962).

---

[1]Viruses might be an exception. They are simpler than cells, yet can reproduce themselves using materials from host cells. Biologists disagree on whether viruses should be considered a form of life.

Consider a bucket that has water pouring in at the top from a tap and flowing out at the same rate through a tap in the bottom. The water level in the bucket stays constant, and a casual observer would call it a "bucket of water." But it is not like an ordinary bucket standing full of water. The water at any instant is not the same water as at any other instant, yet the outward appearance is constant. We are like buckets with water, nutrients, and air flowing through us, but with other, much more complex attributes, such as the ability to reproduce and be affected by genetic changes that let us evolve from generation to generation.

The flow and change of material inside living organisms give us a clue to the kinds of processes involved in the origin of life. We are looking for a process in which complex, carbon-based molecules can enter cell systems and enable them to create new molecules, incorporate these molecules into new structures, eject unused material, and reproduce themselves.

Scientists therefore usually choose to define *life* by these specific carbon-based processes. Often at this point people ask, "What about some unknown form of consciousness? Or what about some unknown chemistry based on other elements, such as silicon, that can form big molecules?" The answer is that we have never observed such life forms, so we can say nothing substantive about them. If they are plausible, we simply increase the probability of life or consciousness in the universe. To judge whether life may exist on other planets, we must find out how those processes got started on Earth and what sort of environments fostered them (Figure 28-1).

THE ORIGIN OF LIFE ON EARTH

Understanding the origin of life requires that we understand the early history of interactions among the elements carbon, hydrogen, oxygen, and nitrogen—the most important elements in living organisms. How did these elements get together into complex molecules such as **amino acids,** the molecules that join to form **proteins,** the huge molecules in cells? And how did proteins aggregate into cells that could absorb surrounding materials, grow, and reproduce?

Production of Complex Molecules

Research in the last few decades has proved that the first steps toward life were surprisingly easy. Chapter

Figure 28-1 Precursors of life on Earth probably formed as a result of the interaction of environmental compounds with lightning as an energy source. Laboratory experiments simulating the early atmosphere and oceans, with sparks representing lightning, produce organic sludges such as the one pictured here floating in the Earth's primitive ocean. Any planet with such an environment would be expected to produce such organic materials. (Painting by Ron Miller.)

18 revealed that complex molecules such as ethyl alcohol (C_2H_5OH) and even the amino acid glycine ($C_2H_5O_2N$) have been discovered in molecular clouds in interstellar space. Extraterrestrial amino acids have also been found inside carbon-rich meteorites. That these molecules originated outside Earth is proved because the structural symmetries of their molecules differ from those of the same types of molecules on Earth.[2] These carbonaceous meteorites once contained liquid water in their carbon-rich soil; the amino acids must have formed in this moist environment in the cometlike or asteroid-like parent bodies of these meteorites.

Amino acids would have quickly formed in Earth's early environment. This was proved by an experiment designed to simulate chemical conditions in Earth's primitive environment. This experiment, called the **Miller experiment,** was first carried out by chemist S. L. Miller (1955), who bottled materials such as methane (CH_4) and ammonia (NH_3), representing Earth's hydrogen-rich primitive atmosphere, together with water, representing the ocean. He then passed sparks through this material to represent lightning. After a short time, Miller found that a sludge containing amino acids, such

as glycine, had formed in the flask. This experiment has been successfully repeated many times with different "primitive atmospheres" and energy sources.

In 1983 researcher C. Ponnamperuma announced that all five of the critical organic compounds responsible for coding genetic information in the DNA and RNA of living cells were found in a single carbonaceous meteorite. At the same time, he and his colleagues were able to synthesize them in a Miller-type experiment in a primitive atmosphere. This work seems to clinch the likelihood that many of life's building blocks arose naturally in extraterrestrial bodies.

In Chapter 12 we saw that chemical interactions involving sunlight have produced a photochemical smog on Saturn's satellite Titan. With its reddish clouds of organic compounds and its possible rains of liquid methane, Titan may give us an intriguing natural laboratory for further study of these processes.

This is not to say that all planets are oozing with organic slime. Sunlight encourages reactions that not only *form* organic molecules but also *break down* organic molecules. When energetic ultraviolet (UV) photons strike organic molecules, the molecules can break apart. Earth has a unique safeguard against this. High in our atmosphere is a layer of air that has a stronger-than-usual concentration of O_3 molecules, or ozone. The ozone molecules are very effective in absorbing UV light. Thus, this so-called **ozone layer** protects life on Earth's sur-

---

[2]The atoms can link into molecules with either "left-hand" or mirror-image "right-hand" symmetry; the ratio of the two types in the meteoritic material is strikingly different from that in terrestrial material.

Figure 28-2 Tidewater pools (foreground) may have facilitated the origin of life. Once organic molecules formed in seawater, evaporation of the water in stranded tidal pools would concentrate the remaining organics, promoting reactions among them and formation of more complex organic substances. Broths rich in amino acids and complex organic molecules resulted and were dumped back into the ocean at subsequent high tides, "fertilizing" the seas. (Photo by author, Baja California.)

face from much damaging UV light (although our suntans are an example of damage done by UV light to the cells of our skin). The protective function of the ozone layer explains recent concern over possible damage to Earth's ozone layer by industrial chemicals.

On worlds like the Moon and Mars, where there is no ozone layer to block UV light, the rate of destruction is high, and gases are insufficient to provide raw materials to form new organics. As noted in Chapter 10, sunlight has apparently destroyed any organic molecules that may have formed on Mars in the past. Indeed, fossil and geochemical evidence suggests life did not emerge on Earth's land surface until ocean plants, which were shielded from sunlight by seawater, emitted enough oxygen to build up an O_3 layer in Earth's atmosphere. This ozone layer then shielded land-based life from solar UV rays.

The Production of Cells

The next step toward life—aggregating complex organic molecules into cell-like structures—is less certain. Although some of the processes can occur in dry environments, liquid water was probably critical to extensive biochemical evolution, because it provided a fluid medium in which materials could move and aggregate. After all, we evolved from the sea, most of our body weight is water, and the earliest fossils are of sea creatures.

Florida biologist Sidney W. Fox has shown that simple heating of dry amino acids (as might happen on a dry planet) can create protein molecules. Once water is added, these proteins assume the shape of round, cell-like objects called **proteinoids,** which take in material from the surrounding liquid, grow by attaching to each other, and divide. Though they are not considered living, they resemble bacteria so much that experts have trouble distinguishing them by appearance.

Possibly related to proteinoids are objects discovered in the 1930s by Dutch chemist H. G. Bungenberg de Jong. When proteins are mixed in water solutions with other complex molecules, both sets of substances spontaneously accumulate into cell-sized clusters called **coacervates.** The remaining fluid is almost entirely free of complex organic molecules.

The next step toward recognizable life forms is still more uncertain. If organic molecules or coacervates are present in a pool of water, they will be left in the pool as the water evaporates. In this way, evaporation of the water in tidewater pools (Figure 28-2) probably produced high, localized concentrations of amino acids, proteins, and other molecules, allowing cell-like structures to form. The cell-like structures in the primeval pools of "organic broth" could have begun reacting with fluids in the pools and with each other, accumulating more molecules and growing more complex, as suggested by Figure 28-3. Eventually these could have evolved into biochemical systems capable of reproducing.

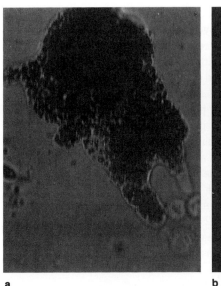
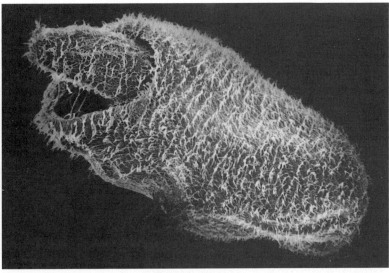

a   b

Figure 28-3 The importance of a fluid medium for primitive evolution is suggested by these photos of single-celled organisms engulfing food from surrounding fluid. **a** An amoeba flows to surround a nearby food particle. (Optical microscope photo; S. L. Wolfe.) **b** The protozoan *Woodruffia* ingests a *Paramecium*. (Electron microscope photo; T. K. Golder.)

Earth's Earliest Life Forms

Whatever the processes, microscopic cellular life must have arisen by 3.5 billion years ago, because the earliest fossil life forms date from that time, in rocks as far apart as Australia and South Africa. One such early life form, which still exists along sea coasts, is **stromatolites;** these are colonies of blue-green algae with cabbagelike size and structure. In the words of paleontologist Stephen Jay Gould (1978), life apparently arose "as soon as it could; perhaps it was as inevitable as quartz . . ."

Whether life evolved smoothly from that time forward is unclear. Some strata dating from 3.5 billion years ago lack fossils. Some researchers think life may have begun in fits and starts, with early species wiped out by giant meteorite impacts or climate changes, only to be replaced as life started again (Walker, 1988).

By about 2.5 billion years ago, life had become common, with stromatolites being a prominent form. Most life remained in the oceans, where water provided a supportive, protective environment. Life forms were mostly soft-bodied, and only rarely produced fossils, so that their development is hard to trace.

It is strange to realize that during most of Earth's history, life would have been hard to find in most *land-scapes.* Even during the Earth's middle age, the land was barren. Some areas must have looked like today's deserts or like Mars. Some areas were moist and washed by rains, but instead of luxurious forests, there were only bare acres, eroded gullies, and grand canyons. Brown vistas stretched to the sea.

Oxygen (O_2) content in the initial atmosphere was much lower than today. Evidence for this includes the fact that sediments older than 2.5 billion years show low oxidation levels, and that even today, stromatolites thrive best in oxygen-poor environments. Plants produce oxygen, however, and the flourishing of the earliest ancient plants drove the O_2 content dramatically upward between 2.5 and 2.0 billion years ago.

Once the O_2 content was high, solar UV broke some O_2 molecules, and the resulting free O atoms joined with O_2 molecules to make O_3 (ozone) molecules. This led to formation of the ozone layer that protected further evolution of life.

Living things changed because of these environmental changes. By about 1.5 billion years ago, new types of organisms appeared in the fossil record; these organisms were adapted to oxygen-rich air, and although they were less resistant to UV damage, were able to flourish because the new ozone layer protected the sur-

Figure 28-4 Lichens, symbiotic combinations of algae and fungi, were probably some of the first organisms on land. As shown here, they grow on rock surfaces, weathering the rock to produce soil. They are widely adaptable, existing in both arctic tundra and hot deserts. Three lichen colonies of red, black, and gray appear in this picture, whose width is about 6 in. (Photo by author, Tsegi Canyon, Arizona.)

face from the sun's UV radiation (Windley, 1977). This was the first big step in expansion of life from the sea onto the land (Figure 28-4).

Biologists agree that the mechanism by which new species evolved to fit the changing environment and the new environmental niches, such as land and air, was **natural selection:** Random changes in the genetic codes (as might be caused by UV irradiation or other processes) occasionally produced offspring with improved survival traits. In turn, these offspring lived longer and had more offspring of their own, promoting retention of the new trait.

A population of organisms on one arid continent, for example, might eventually evolve in a different direction from a population of the same organisms on another, wetter continent. (Witness the different animal populations on Australia and Asia.) A rapid change in global climate toward drier conditions might lead the first group to flourish while the second group died out. The first group would then spread rapidly from their home continent to other regions. The fossil record of a region thus might show the sudden emergence of the first group and extinction of the second. All these steps are probably part of natural selection, though the details of the process are still debated by biologists. Some details are uncertain because we often sample only discontinuous layers of fossil-bearing strata. Nevertheless, many fos-

sil sequences show clear, step-by-step evolution, as with the 50-million-year sequence showing the evolution of a ½-m-tall, three-toed mammal into the modern horse.

There is little mystery about the basic process. As Darwin pointed out long ago, the overall process is like the artificial selection practiced by breeders who develop new strains by producing more offspring between individuals with traits the breeders favor.

The adaptability of life forms is shown by the rapid proliferation of advanced species once they evolved, as indicated in the geological time scale shown in Table 6-1. Although it took about half the available time to go from complex molecules to algae and bacteria, it took only the last 12% of Earth history to go from the first hard-bodied sea creatures (such as trilobites, Figure 28-5) to humans.

PLANETS OUTSIDE THE SOLAR SYSTEM?

From all we have just said, we conclude that if planetary surfaces with the necessary conditions—liquid water and the "CHON" chemicals (carbon, hydrogen, oxygen, and nitrogen)—exist long enough anywhere, life is likely to evolve. Advanced species are likely to appear eventually. But are there such planets?

Marginal Evidence for Planets Outside the Solar System

In view of the hundred billion stars in our galaxy, not to mention the innumerable other galaxies, we should not limit our search for life to the solar system alone. There are four kinds of **evidence for planets near other stars,** but the evidence is not conclusive.

First, most newly formed stars are immersed in cocoon nebulae in which mineral grains have condensed. These grains may aggregate into rocky planetary bodies as a natural by-product of star formation.

Second, some main-sequence stars have systems of dust particles orbiting around them. Vega, for example, has a Jupiter-mass or more of dust particles at least a millimeter across, revealed by infrared satellite telescope. Figure 18-8 showed a photo of a similar dust system around the star Beta Pictoris. The dust may be a by-product of planet or asteroid systems around those stars.

Third, statistics of masses among binary and mul-

Figure 28-5 Fossil trilobite, a hard-bodied sea animal that appeared about 600 million years ago. (Gayle Hartmann and Joyce Rehm.)

tiple stars suggest the likelihood of many stars having low-mass companions that are planets.

Fourth, unseen low-mass companions have actually been detected around a few nearby stars. Some are between the mass of Jupiter and that of the smallest stars; they may be planetlike. (Some of them are discussed in Chapter 21, page 339.)

At the distances of most stars, planets smaller than Jupiter are undetectable by current methods, but only marginally so. Astronomers have proposed new devices and techniques to detect even Earth-sized planets near nearby stars. Among these are space telescopes and sensitive spectrometers that would detect radial-velocity oscillations of stars, caused by planets going around them (O'Leary, 1980). The search for planets near other stars is an exciting enterprise; it may finally clarify whether our own planetary system is commonplace or freakish. For the present, an informed guess might be that between 1% and 30% of all stars have at least one planet nearby.

Habitable Planets?

Even if planets do exist near some other stars, there is no guarantee that they are habitable. Astronomers have proposed several conditions needed to make a planet habitable.

1. The central star should not be more than about 1.5 $M_\odot$, so that it will last long enough for complex life forms to evolve (at least 2 billion years) and so that it will not kill evolving life with too much UV radiation, which breaks down organic molecules.

2. The central star should be at least 0.3 $M_\odot$ to be warm enough to create a reasonably large orbital zone in which a planet could retain liquid water.

3. The central star should not flare violently or emit strong X rays. It should be on the main sequence in order to have been stable long enough to give its planets long-term climatic stability. Conditions 1, 2, and 3 restrict us to main-sequence stars of spectral types F, G, and K.

4. The planet must orbit at the right distance from the star so that liquid water will neither evaporate nor permanently freeze.

5. The planet's orbit must be circular—stable enough to keep it at a proper distance and prevent drastic seasonal changes. Such stable, habitable orbits are less probable, but still possible in common types of binary systems (Harrington, 1977).

6. The planet's gravity must be strong enough to hold a substantial atmosphere.

The American physicist Stephen Dole (1964) reviewed these criteria and speculated that only a few percent of all stars, mostly from 0.9 to 1.0 $M_\odot$, have habitable planets. However, the figure remains highly uncertain.

HAS LIFE EVOLVED ELSEWHERE?

If habitable planets do exist, and if life evolves readily under habitable conditions, shall we immediately conclude that life must be abundant throughout the universe? There are several additional factors to consider. For one thing, planetary environments change with time, so that today's habitable planet may not be habitable tomorrow. This raises the question of the ability of life to adapt to change. The big problem is that we don't know how unusual Earth's history has been. Has life narrowly escaped complete extinction by climatic change in the past? Or was there never any question that natural selection would allow a few species to survive each step of Earth's evolution?

Effects of Astronomical Processes on Biological Evolution

Basic planetary or stellar processes may be involved in encouraging or hindering biological evolution. For example, Earth's fossil record indicates an episode called the "great dying," when as many as 96% of all existing land and sea species became extinct in less than a few million years (Raup, 1979). This happened about 240 million years ago (see Table 6-1, page 97); many paleontologists believe widespread climate changes were involved.

A number of geological and astronomical processes could have caused climate changes so massive that they affected the course of biological evolution. Among these are:

1. Convection of Earth's mantle may have driven plate tectonic crustal splitting, which in turn changed sea levels, ocean currents, wind patterns, and seasonal extremes. These changes affected global climate.

2. Volcanic eruptions may have spewed enough dust into the high atmosphere to reduce the sunlight reaching the surface. For example, widespread sunlight decreases of up to 25% occurred after the 1883 Krakatoa (Sumatra), 1912 Katmai (Alaska), and 1982 El Chichon (Mexico) eruptions. Rarer larger eruptions could cut sunlight for some years, because of dust trapped in the stratosphere, lowering temperatures and causing the decline of some species and ascendancy of new ones.

3. Slight changes in Earth's orbit and the tip of the planetary axis to the plane of the ecliptic, caused by gravitational forces, are believed to have caused major climatic changes, such as the ice ages. (See, for example, Imbrie and Imbrie, 1980; Covey, 1984.)

4. Slight changes in the Sun's radiation may have changed climates.

5. Irradiation from a nearby supernova could have affected the climate or directly damaged organisms.

6. An asteroidal or cometary impact or atmospheric explosion could have damaged the ozone layer, exposing organisms to enhanced radiation. Turco and others (1981) found that the Tunguska asteroid or comet explosion of 1908 (Sekanina and Yeomans, 1984) depleted up to 45%

of the ozone in the Northern Hemisphere, consistent with Smithsonian measures made in 1909–1911.

7. The impact of a large asteroid or comet would also have created a giant explosion crater, kicking up much more dust into the atmosphere than a volcanic eruption. The dust could have obscured the Sun and made midday as dark as night for periods of months.

The first six processes above are somewhat theoretical. If they did not affect the evolution of life on Earth, they might have affected evolution on other, hypothetical planets. But very strong evidence has emerged that the seventh process really did happen, with devastating results. It is worth discussing in more detail.

For many years, geologists have known that a dramatic break in the fossil record, second only to the "great dying," happened 65 million years ago. Some 75% of species (notably dinosaurs) became extinct within a few million years, but no one knew why (Russell, 1982). Around 1980, geochemists astonished the biologists with new evidence: They found that the soil layer exactly at this boundary contained an excess of elements associated with meteorites, and they concluded that a 10-km diameter asteroid struck Earth 65 million years ago (Alvarez and others, 1984). Further studies of the soil layer showed that worldwide fires were touched off, and giant tsunamis, or so-called tidal waves, swept inland from many oceans, probably within a few hours of the impact. Calculations suggested that dust from the impact blocked sunlight for months. The effects were complex but amazing. Fossil records indicate extinction of 44 to 49% of all genera of light-dependent floating marine organisms, but extinction of only 20% of the (less light-dependent) land genera (Ahrens and O'Keefe, 1982). The decimation of CO_2-consuming plankton in the oceans caused the CO_2 concentration in the atmosphere to rise. Although many land animals survived the period of darkness, the rising CO_2 content may have caused a greenhouse warming that led to massive extinctions of land animals during the next 50,000 y or so.

Similar geochemical evidence suggests that a much smaller asteroid, about ½ km in diameter, impacted about 2.3 million years ago in the ocean southwest of South America. This small asteroid caused milder but still dramatic climatic effects, apparently triggering wide glaciation as a result of lowered temperatures caused by clouds (Kyte and others, 1988).

From such evidence, many scientists now believe that the evolution of life on Earth has been influenced by random but dramatic astronomical events.

Adaptability and Diversity of Life

The great variety of ancient and modern species on Earth and the variety of environments in which they have thrived suggest that, given time, life could also have evolved to fit a wide range of conditions on other planets. Even humans have a remarkable adaptability. We thrive from equatorial wet jungles to dry deserts to arctic plains to Andean summits where air pressure is barely half that at sea level.

We can survive a 3% variation in body temperature, from about 303 to 313 K, and simpler organisms can withstand much wider variations. Certain bacteria that appeared about 3 billion years ago can withstand 18% temperature variations, over a range of 70 K (126°F). The environmental range covered by *all* terrestrial species is even greater. For example, some microorganisms live in Antarctic ponds that remain liquid at 228 K (-49°F) because of dissolved calcium salts, and others live in Yellowstone Hot Springs at temperatures of 363 K (194°F).

Habitable pressures range over a factor of 1000. Bacteria exist at altitudes where the atmospheric pressure is only about 20% of normal, whereas more advanced organisms live at ocean depths with pressures of hundreds of atmospheres.

Although sunlight is the ultimate energy source for all familiar organisms (because the ecological food chain leads back to plants that flourish by photosynthesis), recent discoveries show that not even sunlight is necessary to terrestrial life. Oceanographers in 1980 announced the discovery of colonies of strange deep sea animals clustered in the darkness around seafloor volcanic vents. Their metabolism is based on volcanic heat and hydrogen sulfide (H_2S) gas, utilized by bacteria that form the bottom rung of the local food chain (Edmund and Von Damm, 1983). The local pressure is 250 atm, nearly three times that on Venus. Some bacteria recovered from these environments thrive and reproduce at 523 K (482°F), extending the temperature range for known life to a value more than two-thirds that on Venus. The research is still being debated, but the 523 K figure has been suggested as a rough upper limit for life (Brock, 1985), because known amino acids break down at higher temperatures. The seafloor vent life forms show that a pleasant, sunny planet with a clear atmosphere is not the only possible locale for life!

Could any terrestrial organisms withstand the environments of other planets? For simple organisms, probably yes. Grass seed can germinate in a variety of atmospheres composed of carbon, hydrogen, oxygen, and nitrogen compounds. Eight species of insects (relatively complicated creatures) studied at different pressures behaved normally all the way down to 10 to 16% of normal atmospheric pressure on Earth (Siegel, 1972).

For these reasons, spacecraft have been at least partially sterilized to avoid contaminating other planets. Such contamination might not only alter planetary environments but also ruin any chance for us to discover whether simple organisms or complex organic molecules ever formed independently on those planets.

It works the other way, too. Simple organisms from other planets might have devastating effects if they ever reached Earth. There are historical examples of similar events. The plague caused by bacteria introduced into Europe in the 1300s killed about a quarter of all Europeans and as many as three-quarters of the inhabitants in some areas. Diseases introduced into Hawaii after the first European contact in 1778 killed about half of all Hawaiians within 50 years. Some 95% of the natives of Guam were wiped out by disease within a century of continued European contact. For these reasons, early Apollo astronauts were quarantined until it was clear they carried no lunar organisms.

With these facts as well as cultural competition in mind, anthropologist D. K. Stern (1975) made the grim remark: "It is likely that the meeting of two alien civilizations will lead to the subordination of one by the other."

Would They Look Like Us?

Opinion is divided on this point. Species with similar capabilities in similar habitats may evolve to look alike. For example, three different species of large sea creatures "designed" for fast ocean swimming look alike: an extinct reptile, the ichthyosaur; a fish, the shark; and a mammal that returned to the sea, the dolphin. Aliens living on planetary surfaces in gaseous atmospheres and using "intelligence" to manipulate their surroundings with tools might well have bilateral symmetry, appendages used as hands, a pair of eyes designed (like ours but unlike most animals') for stereo vision, and so on.

On the other hand, the famous paleontologist George Gaylord Simpson (1964, 1973) has argued that although life is likely to start, the long chain of environmental changes and evolutionary steps required to produce humans is unlikely to be approximated elsewhere, so there is likely to be a "nonprevalence of humanoids." Simpson criticized the whole attempt to estimate the probabilities of such life as nearly meaningless because of our lack of experiments or examples.

In short, natural selection seems to produce species capable of occupying any habitable environment. Thus we should not be surprised if life has evolved on another planet. But this life may look very strange to us, even if it displays recognizable intelligence. After all, if mushrooms and corals and woolly mammoths and Venus flytraps all evolved on one planet, how much greater may be the differences between life forms on two different planets? Feathers and fur and sex and seeds and symphonies may be the unique products of Earth.

Effects of Technology on Biological Evolution

If life will evolve when the right conditions exist, and if the conditions probably do exist elsewhere, then what can we predict about that life? Should we predict intelligence and civilizations? What do these terms mean? Should we assume that other civilizations will achieve spaceflight or might visit us some day? This raises the question of technology and its role in the evolution of life. Just as cosmologists have only one universe as an example, **exobiology** (the study of possible life on other worlds) has only one inhabited world as an example, so the answers to these questions are uncertain. Many exobiologists have assumed that intelligence by definition involves use of tools to modify the environment.

Whereas limited environmental modification can be helpful, too much or too rapid a change can be fatal! As Pulitzer Prize-winning naturalist René Dubos points out, we are umbilically connected to Earth, and if we alter our planet too much before acquiring the ability to leave it, we are finished. For this reason, the development of technology could actually end civilization on some worlds.

On our own planet, for example, technological "advances" have produced nuclear, biological, and other types of weapons that could involve the whole world in war. For example, the radioactive strontium-90 produced by nuclear explosions has a half-life of 28 y. It was blasted freely into the atmosphere before the Nuclear Test Ban Treaty of 1963. A year after the first H-bomb tests by the United States in the Pacific, strontium-90 deposits in American soil increased soil radioactivity by $\frac{1}{2}$%. A few weeks after a 1976 Chinese nuclear test, airborne radioactive debris fell onto the United States, increasing radiation levels. Obviously, nuclear war could devastate not only civilization but also future forms of life, whose genetic pool would be exposed to high radiation levels for decades. Humanity has thus proved that a planetary culture could wipe itself out by conscious design of weapons, as irrational as that may seem.

This disaster *could* also happen by mistake. As our technology reaches planetary scale, our accidents can also involve the whole planet. The Soviet Chernobyl nuclear power plant accident, for example, affected large regions of northeastern Europe. Some currently planned nuclear power plants will produce radioactive plutonium wastes, among the most toxic of known materials. Our creation of highly radioactive substances presents a technological danger that our society has allowed to develop in spite of knowing about it ahead of time.

The seemingly harmless aerosol can is a different story. In 1974, several planetary scientists realized from theoretical calculations that when Freon (the propellant gas used in such cans) is released into the air, it reaches the Earth's protective ozone layer, high in the stratosphere. Here its breakdown products destroy the ozone layer. Depletion of the ozone would let more solar UV radiation reach the surface, increasing the risks of skin cancer. Even as more Freon was being sold, planetary astronomers and chemists published refined calculations showing that to deplete the Freon already in the atmosphere could take as long as a century. After a 1975 report by the National Academy of Sciences confirmed this danger, moves were made in 1976 and 1977 by American regulatory agencies to phase out Freon aerosol propellants—an example of a technological danger that almost slipped by.[3] Thus, our generation's consumer gimmicks could be creating future health hazards. (See also Schneider, 1989.)

---

[3]But we should be haunted by the words of Harvard planetary physicist Michael McElroy, who performed some of the earliest Freon calculations: "What the hell else has slipped by?"

Our purpose here is not just to toll fashionable bells of doom and gloom. We all hope that by exercising a bit of intelligence and caution, humanity can recognize these dangers and avoid them. Nonetheless, if humanity is any example, long-term survival of a planetary culture is not assured. Although we have been around for less than 1% of the age of our planet, we are already beginning to have brushes with global disaster. Thus we can speculate that if evolution produces intelligent societies that remain tied to one planet, some may last less than a percent of the age of the universe—reducing the chance that a nearby culture will be around at the same time that we are.

On the plus side, we have succeeded in identifying the **cultural hurdle** that we (and perhaps intelligent species on any planet) must surmount: the transition from scattered, competing nation-states with the capability to damage the planet to stable global or interplanetary societies of intelligence and imagination. Perhaps some alien cultures have crossed this cultural hurdle and spread across many planets, thus ensuring their survival against ecological disaster on any one planet. Such cultures might last and be detectable for millions of years instead of thousands.

Our conclusion so far is that although some fraction of the stars have planets, most of these planets may be inclement for life, as symbolized by Figure 28-6a and b. But a fraction of the planets may have produced life, as symbolized in Figure 28-6c. Of these, a fraction may have produced civilizations that now exist or have been destroyed. The next question is whether any actual evidence for extraterrestrial carbon-based life as we know it exists today.

Alien Life: In the Solar System or Among the Stars?

Spacecraft exploration of the planets has virtually disproved advanced extraterrestrial **life in the solar system** and has probably ruled out any extraterrestrial life whatsoever in our system. The terrestrial planets except for Earth seem devoid of life, according to our space probes, and the outer planets seem too cold.

There is no direct evidence of life among the stars, but American, Soviet, and other scientists in international meetings have put together a method for considering the possibilities (Sagan, 1973). The logic is to try to estimate the various fractions—of stars having planets,

of those planets that are habitable, of those where conditions remain favorable long enough for life to evolve, of those where life does evolve, of those where intelligence evolves, and of the planet's life during which intelligence lasts. The product of all these fractions is the estimated fraction of all stars harboring intelligence. This statement is sometimes called the **Drake equation,** after radio astronomer Frank Drake, who first formalized this scheme for discussing the possibility that alien intelligence exists.

Table 28-1 on page 447 shows "optimistic" and "pessimistic" estimates of the various fractions, and consequent estimates of the upper and lower limits on the fraction of stars that might have civilizations today. In the optimistic case, the answer comes out to be about one star in 50; in the pessimistic case, only one star in 100 million million. These are little more than educated guesses, because we lack the data to make better scientific estimates. But this need not stop us from considering the consequences of our conjectures.

In the optimistic case, the nearest civilization might be only 15 light-years away—amazingly close. At the speed of light it might take only one generation to reach it. In the pessimistic case, the closest civilization would probably not be in our own galaxy, but roughly 10 million light-years away (a few megaparsecs) in a distant galaxy. Nonetheless, there are so many galaxies that it is hard to avoid the conclusion—even with the pessimistic view—that **life outside the solar system** is likely, with millions or billions of technological civilizations. If so, then at this moment intelligent creatures may be pursuing their own ends in unknown places under unknown suns.

Where Are They?

Radio astronomers have listened for radio messages from alien civilizations and heard none. Nor do our skies seem to be overrun with alien visitors trying to contact us. Nonetheless, theorists have recently pointed out that if *any one* civilization began sending out multigeneration slower-than-light starships that set up colonies, which in turn sent out new starships in a few generations to other stars, that civilization could populate all the habitable planets in the galaxy on a timescale short compared to the galaxy's age. (The time required would be long compared to the timescale of evolution, so that the original civilization might produce many new species as it went.) So why haven't we heard from aliens?

a

b

c

Figure 28-6 Too hot, too cold, and just right. Scenes on imaginary planets emphasize that even if extrasolar planets do exist and have atmospheres and potential for life, establishing the right climate for life to evolve may be chancy. **a** The surface of this planet, too close to a pair of hot, massive stars, is a sun-blasted desert. **b** On this world, far from a red and white star pair, water is all frozen. **c** Some worlds may exist where liquid water persists and life forms have evolved. (Paintings by Don Dixon, Ron Miller, and the author, respectively.)

One answer could be that there are no other aliens—intelligent life on Earth is some sort of unique accident. Another answer might be that we *are* being visited—witness the "flying saucer" reports. Some UFO reports are intriguing, but the verifiable evidence for actual alien spaceships is abysmally poor. (For further discussion of UFOs, see Enrichment Essay A.) If any alien visits have actually occurred, we have no proof of them.

A third possible answer is that the pessimistic figures are right and the nearest civilizations are in distant galaxies. Even their radio messages, if any, could be 10 million years old by the time we receive them, and their spaceships would be unlikely to reach Earth if limited to speeds less than that of light, as current physics requires.

A fourth possible answer is that biological evolution need not produce creatures who have a desire to build "civilizations" or travel through space. Not even all human societies necessarily evolve toward technology. Are humans fated to be explorers, bridgebuilders, and businesspeople rather than artists, athletes, or daydreamers? Perhaps the aggressive technocracy that we call

TABLE 28·1

Estimated Fraction of Stars with Planets Having Intelligent Life

	Plausible Lower Limit	Plausible Upper Limit
Stars having planets	0.01	0.3
Those stars ever having habitable conditions on at least one planet	0.1	0.7
Those planets on which such conditions last long enough for life to evolve	0.1	1
Those planets on which life does evolve	0.1	1
Those planets on which habitable conditions last long enough for intelligence to evolve	0.1	0.9
Those planets on which intelligence does evolve	0.1	1
Those planets on which intelligent life endures	10^{-7}	0.1
Product of fractions in column: Fraction of all stars with planets having intelligent life	10^{-14}	0.02
Implication: Distance to nearest civilization	10^7 ly	15 ly

modern civilization is just one type of *cultural* activity rather than a universally achieved stage of *biological* evolution. Historically, patterns we once assumed to be biological have turned out to be merely cultural. (Confusion of these two has led to the racist and sexist biases we are still trying to overcome.) We certainly cannot expect the common science fiction movie scenario, in which humanlike aliens walk out of saucers and invite us to join their democratically constituted United Planets—a galactic organization structured by documents that are curiously reminiscent of the U.S. Constitution. So why assume that other civilizations would even try to visit us?

This brings us to a fifth and perhaps most significant answer: We may be farther from aliens in evolutionary time than in physical space. Even if another planet started evolving at exactly the same time as ours, and even if its biochemistry produced creatures like us, those creatures are not likely to be in a phase of evolution similar to ours. If the evolutionary "clocks" on the two planets got only 1% out of synchronization, they would be 50 million years ahead of us or behind us—as far from us evolutionarily as we are from early mammals. Thus even if other planets produced civilizations we could recognize, contact might have to come in a very narrow time

interval in order for us to see any recognizable common interests. This is suggested by Figure 28-7.

Radio Communication

Possibly our period of isolation may be nearing an end. For half a century we have been broadcasting radio communications among ourselves. Already our unintentional but weak alert is more than 60 ly out from Earth. Aliens may one day pick up our signals and send radio messages (or an expedition?) in return. Radio astronomers in both the United States and the Soviet Union are therefore still conducting modest programs to listen for such messages with large radio telescopes. In one such search, radio astronomers listened for broadcasts near 1420 MHz (the 21-cm wavelength that marks an astronomically important radiation from neutral hydrogen atoms in space), targeting on all solar-type stars (185 in all) within 25 pc and known to be members of multiple star systems. No artificial broadcasts were identified (Horowitz, 1978). Goldsmith and Owen (1979) summarize other searches involving more than 600 stars. All were negative. But in terms of total listening time and the total number of possible wavelengths to tune in, the search has hardly begun. Larger listening instru-

Figure 28-7 Even if 10,000 alien expeditions have visited Earth at random times during our planet's history, visits would have averaged hundreds of thousands of years apart. Even the most recent expedition in such a scenario would be unlikely to have left any readily recoverable historical or archaeological traces. (Painting by Jim Nichols.)

ments have been proposed in both the United States and the Soviet Union.

Meanwhile, we have sent a few more deliberate messages. Several space vehicles, such as Pioneer 10 (which flew by Jupiter and left the solar system in 1973), carry plaques designed to convey our appearance and location to possible alien discoverers of the derelict spacecraft sometime in the future. The first radio message was sent from the large radio telescope at Arecibo, Puerto Rico, in 1974, beamed toward globular star cluster M 13, which is 27,000 ly away.[4]

SUMMARY

Discovery of firm evidence for alien civilizations or even alien life forms could be a pivotal development in our view of ourselves, as suggested in Figure 28-8. Yet questions of exobiology, especially of intelligent life on other worlds,

leave us with a mystery. Experimental evidence indicates rather strongly that life should start on other planets if liquid water, energy, and the right chemicals are present. Astronomical evidence suggests, but does not prove, that habitable planets ought to exist elsewhere in the universe. Biological evidence shows that life is adaptable and species can evolve to fit different environments, from ocean depths to low-pressure atmospheres of different composition. Yet advanced life has not evolved to survive on Mars.

Although the limited evidence indicates that other life forms should exist, there is no evidence that they do or that they have tried to communicate with us. We can only speculate about the reasons. Perhaps they are too far away. Perhaps most civilizations destroy themselves before successfully expanding into the universe. Perhaps evolution carries them beyond a stage where they would care to communicate with us. Perhaps they are unrecognizable.

Arthur C. Clarke has remarked that any technology much advanced beyond your own looks like magic. Perhaps we are too limited by our own concept of civilization. After all, one creature's civilization may be another's chaos, as shown by Mohandas Gandhi's remark when asked what he thought of Western civilization: He said it wouldn't be such a bad idea. It seems likely that our first contact with aliens (if they exist) might be as incomprehensible as the dramatized contact in the closing segment of Clarke's novel and the Kubrick–Clarke film, *2001*.

Clearly we have been reduced to speculation by a lack

---

[4]Astronomers in these projects have been deluged with letters ranging from support to complaints about the nudity of the human figures on the Pioneer 10 plaque. One telegram read: "Message received. Help is on the way—M 13." Its authenticity might be questioned, because the round-trip message time to M 13 is 54,000 y!

a

b

Figure 28-8 Discovery of positive evidence for extraterrestrial civilizations could be a pivotal and dramatic event. It could occur **a** by our going out and finding it (discovery of an alien artifact on the Moon, as dramatized in *2001—A Space Odyssey*, MGM Studios) or **b** by its arriving on Earth (arrival of mother ship as dramatized in *Close Encounters of the Third Kind*, ©1977 Columbia Pictures Industries, Inc.).

of facts. Indeed, the whole field of exobiology has been criticized as a science without any subject matter. Exobiology recalls Mark Twain's comment "There is something fascinating about science. One gets such wholesale returns of conjecture from such a trifling investment of fact." The only way to reduce the conjecture and increase the proportion of fact is to pursue research in many related fields—physics, chemistry, geology, meteorology, and biology—and listen to the skies with radio telescopes. There may be surprises waiting out there.

CONCEPTS

organic molecule	natural selection
life	evidence for planets near other stars
cell	
organic chemistry	exobiology
amino acid	cultural hurdle
protein	life in the solar system
Miller experiment	Drake equation
ozone layer	life outside the solar system
proteinoid	
coacervate	
stromatolite	

PROBLEMS

1. Compare the probability of detecting radio broadcasts from intelligent creatures native to hypothetical planets near Alpha Centauri, Gamma Cephei, Sirius, and Betelgeuse. See data mentioned in this and other chapters.

2. Describe several ways in which Earth's geological evolution has affected the evolution of life.

3. Describe several ways in which extraterrestrial events, such as solar or stellar evolution, might have affected the evolution of life on Earth or other planets.

4. Describe ways technology could affect the survival of intelligent life on Earth or other planets. Construct scenarios of (*a*) the possible destruction of life and (*b*) the assured survival of life. (*Hint:* In case *b,* consider the impact of space travel.)

5. In your own opinion, what would be the long-range consequences of the following:

a. Arrival of an alien spacecraft and visitors at a prominent place, such as the UN Building

b. Discovery of radio signals arriving from a planet about 10 ly distant and asking for two-way communication

c. Proof (by some unspecified means) that life existed *nowhere else* in the observable universe

6. Given the assumption that our technology has the potential for creating planetwide changes in environment, defend the proposition that *if* life on other worlds produces technologies like ours, then that life is likely either to have become extinct or to be widely dispersed among many planets by means of space travel.

7. In view of the devastation wrought on many terrestrial cultures by contact with more technologically advanced cultures, which would you say is the safest course: (1) active broadcasting of radio signals to show where we are in hopes of attracting friendly contacts; (2) careful listening with large radio receivers to see if there are any signs of intelligent life in space; or (3) neither broadcasting nor listening, but just waiting to see what happens?

a. How would the results of our listening program be affected if other intelligent species had reached the first, second, or third conclusion?

b. If we listen at many frequencies and pick up no artificial signals, does that prove that life has not evolved elsewhere in the universe?

8. Why would native life be unlikely on a planet associated with:

a. An O or B star?

b. A red giant star?

c. A white dwarf?

d. A pulsar?

9. For each known extraterrestrial planet in the solar system, list several environmental factors believed to be adverse to the existence of advanced life forms.

10. According to present astronomical theory, the Sun should eventually use all its hydrogen and turn into a red giant.

a. When will this happen (see Chapter 19)?

b. How does this compare with the time scale of biological evolution?

c. Would you expect the human species to be recognizable by the time this happens?

d. In what way might life that originated on Earth conceivably escape such an event?

The Cosmic Perspective

Just as a vacation in an interesting place can help us view our own workaday cares from a clearer perspective, our astronomical explorations to the ends of the known universe help us view our earthly cares from a clearer, cosmic perspective. In this epilogue I will give some personal thoughts about how our cosmic journey reflects on our daily lives.

This cosmic perspective has had a deep impact on humanity during the last few centuries. Astronomical experience has outmoded once popular ideas, such as:

1. The idea that there are physical heavens and hells directly above and below Earth

2. The idea that Earth is the center of the universe, around which all other bodies revolve

3. The idea that humans are the inevitable lords of creation

4. The idea that the heavens are unchanging, or that they are constant except for occasional sudden creation of new features

5. The idea that Earth is the sole and inexhaustible reservoir of raw materials

What has the cosmic perspective given us in exchange for these cherished ideas? Space extends without known limit around us—a new frontier. Planets circle suns; suns circle mysterious galactic cores. Humans are a special part of a great chain of life, in which many life forms have been linked. We don't know if this chain of life is strong and stretches among many planets in our galaxy, or if it is weak and exists only in very rare, scattered segments. The heavens evolve constantly. Humans have watched new starlike objects appear and old stars explode. Even Earth itself wobbles so that Polaris will be the North Star for only a few centuries. Although we are exhausting Earth's finite resource-rich surface layer, we may be able to utilize cosmic sources of energy and material with less damage to Earth than is caused by our present resource-seeking and resource-consuming activities.

Some critics have said that science has only demeaned humans by putting us on a satellite of an average star on one side of an average galaxy. Certainly these ideas shocked people during the Renaissance. But as astronomer Harlow Shapley asked:

Are we debased by the greater speed of the sparrow, the larger size of the hippopotamus, the keener hearing of the dog, the finer detectors of odor possessed by insects? We can easily get adjusted to all of these. . . . We should also take the stars in stride. We should adjust ourselves to the cosmic facts.

Science has also been criticized for replacing outmoded ideas only with machines, which have brought us the evils of pollution, nuclear weapons, and a dehumanized existence. This criticism seems to me to confuse science (knowing) with technology (using knowledge and material). There is a difference between knowing and using. To me there is something appealing about knowing as much as possible and using as little as necessary. The popular criticism of "science and technology" (lumped together as if one) is really a criticism of our *social* mechanisms—how we have *used* knowledge and materials.

And it is justified. There is great irony in the fact that despite 10,000 years of effort to produce labor-saving devices, many people have trouble getting by on 40-hour-per-week jobs where they labor to make products that do not even interest them and that they themselves may label as "consumer gimmicks." This is a failure of our overall culture, rather than a failure of science or technology. The cosmic perspective is making this clearer, as more people begin to think in terms of our total planetary budget of materials, energy, and creative talent. The cosmic perspective helps us see that the political and social difficulties of maintaining a stable, workable civilization are at least as challenging as the problems of scientific exploration.

Besides giving us a new view of our place in the universe and our planetary society, the cosmic perspective gives us facts on which to base our actions. Facts, of course, are merely conclusions drawn from observations, usually verified by generations of observers. They provide a foundation for practical life in the real world. They grade indistinguishably into hypotheses about how the universe acts and how it is put together.

If a philosopher–critic argues that we can never really know facts, a scientist can agree and argue that acting on hypotheses is like betting. If you are forced to put your money on one thing or another, you try to put it on what is most likely to be successful. The world forces us to act, and we want to act on the basis of the ideas that are most likely to be true. So we *act* on our understanding of physical laws when we design bridges, and we *act* on our understanding of the solar system when we choose not to waste time or money on astrological advice. The scientific method of learning about the universe is simply a scheme for accumulating hypotheses on which to act until better information comes along. Hypotheses are useful because they suggest possibilities of new discoveries, new principles, and new places to go. They suggest goals, tests, and experiments.

This brings up the attempts, mentioned in Chapter 14, to legislate a retreat from scientific debate about our cosmic environment. According to some critics of science, the last century of scientific exploration has demeaned human and religious values, and society would be better served if the theories resulting from scientific research could be suppressed or force-fitted into one ideological mold or another. Key among the theories criticized is the theory of biological evolution.

In the United States, for example, lawsuits and legislation in several states have called for bans on public school teaching of evolution, or "equal time" presentations of some alternative hypothesis, usually a "creationist" hypothesis involving the sudden, one-time creation of the universe, the Earth, geological features, and species. The proposed creationist hypothesis is generally derived from fundamentalist (literal) interpretations of the Judeo–Christian Bible, raising the question of why a science teacher (if he or she *must* teach a tradition outside the scientific method) should choose this particular creation tradition over the creation traditions of other cultures, such as Hindu, Buddhist, or native American Indian. Legal attacks on the teaching of biological evolution continued in the 1980s. Nonetheless, "equal-time" creationist laws have been struck down by U.S. courts on the constitutional grounds that the state is not to be involved in teaching religious traditions.

These trends leave unclear whether only the theory of biological evolution of species will come under legal attack, or whether the attack will one day spread to concepts of stellar evolution, planet growth, and the evolution of the Earth's atmospheric and geological environment, since some creationist traditions propose instantaneous or seven-day formation of all present features in the physical universe. Many backers of the current "creationist" legal cases reportedly believe the world was created only 10,000 years ago.

The popular press has presented this as an argument of "science versus religion," but I fail to see it in quite such an adversary way. It seems to me that real religious awe and scientific curiosity are very similar; both involve the recognition of something "bigger than we are," about which we don't have final answers. When we ask the right questions, nature suggests answers that we can act on. To turn away from this dialogue with the universe and insist that science courses include someone's interpretation of old writings seems both an antiintellectual and antireligious abandonment of the talents we've been given.

Earlier chapters showed examples of civilization getting into trouble in this way before. For example, Chapter 3 described how medieval scholars accepted the views of Aristotle and Ptolemy on the Earth's place among the planets, arresting and even executing dissenters who later turned out to be correct. The Renaissance came as scholars began to debate these questions not with old manuscripts but with *new observational evidence*.

Science (and science courses) involve the discussion of *evidence*—that is, nature. Any hypothesis—about evolution of species, evolution of stars, the sudden appearance of humanity in the Garden of Eden, the sudden creation of everything in modern form 10,000 years ago, the big bang, relativity—can be suggested and discussed scientifically as long as the discussion centers around observation, experiment, calculation, and prediction. The answer to the current attack on science is not, one hopes, to legislate that favorite theories must be taught, but rather to continue the most open, rich, vigorous discussions of hypotheses as tested against evidence. Out of such a free discussion of evidence, one is free to shape one's philosophical beliefs—a practical cosmic perspective. (See Enrichment Essay A for further thoughts on separating science and nonscience.)

Does the cosmic perspective have further bearing on economic, political, social, or philosophical problems of our civilization? Each person brings his or her own experience into answering such a question, but here are some loosely connected thoughts that might be considered:

1. A classic science fiction plot deals with a vast interstellar spaceship, kilometers long, in which generations of men and women live and die, maintaining a stable environment and culture while en route from one part of the galaxy to another. This plot turns out to be the truth. We have only recently realized that we are on board just such a vehicle—Spaceship Earth—with its finite area and finite resources.

2. There must be few places in the universe where we can walk naked away from our machines, breathe in the atmosphere, let the light of the nearest star fall on us, and find water to drink on the surface. In other words, the Earth may be a Hawaii in a universe of Siberias. Although we can take technology with us to adapt to new places (as we have adapted to the extremes of Earth), it would be a good idea to take care of the one known place in the universe where we can do these things.

3. Twentieth-century global society is not living within its planetary means. Certain resources, such as metals, coal, and natural gas, are being consumed at rates hundreds of thousands of times faster than their replacement rates. If the present growth in consumption rates continues, most of those resources will be exhausted during the next century (between 2000 and 2100).

4. It seems technologically feasible to engineer a stable society, but it also seems possible to destroy civilization and all life on Earth either by design or by accident.

5. Pressures on humanity and the environment will increase the longer we draw on Earth's finite resources. Even an era of conservation only stretches the time available unless new materials, energy, or ideas to solve the underlying problem of consumption in a finite world are developed. Pursuit of human capabilities in space may offer a way out of these problems. Gradually shifting heavy industry into space, where solar energy could be used, might help reverse the polluting trends on Earth today; it could reduce the mining of fossil fuels and some metals, reduce the emission of pollutants into the atmosphere, and foster an ideal of Earth stewardship. Resulting pollution of the space environment would be minimal, both because of its vastness (the volume of the solar system is about 100 billion billion times that of our atmosphere) and because gases and dust are flushed out of the solar system by the solar wind.

6. We can describe the arrangement and history of the planetary and galactic systems in which we find ourselves, but we do not really understand the arrangement or history of the largest-scale clusters of galaxies.

The most fundamental structure of the universe is thus still a mystery.

7. We may be prohibited, by the Hubble recession and the finite speed of light, from ever obtaining information about places beyond about 10,000 Mpc.

8. To some philosophers, the only reality is the reality directly witnessed by human observers. Yet three centuries of repeated, reliable observations give us the right to talk about real events that happened over a billion years ago and real places where no person has ever walked. Time and distance can be objectively probed, giving results that remain consistent (within our limits of measurement), regardless of who does the measuring. Certain places on Mars have looked the same whether photographed in 1971 by an American spacecraft or in 1973 by a Russian spacecraft. Thus it is reasonable to suppose that there are real places where things happen, even though there are no humans to see them. In a way, it is astonishing to think that rocks roll down hills on the far side of the Moon when no one is there; and that waves crashed on Earth's beaches for billions of years before there were land animals to see them. . . .

9. No place in the universe is permanent. Probably the Sun will run out of hydrogen within the next 6 to 10 billion years, expanding to the red giant state, engulfing Earth, and finally collapsing to a dwarf. New stars will form near other stars, and perhaps other Earthlike planets will form around some of these stars.

10. As the physical universe changes, the biological universe also changes. In relatively brief periods of millions of years, new creatures may evolve, just as humanity itself appeared only a few million years ago. A philosophy that sees humanity as a permanent fixture in an unchanging environment may be practical for short periods such as human lifetimes or centuries, but it cannot be defended on a cosmic scale.

11. In spite of our seeming sophistication and ability to modify our local surroundings, the universe remains implacable. We humans live on a small planet in the midst of forces and processes, some far greater than any we can create or modify. Beyond our mortal processes are greater processes of death and birth that we have to accept as part of nature. Whole worlds are demolished and whole worlds are created.

As the American naturalist Henry David Thoreau commented with regret, most modern people have forgotten that the Greek word for the universe, κόσμος (cosmos), means beauty or order. We need a sense of beauty and majesty that not only responds to a pleasant sunset but can also encompass that more permanent cosmic sunset: a supernova. Such chaotic explosions are also part of the *order of things* in the cosmos.

The question of creators is a mystery—exciting, puzzling, beyond our abilities to explain. Here astronomy establishes a point of contact with both old and new religious perceptions. Beyond this point, astronomers are no better equipped than anyone else to make definite statements.

12. Many mysteries remain, and much space is left still to explore. Such explorations seem worthy of human energy—worthier, for example, than warfare. It is good to admit that there is still mystery in the universe. There is more to life than what we see on this particular planet at this particular time.

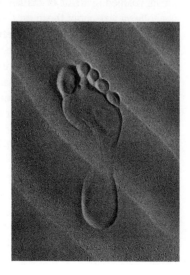

Pseudoscience and Nonscience

When confronted with a hypothesis that sounds far-fetched, some scientists may immediately label it pseudoscience or even nonscience. This is often unfair. The true scientific method is to *test* hypotheses. A good hypothesis must predict some things about nature, and if the predictions are wrong, the hypothesis must be rejected or modified. Though a hypothesis can never be proved "ultimately true," if experiments keep turning up consistent results, it is considered more and more reliable and comes to be an accepted theory or even a law.

A **pseudoscience** is a body of hypotheses treated as if true (usually with commercial intent) but without any consistent supporting experimental or observational evidence. Pseudoscience may appear to be backed by the trappings of real science, such as quotations of evidence, but the evidence is often hearsay, the references are often to other poorly researched commercial books, and the work is rarely reviewed by other professional researchers before publication. Note that pseudoscience is not determined by subject matter but by its method of dealing with *evidence*.

The dangers of pseudoscience are twofold. First, its practitioners often bilk consumers of their money by falsely promising new discoveries or mystic knowledge. Second, it misrepresents real scientific discovery and often contributes to antiintellectual attitudes that exchange mysticism and magic for exploration and discovery. The exchange is a poor one because, as I hope I have shown in this book, real discoveries about the universe are as exciting as the erroneous claims of pseudoscience.

Astrology is a classic example of pseudoscience. One of the ancient systems for attempting to predict events from patterns in nature (such as patterns in tea leaves or flocks of migrating geese), astrology tries to predict events by using the patterns made by the moving planets against the background of starry constellations. (Chapter 1 traces its roots and its shortcomings in more detail.) During the tortuous course of history, many superstitions have been abandoned but others survive. Few Americans use tea leaves to plan their day, but many pore over the astrology column in the morning paper. Old traditions (whether sensible or not) die hard. New knowledge comes slowly, but faster in our century than ever before.

PSEUDOSCIENCE BASED ON EXTRAORDINARY REPORTS

Another form of pseudoscience has grown up around alleged eyewitness reports of strange events. It is perfectly permissible to hypothesize that strange events, such as an alien visit to Earth, will sometimes occur. We cannot dismiss the hypothesis, but we can ask if any *verifiable* evidence exists. To answer this question, we need to understand the processes that lead to eyewitness reports.

Perception, Conception, and Reporting

One-time events are hard to confirm. It is easier to learn from repeatable experiments in the laboratory than from sporadic views of celestial phenomena, which is one reason why less occult lore accumulates around a science like hydraulics than around astronomy.

The first step in generating an eyewitness observation is *perception*—the observer's intake of external sensory stimuli. This perception is converted in the observer's brain into a *conception*. This second step involves subjective factors, such as the association the person may make between the object and concepts prevailing in the culture. For example, witnesses almost invariably conceive of meteors in terms of the distances of aircraft. They say, "It landed just behind the barn," when the meteor may in fact have been hundreds of miles away.

Reporting, the third step, transmits conceptions to other people. Throughout most of history, this process has been by word of mouth, with secondhand reports and hearsay blending into long-lived oral traditions. Even today, it is hard to transmit conceptions accurately by words. There is a UFO joke about a man seeing a large orange object on the ground, with flashing red lights, rows of windows, and small people inside. Does this account make you visualize a flying saucer? The object was a school bus. This joke shows how, in the appropriate context, simple, accurate words can lead many people to the wrong conception.

The fact that a report appears in print or on TV can affect the beliefs of millions of people. Astronomer Carl Sagan (Sagan and Agel, 1975) recounts an example of public overacceptance. He saw a woman reading a pseudoastronomy best-seller that contained many errors and flagrantly misquoted Sagan himself. Sagan asked the woman if she knew the book contained inaccuracies. "It couldn't," she said, "because they wouldn't let him publish it if it weren't true." Not so. Many news and entertainment media thrive on sensational stories with minimal documentation, so readers must maintain a healthy skepticism.

The processes of perception, conception, and reporting of rare events can in principle be a source of scientific information, but in practice they often produce misleading data. Reports of weird events are likely to circulate whether the events really occurred or not.

UFOs

UFOs are unidentified objects in the sky. Undeniably, there have been thousands of reports of different kinds of UFOs, ranging from alleged metal disks to amorphous glows to unusual objects detected by radar. The previous discussion shows why they are hard to study scientifically. The real question is whether *any* of these represent unknown phenomenon.

Most of the literature on UFOs is pseudoscience—not because of the subject, but because most UFO books abound in exaggerated claims and distortions. A 1968 University of Colorado study (in which I personally participated in hopes of discovering exciting evidence of unusual phenomena) established that many of the classic UFO photos are either fakes or photos of known natural phenomena. Nevertheless, these photos reappear, year after year, in new UFO publications.

There is a faddish element to UFO reports: waves of "sightings" followed space events such as the first Sputniks and the first photos from Mars, and many UFO hoaxes followed within weeks of the first "flying saucer" report in 1947. Waves of UFO reports have even occurred in earlier eras, when the UFOs were reported to look not like the saucers popularized in our movies but like images popular in those times. A UFO wave in the 1890s reported "airships" looking like Victorian dirigibles; this image came from Jules Verne's novel *Robur the Conquerer,* in which a mad scientist terrifies the world from . . . you guessed it . . . his newfangled dirigible.

When a satellite reentering the atmosphere broke into burning pieces in the night sky over the eastern United States in 1968, many people witnessed the phenomenon. Some correctly identified it, but many others misconceived it. Of 30 extensive reports collected by the Air Force, 57% said that the objects were flying in formation, implying intelligent control. About 17% conceived that the glowing objects must be attached to an unseen black object, and they reported a nonexistent "cigar-shaped" or "rocket-shaped" object, sometimes with glowing windows. One particularly elaborate report spuriously declared that it had . . . "square shaped windows along the side . . . the fuselage was constructed of many pieced or flat sheets . . . (Condon and others, 1969).

The abundance of such misconceptions and misreporting does not, of course, prove that "flying saucers" or alien spaceships do not exist. But it does show that proof of their existence is nearly impossible to establish from the kinds of reports available. While a few unsolved UFO cases may suggest unusual atmospheric phenomena, none give good evidence for spaceships. If there really are any spaceships around, one might predict, in view of the increasing numbers of surveillance devices and tourist cameras, that good evidence should appear. So far, none has.

Astrocatastrophes

A very interesting case was a study of ancient myths by naturalist and psychiatrist Immanuel Velikovsky. He used the pseudoscientific method of interpreting ancient documents literally. For example, where the Book of Joshua says the Sun stood still (10:13), Velikovsky assumed that Earth indeed suddenly stopped turning and then started again, ignoring modern evidence that this is unlikely. Compiling his results, Velikovsky (1950) concluded that two astronomical supercatastrophes occurred around 1500 B.C. and 750 B.C., during which Venus first entered the solar system, passed near Mars, and then passed near Earth, reaching its present orbit only after 750 B.C.

Velikovsky's work was an interesting compilation of strange myths, but when astronomers criticized his work, he was celebrated as an antiestablishment underdog and his books became best-sellers. His conclusions have been solidly refuted by scientists, however. For example, ancient records of Venus' motions and of old eclipses show no signs of these disturbances; nor do computer calculations that can trace planetary motions back millions of years. Moreover, the lunar lava flows that Velikovsky attributed to the catastrophes date from 3 billion years ago, not 3000.

Although Velikovsky was far too literal in interpreting old records, astronomical events such as meteorite falls may indeed have influenced early cultures and myths. For example, geologist Cesare Emiliani and his coworkers (1975) described physical evidence of worldwide coastal flooding around 9600 B.C., which they suggest may be the basis of the worldwide flood myth.

"SCIENTIFIC CREATIONISM"

Renewed attacks on the theory of evolution raise the issue of a new kind of pseudoscience, often called *scientific creationism.* The history of this movement is traced in some detail by Numbers (1982). It grew out of a feeling by some parents that scientific evidence (usually being taught in biology classes) conflicts with the family's religious faith. This faith may include the idea that humans did not evolve from other animals, and that Earth was created only a few thousand years ago. In its most extreme form, "scientific creationism" proposes that humanity (or the solar system or whatever) was created all at once along with fossils in the ground, radioisotopes in their present distribution, and so on. In this view, the fossils that suggest evolution were put there to fool us. While this is a logically consistent hypothesis in which one might choose to put one's faith, it cannot be dealt with scientifically.

This point can be understood by considering the theory of "fairies in my rose garden." Suppose I hypothesize that shy fairies live in my rose garden and help tend the roses, but are so clever that they always know if I put a camera or recorder out to document them; thus they avoid leaving any record. This explains why no evidence of fairies exists. This is a fine hypothesis, but it is not a part of rational science, because it has the built-in proviso that no meaningful evidence can be gathered. Once the assumption is made that fairies exist, many books could be published about where they might live, how they might dress, and so on. These books could be fine literature and brilliant exercises in logic, but they would be speculation, not science. (Some of the more enthusiastic books about life on other worlds have been criticized on these same grounds; but at least we have evidence of amino acids in meteorites, biochemical evidence about laboratory reactions, and astronomical evidence about the environments of other star systems!)

What is the main danger of these ideas? It does not lie with the hypotheses, whose merits should always be tested against natural evidence. The main problem is with the underlying premise that the answers are already known from ancient sources and that science merely studies nature to verify those answers. Instead, the business of science is an open, two-way dialogue with nature. If we ask the right questions, nature not only shows us the answers but may lead us on to the next question. Our discoveries may not only be interesting in themselves but also lead to practical applications. These results are less likely in a science that is conducted only to verify conclusions that are already "known" from ancient authorities. We commit a sin of pride if we presume to tell nature what the answer is, before we even attempt our observations!

Ideally, a science course should be a synthesis of evidence about the nature of the universe derived from observation, experimentation, and calculation. How this evidence is organized into a system of philosophical or religious thought is the subject of philosophy and religion courses—which are exciting enterprises in themselves, but not science. Yet in the 1980s lawsuits have attempted to force the teaching of scientific creationism under the guise of science in science courses. In the opinion of most educators, scientific creationism does not belong in a science classroom because it does not deal in a testable way with evidence. It is not part of a dialogue with nature; it is not science. It might be judged on its merits in a course on comparative religion.

HOW TO REACT TO PSEUDOSCIENCE

When considering how generations of people around the world have wasted their resources on dubious cures, astrological advice, and superstitious cults, we may grow impatient and demand that such activities be abolished. Yet, one of the beauties of Western democracy is that we attempt, at least, to allow freedom of speech—freedom of competing ideas—in a hope that the right ideas will emerge as the incorrect ideas are found to be flawed. We try to encourage healthy debate, and hope to find what seems closest to the truth.

Perhaps in this approach we do not pay enough attention to the fact that the human mind has an amazing ability to conceive self-consistent systems of thought that in the final analysis have little to do with reality. Sometimes it is labeled fiction: Tolkien's novels of hobbits, for example, create an intricate world of gnomes with their own language, mythology, and history—but no one claims they are real. Sometimes we think our thought system *may* reflect reality and we label it a scientific hypothesis—but then we test it repeatedly against observation and experiment to see if we can disprove it. If so, we replace it with something *better*. Sometimes we simply claim that our thought system *is* reality and urge people to give their hearts to it. If these ideas are untestable, uncritical acceptance may be claimed as a virtue. It is these thought systems—pseudoscience and superstition—that threaten to drain away energy, resources, and imagination.

What we can do with any thought system—fiction, scientific hypothesis, or superstition—is to use *evidence* to test it against experience and reality. In the meantime, scientists and teachers must avoid being dogmatic about interpreting natural evidence, and people everywhere should cheerfully anticipate continuing refinement of scientific knowledge with new discoveries that deepen our understanding of the universe and our relation to it.

Astronomical Coordinates and Timekeeping Systems

Two kinds of astronomical systems are related to the rotation of Earth. The first is the system of astronomical coordinates, whereby astronomers map positions of objects in the sky. The second is the timekeeping system employed around the world—an outgrowth of the early timekeeping and calendar systems described in Chapter 1.

RIGHT ASCENSION AND DECLINATION COORDINATES

We use Earth's spin to define coordinate systems for locating objects both on the ground and in the sky. In the ground system—the system of latitude and longitude—latitude lines are perpendicular to the axis of rotation, whereas longitude lines intersect at the poles of rotation.

A nearly identical system describes locations of stars and other celestial objects. It involves coordinates called *right ascension* and *declination* (often abbreviated "R.A." and "Dec."). *Right ascension and declination are merely longitude and latitude, respectively, projected from the center of Earth through its surface onto the sky and fixed with respect to the stars.* The celestial equator (see page 11) is the line of 0° declination, and the north celestial pole (see page 10) is at declination 90°N, usually written +90°; the south celestial pole is at −90°.

To visualize this system, imagine Earth as a giant nonrotating glass sphere with only the longitude and latitude lines engraved on it, and imagine yourself at the center of the sphere. You look out and see the longitude and latitude lines, by which you can locate features on Earth's surface. But these lines are also the lines of right ascension and declination, and you see them projected onto the stars. Earth's equator overlaps the celestial equator. Earth's North Pole overlaps the north celestial pole and almost obscures the star Polaris.

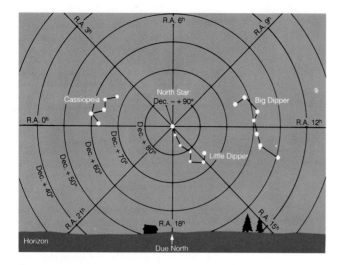

a

Figure B-1 **a** Schematic sky view looking north from the mid–United States on a February evening. Prominent constellations and the right ascension–declination system of imaginary coordinates are shown. Note that the leading star in Cassiopeia is nearly on right ascension 0ʰ and declination 60°. **b** The same view about an hour later. Note how the right ascension–declination system moves with the stars as they rise in the east and set in the west.

ascension places it 1^h below the eastern horizon, for example, astronomers know that it will rise in an hour. (Earth turns 15° in an hour, so 1 h = 15°.)

Stars are cataloged by their right ascension and declination. Sirius, for example, lies at R.A. 6^h41^m, dec. −17°; Polaris, which is virtually on the north celestial pole, is at R.A. 1^h23^m, dec. +89°.

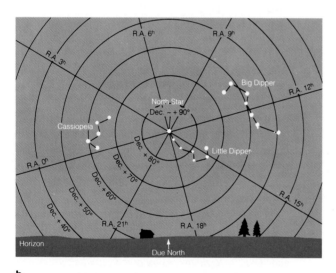

b

THE EFFECT OF PRECESSION ON RIGHT ASCENSION AND DECLINATION

Now we can more clearly describe the effects of precession (page 33). Because the right ascension–declination system is anchored on Earth, and because Earth wobbles with respect to the stars every 26,000 y, the system shifts among the stars. The coordinates of stars shift slightly every year, and astronomers must continually update the right ascension and declination positions listed in catalogs. Catalogs generally specify "epoch 1950" or "epoch 2000" as the year for which the positions are listed. The same shift of the celestial equator and vernal equinox invalidated the original system of astrology by shifting the right ascension–anchored astrological signs away from their original constellations, as described in Chapter 1.

ANOTHER CELESTIAL COORDINATE SYSTEM

A system useful for describing the position of objects in the sky during both day and night is the *altitude–azimuth system*. *Altitude* is the number of degrees above the horizon. *Azimuth* is the number of degrees around the horizon, starting from the north point (0°) and moving through the east (90°). Since azimuth is counted from the north to the east, an azimuth of, say, 60° is often written N60°E for maximum clarity. Thus an object on the horizon due east would have altitude

Figures B-1 and B-2 help in visualizing the right ascension–declination system of sky coordinates. Figure B-1 shows two views of the northern sky an hour apart as the stars wheel around the north celestial pole Figure B-2 shows two similar views of the eastern sky as Orion rises there.

As zero right ascension, astronomers have chosen the right ascension line running through the vernal equinox: the point where the Sun appears to cross the celestial equator going north. Instead of measuring right ascension in degrees, astronomers measure it in hours, because they are interested in the time required for apparent motions of stars across the sky. If a star's right

Figure B-2 a Schematic sky view looking east on a November evening. Prominent constellations and the right ascension–declination system of imaginary coordinates. Note that Orion lies on the celestial equator. **b** The same view an hour later. The right ascension–declination system moves with the stars as they rise in the east.

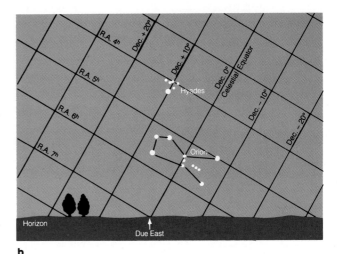

0°, azimuth N90°E. An object halfway up the sky in the west would have altitude 45°, azimuth N270°E. This system is clarified in Figure B-3, which shows two views of Orion rising in the east with altitude and azimuth lines superimposed.

An object at the *zenith* has altitude 90°. The most important azimuth line is the *meridian* (see page 11).

SYSTEMS OF TIMEKEEPING

Before clocks were invented, time was kept by watching the apparent motion of the Sun around Earth, caused by Earth's rotation. This is called *apparent solar time.* Noon was the moment when the Sun crossed the meridian. *A.M. (antemeridiem)* is the period before the Sun crosses the meridian; *P.M. (postmeridiem),* the period after. Once clocks were invented, it became clear that the Sun's apparent motions were not exactly uniform, because the combination of Earth's rotation and orbital revolution produces days of varying length. Later, more careful measurements revealed slight, sudden speedups and slowdowns in Earth's rotation due to earthquakes, tides, and winds.

For these reasons, today's clocks run on *mean solar time,* a time system based on the uniform average rate of the Sun's motion rather than its actual position in the sky. The most accurate calibration of clocks for scientific work is now determined by an internationally accepted definition of the second based on constant atomic processes: 1 s is 9,192,631,770 cycles of a certain type of radiation from the cesium-133 atom. This standard is called *atomic time.*

For long-term prediction of planetary positions or eclipses, astronomers have defined an additional time system based on the motions of the planets, called *ephemeris time.*

Astronomers, who sight their telescopes on the stars every night, need a star-based time system rather than the Sun-based system of mean solar time by which most clocks run. This system, called *sidereal time,* is simply a clock reading equal to the right ascension crossing the local meridian at any moment.

TOWARD THE MODERN CALENDAR

Just as timekeeping based on Earth's rotation has been continually refined, so has the calendar system based on Earth's orbital revolution. Chapter 1 described how early people developed calendar systems and estimated the length of the year. The year was often officially started on an astronomically determined date, such as the summer solstice. One problem with early calendars was accurately determining the number of days in the year. We now know that the sidereal year (Earth's motion around the Sun with respect to the stars) is 365.256 ephemeris days, and the tropical year (cycle of seasons) is 365.242 ephemeris days.

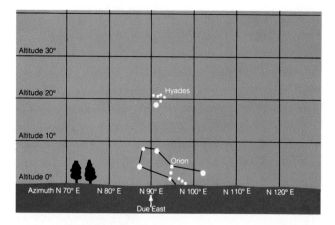

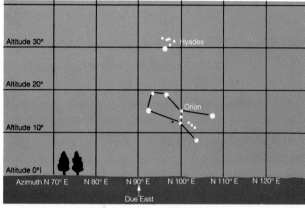

b

Figure B-3 **a** This figure is the same as Figure B-2a, but it uses the altitude–azimuth system of coordinates instead of the right ascension–declination system. In this view, the trees reach an altitude of about 5° and the Hyades are at an altitude of about 20°. **b** The same view about an hour later. Note that the altitude–azimuth system of coordinates is tied to the horizon and does not move with the stars. The trees still reach an altitude of about 5°, while the Hyades have risen to an altitude of 30°.

intercalated at the end of the last month, February, where we still insert our extra day every leap year.

By 46 B.C., however, local officials in different parts of the empire were inserting extra days as they pleased, and the calendar became confused. Julius Caesar had the calendar revised in 44 B.C. under the direction of the Alexandrian astronomer Sosigenes, and the modern leap year scheme was formalized. The fifth month of the new *Julian* calendar (Quintilis) was renamed July in honor of Julius Caesar. August was named after Augustus Caesar, who made some additional improvements in 7 B.C.

The Julian calendar worked well for more than a thousand years, but since the year is not exactly $365\frac{1}{4}$ days long, new errors crept in. In the 1200s, European scientists noted that there was a full week discrepancy in the date of the vernal equinox. To fix this, Pope Gregory XIII commissioned a new revision of the calendar during the period from 1576 to 1603.

This *Gregorian* calendar was adopted quickly in Roman Catholic countries, more slowly in Protestant countries, and still more slowly in Eastern Orthodox and other countries. England and the colonies converted in 1752, dropping 11 days. (Riots and legislation against rent abuse followed because of the resulting 19-day month.) Russia converted during the 1917 revolution and had to drop 13 days. Because of the different dates of conversion to the Gregorian calendar in different countries, scholars must beware of historical records of dates, and not compare dates in different countries without considering the calendar in effect at the time that records were made.

To clear up historical confusion about dates, astronomers have devised a system called the *Julian day count*, which designates any date in history, anywhere in the world, by the number of days since January 1, 4713 B.C. On July 4, 1976, to pick a nonrandom date, the Julian day count was 2,443,964. Computer-calculated dates of ancient eclipses and other astronomical events are expressed in this system.

If early timekeepers adopted a calendar with, say, 365 days in a year, they discovered that the astronomically determined New Year's Day was one day in error after four years. This problem was resolved by a process called *intercalation:* Astronomers, government officials, or priests announced extra days, often celebrated as holidays, every few years to keep the calendar in step with astronomical events.

As long as society was stable, this process could be kept up for many years. The Romans, for example, developed the ancestor of our calendar, beginning the year at the spring equinox. (The Latin-derived names *September* through *December* represent month positions 7 through 10 in that calendar.) The extra days were

Powers of 10

It is no accident that the word *astronomical* has become a synonym for "enormous, almost beyond conception." Astronomy is full of extraordinary numbers designating great ages and distances. In astronomy (and other sciences), therefore, a convenient shorthand system of writing numbers is favored. In this system, an exponent, or superscript, designates the number of factors of 10 that have to be multiplied together to give the desired quantity—that is, the number of zeros in the quantity. For example:

$$1 = 10^0$$

$$10 = 10^1$$

$$100 = 10^2$$

$$1,000 = 10^3$$

$$10,000 = 10^4 \text{ and so on}$$

The most often used of these large numbers are

$$\text{one thousand} = 1,000 \qquad\qquad = 10^3$$

$$\text{one million} = 1,000,000 \qquad = 10^6$$

$$\text{one billion}^1 = 1,000,000,000 \quad = 10^9$$

$$\text{one trillion} = 1,000,000,000,000 = 10^{12}$$

The usefulness of this system is illustrated by expressing the age of Earth

$$4,600,000,000 \text{ y} = 4.6 \times 10^9 \text{ y}$$

or the distance that light travels in a year

$$6,000,000,000,000 \text{ mi} = 6 \times 10^{12} \text{ mi}$$

A similar system is used for very small numbers.

---

[1]Beware—the British use the term *billion* to refer not to 10^9 but to 10^{12}. This book uses *billion* to mean 10^9.

Here the exponent is negative and refers to the number of decimal places. Thus:

$$0.0001 = 10^{-4}$$
$$0.001 \ = 10^{-3}$$
$$0.01 \ \ = 10^{-2}$$
$$0.1 \ \ \ = 10^{-1}$$
$$1.0 \ \ \ = 10^{0}$$

The density of gas in interstellar space, which is

$$0.000,000,000,000,000,000,000,002 \text{ g/cm}^3$$

can be more conveniently written as

$$2 \times 10^{-24} \text{ g/cm}^3$$

One of the beauties of the metric system (discussed further in Appendix 2) is that the units are related by powers of 10. For instance, a kilometer is 10^3m (as opposed to our own confusing English system, where 1 mi is 5280 ft). To make the system easier to use, the International System of units (known as the SI metric system) has a standardized set of prefixes, such as *kilo*, that represent multiples of 10^{-3}, 10^3, 10^6, and so forth. These prefixes are shown in Table A1-1.

TABLE A1-1		
Prefixes Used in the SI Metric System of Units		
Prefix	**Symbol**	**Multiple**
Tera- (TER-a)	T	10^{12}
Giga- (JIG-a)	G	10^{9}
Mega- (MEG-a)*	M	10^{6}
Kilo- (KILL-low)*	k	10^{3}
Hecto- (HECK-toe)	h	10^{2}
Deka- (DECK-a)	da	10^{1}
Deci- (DESS-ee)	d	10^{-1}
Centi- (SEN-tee)*	c	10^{-2}
Milli- (MILL-ee)*	m	10^{-3}
Micro- (MY-crow)*	μ	10^{-6}
Nano- (NAN-oh)	n	10^{-9}
Pico- (PEE-koh)	p	10^{-12}
Femto- (FEM-toe)	f	10^{-15}
Atto- (AT-oh)	a	10^{-18}

Note: Starred items are the most important in ordinary applications.

APPENDIX 2

Units of Measurement

Most scientists express distances and other measurements in metric units, and the world is in the process of converting to this system. This conversion will save time and money, since publications, quantities of goods, and machine parts will be measured in one uniform system. The metric system is also easier to learn and use, since it expresses units in multiples of 10 (like our money system), instead of in unpredictable multiples such as 12 inches per foot, 5280 feet per mile, and 16 ounces per pound.

Table A2-1 gives some common units in both systems. This book has emphasized the International System of metric units (abbreviated "SI" for *Système Internationale*), which uses the meter, kilogram, and second as the basic measures of length, mass, and time, respectively. When using the SI system to solve problems involving complex quantities or constants, such as energy or the gravitational constant G, one must make sure that these quantities are expressed in terms of the basic units kg, m, and s. Equivalents in the centimeter–gram–second (cgs) system and in larger units are also given in Table A2-1.

Table A2-2 gives some of the more useful physical constants expressed in the SI system.

TABLE A2·1

Conversion Factors for Units of Measurement

	SI (m-kg-s)	cgs System
1 inch	0.025 m	2.54 cm
1 foot	0.305 m	30.5 cm
1 yard	0.914 m	91.4 cm
1 mile	1609 m	1.609×10^5 cm
1 pound	0.454 kg	454 g
1 astronomical unit (AU)	1.50×10^{11} m	1.50×10^{13} cm
1 light-year (ly)	9.46×10^{15} m	9.46×10^{17} cm
1 parsec (pc) = 3.26 ly	3.08×10^{16} m	3.08×10^{18} cm
1 kiloparsec (kpc)	3.08×10^{19} m	3.08×10^{21} cm
1 megaparsec (Mpc)	3.08×10^{22} m	3.08×10^{24} cm
1 solar mass (1 $M_\odot$)	1.99×10^{30} kg	1.99×10^{33} g

TABLE A2·2

Some Useful Physical Constants (SI system of metric units)

Constant	Numerical Value
Seconds of arc in one radian	206,265
G (Newton's universal gravitational constant)	6.67×10^{-11} N · m^2/kg^2
Wien's constant	2.90×10^{-3} m · K
k (Boltzmann constant)	1.38×10^{-23} J/K
c (velocity of light)	3.00×10^8 m/s
σ (Stefan–Boltzmann constant)	5.67×10^{-8} W/m^2 · K^4
Mass of a proton	1.67×10^{-27} kg
Mass of an electron	9.11×10^{-31} kg
Solar constant	1390 W/m^2
$L_\odot$ (solar luminosity)	8×10^{26} W = 8×10^{26} J/s

Supplemental Aids in Studying Astronomy

Astronomical Calendar. Published annually by Guy Ottewell; sponsored by Department of Physics, Furman University, Greenville, SC 29613, (803) 294-2207. $12.00 each; discounts for volume purchases (1988 price including tax and postage). Star maps for each month; lists of astronomical events for each month; explanation of phenomena and various terms; astronomical glossary.

Astronomy Magazine. Published monthly; 411 East Mason St., 6th Floor, Milwaukee, WI 53202. $24/year (12 issues; 1988 price). Popular articles on astronomical phenomena and observing techniques; astronomical paintings and photographs; monthly star charts.

Griffith Observer. Published monthly; Griffith Planetarium, 2800 E. Observatory Rd., Los Angeles, CA 90027. $10/year (12 issues; 1988 price). Popular articles about astronomy.

Journal of the Association of Lunar and Planetary Observers. Published bimonthly; Astronomical Society of Las Cruces, Box 921, Las Cruces, NM 88004. $8.00/year membership (6 issues; 1981 price). Journal of a premier association of amateur astronomers devoted to reports of amateur research and observing projects concerning the Moon, planets, and comets. Publishes drawings and reports by members and coordinates research projects.

Mercury Magazine (Journal of the Astronomical Society of the Pacific). Published bimonthly; Astronomical Society of the Pacific, 1290 24th Avenue, San Francisco, CA 94122. $24.00/year membership (6 issues; 1988 price). Popular articles about astronomy and the astronomical scene; news of the society, which encourages astronomy and holds occasional public functions.

Minor Planet Bulletin. Contact Prof. R. G. Hodgson, Dort College, Sioux Center, IA 51250, for information. $7.00/year. Bulletin of the Minor Planet Section of the Association of Lunar and Planetary Observers with news of recent discoveries, predicted visibilities, and observing projects in which amateur observers can make scientific contributions. Mostly useful for advanced telescope observers.

The Planetary Report. Published bimonthly; 116 S. Euclid Ave., Pasadena, CA 91101. Journal of a society started in 1980 by Carl Sagan and others to promote popular and congressional support for space exploration programs; news of NASA mission plans, recent discoveries about the solar system; recent spacecraft photos; paintings. Membership ($20, 1988 price) includes journal subscription.

Sky and Telescope Magazine. Published monthly; 49-50-51 Bay State Rd., Cambridge, MA 02138. $21.95/year (12 issues; 1988 price). Popular and semitechnical articles on astronomical phenomena, history, telescope making, and observing techniques; photographs; monthly star charts.

Space World. Published monthly. Box 296, Amherst, WI 54406. $26/year (12 issues, 1988 price). Magazine of a space advocacy group reporting national and international space progress.

Glossary

AU: (See *astronomical unit.*)

absolute: Intrinsic; not dependent on the position or distance of the observer.

absolute brightness: Any measure of the intrinsic brightness or luminosity of a celestial object.

absolute magnitude: The absolute brightness (luminosity) of a star expressed in the magnitude system. The Sun's absolute magnitude is $+5$.

absorption: The loss of photons as light passes through a medium. A photon is lost when it strikes an electron, and the photon's energy is consumed in knocking the electron to a higher energy level.

absorption band: A dark or dim region of the spectrum, caused by absorption of light over a moderate range of wavelengths, typically about 0.1 nm, usually by molecules or crystals.

absorption line: In a spectrum, a reduction in intensity in a narrow interval of wavelength, caused by absorption of the light by atoms between the source and the observer.

accretion disk: A disk of gas and dust, usually hot, surrounding a star. Usually used in the context of material that has been thrown off a nearby star and is falling into a companion star.

achondrite: A type of stony meteorite in which chondrules have been destroyed, probably by heating or melting.

age of the Earth: The period since the Earth's formation from planetesimals, estimated to be 4.5 billion years.

age of globular clusters: About 14 ± 4 billion years.

age of the (Milky Way) galaxy: Estimated to be roughly 14 ± 3 billion years.

age of open clusters: Time since formation of open clusters, judged by their H–R diagrams. Most are less than 100 million years. Ages from 1 million to a few billion years have been reported.

age of stars: Time since star formation, typically billions of years for smaller stars, but less than a million years for some massive stars in recently formed clusters. Age is difficult to measure for individual stars, but possible to measure for clusters of stars.

age of the universe: The time since the big bang, 10 to 18 billion years ago (probably close to 14 billion years ago).

airglow: Visible and infrared glow from the atmosphere produced when air molecules are excited by solar radiation.

Airy disk: In the telescopic image of a star, a small disk caused by optical effects.

Alexandrian library: The research institution and collection of ancient works preserved after the fall of Rome at Alexandria, Egypt. Alexandrian knowledge passed to the Arabs with the Arab conquest of Alexandria, and eventually back into Europe around A.D. 100–1500.

Alpha Centauri: (1) The nearest star system, composed of three members; (2) the brightest of these three.

amino acid: A complex organic molecule important in composing protein and called a "building block of life."

Andromeda galaxy: The nearest spiral galaxy comparable to our own, about 660 kpc away.

angular measure: Any measure of the size or separation of two objects as seen from a specified point, expressed in angular units (degrees, minutes of arc, or seconds of arc), but not linear units (such as kilometers, miles, or parsecs).

angular size: The angle subtended by an object at a given distance.

annular solar eclipse: An eclipse in which the light source is almost, but not quite, covered, leaving a thin ring of light at mid-eclipse.

aperture: The diameter of the light-gathering objective in a telescope.

apogee: The point in an orbit around the Earth that is farthest from the Earth.

Apollo asteroids: Asteroids that cross the Earth's orbit.

Apollo program: The U.S. program to land humans on the Moon, 1961–1972; first landing July 20, 1969.

apparent: Not intrinsic, but dependent on the position or distance of an observer.

apparent brightness: The brightness of an object as perceived by an observer at a specified location (but not measuring the object's intrinsic, or absolute, brightness).

apparent magnitude: Apparent brightness of one star relative to another as expressed in the magnitude system (see examples in Table 14-2).

apparent solar time: Time of day determined by the Sun's actual position in the sky. Apparent solar noon occurs as the Sun crosses the meridian. Apparent solar time is different at each different longitude.

apparition: The period of a few weeks during which a planet is most prominent or best placed for observation from Earth.

association: A loosely connected grouping of young stars.

asteroid: A rocky or metallic interplanetary body (usually larger than 100 m in diameter).

asteroid belt: The grouping of asteroids orbiting between Mars and Jupiter.

asthenosphere: In a planetary body, a subsurface layer that is more plastic than adjacent layers because the combination of pressure and temperature places it near (or slightly above) the melting point. Asthenospheric movements may disrupt the planet's surface.

astrology: The superstitious belief that human lives are influenced or controlled by the positions of planets and stars; this belief is rejected by modern astronomers and other scientists.

astrometric binary: A binary star system detectable from the orbital motion of a single visible component.

astrometry: The study of positions and motions of the stars.

astronomical unit (AU): The mean distance from the Earth to the Sun, about 150 million kilometers.

astronomy: The study of all matter and energy in the universe.

atom: A particle of matter composed of a nucleus surrounded by orbiting electrons.

aurora: Glowing, often moving colored light forms seen near the north and south magnetic poles of the Earth; caused by radiation from high-altitude air molecules excited by particles from the Sun and Van Allen belts.

B-type shell star: A star of spectral type B that occasionally blows off a cloud of gas, forming a gaseous shell around the star.

Balmer alpha line: A brilliant red spectral line at wavelength 656.3 nm, caused by transition of the electron in the hydrogen atom from the third-level orbit to the second-level orbit.

barred spiral galaxy: A spiral galaxy whose spiral arms attach to a barlike feature containing the nucleus.

basalt: A type of igneous rock, often formed in lava flows, common on the Moon and terrestrial planets.

basaltic rock: Igneous rock (including basalt) with a composition resembling basalt and a relatively low content of quartz (SiO_2).

basin: Large impact crater on a planet, usually several hundred kilometers across, flooded with basaltic lava and surrounded by concentric rings of faulted cliffs.

belts: Dark cloud bands on giant planets.

big bang theory: The theory that an explosionlike event that initiated the universe as we know it, probably between 10 and 18 billion years ago. Increasing evidence supports this theory.

binaries, classes of: Three categories of binary stars, depending on whether neither, one, or both fill their Lagrangian lobes.

binary star system: A pair of coorbiting stars.

bipolar jetting: A phenomenon in which narrow streams of gas are ejected at very high speed in opposite directions from the centers of some disks of gas, perpendicular to the disk; bipolar jetting is seen both in accretion disks around individual stars and in galactic disks. The mechanism is uncertain.

black hole: An object whose surface gravity is so great that no radiation or matter can escape from it. Some black holes discussed in astronomy are collapsed stars, but much smaller ones are theoretically possible.

blue shift: A Doppler shift of spectral features toward shorter wavelengths, indicating approach of the source.

Bode's rule: A convenient memory aid for listing the planets' distances from the Sun.

body tide: A tidal bulge raised in the solid body of a planet.

Bok globule: A relatively small, dense, dark cloud of interstellar gas and dust, usually silhouetted against bright clouds.

bolometric luminosity: The total energy radiated by an object at all wavelengths, usually given in joules per second (identical to watts).

Tycho Brahe (1546–1601): Danish astronomer who recorded planetary positions, ultimately enabling Kepler to deduce laws of planetary orbits.

breccia: A rock made from angular fragments cemented together.

brecciated meteorite: A meteorite formed from cemented fragments of one or more meteorite types.

brown dwarf: A starlike object too small to achieve nuclear reactions in its center; any stellar object smaller than about $0.08 \, M_\odot$.

bubble: A roughly spheroidal shell of interstellar gas blown outward from a star by a stellar explosion or strong stellar wind.

burster: A stellar object (neutron star?) that emits flashes of light lasting only seconds. Many emit gamma rays, but visible light bursters are also suspected.

Callisto: Outermost of the four large Moons of Jupiter, and most heavily cratered of the four.

Cambrian period: A period from 570 to 500 million years ago during which fossil-producing species of plants and animals first proliferated.

canals: Alleged straight-line markings on Mars found not to exist by spacecraft visits to Mars.

capture theory: A theory of origin of a planet–satellite or binary star system in which one body captures another body by gravity.

carbonaceous chondrite: A type of carbon-rich and volatile-rich meteorite, believed to be a nearly unaltered example of some of the earliest-formed matter in the solar system.

carbonaceous material: Black material rich in carbon and carbon compounds, found in carbonaceous chondrites and believed to color many black comets and asteroids of the outer solar system.

carbon cycle: A series of nuclear reactions in which hydrogen is converted to helium, releasing energy in stars more massive than about 1.5 $M_\odot$. Carbon is used as a catalyst.

Cassini's division: The most prominent gap in Saturn's rings.

catastrophic theory: A theory invoking sudden or very short (cosmically or geologically speaking) energetic events to explain observed phenomena.

catastrophism: An early scientific school which held that most features of nature formed in sudden events, or catastrophes, instead of by slow processes.

cause of eclipses: The falling of a shadow of one body onto another body.

cause of the seasons: The tilt (obliquity) of the Earth's axis to its orbit plane causes first the North Pole and later the South Pole to be tipped toward the Sun during the course of a year.

celestial equator: The projection of the Earth's equator onto the sky.

celestial poles: The projections of the two poles of the Earth's rotation onto the sky.

cell: The unit of structure in living matter.

center of mass: The imaginary point of any system or body at which all the mass could be concentrated without affecting the motion of the system as a whole; the balance point.

Cepheid variable: Any of a group of luminous variable stars with periods of 5 to 30 d (depending on their population). The periods are correlated with luminosity, allowing distance estimates out to about 3 mpc.

cgs system: Metric system of measurement using centimeters, grams, and seconds as the fundamental units.

chain reaction theory: A theory of the cause of spiral arms in spiral galaxies. Star formation in a region leads to a chain reaction of adjacent star formation, and the region of young stars gets sheared into a spiral arm.

Chandrasekhar limit: A mass of about 1.4 $M_\odot$, the maximum for white dwarfs; stars of greater mass have too great a central pressure, causing formation of a star type denser than a white dwarf.

channel: One of the river-bed-like valleys on Mars, which are possible sites of ancient Martian rivers.

chaotic rotation: A form of rotation in which dynamical forces cause the rotation period to change irregularly from one rotation to the next. Applies to Saturn's moon Hyperion.

Charon: Pluto's satellite.

chemical reaction: Reaction between elements or compounds in which electron structures are altered; atoms may be moved from one molecule to another, but nuclei are not changed and thus no element is changed to another.

chondrite: Stony meteorite containing chondrules, believed to be little altered since their formation 4.6 billion years ago.

chondrule: BB-sized spherule in certain stony meteorites, believed among the earliest-formed solid materials in the solar system.

chromosphere: A reddish-colored layer in the solar atmosphere, just above the photosphere.

circular velocity: Velocity of an object in circular orbit:

$$V = \sqrt{\frac{GM}{R}}$$

where G = Newton's gravitational constant. In SI units it is 6.67×10^{-11} N·m²/kg²
M = mass of central body
R = distance of orbiter from center of central body

circumpolar zone: The zone of stars, centered on a celestial pole, that never sets, as seen from a given latitude.

circumstellar nebula: Gas and dust surrounding a star.

classes of binaries: See *binaries, classes of*.

closed space: Space that is curved in such a way that the total volume is finite.

closed universe: A theoretical model of the universe with finite volume and curved space. In most cosmological models, the universe, if closed, will eventually recollapse.

closure density: A critical density of matter in the universe, above which the universe would be closed.

cluster of galaxies: A relatively close grouping of galaxies, often with some members coorbiting or interacting with each other.

co-accretion theory: A class of theory in which two coorbiting objects form by accreting at the same time, side by side.

coacervate: Cell-sized, nonliving globule of proteins and complex organic molecules that forms spontaneously in a water solution.

Coal Sack Nebula: A prominent dark nebula about 170 kpc away silhouetted against the Milky Way.

cocoon nebula: A dust-rich nebula enclosing and obscuring a star during its formation, but later shed.

collapse: Rapid contraction, especially of a cloud of gas and dust during star formation.

colliding galaxies: Galaxies undergoing interpenetration or close enough to cause major gravitational distortion of each other.

coma: (1) The diffuse part of the head of a comet surrounding the nucleus; (2) a type of distortion in some telescopes and optical systems.

comet: An ice-rich interplanetary body that, when heated by the Sun in the inner solar system, releases gases that form a bright head and diffuse tail. (See also *coma*.)

comet head: The coma and nucleus regions of a comet.

comet nucleus: The brightest starlike object near the center of a comet's head; the physical body (believed to be icy and a few kilometers across) within a comet.

comet tail: Diffuse streamers of gas and dust released from a comet and blown in the direction away from the Sun by the solar wind.

comparative planetology: An interdisciplinary field of astronomy and geology attempting to discover and explain differences between planets in properties such as climate and interior structure.

composition of Population II stars: About $\frac{3}{4}$ hydrogen and $\frac{1}{4}$ helium, by mass, with virtually no heavier elements.

compound telescope: A telescope that combines lenses and mirrors in the light-gathering system.

condensation sequence: The sequence in which chemical compounds condense to form solid grains in a cooling, dense nebula.

conditions in the galactic center: A dense concentration of stars (mostly Population II), dust, and gas, possibly with one or more large black holes marking the nucleus.

conduction: One of three processes that transfers heat from hot to cold regions; occurs as fast-moving molecules in the hot region agitate adjacent molecules.

conjunction: The period when a planet lies at zero or minimum angular distance from the Sun, as seen from the Earth.

conservation of angular momentum: A useful physical rule which states that the total angular momentum in an isolated system remains constant.

constellation: Imaginary pattern found among the stars, resembling animals, mythical heroes, and the like; different cultures map different constellations.

contact binary: A coorbiting pair of stars whose inner atmospheres or surfaces touch.

continental drift: The motion of continents due to (convective?) motion of underlying material in the Earth's mantle.

continental shield: A stable, ancient region, usually flat and oval-shaped, in a continent.

continuous creation theory: The hypothesis that matter is being created in interstellar or intergalactic space during the current era (and always).

continuum: In a spectrum with absorption or emission lines, the background continuous spectrum.

convection: One of three modes of transmission of heat (energy) from hot regions to cold regions; involves motions of masses of material.

Copernican revolution: The intellectual revolution associated with adopting Copernicus' model of the solar system, which displaced the Earth from the center of the universe.

core: The densest inner region of the Earth, probably of nickel–iron composition; in other planets, similar high-density central regions; in the Sun or stars, a dense central region where nuclear reactions occur; in galaxies, the densest, brightest central regions.

Coriolis drift: A departure from a straight-line trajectory, perceived by an observer in a rotating system; Coriolis effects in clouds were early evidence of the Earth's rotation.

corona: (1) The outermost atmosphere of the Sun, having a temperature of about 1 to 2 million Kelvin; (2) circular tectonic structure on the surface of Venus, typically a few hundred kilometers in diameter.

coronograph: An instrument permitting direct observation of the solar corona without an eclipse.

cosmic fuels: Nonfossil energy sources provided by cosmic processes; for example, solar and geothermal energy.

cosmic rays: High-energy atomic particles (85% protons) that enter the Earth's atmosphere from space. Many may originate in supernovae and pulsars.

cosmogony: Any theory of the origin of the universe or one of its component systems, such as galaxies or the solar system.

cosmological principle: The assumption (unproven) that the universe is homogeneous and isotropic.

cosmological red shift: Any redward Doppler shift attributed to the mutual recession of galaxies or the expanding universe.

cosmology: The study of the structure of the universe. The term is often broadened to include the origin of the universe as well.

Cretaceous-Tertiary extinctions: Extinctions of many species 65 million years ago, probably triggered by the impact of an asteroid.

crust: The outermost, solid layer of a planet, with composition distinct from the mantle and differentiated by a seismic discontinuity.

cultural hurdle: The hypothetical survival requirement for a planetary culture between the time it achieves technology capable of quickly altering its planetary environment and the time it can establish viable bases off its planet; the uncertainty of the probability of survival affects our estimates of the probability of intelligent life elsewhere in space.

curved space: Space in which Euclidean solid geometry is not valid.

daughter isotope: An isotope resulting from radioactive disintegration of a parent isotope.

declination: Angular distance north or south of the celestial equator. (Abbreviation: Dec.)

degenerate matter: Matter in a very high density state in which electrons are freed from atoms and pressure is a function of density but not temperature.

degree: An angle equaling $\frac{1}{360}$ of a circle.

density—wave theory: The leading theory for explaining the formation of spiral arms in galaxies, which posits periodicities in star, dust, and gas motions.

deposition: The accumulation of eroded materials in one place.

desert: Any of the brighter regions of Mars.

differential rotation: The differences in speed for stars at different distances from the center of the galaxy. Orbital velocities are actually slower at 5000 pc from the center than at the Sun's distance, which is 8000 to 10,000 pc from the center.

differentiation: Any process that tends to separate different chemicals from their original mixed state and concentrate them in different regions.

diffraction: The slight bending of light rays as they pass edges, producing spurious rays and rings in telescopic images of stars.

dimensions of the galaxy: As early as 1935 astronomers agreed that the Sun is about 8000 to 10,000 pc from the center, and the overall diameter of our galaxy is about 30,000 pc.

Dione: (See *Tethys, Dione, and Rhea*)

dirty iceberg model: A theoretical description of a comet nucleus as a large icy body with bits of silicate "dirt" embedded in it.

disruption of open clusters: Gradual dispersion of stars from an open cluster as the cluster is sheared by differential galactic rotation, and as high-speed stars escape. Disruption time is usually a few hundred million years.

distance limit for reliable parallaxes: 20 parsecs.

Doppler shift: The shift in wavelength of light or sound as perceived by the observer of an approaching or receding body. For speeds well below that of light, the shift is given by the equation

$$\text{Original wavelength} \times \frac{\text{radial velocity}}{\text{velocity of light}}$$

Drake equation: The statement that the fraction of stars harboring intelligent life equals the number of all stars times a sequence of fractions, such as the fraction of all stars having planets, the fraction of planets that are habitable, and so on. Named after radio astronomer Frank Drake.

duration of element formation: For each element, the time interval during which it formed. Most atoms of light elements, like H, formed in the first minutes of the big bang, but heavy elements have been synthesized inside stars for billions of years.

dust trails: Toroidal lanes of dust stretching along elliptical orbits around the solar system; probably due to asteriod collisions or comet dust ejection.

dwarf elliptical galaxy: An ellipsoidal galaxy resembling a globular cluster but usually at least a few times larger.

Earth: The third planet from the Sun.

earthquake: Vibration or rolling motion of the Earth's surface accompanying the fracture of underground rock.

eclipse: An event in which the shadow of one body falls on another body.

eclipsing binary: A binary star system seen virtually edge-on so that the stars eclipse each other during each revolution.

eclipsing—spectroscopic binary: An eclipsing binary whose motions are measurable from spectral Doppler shifts. The most informative type of binary star.

ecliptic: (1) The plane of the Earth's orbit and its projection in the sky as seen from Earth; (2) approximately, the plane of the solar system.

effective temperature: Temperature of an object as calculated from the properties of the radiation it emits.

ejecta blanket: A layer of debris thrown out of a crater onto a planet's surface.

electromagnetic radiation: Light, radio waves, X rays, and other forms of radiation that propagate as disturbances in electric and magnetic fields, travel at the speed of light, and combine to make up the electromagnetic spectrum.

electron: Negatively charged particle orbiting around the atomic nucleus, with mass 9.1×10^{-31} kg.

element: A chemical material with a specified number of protons in the nucleus of each atom. Atoms with one proton are hydrogen; with two protons, helium; and so on.

ellipse: A closed, oval-shaped curve (generated by passing a plane through a cone) describing the shape of the orbit of one body around another.

emission: Release of electromagnetic radiation from matter.

emission band: Narrow wavelength interval in which molecules emit light.

emission line: Very narrow wavelength intervals in which atoms emit light.

Enceladus: One of the inner moons of Saturn, notable for its very bright, fissured surface of water ice, with several young, sparsely cratered areas.

energy: In physics, a specific quality equal to work or the ability to do work. Energy may appear in many forms, including electromagnetic radiation, heat, motion, and even mass (according to the theory of relativity).

energy level: The orbit of an electron around the nucleus of an atom.

English system: A nondecimal system of units using pounds, inches, and seconds, now being replaced by the more convenient metric system.

ephemeris: A table of predicted positions of a planet, asteroid, or other celestial body.

ephemeris time: A timekeeping system based on the motions of planets; more regular than conventional systems based on the Earth's rotation.

epicycle: A small circular motion superimposed on a larger circular motion.

epicycle theory: An early theory by Ptolemy that the planets move around the Earth in epicycles.

equatorial zone: On Jupiter, Saturn, and possibly other great planets, a bright cloud zone near the equator.

equinox: The date when the Sun passes through the Earth's equatorial plane (occurs twice annually).

erg: The unit of energy in the cgs metric system.

erosion: Removal of rock and soil by any natural process.

escape velocity: The minimum speed needed to allow a projectile to move away from a planet and never return to its point of launch. It equals $\sqrt{2} \times$ circular velocity. (See *circular velocity*.)

Eta Carinae Nebula: A nebula around a peculiar novalike variable star about 2 kpc away.

Europa: One of Jupiter's four large moons, notable for its smooth, bright, icy surface.

event horizon: The theoretical surface around a black hole from which matter and energy do not escape.

evidence for planets near other stars: Although planets the size of Jupiter or smaller near other stars are beyond our current detection capability, new techniques of imaging and astrometry will allow their detection within a few years, if they exist. Already some objects larger than Jupiter but smaller than stars have been found.

evidence for present-day star formation: The fact that the solar system is much younger than the galaxy; existence of young open clusters; existence of short-lived massive stars.

evidence of spiral structure (in Milky Way): Spiral distribution of hydrogen gas mapped by 21-cm radio line; spiral distribution of open clusters; spiral patterns observed in other galaxies.

evolution of galaxies: Changes in form and stellar populations of galaxies as a result of consuming gas and dust during star formation and production of heavy elements during star evolution.

evolutionary theory: A theory in which changes occur by relatively slow processes or processes commonly growing out of the initial conditions, rather than by sudden or unusual processes.

evolutionary track: The sequence of points on the H–R diagram occupied by a star as it evolves.

excitation: The process of causing an atom or molecule to go into an excited state, that is, having some electrons in elevated energy levels; the state of being excited.

excited atoms and molecules: Atoms or molecules in which electrons are not all in the lowest possible energy levels.

excited state: The state of an atom or molecule when not all electrons are in the lowest possible energy levels.

exobiology: Study of life beyond the Earth.

expanding universe: A term popularized by Eddington to describe the mutual recession of galaxies.

explorative interval: The hypothetical interval of time during which a species actively engages in the exploration of other planets.

extragalactic standard of rest: An assumed stationary frame of reference defined by using the nearby galaxies as reference objects.

fault: A fracture along which displacement has occurred on the solid surface of a planet or other celestial body.

fields: Entities dispersed in space but having a measurable value or magnitude that can be measured at any point in space. Examples are gravitational, electric, and magnetic fields.

fireball: An unusually bright meteor, which may yield meteorites.

first quarter: The phase of the Moon when it is one-fourth of the way around its orbit from new moon; the first quarter moon is seen in the evening sky with a straight terminator and half the disk illuminated.

fission theory: A theory of origin of a planet–satellite or binary star system by breakup of a single original body.

flare: (1) On the Sun, a sudden, short-lived, localized outburst of energy, often ejecting gas at speeds exceeding 1000 km/s; (2) an outburst from certain types of variable stars, sometimes called *flare stars.*

focal length: The distance from a lens or mirror to the point where it focuses the image of a very distant object, such as the Moon.

focus: One of the two interior points around which planets or stars move in an elliptical orbit.

forbidden line: Spectral line arising from a metastable state in atoms.

force: In physics, a specific phenomenon producing acceleration of mass. Forces can be generated in many ways, such as by gravity, pressure, and radiation.

formation of galaxies: Processes that led to subdivision of the universe's gas after the big bang and its collapse into individual galaxies.

free-fall: Motion under the influence of gravity only, without any other force or acceleration, such as rocket firing.

free-fall contraction: Contraction of a cloud or system of particles by gravity only, unresisted by any other force.

frequency: Number of electromagnetic oscillations per second corresponding to electromagnetic radiation of any given wavelength.

full moon: The phase of the Moon when it is closest to 180° from the Sun and therefore fully illuminated.

funneling effect: Concentration of all evolutionary tracks into the giant region of the H–R diagram.

galactic cannibalism: The absorption of one galaxy by another during a collision, forming a new, larger galaxy.

galactic equator: The plane of the Milky Way galaxy projected on the sky.

galactic halo: A spherical swarm of globular clusters above and below the galactic disk, centered on a point in the direction of the constellation Sagittarius.

galactic jets: Narrow beams of extremely hot, ionized gas shooting out of some active galaxies in two opposite directions, above and below the plane of the disk. They radiate enormous amounts of energy and their origin is uncertain.

galactic latitude: Angular distance around the galactic equator from the galaxy's center.

galactic longitude: Angular distance from the galactic equator.

galactic nucleus: The center of a galaxy.

galaxies with active nuclei: Galaxies whose centers emit more energy than other, normal galaxies. They are generally strong radio sources.

galaxy: Any of the largest groupings of stars, usually of mass 10^8 to 10^{13} $M_\odot$.

Galilean satellites: The four large satellites of Jupiter, discovered by Galileo.

Galileo Galilei (1564–1642): Italian scientist who first applied the telescope to observe other planets, discovering lunar craters, Jupiter's moons, and other celestial phenomena.

Galileo program: A NASA mission to put an unmanned probe in orbit around Jupiter in the 1990s.

Ganymede: Largest of the four Galilean satellites of Jupiter, with a fractured and cratered ice surface.

geological time scale: The sequence of events in the history of the Earth.

giant elliptical galaxy: Galaxies with diffuse elliptical form, somewhat resembling globular clusters but much larger.

giant planet: (1) Jupiter, Saturn, Uranus, or Neptune; (2) any planet much more massive than the Earth.

giant star: Highly luminous star larger than the Sun. O and B main-sequence stars are sometimes called *blue giants*; evolved stars of extremely large radius are called *red giants*.

gibbous: A phase between half-illuminated and fully illuminated, with a convex terminator (pronounced with hard g, as in "give").

globular cluster X radiation: X-ray radiation from globular clusters, discovered unexpectedly in the 1970s; it indicates energetic environments somewhere within them.

globular star cluster: A dense spheroidal cluster of stars, usually old, with mass of 10^4 to 10^6 $M_\odot$.

grain: Small (usually microscopic) solid particle in space.

granite: A rock type of modest density and high silica content, formed in association with differentiation processes and, therefore, found primarily on the Earth.

granitic rock: A silica-rich, light-colored rock type common in the Earth's continents. Granites, being low in density, tend to float to the surfaces of planets that have had extensive melting in the outer layers.

granules: Convection cells 1000 to 2000 km across, rising from the subphotospheric layers of the Sun. Each granule rises at a speed of 2 to 3 km/s and lasts for a few minutes.

gravitational contraction: Slow contraction of a cloud, star, or planet due to gravity, causing heat and radiation.

gravitational lensing: The creation of a distorted image of a distant quasar or galaxy when its light is focused by the gravity of a galaxy between it and us.

gravitational red shift: A red shift caused by light emitted from a region of very high gravity, such as a neutron star or the environs of a black hole.

gravity: The force by which all masses attract all other masses.

great circle: Any circle on the surface of a sphere (especially the Earth or sky) generated by a plane passing through the center of the sphere; the shortest distance between two points on a sphere.

Great Red Spot: A large, reddish, oval, semipermanent cloud formation on Jupiter.

greenhouse effect: Heating of an atmosphere by absorption of outgoing infrared radiation.

Gregorian calendar: Essentially the modern calendar system, introduced around A.D. 1600 under Pope Gregory XIII, and containing the modern system of reckoning leap years.

ground state: The lowest energy state of an atom, in which all electrons are in the lowest possible energy levels.

Gum Nebula: A large, relatively nearby nebula in the Southern Hemisphere sky, detected in hydrogen alpha light and formed by a supernova explosion estimated to have occurred around 9000 B.C.

HI region: Interstellar region in which hydrogen is predominantly neutral.

HII region: Interstellar region in which hydrogen is predominantly ionized.

half-life: In any phenomenon, the time during which the main variable changes by half its original value; often used loosely to indicate the characteristic time scale of a phenomenon. In radioactive decay, the time for half the atoms in a system to disintegrate.

Halley's comet: The most famous comet, which visits the inner solar system every 76 years, most recently in 1986 when close-up photos were made by space probes.

Hayashi track: A sharply descending evolutionary track in the H–R diagram covering the early period of stellar evolution from the high-luminosity phase to the main sequence.

heliacal rising: A star's first visible rising during the yearly cycle.

heliacal setting: A star's last visible setting during the yearly cycle.

helium flash: Runaway helium "burning" inside a star as it evolves off the main sequence and into the giant phase of evolution. It occurs when degenerate gas at the star's center reaches a temperature of about 10^8 K.

Helmholtz contraction: Slow contraction of a cloud or system of particles by the force of gravity, which is retarded by outward gas pressure and the limited rate at which radiation can escape.

hierarchical universe: A theoretical model of the universe organized in ever-larger clusters of galaxies, plausibly with a density that approaches zero as larger volumes are considered.

high-luminosity phase: A star's short-lived stage of maximum brightness during pre-main-sequence evolution.

high-velocity star: A star with a high velocity relative to the Sun; generally associated with the galactic halo.

homogeneous: Uniform in composition throughout the volume considered.

hour angle: The number of hours since a star (or other body) last crossed the local meridian.

H–R diagram: A technique for representing stellar data by plotting spectral type (or color or temperature) against luminosity (or absolute magnitude), named after its early proponents, Hertzsprung and Russell.

Hubble Space Telescope: A large telescope with 2.4-m (94-in.) mirror originally designed to be orbited by the Space Shuttle in 1985. Due to the Challenger disaster, its launch has been delayed.

Hubble's constant: The ratio between a galaxy's recession speed and its distance, measured to be about 50 to 100 km/s/mpc.

Hubble's relation: Expression indicating how the recession velocity of a galaxy increases with distance.

Hydrogen alpha line: The designation of hydrogen's red spectral line at 656.3 nm, more properly called *hydrogen Balmer alpha.*

hydrogen Balmer series: The series of all hydrogen spectral lines from 364.6 to 656.3 nm, caused by electron transitions between the second and higher energy levels.

hyperbola: The orbital curve followed by any free-falling body moving faster than escape velocity.

Hyperion: An irregularly shaped outer moon of Saturn, noted for its chaotic rotation.

hypothesis: A proposed explanation of an observed phenomenon or a proposal that a certain observable phenomenon occurs.

Iapetus: An outer moon of Saturn noteworthy because one hemisphere has a bright, icy surface, and the other, a black carbonaceous surface.

IC number: The catalog number of a cluster, nebula, or galaxy in the *Index Catalog.*

igneous rock: Rock crystallized from molten material.

impact crater: A roughly circular depression of any size (known examples range from microscopic size to diameters greater than 1000 km) caused by a meteorite impact.

impact-trigger theory: The leading theory of the moon's origin, in which material was blasted off Earth's mantle and then reaccumulated to form the moon.

incidence of multiplicity: Among stars, the fraction of systems that contain more than one star. Probably 50 to 70% of systems have companion stars, many of these having more than one companion.

inferior planet: Mercury or Venus.

inflation: The period of extremely rapid expansion of the scale of space in the first fraction of a second after the big bang.

inflationary big-bang theory: Model of the big bang calling for a super-rapid expansion of space in the first fraction of a second after the big bang.

infrared light: Radiation of wavelength too long to see, usually about 1 to 100 μm.

infrared star: A star detected primarily by infrared light.

initial conditions: (1) In any scientific problem, the conditions that define the beginning of the process being studied; (2) in cosmogony, the conditions at the moment of, or immediately after, the big bang.

intense early bombardment: The very intensive bombardment of planetary bodies by meteorites, from 4.6 to about 4 billion years ago, following plant formation.

interferometry: A system for obtaining high-resolution astronomical observations by linking several physically separated telescopes electronically, in effect creating a single, much larger telescope.

intergalactic globular cluster: A globular star cluster or a small galaxy resembling same, found in intergalactic space.

interstellar atom: Atom of gas in interstellar space.

interstellar grain: Microscopic solid grain in interstellar space; interstellar dust.

interstellar molecule: Molecule of gas in interstellar space.

interstellar obscuration: Absorption of starlight by interstellar dust, causing distant objects to appear fainter.

interstellar reddening: Loss of blue starlight due to interstellar dust, causing distant objects to appear redder and fainter.

interstellar snowball: Hypothetical interstellar particle larger than an interstellar grain.

inverse square law: The relation describing any entity, such as radiation or gravity, that varies as $1/r^2$, where r is the distance of the entity from the source.

Io: The innermost Galilean moon of Jupiter, famous for its active volcanism, which is unique among moons.

ion: Charged atom or molecule.

ionization: The process of knocking one or more electrons off a neutral atom or molecule.

iron meteorite: Meteorite composed of a nearly pure nickel–iron alloy.

irregular galaxy: A galaxy of amorphous shape. Most have relatively low mass (10^8–10^{10} $M_\odot$).

irregular variable: A star that fluctuates in brightness irregularly.

isotope: A form of an element with a specified number of neutrons in the nucleus. Each element may have many possible isotopic forms, but only a few are stable.

isotropic: Appearing uniform no matter what the direction of view.

joule: The unit of energy in the SI metric system of units.

Julian date: The date based on a running tabulation of days, starting January 1, 4713 B.C.

Jupiter's atmospheric composition: Mostly hydrogen, with additional helium, hydrogen-based compounds, and other gases.

Jupiter's infrared thermal radiation: Infrared radiation from Jupiter due to slow contraction of its interior and exceeding the incoming solar radiation.

Jupiter's interior: Beneath Jupiter's atmosphere, a high-pressure region of liquid hydrogen, liquid metallic hydrogen, and a small central core of rocky material.

Jupiter's temperature: Temperatures around $-200°F$ are measured at and above the cloud tops in Jupiter's atmosphere, but the air temperature increases below the clouds to values around room temperature and even warmer.

Kelvin scale: The absolute temperature scale, with 0 K = absolute zero. A Kelvin degree is the same size (same temperature difference) as a centigrade degree.

Johannes Kepler (1571–1630): Astronomer who first deduced the shapes and relations of planets' elliptical orbits.

Kepler's laws: The three laws of planetary motion that describe how the planets move, show that the Sun is the central body, and allow accurate prediction of planetary positions.

kiloparsec (kpc): 1000 parsecs.

Kirchhoff's laws: Laws describing conditions that produce emission, absorption, and continuous spectra.

$L_\odot$: The luminosity of the Sun, 4×10^{26} watts.

LSR: (See *local standard of rest*.)

Lagrangian points: In an orbiting system with one large and one small body, an array of five points where a still smaller body would retain a fixed position with respect to the other two.

Lagrangian surface: An imaginary surface with a figure-8 cross section surrounding two coorbiting bodies in circular orbits and constraining motions of particles within the system.

lava: Molten rock on the surface of a planet.

life: A process in which complex, carbon-based materials organized in cells take in additional material from their environment, replicate molecules, reproduce, and do other weird things like writing books.

life on Mars: Although long sought and believed possible by many scientists, biological processes on Mars were apparently ruled out in 1976 when Viking landers found no organic material there.

life in the solar system: Hypothetical biological activity on other planetary bodies, such as on Mars, under Europa's ice, or in Jupiter's atmosphere. Now regarded as unlikely.

life outside the solar system: Hypothetical biological activity beyond the outskirts of our solar system, as on hypothetical planets around other stars. Such life is regarded as plausible, even if sparsely scattered, by most astronomers.

light curve: A plot of brightness of a star (or other object) versus time.

light-gathering power: The ability of a telescope or binoculars to gather light. It is proportional to the area of the objective, that is, the square of the aperture.

light-year: The distance light travels in one year, 9.46×10^{12} km.

limb: The apparent edge of a celestial object.

line of nodes: A line formed by the intersection of an orbit and some other reference plane, such as the plane of the solar system.

linear measure: Measurement involving linear distances, as opposed to angles or angular distances.

lithosphere: The solid rocky layer in a partially molten planet.

Local Group: The cluster of galaxies to which the Milky Way and 26 nearby galaxies belong.

local standard of rest (LSR): A frame of reference moving with the average velocity of the nearby stars (out to about 50 pc from the Sun).

luminosity: The total energy radiated by a source per second. The luminosity of the Sun ($L_\odot$) is 4×20^{26} watts.

lunar eclipse: Dimming of the Moon as it passes into the Earth's shadow.

$M_\odot$: The mass of the Sun, 2×10^{30} kg.

M giant: A giant star of spectral class M.

Magellanic clouds: The two galaxies nearest the Milky Way, irregular in form and visible to the naked eye in the Southern Hemisphere.

Magellanic stream: Gas filaments connecting the Magellanic clouds to the Milky Way.

magma: Underground molten rock.

magma ocean: Primordial layer of molten lava on the initial surface of the Moon and (by inference) planets.

magnetic braking: The slowing of rotation of a star or planet by interaction of its magnetic field with surrounding ionized material.

magnetic field: Region of space in which a compass (or other detector) would respond to magnetism of some body.

magnification: Apparent angular size of a telescope image divided by the angular size of the object seen by naked eye.

main-sequence star: One of the group of stars defined on the H–R diagram that have a relatively stable interior configuration and are consuming hydrogen in nuclear reactions; a star on the main sequence.

mantle: A region of intermediate density surrounding the core of planets.

mare (pl. *maria*): A dark-colored region on a planet or satellite; a region of basaltic lava flow on the Moon.

Mars-crossing asteroids: Asteroids whose orbits cross that of Mars.

Mars' rotation period: The day on Mars—$24^h 37^m$, only slightly longer than the Earth's.

Martian air pressure: The pressure exerted by the very thin Martian air. On Mars' surface, the air pressure is only about 0.7% of that at the Earth's surface.

Martian air temperature: Soil temperatures approach or exceed freezing in the day; the air is colder. Night temperatures around $-123°F$ are recorded.

Martian atmosphere: The thin gases around Mars, composed almost entirely of carbon dioxide (CO_2).

Martian meteorites: A handful of meteorites about 1.3 billion years old, believed to have been blasted off Mars about 0.2 billion years ago.

maser (*microwave amplification by simulated emission of radiation*): (1) A device that amplifies microwave radio waves through special electronic transitions in atoms; (2) an interstellar cloud that acts in this way.

mass: (1) Material; (2) the amount of material.

mass/luminosity ratio: The mass per unit of light or total radiation emitted from an object such as a galaxy.

mass–luminosity relation: The relation between the mass of a main-sequence star and its total radiation rate; the more massive, the greater the luminosity.

mass of the galaxy: About 4×10^{41} kg, or $2 \times 10^{11} M_\odot$, as estimated from the circular velocity equation.

Maunder minimum: The interval from 1645 to 1715, when solar activity was minimal.

mean density: Mass of an object divided by its volume.

mean solar time: Time shown by conventional clocks, determined by the Sun's mean rate averaged over the year.

megaparsec: One million parsecs.

meridian: (1) A north–south line on a planet, moon, or star; (2) a great circle through the celestial pole and the zenith.

Messier number: The catalog number of a nebula, star cluster, or galaxy in *Messier's Catalog*.

metallic hydrogen: A high-pressure form of hydrogen with free electrons.

metastable state: In an atom, a configuration of electrons that is relatively long-lived, but is rarely found on the Earth because it is disrupted by collisions with other atoms; it may be found in interstellar atoms, creating forbidden spectral lines.

meteor: A rapidly moving luminous object visible for a few seconds in the night sky (a "shooting star").

meteorite: An interplanetary rock or metal object that strikes the ground.

meteorite impact crater: Circular depression in planetary surfaces, caused by explosions as meteorites crash into the surfaces at high speeds.

meteoroid: A particle in space, generally smaller than a few meters across.

meteor shower: A concentrated group of meteors, seen when the Earth's orbit intersects debris from a comet.

meter: 39.4 in.

Milky Way galaxy: The spiral galaxy in which we live.

Miller experiment: An experiment in which amino acids are created in laboratory conditions simulating the conditions of the early Earth.

Mimas: The innermost large moon of Saturn, icy and heavily cratered.

minerals: Chemical compounds, usually in the form of crystals, that constitute rocks.

minute of arc: An angle equaling $\frac{1}{60}$ of a degree.

Miranda: The innermost of five large moons of Uranus, noted for puzzling fractured and grooved terrain.

missing mass: Mass suspected to exist distributed through the universe (in intergalactic space?) that, when added to the currently observed mass, would bring the universe closer to closure density.

mks system: A metric system of units expressing length in meters, mass in kilograms, and time in seconds. (See also *cgs system.*)

molecular cloud: An interstellar cloud of gas and dust, with greater than average density, dust content, and high concentration of molecules.

Moon: (1) The Earth's natural satellite; (2) any satellite.

multiple scattering: Redirection of electromagnetic radiation (such as light waves) by repeated interaction with atoms, molecules, or dust grains in space or in an atmosphere.

multiple star system: A system of three or more stars orbiting around each other.

mutual recession of galaxies: The phenomenon that all distant galaxies are moving away from us, and the farther away they are, the faster they are receding.

natural selection: The theory that states that those individuals best adapted to the ever-changing environment produce a greater number of offspring.

nebula: A cloud of dense gas and/or dust in interstellar space or surrounding a star.

negative hydrogen ions: Hydrogen atoms that have temporarily captured an extra electron, responsible for opacity in the photosphere of the Sun and many stars.

Neptune: The outermost gas giant planet in the outer solar system.

neutrino: A subatomic particle created in certain nuclear reactions inside stars. It can pass through most matter. Its mass is uncertain, being either zero or a tiny fraction of an electron's mass.

neutron: One of the two major particles constituting the atomic nucleus; it has zero charge and mass 1.6749×10^{-27} kg.

neutron star: A star with a core composed mostly of neutrons, with density 10^{16} to 10^{18} kg/m^3. Many or most neutron stars are pulsars.

new moon: The phase of the moon when it is nearest the Earth–Sun line, hence invisible from Earth because of the Sun's glare.

Isaac Newton (1642–1727): English physicist who discovered the spectrum and laws of gravitation and motion; he also developed calculus and made other discoveries. Possibly the greatest physicist of history.

Newtonian–Euclidean static cosmology: A hypothetical model of the universe with infinite volume, no expansion, and Euclidean geometry.

Newton's laws of motion: Three rules describing motion and forces. Briefly, (1) a body remains in its state of motion unless a force acts on it; (2) force equals mass times acceleration; (3) for every action there is an equal and opposite reaction.

NGC number: The catalog number of a nebula, cluster, or galaxy in the *New General Catalog*.

node: One of two points where an orbit crosses a reference plane.

noncosmological red shift: A hypothetical red shift of distant galaxies *not* caused by the Doppler effect.

non-Euclidean geometry: A hypothetical geometry in which Euclid's relations are not true; a geometry of curved space.

nonthermal radiation: Radiation *not* due to the heat of the source; for example, synchrotron radiation.

North Star: (1) Polaris; (2) any bright star that happens to be within a few degrees of the north celestial pole during a given era.

nova: A type of suddenly brightening star (from the Latin for "new") resulting from explosive brightening when gas is dumped from one member of a binary star pair onto the other.

nuclear reaction: Reaction involving the nuclei of atoms in which a nucleus changes mass.

nuclear winter: A sudden drop of temperature on Earth hypothesized to occur if extensive nuclear explosions (or an asteroid or comet impact) ejected enough dust and smoke into the atmosphere to block sunlight.

nucleus: (1) The matter at the center of an atom, composed of protons and neutrons; (2) the bright central core (or solid body) of a comet; (3) the bright central core of a galaxy.

O association: An association of O-type stars.

objective: The major light-gathering element of a telescope; the mirror in a reflector and the lens in a refractor.

oblate spheroid: The shape assumed by a sphere deformed by rotation.

obliquity: The angle by which a planet's rotation axis is tipped to its orbit.

Occam's razor: The principle that the simplest hypothesis, with the fewest assumptions, is most likely to be correct. Named after its use to cut away false hypotheses.

ocean tide: The Moon's tidal stretching of the Earth, as observed in the ocean surface (as opposed to body tide).

Olbers' paradox: The problem of why the sky is dark at night if the universe is filled with stars.

Oort cloud: The swarm of comets surrounding the solar system.

opacity: The extent to which gaseous (or other) material absorbs light.

open space: Space that is uncurved or curved in such a way as to have infinite volume and no boundaries.

open star cluster: A grouping of relatively young Population I stars (usually 10^2–$10^3 M_\odot$), sometimes called a *galactic cluster.*

open universe: A universe with infinite volume and no boundaries. In most cosmological models, the universe, if open, will continue expanding forever.

opposition: The period when a superior planet lies in an opposite direction from the Sun as seen from the Earth. At opposition, a planet appears in the midnight sky, well placed for observation.

optical double star: A pair of stars that have small angular separation but are not coorbiting.

orbital precession: A slow, cyclical change in the orientation of the plane of an orbit.

organic chemistry: Chemistry involving organic molecules.

organic molecule: Molecule based on the carbon atom, usually large and complex, but not necessarily part of living organisms.

origin of the heavy elements: The process by which light elements fused to form nuclei of heavy elements, primarily inside massive stars.

Orion Nebula: Several hundred solar masses of gas and dust composing the core of a star-forming region about 460 parsecs away. One of the most prominent nebulae in our sky, it forms the central "star" in Orion's sword.

Orion star-forming region: The larger region in which stars are forming around the Orion Nebula.

oscillating universe: A hypothetical model of the universe with cycles of contraction and expansion.

ozone layer: An atmospheric layer rich in ozone (O_3), created by the interaction between oxygen molecules (O_2) and solar radiation. On Earth, its altitude is about 20–60 km.

parabola: (1) The curved trajectory followed by a particle moving at escape velocity; (2) the curve of a Newtonian telescope's primary mirror.

parallax: An angular shift in apparent position due to an observer's motion; more specifically, a small angular shift in a star's apparent position due to the Earth's motion around the Sun. *Stellar parallax,* used to measure stellar distance, is defined as the angle subtended by the radius of the Earth's orbit as seen from the star.

parent body: A body from which a meteorite formed and later broke off as a fragment.

parent isotope: A radioactive isotope that disintegrates and forms a daughter isotope.

parsec: A distance of 206,265 AU, 3.26 ly, or 3.09×10^{13} km; defined as the distance corresponding to a parallax of 1 second of arc.

partial solar eclipse: An eclipse in which the light source is not totally obscured from an observer.

particlelike properties of light: Characteristics of light, such as concentration of energy and momentum in discrete microscopic packets (photons) that mimic the properties of particles.

Pauli exclusion principle: A principle of subatomic physics specifying that no two electrons in a very small volume have exactly the same properties of energy, motion, and so on.

peculiar velocity: A star's velocity with respect to the local standard of rest.

penumbra: (1) The outer, brighter part of a shadow, from which the light source is not totally obscured; (2) the outer, lighter part of a sunspot.

perigee: The point in an orbit around the Earth that is closest to the Earth.

permafrost: Semipermanent underground ice.

phase: The apparent shape of an illuminated body, varying with the "phase angle" from observer to body to illumination source.

Phoebe: The outermost moon of Saturn, a small dark moon believed to be captured.

photometry: The measurement of the amount of light, either total or in different specified colors, coming from an object.

photon: The quantum unit of light, having some properties of a wave. For each wavelength of radiation, the photon has a different energy.

photosphere: The light-emitting surface layer of the Sun.

physical binary stars: Two stars orbiting around a common center of mass.

Planck's law: A formula that describes the energy associated with each wavelength (color) in the spectrum. It shows that photons of blue light are more energetic than photons of red light.

planet: A solid (or partially liquid) body orbiting around a star but too small to generate energy by nuclear reactions.

"Planet X": A term sometimes used for a hypothetical tenth planet in the solar system, beyond Pluto.

planetary nebula: A type of circumstellar gas cloud that has spheroidal shape and often appears as a faint disk in telescopes; it has nothing to do with planets except for the rough resemblance to a planet's shape when seen telescopically.

planetesimal: One of the small bodies from which planets formed, usually ranging from micrometers to kilometers in diameter.

planetology: The study of the planets' origins, evolution, and conditions.

plasma: A high-temperature gas consisting entirely of ions, instead of neutral atoms or molecules. Because of the high temperature, the atoms strike each other hard enough to keep at least the outer electrons knocked off.

plate: Moving unit of the Earth's lithosphere, typically of continental scale.

plate tectonics: Motions of a planet's lithosphere, causing fracturing of the surface into plates. Primary example occurs on the Earth.

Pluto: Cataloged as the outermost and smallest planet in the solar system, but possibly one of many small worldlets on the fringe of the solar system.

Polaris: The North Star.

Population I: Stars with a few percent heavy elements (heavier than helium), found in the disks of spiral galaxies and in irregular galaxies.

Population II: Stars composed of nearly pure hydrogen and helium, found in the halo and center of spiral galaxies, in elliptical galaxies, and to a limited extent in irregulars.

powers of 10: The number of times 10s must be multiplied together to give a specific number; the exponent of 10. (Example: $10^2 = 100$; the power, or exponent, is 2; see also Appendix 1.)

precession: The wobble in the position of a planet's rotation axis caused by external forces. Also, the change in a coordinate system (tied to any planet) caused by such a wobble.

pre-main-sequence star: Evolutionary state of stars prior to arrival on the main sequence, especially just before the main sequence is reached.

primary atmosphere: The atmosphere of a planet (if any) just after planet formation. (See also *secondary atmosphere.*)

primeval fireball: The hypothetical, expanding, initial cloud of high temperature plasma during the big bang.

principle of relativity: The principle that observers can measure only relative motions, since there is no absolute frame of reference in the universe by which to specify absolute motions.

problem of the missing mass: Motions of stars around galaxies imply that these galaxies have more mass than accounted for by the visible stars. Thus there must be unseen mass in some uncertain form. Astronomers are still searching for this "missing" mass.

prograde and retrograde satellite orbits: Satellite orbits in which motion is in the same or opposite direction, respectively, as the planet's rotation.

prograde rotation: Spinning on an axis from west to east, or counterclockwise as seen from the North Pole (as in the case of the Earth).

prominence: A radiating gas cloud extending from the solar surface into the thinner corona.

prominent star: One of the brightest stars in our sky, but not necessarily one of the nearest.

proper motion: The angular rate of motion of a star or other object across the sky. (Most stars have proper motions less than a few seconds of arc per year.)

protein: Any of several types of complex organic molecules made from amino acids inside plants and animals, which are essential in living organisms.

proteinoid: Cell-like, nonliving spheroid of protein molecules created in the laboratory by heating amino acids and adding water; a possible step in the evolution of life.

protogalaxy: A gravitationally stable cloud of galactic mass contracting toward galactic dimension.

proton: One of the two basic particles in an atomic nucleus, with positive charge and mass $1.6726 = 10^{-27}$ kg.

proton–proton chain: A series of thermonuclear reactions that convert hydrogen nuclei to helium nuclei, converting a tiny amount of mass into energy.

protoplanet: A planet shortly before its final formation. Sometimes hypothesized to have a massive atmosphere and greater mass than in its present state.

protostar: A gravitationally stable cloud of stellar mass contracting in an early pre-main-sequence evolutionary state.

pseudoscience: Research that has the trappings of science but does not follow the scientific method, usually lacking review and repetition of observations by independent researchers.

Ptolemaic model: The ancient Earth-centered model of the solar system, with the Sun, Moon, and other planets moving in epicycles.

pulsar: A rapidly rotating neutron star with a strong magnetic field, observed to emit pulses of radiation.

QSO: Quasi-stellar object; faint bluish star with a spectrum similar to that of a quasar.

quantum: A small indivisible unit of some quantity such as energy or mass.

quasar: Any of a group of starlike, faint celestial objects with very large red shifts. Many astronomers believe they are extremely distant galaxies of unusually energetic form.

r-process reactions: Rapid reactions, probably occurring inside supernovae, in which heavy elements are formed as atomic nuclei capture neutrons. (See also *s-process reactions.*)

radial velocity: The velocity component along the line of sight toward or away from an observer. Recession is positive; approach is negative.

radiation: (1) Any electromagnetic waves or atomic particles that transmit energy across space; (2) one of three modes of heat (energy) transmission through stars or planets from warm regions to cool regions.

radiation pressure: An outward pressure on small particles exerted by electromagnetic radiation in a direction away from the light source.

radioactive atom: Any atom whose nucleus spontaneously disintegrates.

radio galaxy: A galaxy that emits unusually large amounts of radio radiation.

radioisotopic dating: Dating of rock or other material by measuring amounts of parent and daughter isotopes.

ray: A bright streak of material ejected from a crater on the Moon or other planet.

Rayleigh scattering: Scattering of light by particles smaller than the light's wavelength. This process favors scattering of blue light.

red giant: A post-main-sequence star whose surface layers have expanded to many solar radii and have relatively low temperatures.

red shift: A Doppler shift of spectral features toward longer wavelengths, indicating recession of the source.

red shift of galaxies: The shift toward longer wavelengths in light of distant galaxies, due to their recession from the solar system. It increases with galaxies' distances.

reflected radiation: Radiation that has arrived from outside a body and bounced off, as opposed to thermal radiation.

reflector: A type of telescope using a mirror as the light collector.

refractor: A type of telescope using a lens as the light collector.

refractory element: An element least likely to be driven out of a material by heating. These elements are usually concentrated in the last components to melt when a material such as rock is heated.

regolith: A powdery soil layer on the Moon and some other bodies caused by meteorite bombardment.

regression of nodes: A shifting of the R.A. = Dec. coordinate system, relative to the stars, as a result of the 26,000-y wobble of the Earth's rotation axis.

relativistic: Moving at speeds near that of light.

relativity: (See *principle of relativity.*)

representative stars: A sample of stars randomly drawn from the total population of stars in space.

resolution: The smallest angle that can be discerned with an optical system; for example, the eye can resolve about 2 minutes of arc.

retrograde motion: Revolution or rotation from east to west contrary to the usual motion in the solar system.

retrograde rotation: Spinning on an axis from east to west, or clockwise as seen from the North Pole (opposite to the spin direction of Earth).

Rhea: (See *Tethys, Dione, and Rhea.*)

rift: A major split in a planet's lithosphere due to active or incipient plate tectonic stresses.

right ascension: Longitude lines projected onto the celestial sphere. (Abbreviation: R.A.)

rille: A type of lunar valley.

Roche's limit: The distance from a large body within which tidal forces would disrupt a satellite.

rock: Solid aggregation of minerals.

rotation curve: Orbital velocity as a function of distance from the center of a galaxy.

rotational line broadening: Broadening of spectral lines due to rotation of the source.

RR Lyrae star: A type of variable star similar to the Cepheids that has been found associated with Population II and not Population I.

runaway star: A star rapidly moving away from a region of recent star formation.

Russell–Vogt theorem: The theorem stating that the equilibrium structure of a star is determined by its mass and chemical composition.

s-process reactions: Slow reactions in giant stars in which heavy elements are built up as atomic nuclei capture neutrons. (See also *r-process reactions.*)

Saha equation: An equation derived in 1920, by the Indian physicist Saha, that tells the percent of atoms in each different excited state, given the conditions in a gas. This in turn controls what spectral lines are emitted or absorbed by that gas.

saros cycle: An interval of 6585 d (about 18 y) separating cycles of similar eclipses, used by ancient people to predict eclipses.

satellite: Any small body orbiting a larger body.

Saturn: The sixth planet out from the Sun, famous for its prominent rings.

Saturn's atmosphere: The thick, cloudy gases around Saturn, composed mostly of hydrogen.

Saturn's ring system: A system of innumerable icy particles orbiting Saturn.

Saturn's satellite system: A family of at least 17 moons orbiting Saturn, ranging from 20 km diameter up to a size slightly exceeding that of the planet Mercury.

science: Study of nature using the scientific method. (See *scientific method.*)

scientific method: The method of learning about nature from making observations, formulating hypotheses, and constructing observational or experimental tests to see if the hypotheses are accurate.

seasonal changes on Mars: Changes in shape and darkness of the dusky patches on Mars from summer to winter and year to year. Once thought to indicate Martian vegetation, the changes are now known to result from blowing dust deposits.

second of arc: An angle equaling $\frac{1}{3600}$ of a degree.

secondary atmosphere: A planet's atmosphere after modification by outgassing and other processes. (See also *primary atmosphere*.)

sedimentary rock: Rock formed from sediments.

seeing: The quality of stillness or lack of shimmer in a telescopic image, associated with atmospheric conditions. If the atmosphere is very still and the image is sharp, the seeing is said to be good.

seismic waves: Waves passing through the interior or surface layers of a planet due to a seismic disturbance, such as an earthquake or large meteorite impact.

seismology: Study of vibrational waves passing through planets, revealing internal structure.

selection effect: Any effect that systematically biases observations or statistics away from a correct understanding.

Seyfert galaxy: A type of galaxy with a bright, bluish nucleus, possibly marking a transition between ordinary galaxies and quasars.

shepherd satellites: Satellites that move near planetary rings and act to confine the ring particles onto certain orbits.

short-period comet: A comet with revolution period less than about 100 y.

SI metric system: The internationally standardized scientific system of units, in which length is given as meters, mass as kilograms, and time as seconds.

sidereal: Referring to stars.

sidereal period: A period of rotation or revolution where the movement is measured relative to the stars.

sidereal time: Time measured by the apparent motion of the stars (instead of the Sun), used by astronomers to point telescopes toward celestial targets; it is the right ascension that is on the meridian at any given location.

significant figures: The number of digits known for certain in a quantity.

small-angle equation: The equation giving the relation between the distance D of an object, its diameter d, and its angular size α (expressed in seconds of arc):

$$\frac{\alpha}{206,265} = \frac{d}{D}$$

solar apex: The direction toward which the Sun is moving relative to nearby stars.

solar constant: The amount of energy reaching a Sun-facing square meter at a given planet's (usually the Earth's) orbit per unit time; for the Earth, it is 1390 W/m^2.

solar core: The Sun's central region of high-pressure gases where nuclear energy is produced.

solar cycle: 22-y cycle of solar activity.

solar eclipse: Partial or total blocking of the Sun's light by an astronomical body (in most usages, by the Moon).

solar nebula: The cloud of gas around the Sun during the formation of the solar system.

solar rotation: Turning of the Sun on its axis in 25.4 d.

solar system: The Sun and all bodies orbiting around it.

solar wind: An outrush of gas past the Earth and beyond the outer planets. Near the Earth, the solar wind travels at velocities near 600 km/s, sometimes reaching 1000 km/s.

solstice: The date when the Sun reaches maximum distance from the celestial equator (occurs twice annually).

solstice principle: According to this principle, the sunrise and sunset positions of the Sun on the eastern and western horizons (respectively) shift positions according to season and the observer's latitude. The principle can be used to optimize passive solar energy input into a home or other building.

space velocity: A star's velocity with respect to the Sun.

spectral class: A class to which a star belongs because of its spectrum, which in turn is determined by its temperature. The spectral classes are O, B, A, F, G, K, and M, from hottest to coolest.

spectral line strength: Measure of the total energy absorbed or emitted in a spectral line.

spectrograph: An instrument for recording a photographic image of a spectrum.

spectroheliograph: An instrument for observing the Sun in certain specified wavelengths.

spectrometer: An instrument for tracing the intensity of a spectrum at different wavelengths; the result is a graph.

spectrophotometry: The study of the amount of radiation at each wavelength in the spectrum.

spectroscopic binary: A binary star revealed by varying Doppler shifts in spectral lines.

spectroscopy: Study of spectra, especially as revealing the properties of the light source.

spectrum: Light from an object arranged in order of wavelength; specifically, the colors of visible light, arranged in this order.

spectrum binary: A binary revealed by mixture of two spectral classes in the spectrum.

spectrum—luminosity diagram: The H–R diagram.

speed of light: Designated as c, the speed of light is about 300,000 km/s and is constant as perceived by all observers.

spicule: Narrow jet of gas extending out of the solar chromosphere, with a lifetime of about 5 min.

spiral arm: In spiral galaxies, one of the arms lying at an angle to the Sun–center line. The arms contain open clusters, O and B stars, and nebulae.

spiral galaxy: A disk-shaped galaxy with a spiral pattern, typically containing 10^{10}–$10^{12}\,M_\odot$ of stars, dust, and gas.

standard time: Solar time appropriate to the given local time zone.

star: A mass of material, usually wholly gaseous, massive enough to initiate (or to have once initiated) nuclear reactions in its central region.

starburst: A relatively sudden and rapid episode of star formation in a galaxy, probably triggered in some cases by collision with another galaxy.

Stefan–Boltzmann law: A law giving the total energy E radiated from a surface of area A and temperature T per second: $E = \sigma T^4 A$. Sigma (σ), the Stefan–Boltzmann constant, equals $5.67 \times 10^{-8}\,\mathrm{W/m^2 \cdot K^4}$ in SI units.

stellar evolution: Evolution of every star from one form to another forced by changes in composition as nuclear reactions proceed.

stellar populations in galaxies: Groupings of stars with different composition and age. Population I consists of young stars with a few percent of heavy elements; stars near the Sun are Population I. Population II includes older stars with virtually no heavy elements.

Stonehenge: A prehistoric English ruin with built-in astronomical alignments.

stony–iron meteorite: Stony meteorite that probably comes from deep within the parent body, where melted stony and metallic material coexisted.

stromatolite: A primitive life form, colonies of blue-green algae that were among the first to appear along shorelines. They are among the most abundant fossils from 3.5 to 2 billion years ago.

subfragmentation: Breakup of a contracting cloud into smaller condensations.

subfragmentation theory: A theory of formation of binary and multiple stars by breakup of the protostellar cloud into two or more components during its collapse from nebular to stellar dimensions.

subtend: To have an angular size equal to a specified angle. For example, the Moon subtends to $\frac{1}{2}°$.

Sun: The star orbited by the Earth.

Sun's composition: Almost 75% hydrogen and 25% helium.

sunspot: A magnetic disturbance on the Sun's surface that is cooler than the surrounding area.

Sun's revolution period: The 230-million-year period taken by the Sun to complete its orbit around the Milky Way galaxy.

superbubble: A large volume of hot gas in interstellar space, formed by coalescence of bubbles blown around supernovae.

supercluster of galaxies: Cluster of clusters of galaxies.

supergiant star: An extremely luminous star in the uppermost part of the H–R diagram.

supergranulation: Large-scale (15,000–30,000 km in diameter) convective cell patterns in the solar photosphere.

superior planet: Any planet with an orbit outside the Earth's orbit.

supernova: A very energetic stellar explosion expending about 10^{42} to 10^{44} joules and blowing off most of the star's mass, leaving a dense core.

symbiotic stars: A pair of stars whose evolutions are affecting each other, especially by mass transfer.

synchronous rotation: Any rotation such that a body keeps the same face toward a coorbiting body.

synchrotron radiation: Radiation emitted when electrons move at nearly the speed of light in a magnetic field.

synodical month: One complete cycle of lunar phases, 29.53 d.

tangential velocity: The velocity component perpendicular to the line of sight.

Tarantula Nebula (30 Doradus): A huge HII emission nebula in the Large Magellanic Cloud.

T association: An association of T Tauri stars.

tectonics: Disruption of planetary or satellite surfaces by large-scale mass movements, such as faulting.

telescope: An instrument for collecting electromagnetic radiation and producing magnified images of distant objects.

temperature: A measure of the average energy of a molecule of a material.

terminator: The dawn or dusk line separating night from day on a planet or satellite.

terrestrial planet: (1) Mercury, Venus, Earth, or Mars; (2) a planet composed primarily of rocky material.

Tethys, Dione, and Rhea: Intermediate-sized icy moons of Saturn.

theory: A body of hypotheses, often with mathematical backing and having passed some observational tests; often implying more validity than the term *hypothesis*.

theory of cosmological red shifts: The theory that galaxies' red shifts are all due to recessional motion, increase with distance, and thus give an indicator of distance.

theory of noncosmological red shifts: The theory that at least some galaxies' red shifts are not Doppler shifts due to recession, but are due to some other cause.

theory of star formation: The theory that describes how stars form by the gravitational collapse of interstellar clouds of dust and gas.

thermal escape: Escape of the fastest-moving gas atoms or molecules from the top of a planet's atmosphere by means of their thermal motion.

thermal motion: Movement of atoms and molecules associated with the temperature of the material; they grow faster as the temperature increases.

thermal radiation: Electromagnetic radiation emitted by a body and associated with an object's temperature; it grows greater and bluer in color as the temperature increases.

third quarter: The phase of the Moon when it is three-fourths of the way around its orbit from new moon; the third quarter moon is seen in the dawn sky with a straight terminator and half the disk illuminated.

3-kpc arm: An inner arm of our galaxy expanding from the center at about 53 km/s.

3-K radiation: Radio radiation coming uniformly from all over the sky, believed to be a red-shifted remnant of the big bang radiation.

thrust: The force generated by a high-speed discharge, as from a rocket or airplane.

tidal recession: Recession of the Moon (or other satellite) from the Earth (or other planet) caused by tidal forces.

tide: A bulge raised in a body by the gravitational force of a nearby body.

Titan: Saturn's largest moon, famous for its thick, smoggy-orange nitrogen atmosphere.

total solar eclipse: (1) An eclipse in which the light source is totally obscured from a specified observer; (2) an eclipse in which a body is entirely immersed in another's shadow. (See also *eclipse.*)

transit: (1) Passage of a planet across the Sun's disk; (2) any passage of a body with a small angular size across the face of a body with a large angular size.

triple-alpha process: A nuclear reaction in which helium is transformed into carbon in red giant stars.

Triton: Neptune's largest moon.

Trojan asteroids: Asteroids caught near the Lagrangian points in Jupiter's orbit, 60° ahead of and 60° behind the planet.

tsunami: A large ocean wave generated by earthquake or volcanic activity (the correct name for a tidal wave).

T Tauri star: A type of variable star, often shedding mass, believed to be still forming and contracting onto the main sequence.

"tuning fork" diagram: A classification scheme for galaxies.

21-cm emission line: The important radio radiation at 21-cm wavelength from interstellar neutral atomic hydrogen.

21-cm radio waves: Produced by neutral hydrogen, these waves are especially useful for galactic mapping because they allow us to detect HI clouds, which are concentrated in the spiral arms.

ultrabasic rock: A rock of high density, low silica content, and high iron content, often derived from the upper mantle of a planet or satellite.

ultraviolet light: Radiation of wavelength too short to see, but longer than that of X rays.

umbra: (1) The dark inner part of a shadow, in which the light source is totally obscured; (2) the dark inner part of a sunspot.

universe: Everything that exists.

Uranus: The seventh planet outward from the Sun.

Uranus' rings: A system of very narrow rings of dark particles around Uranus, scarcely detectable from Earth.

Uranus' rotation axis: Notable for its almost right-angle tilt (obliquity) of 97° to Uranus' orbital plane.

Uranus' satellites: A system of five large moons discovered from Earth, and another ten discovered by Voyager 2 in 1986.

Urey reaction: Reaction by which the Earth's carbon dioxide was concentrated in carbonate rocks after dissolving in seawater.

Van Allen belts: Doughnut-shaped zones around the Earth (or another planet with a strong magnetic field) that traps energetic ions from the Sun.

variable star: A star that varies in brightness.

Venera 7: First spacecraft to land successfully on another planet; it transmitted data from the surface of Venus in 1970.

Viking 1: The first successful probe to land on Mars (July 20, 1976). It made the first surface photos and measures of soil composition.

Viking 2: The second successful probe to land on Mars (September 3, 1976).

visible light: Electromagnetic radiation at wavelengths that can be perceived by the eye.

visual apparent magnitude: An apparent magnitude estimate based only on visual radiation from an object (excluding infrared, ultraviolet, X rays, and so on).

visual binary: A binary in which both components can be seen.

volatile element: Element easily driven out of a material by heating.

volcanic crater: A circular depression caused by volcanic processes such as explosion or collapse.

volcanism: Eruption of molten materials at the surface of a planet or satellite.

volcanoes: Sites where molten materials erupt from inside a planet or satellite.

wavelength: (1) The length of the wavelike characteristic of electromagnetic radiation; (2) in any wave, the distance from one maximum to the next.

wavelike properties of light: Characteristics of light, such as frequency and diffraction, that mimic properties of waves.

white dwarf star: A planet-sized star of roughly solar mass and very high density (10^8 to 10^{11} kg/m^3) produced as a terminal state after nuclear fuels have been consumed.

white light: A mixture of light of all colors in proportions as found in the solar spectrum.

Wien's law: A formula giving the wavelength W at which the maximum amount of radiation comes from a body of temperature T. The formula is $W = 0.00290/T$.

Wolf–Rayet star: A type of very hot star ejecting mass.

W Ursae Majoris stars: Contact binary stars.

X ray: Electromagnetic radiation of wavelength about 0.01 to 10 nm.

X-ray source: Celestial object emitting X rays; many are probably binary systems where mass is transferred.

zenith: The point directly overhead.

zero-age main sequence: The main sequence defined by a population of stars all of which have just evolved onto the main sequence. (Further evolution modifies the main sequence shape on the H–R diagram slightly.)

zodiac: A band around the sky about 18° wide, centered on the ecliptic, in which the planets move.

zodiacal light: A glow, barely visible to the eye, caused by dust particles spread along the ecliptic plane.

zone: Light cloud band on a giant planet.

zone of avoidance: A band around the sky, centered on the Milky Way, in which galaxies are obscured by the Milky Way's dust.

References

Asterisked references are less technical, more readily available, and recommended for general reading. They are good general references for preparation of term papers.

Chapter 1

*Ashbrook, J. 1973. "Astronomical Scrapbook." *Sky and Telescope 46:* 300.

*Bok, B. 1975. "A Critical Look at Astrology." *The Humanist,* September/October, p. 5.

*Carlson, J. B. 1975. "Lodestone Compass: Chinese or Olmec Primacy?" *Science 189:* 753.

*Gingerich, O. 1967. "Musings on Antique Astronomy." *American Scientist 55:* 88.

*———. 1984. "The Origin of the Zodiac." *Sky and Telescope 67:* 218.

*Hoyle, F. 1972. *From Stonehenge to Modern Cosmology.* San Francisco: W. H. Freeman.

Humphreys, C. J., and W. G. Waddington. 1983. "Dating the Crucifixion." *Nature 306:* 743.

*Jerome, L. E. 1975. "Astrology: Magic or Science?" *The Humanist,* September/October, p. 10.

Lockyer, J. N. 1894. *The Dawn of Astronomy.* Cambridge, Mass.: M.I.T. Press. (Reprint of edition copyrighted 1893, published 1894.)

Lowe, D. R., G. Byerly, F. Asaro, and F. Kyte. 1989. "Geological and Geochemical Record of 3400-Million-Year-Old Meteorite Impacts." *Science 245:* 959.

Lowell, P. 1906. *Mars and Its Canals.* 2nd ed. New York: Macmillan.

Luce, G. G. 1975. "Trust Your Body Rhythms." *Psychology Today,* April, p. 52.

Ovenden, M. 1966. "The Origin of Constellations." *Philosophical Journal 3:* 1.

*Pannekoek, A. 1961. *A History of Astronomy.* London: Allen and Unwin.

Schaefer, B. E. 1989. "Dating the Crucifixion." *Sky and Telescope,* April 1, p. 374.

*Stephenson, F. R. 1982. "Historical Eclipses." *Scientific American,* October, p. 170.

Chapter 2

*Gingerich, O. 1986. "Islamic Astronomy." *Scientific American,* April, p. 74.

Needham, J. 1959. *Science and Civilization in China.* Vol. 3. London: Cambridge University Press.

Pritchard, J. B., ed. 1955. *Ancient Eastern Texts Relating to the Old Testament.* 2nd ed. Princeton, N.J.: Princeton University Press.

*Thomsen, D. E. 1984. "Calendric Reform in Yucatan." *Science News 126:* 282.

Zeilik, M. 1985. "The Ethnoastronomy of the Historic Pueblos, I: Calendrical Sun Watching." *Archaeoastronomy 8:* S1.

Chapter 3

*Gingerich, O. 1973a. "Copernicus and Tycho." *Scientific American,* December, p. 86.

*———. 1973b. *Crisis Versus Aesthetic in the Copernican Revolution.* Cambridge, Mass.: Smithsonian Astrophysical Observatory.

*Hartmann, W. K. 1983. *Moons and Planets.* Belmont, Calif.: Wadsworth.

*Lerner, L. S., and E. A. Gosselin. 1986. "Galileo and the Specter of Bruno." *Scientific American 255,* November, p. 126.

Chapter 4

*Arnold, J. R. 1980. "The Frontier in Space." *American Scientist 68:* 299.

*Banks, P., and D. Black. 1987. "The Future of Science in Space." *Science 236:* 244.

Dessler, A. T. 1984. "The Vernov Radiation Belt (Almost)." *Science 226,* no. 4677, editorial page.

*Dyson, F. 1969. "Human Consequences of the Exploration of Space." *Bulletin of Atomic Scientists,* September, p. 8.

*Hartmann, W. K., R. Miller, and P. Lee. 1984. *Out of the Cradle.* New York: Workman Publishing.

*Lewis, R. S. 1969. *Appointment on the Moon.* New York: Viking Press.

National Commission on Space. 1986. *Pioneering the Space Frontier.* New York: Bantam Books.

Newton, I. 1962. *Principia.* A. Motte, trans. Berkeley: University of California Press. (Originally published 1687.)

*O'Neill, G. K. 1977. *The High Frontier.* New York: William Morrow.

Chapter 5

Crawford, D. S., and T. B. Hunter. 1990. "The Battle Against Light Pollution," *Sky and Telescope 80:* 23.

*Hjellming, R., and R. Bignell. 1982. "Radio Astronomy with the Very Large Array." *Science 216:* 1279.

Soifer, B., D. Beichman, and D. Sanders. 1989. "An Infrared View of the Universe." *American Scientist 77:* 46.

Chapter 6

Alvarez, L. W., and others. 1980. "Extraterrestrial Cause for the Cretaceous-Tertiary Extinction." *Science 208:* 1095.

Ganapathy, R. 1980. "A Major Meteorite Impact on the Earth 65 Million Years Ago: Evidence from the Cretaceous-Tertiary Boundary Clay." *Science 209:* 921.

*Hurley, P. M. 1968. "The Confirmation of Continental Drift." *Scientific American,* April, p. 52.

Kerr, R. A. 1987. "Milankovitch Climate Cycles Through the Ages." *Science 235:* 973.

Kozlovsky, Y. A. 1984. "The World's Deepest Well." *Scientific American,* p. 436.

*Levi, B., and T. Rothman. 1985. "Nuclear Winter: A Matter of Degrees." *Physics Today,* September, 58.

*Morris, S. C. 1987. "The Search for the Precambrian–Cambrian Boundary." *American Scientist 75:* 157.

Pepin, R. O. 1976. "The Formation Interval of the Earth." *Abstracts Seventh Lunar Science Conference,* Houston: Lunar and Plantary Science Institute.

*Schneider, S. H. 1987. "Climate Modeling." *Scientific American 256:* 72.

Schopf, J. W. 1975. "The Age of Microscopic Life." *Endeavour 34:* 51.

*Silberner, J. 1987. "Predicting Parkfield." *Science News 131:* 268.

Toon, O. B., and others. 1982. "Evolution of an Impact-Generated Dust Cloud and Its Effects on the Atmosphere." In *Geological Sociey of America Special Paper 190,* ed. L. Silver and P. Schultz. Boulder, Colo.: Geological Society of America.

*Turco, R. P., and others. 1984. "The Climatic Effects of Nuclear War." *Scientific American 251,* August, p. 33.

Wolbach, W., R. Lewis, and E. Anders. 1986. "Cretaceous Extinctions: Evidence for Wildfires and Search for Meteoritic Material." *Science 230:* 167.

Chapter 7

Cameron, A., and W. Ward, 1976. "The Origin of the Moon" (Abstract). In *Lunar Science VII.* Houston: Lunar Science Institute, pp. 120–122.

*Goldreich, P. 1972. "Tides and the Earth–Moon System." *Scientific American,* April, p. 42.

Hartmann, W. K., and D. Davis. 1975. "Satellite-Sized Planetesimals and Lunar Origin." *Icarus 24:* 504.

Hartmann, W. K., R. Phillips, and J. G. Taylor. 1986. *Origin of the Moon.* Houston: Lunar and Planetary Institute.

Lin, T. D. 1986. "Lunar Concrete." *Lunar News* (from Lunar and Planetary Institute, Houston), no. 47, p. 2.

Taylor, S. R. 1975. *Lunar Science: A Post-Apollo View.* New York: Pergamon Press.

*———. 1982. *Planetary Science: The Lunar Perspective.* Houston: Lunar and Planetary Institute.

*Thomsen, D. E. 1986. "Man in the Moon." *Science News 129:* 154.

Chapter 8

Meech, K., and M. Belton. 1989. "(2060) Chiron." International Astronomical Union Circular No. 4770, April 11.

Tholen, D., W. K. Hartmann, and D. Cruikshank, 1988. "(2060) Chiron." International Astronomical Union Circular No. 4554, February 24.

Thomas, P., C. Weitz, and J. Veverka, 1989. "Small Satellites of Uranus: Disk-Integrated Photometry and Estimated Radii." *Icarus 81:* 92.

Chapter 9

Grinspoon, D. H. 1987. "Was Venus Wet? Deuterium Reconsidered." *Science 238:* 1702.

Kerzhanovich, V. V., and M. Marov. 1983. "The Atmospheric Dynamics of Venus According to Doppler Measurements by the Venera Entry Probes." In *Venus,* eds. D. M. Hunten and others. Tucson: University of Arizona Press.

Oyama, V. I., and others. 1979. "Venus Lower Atmospheric Composition: Analysis by Gas Chromatography." *Science 203:* 802.

Pieters, C. M., and others. 1986. "The Color of the Surface of Venus." *Science 234:* 1379.

*Schubert, G., and C. Covey. 1981. "The Atmosphere of Venus." *Scientific American,* June, p. 66.

Von Zahn, and others. 1983. "Composition of the Venus Atmosphere." In *Venus,* eds. D. M. Hunten and others. Tucson: University of Arizona Press.

Walker, J. C. C. 1977. *Evolution of the Atmosphere.* New York: Macmillan.

Chapter 10

Greeley, R. 1987. "Release of Juvenile Water on Mars: Estimated Amounts and Timing Associated with Volcanism." *Science 236:* 1653.

*Haberle, R. M. 1986. "The Climate of Mars." *Scientific American,* May, p. 54.

Kerr, R. A. 1986. "Mars Is Getting Wetter and Wetter." *Science 233:* 936.

*Lowell, P. 1895. *Mars and Its Canals.* 2nd ed. New York: Macmillan.

McSween, H.Y., Jr. 1985. "SNC Meteorites: Clues to Martian Petrologic Evolution?" *Reviews of Geophysics 23:* 391–416.

Owen, T., and others. 1977. "The Composition of the Atmosphere at the Surface of Mars." *Journal of Geophysical Research 82:* 4635.

Sagan, C., O. Toon, and P. Gierasch. 1973. "Climatic Change on Mars." *Science 181:* 1045.

Warren, P. H. 1987. "Mars Regolith vs. SNC Meteorites: Possible Evidence for Abundant Crustal Carbonates." *Icarus 70:* 153.

*Wells, H. G. 1898. *The War of the Worlds.* London: W. Heinemann.

Wright, I., M. Grady, and C. Pillinger. 1989. "Organic Materials in a Martian Meteorite." *Nature 340:* 220.

Chapter 11

*Chapman, C. R. 1968. "The Discovery of Jupiter's Red Spot." *Sky and Telescope 35:* 276.

*Hartmann, W. K. 1983. *Moons and Planets.* Belmont, Calif.: Wadsworth.

Peale, S. J., P. Cassen, and R. Reynolds. 1979. "Melting of Io by Tidal Dissipation." *Science 203:* 892.

Chapter 12

Binzel, R. P. 1990. "Pluto." *Scientific American 262:* 50.

Cruikshank, D. P., and P. Silvaggio. 1979. "Triton: A Satellite with an Atmosphere." *Astrophysical Journal 233:* 1016.

Cruikshank, D. P., R. Brown, and R. Clark. 1983. "Nitrogen on Triton." *Bulletin of the American Astronomical Society 15:* 857.

*Grosser, M. 1962. *The Discovery of Neptune.* Cambridge, Mass.: Harvard University Press.

Lebofsky, L., T. Johnson, and T. McCord. 1970. "Saturn's Rings: Spectral Reflectivity and Compositional Implications." *Icarus 13:* 226.

Littmann, Mark. 1989. "Where Is lanet X?" *Sky and Telescope 78:* 596.

Luu, Jane, and D. Jewitt. 1988. "A Two-Part Search for Slow-Moving Objects." *Astronomical Journal 95:* 1256.

Marcialis, R., G. Rieke, and L. Lebofsky. 1987. "The Surface Composition of Charon: Tentative Identification of Water Ice." *Science 237:* 1349.

Tholen, D. J., and M. W. Buie. 1988. "Circumstances for Pluto—Charon Mutual Events in 1989." *Astronomy Journal 96:* 1977–1982.

Chapter 13

Brandt, J. C., and M. Niedner. 1986. "The Structure of Comet Tails." *Scientific American,* January, p. 49.

Gaffey, M. J., and T. McCord. 1977. "Mining the Asteroids." *Mercury 6:* 1.

*Hartmann, W. K. 1975. "The Smaller Bodies of the Solar System." *Scientific American,* September, p. 143.

*———. 1982. "Mines in the Sky Are Not So Wild a Dream." Smithsonian, September, p. 70.

Kamoun, P. G., and others. 1982. "Comet Encke: Radar Detection of Nucleus." *Science 216:* 293.

Lebofsky, L. A., and others. 1981. "The 1.7 to 4.2-m Spectrum of Asteroid 1 Ceres: Evidence for Structural Water in Clay Minerals." *Icarus 48:* 453.

O'Leary, B. 1983. "Mining the Earth-Approaching Asteroids for Their Precious and Strategic Metals." *Advances in Astronautical Sciences 53:* 375.

Ostro, S. J., and others. 1990. "Radar Images of Asteroid 1989 PB." *Science 248:* 1523.

Sekanina, Z. 1983. "The Tunguska Event: No Cometary Signature in Evidence." *Astronomy Journal 88:* 1382.

Sykes, M., L. Lebofsky, D. Hunten, and F. Low. 1986. "The Discovery of Dust Trails in the Orbits of Periodic Comets." *Science 232:* 1115.

Chapter 14

Anders, E., and M. Ebihara. 1982. "Solar-system Abundances of the Elements." *Geochimica et Cosmochimica Acta 46:* 2365.

*Grossman, L. 1975. "The Most Primitive Objects in the Solar System: Carbonaceous Chondrites." *Scientific American,* February, p. 30.

Hubbard, W., and D. Stevenson. 1986. "Interior Structure of Saturn." In *Saturn,* eds. T. Gehrels and M. Matthews. Tucson: University of Arizona Press.

*Lewis, J. S. 1974. "The Chemistry of the Solar System." *Scientific American,* March, p. 50.

McSween, H. Y., Jr. 1989. "Chondritic Meteorites and the Formation of Planets." *American Scientist 77:* 146.

Shapley, H., and H. Howarth, eds. 1929. *A Source Book in Astronomy.* New York: McGraw-Hill.

*Stone, E., and others. 1979. "The Voyager 2 Encounter with the Jupiter System." *Science 206:* 925.

*———. 1982. "The Voyager 2 Encounter with the Saturn System." *Science 212:* 499.

*———. 1986. "The Voyager 2 Encounter with the Uranus System." *Science 233:* 39.

*Wetherill, G. W. 1985. "Occurrence of Giant Impacts During the Growth of the Terrestrial Planets." *Science 228:* 877.

———. 1990. "Formation of the Earth." *Annual Reviews of Earth and Planetary Science 18:* 205.

Chapter 15

Anders, E. and N. Grevesse. 1988. "Abundances of the Elements: Meteoritic and Solar." Preprint submitted to *Geochimica Cosmochimica Acta.*

Bahcall, John N. 1990. "The Solar-Neutrino Problem." *Scientific American 262:* 54.

Foukal, Peter V. 1990. "The Variable Sun." *Scientific American 262:* 34.

Gibson, E. G. 1973. *The Quiet Sun.* Publication no. SP-303. Washington, D.C.: National Aeronautics and Space Administration.

*Kreith, F., and R. T. Meyer. 1983. "Large-Scale Use of Solar Energy with Central Receivers." *American Scientist 71:* 598.

*Meadows, J. 1984. "The Origins of Astrophysics." *American Scientist 72:* 269.

*Pasachoff, J. M. 1980. "Our Sun." In *Astronomy Selected Readings,* ed. M. A. Seeds. Menlo Park, Calif.: Benjamin/Cummings.

*Snell, J. E., P. Achenbach, and S. Peterson. 1976. "Energy Conservation in New Housing Design." *Science 192:* 1305.

*Thomsen, D. E. 1985. "Solar News: Convection and Magnetism." *Science News 127:* 326.

*Wolfson, R. 1983. "The Active Solar Corona." *Scientific American,* February, p. 104.

Chapter 16

Allen, C. W. 1973. *Astrophysical Quantities.* 3rd ed. London: Athlone Press.

Cannon, A. J. 1924. *The Henry Draper Catalogue of Stars.* Cambridge: The Harvard Annals.

Chapter 17

Baum, W. A. 1986. "The Role of Space Telescopes in the Detection of Brown Dwarfs." In *Astrophysics of Brown Dwarfs,* eds. M. Kafatos, R. Harrington, and S. Maran. Cambridge: Cambridge University Press.

*Burnham, R. 1978. *Burnham's Celestial Handbook.* New York: Dover Publications.

*Humphreys, R., and K. Davidson. 1984. "The Most Luminous Stars." *Science 223:* 243.

Iben, I. 1967. "Stellar Evolution: Within and Off the Main Sequence." *Annual Review of Astronomy and Astrophysics 4:* 171.

Kamper, K., and A. J. Wesselink. 1978. "Alpha and Proxima Centauri." *Astronomical Journal 83:* 1653.

van de Kamp, P. 1981. *Stellar Paths.* Dordrecht, Netherlands: Reidel.

Westbrook, C., and C. B. Tarter. 1975. "On Protostellar Evolution." *Astrophysical Journal 200:* 48.

Chapter 18

Bailly, J., and C. Lada. 1983. "The High-Velocity Molecular Flows Near Young Stellar Objects." *Astrophysical Journal 265:* 824.

*Boss, A. P. 1985. "Collapse and Formation of Stars." *Scientific American,* January, p. 40.

Cohen, M., and L. Kuhi. 1979. *Astrophysical Journal,* Supplement Series, December.

D'Antona, F. 1987. "Evolution of Very Low Mass Stars and Brown Dwarfs II." *Astrophysical Journal 320:* 633.

Hartmann, L., and J. Raymond. 1984. "A High-Resolution Study of Herbig-Haro Objects 1 and 2." *Astrophysical Journal 276:* 560.

Hayashi, C. 1961. "Stellar Evolution in the Early Phases of Gravitational Contraction." *Publications of the Astronomical Society of Japan 13:* 450.

Henyey, L., R. LeLevier, and R. Levée. 1955. "The Early Phases of Stellar Evolution." *Publications of the Astronomical Society of the Pacific 67:* 154.

Iben, I. 1965. "Stellar Evolution I. The Approach to the Main Sequence." *Astrophysical Journal 141:* 993.

*Lada, C. J. 1982. "Energetic Outflow from Young Stars." *Scientific American,* June, p. 82.

Lunine, J., W. Hubbard, and M. Morley. 1986. "Evolution and Infrared Spectra of Brown Dwarfs." *Astrophysical Journal 310:* 238.

Mendoza, V. E. 1968. "Infrared Excesses in T Tauri Stars and Related Objects." *Astrophysical Journal 143:* 1010.

Schild, R. E. 1990. "A Star Is Born." *Sky and Telescope 80:* 600.

Schmelz, J. 1984. "An Investigation of T Tauri Variability." *Astronomical Journal 89:* 108.

Schwartz, R. D. 1983. "Herbig–Haro Objects." *Annual Reviews in Astronomy and Astrophysics 21:* 209.

*Scoville, N., and J. Young. 1984. "Molecular Clouds, Star Formation and Galactic Structure." *Scientific American,* April, p. 42.

Walker, M. 1972. "Studies of Extremely Young Clusters. VI." *Astrophysical Journal 175:* 89.

*Welch, W., and others. 1985. "Gas Jets Associated with Star Formation." *Science 228:* 1389.

Chapter 19

Baade, W., and F. Zwicky. 1934. "Cosmic Rays from Super-Novae." *Proceedings of the National Academy of Science 20:* 259.

Baize, P. 1980. "Les Masses des Étoiles Variables a Longue Période." *L'Astronomie 94:* 71.

Bonneau, D., and others. 1982. "The Diameter of Mira." *Astronomy and Astrophysics 106:* 235.

*Bova, B. 1973. "Obituary of Stars: A Tale of Red Giants, White Dwarfs, and Black Holes." *Smithsonian 4:* 54.

*Burnham, Robert. 1978. *Burnham's Celestial Handbook.* New York: Dover Publications.

*Burrows, A. 1987. "The Birth of Neutron Stars and Black Holes." *Physics Today,* September, p. 28.

*Hawking, S. W. 1977. "The Quantum Mechanics of Black Holes." *Scientific American,* January, p. 34.

*Helfand, D. 1983. "Theory Points to Pulsating White Dwarfs." *Physics Today,* January, p. 21.

*———. 1987. "Bang: The Supernova of 1987." *Physics Today,* August, p. 25.

*Kaler, J. B. 1986. "Planetary Nebulae and the Death of Stars." *American Scientist 74:* 244.

Leavitt, H. S. 1912. *Periods of 25 Variable Stars in the Small Magellanic Cloud.* Harvard College Observatory Circular no. 173, p. 1.

*Liebert, J. 1980. "White Dwarf Stars." *Annual Review of Astronomy and Astrophysics 18:* 363.

Nolan, P., and J. Matteson. 1983. "A Feature in the X-ray Spectrum of Cygnus X-1: A Possible Positron Annihilation Line." *Astrophysical Journal 265:* 389.

Oppenheimer, J. R., and R. Serber. 1938. "On the Stability of Stellar Neutron Cores." *Physics Review 54:* 540.

Pines, D. 1980. "Accreting Neutron Stars, Black Holes, and Degenerate Dwarf Stars." *Science 207:* 597.

Reddy, F. 1983. "Supernovae: Still a Challenge." *Sky and Telescope 66:* 485.

Shapley, H., ed. 1960. *Source Book in Astronomy, 1900–1950.* Cambridge, Mass.: Harvard University Press.

Shu, Frank H. 1982. *The Physical Universe.* Mill Valley, Calif.: University Science Books.

Spergel, D., and others. 1987. "A Simple Model for Neutrino Cooling of the Large Magellanic Cloud Supernova." *Science 237:* 1471.

Waldrop, M. M. 1983. "The 0.001557806449023-Second Pulsar." *Science 219:* 831.

Wheeler, J., and K. Nomoto. 1985. "How Stars Explode." *American Scientist 73:* 240.

Woosley, S., and T. Weaver. 1989. "The Great Supernova of 1987." *Scientific American,* August, p. 32.

Chapter 20

Allen, C. W. 1973. *Astrophysical Quantities.* 3rd ed. London: Athlone Press.

*Blitz, L. 1982. "Giant Molecular-Cloud Complexes in the Galaxy." *Scientific American,* April, p. 84.

Brackenridge, G. R. 1981. "Terrestrial Paleoenvironmental Effects of a Late Quaternary-Age Supernova." *Icarus 46:* 81.

Greenberg, J. M. 1984. "The Structure and Evolution of Interstellar Grains." *Scientific American,* June, p. 124.

Hartmann, J. F. 1904. "Investigations on the Spectrum and Orbit of σ Orionis." *Astrophysical Journal 19:* 268.

Maran, S., J. Brandt, and T. Stecher, eds. 1973. *The Gum Nebula and Related Problems.* Publication no. SP-332. Washington, D.C.: National Aeronautics and Space Administration.

*Scoville, N., and J. Young. 1984. "Molecular Clouds, Star Formation and Galactic Structure." *Scientific American,* April, p. 42.

Shu, F. H. 1982. *The Physical Universe: An Introduction to Astronomy.* Mill Valley, Calif.: University Science Books.

Chapter 21

Abt, H. A. 1979. "The Frequencies of Binaries on the Main Sequence." *Astronomical Journal 84:* 1591.

———. 1983. "Normal and Abnormal Binary Frequencies." *Annual Review of Astrophysics 21:* 343.

Abt, H., and S. Levy. 1976. "Multiplicity Among Solar-Type Stars." *Astrophysical Journal,* Supplement series 30: 273.

Abt, H., Ana Gomez, and S. Levy. 1990. "The Frequency and Formation Mechanism of B2–B5 Main-Sequence Binaries." *Astrophysical Journal Supplements 74:* 551.

Arny, T., and P. Weissman. 1973. "Interaction of Proto-Stars in a Collapsing Cluster." *Astronomical Journal 78:* 310.

*Batten, A. H. 1973. *Binary and Multiple System of Stars.* Oxford: Pergamon Press.

Dyck, H., T. Simon, and B. Zuckerman. 1982. "Discovery of an Infrared Companion to T Tauri." *Astrophysics Journal Letters 225:* L103.

Fekel, F. C. 1987. "Multiple Stars: Anathemas or Friends?" *Vistas in Astronomy 30:* 69.

Grindlay, J. E., and others. 1984. "The Central X-ray Source in SS 433." *Astrophysical Journal 277:* 286.

Hansen, R., B. Jones, and D. Lin. 1983. "The Astrometric Position of T Tauri and the Nature of Its Companion." *Astrophysics Journal Letters 270:* L27.

*Heintz, W. D. 1978. *Double Stars.* Dordrecht, Netherlands: Reidel.

*Margon, B. 1980. "The Bizarre Spectrum of SS 433." *Scientific American,* October, p. 54.

*———. 1982. "Relativistic Jets in SS 433." *Science 215:* 247.

McMillan, S., P. Hut, and J. Makino. 1990. "Star Cluster Evolution with Primordial Binaries 1. A Comparative Study." *Astrophysical Journal 362:* 522.

Popper, D. M. 1980. "Stellar Masses." *Annual Review of Astronomy and Astrophysics 18:* 115.

Thomsen, D. 1986. "A 'Brickbat' in the Sky." *Science News 127:* 154.

Ventura, J., and others. 1983. "Can X-ray Bursts Originate from Low-Mass Binaries?" *Nature 301:* 491.

Wanner, J. F. 1967. "The Visual Binary Krüger 60." *Sky and Telescope 33:* 16.

Chapter 22

Jones, K., and P. Demarque. 1983. "The Ages and Composition of Old Clusters." *Astrophysical Journal 264:* 206.

King, I. R. 1985. "Globular Clusters." *Scientific American,* July, p. 79.

Peterson, C. 1987. *Publications of the Astronomy Society of the Pacific 99:* 1153.

***Shapley,** H. 1930. *Star Clusters.* Cambridge, Mass.: Harvard University Press.

Trumpler, R. 1930. "Absorption of Light in the Galactic System." *Publications of the Astronomical Society of the Pacific 42:* 214.

Chapter 23

***Geballe,** T. R. 1979. "The Central Parsec of the Galaxy." *Scientific American,* July, p. 60.

Genzel, R., and others. 1984. "Far-infrared Spectroscopy of the Galactic Center: Neutral and Ionized Gas in the Central 10 Parsecs of the Galaxy." *Astrophysical Journal 276:* 551.

Hawkins, M. R. S. 1983. "Direct Evidence for a Massive Galactic Halo." *Nature 303:* 406.

Jansky, K. 1935. "A Note on the Source of Interstellar Interference." *Proceedings of the Institute of Radio Engineers 23:* 1158.

Lacy, J. H., and others. 1980. "Observations of the Motion and Distribution of the Ionized Gas in the Central Parsec of the Galaxy. II." *Astrophysical Journal 241:* 132.

Leventhal, M., and others. 1982. "Time-variable Positron Annihilation Radiation from the Galactic Center Direction. *Astrophysical Journal Letters 260:* L1.

Lo, K., and M. Claussen. 1983. "High-resolution Observations of Ionized Gas in Central 3 Parsecs of the Galaxy: Possible Evidence for Infall." *Nature 306:* 647.

Reber, G. 1944. "Cosmic Static." *Astrophysical Journal 100:* 279.

Sanders, R., and K. Prendergast. 1974. "The Possible Relation of the 3-kpc Arm to Explosions in the Galactic Nucleus." *Astrophysical Journal 188:* 489.

Shapley, H., and H. Howarth, eds. 1929. *A Source Book in Astronomy.* New York: McGraw-Hill.

Waldrop, M. M. 1985. "The Core of the Milky Way." *Science 230:* 158.

Chapter 24

Allen, C. W. 1973. *Astrophysical Quantities.* London: Athlone Press.

Freedman, W., and B. Madore. 1983. "Time Evolution of Disk Galaxies Undergoing Stochastic Self-propagating Star Formation." *Astrophysical Journal 265:* 140.

Gunn, J. E. 1979. "Feeding the Monster: Gas Discs in Elliptical Galaxies." In *Active Galactic Nuclei,* eds. G. Hazard and S. Mitton. London: Cambridge University Press.

***Hirshfeld,** A. 1980. "Inside Dwarf Galaxies." *Sky and Telescope 59:* 287.

Hodge, P. 1987. "The Local Group." *Mercury 16:* 2.

***Humphreys,** R., and K. Davidson. 1984. "The Most Luminous Stars." *Science 223:* 243.

Humphreys, Roberta, M. Massey, and Wendy Freedman. 1990. "Spectroscopy of Luminous Blue Stars in M31 and M33." *Astronomical Journal 99:* 84.

Hunter, D., and J. Gallagher, III. 1989. "Star Formation in Irregular Galaxies." *Science 243:* 1557.

Irwin, M. J., and others. 1990. "A New Satellite Galaxy of the Milky Way in the Constellation of Sextans." *Monthly Notices of the Royal Astronomical Society 244 (2):* 16P–19P.

Mathis, J., B. Savage, and J. Cassinelli. 1984. "A Superluminous Object in the Large Cloud of Magellan." *Scientific American,* August, p. 52.

Ostriker, J. P. 1981. "Some Thoughts on Galaxy Formation." *Bulletin of the American Astronomy Society 12:* 770.

***Rubin,** V. 1983. "Dark Matter in Spiral Galaxies." *Scientific American,* June, p. 96.

Sandage, A., and G. A. Tammann. 1974. "Steps Toward the Hubble Constant. III." *Astrophysical Journal 194:* 223.

***Schweizer,** F. 1986. "Colliding and Merging Galaxies." *Science 231:* 227.

***Shapley,** H. 1957. *The Inner Metagalaxy.* New Haven, Conn.: Yale University Press.

***Silk,** J. 1987. "The Formation of Galaxies." *Physics Today 40,* no. 4, p. 28.

Simkin, S., and others. 1987. "Markarian 348: A Tidally Disturbed Seyfert Galaxy." *Science 235:* 1367.

Spillar, E. J., and others. 1990. "Infrared Speckle Observations of the Nucleus of M31." *Astrophysical Journal Letters 349:* L13.

***Strom,** K., and S. Strom. 1982. "Galactic Evolution: A Survey of Recent Progress." *Science 216:* 571.

Thackeray, A. D. 1971. "Survey of Principal Characteristics of the Magellanic Clouds." In *The Magellanic Clouds,* ed. A. Muller. Boston: Reidel.

***Toomre,** A., and J. Toomre. 1973. "Violent Tides Between Galaxies." *Scientific American,* December, p. 38.

van den Bergh, S. 1972. "Search for Faint Companions to M31." *Astrophysical Journal 171:* L35.

Chapter 25

***Anderson,** P. 1987. "Massive Objects in Galactic Nuclei May Be Black Holes." *Physics Today,* October, p. 22.

Balick, B., and T. Heckman. 1983. "Spectroscopy of the Fuzz Associated with Four Quasars." *Astrophysical Journal 265:* L1.

Bartel, N., and others. 1985. "Hubble's Constant Determined Using Very-long Baseline Interferometry." *Nature 318:* 25.

Burbidge, G., and others. 1990. "Associations Between Quasistellar Objects and Galaxies." *Astrophysical Journal Supplements 362:* 759.

Chambers, K., G. Miley, and W. van Breugel. 1990. "4C 41.17: A Radio Galaxy at a Redshift of 3.8." *Astrophysical Journal 363:* 21.

Chincarini, G., and H. Rood. 1980. "The Cosmic Tapestry." *Sky and Telescope 59:* 364.

Djorgovski, S., and others. 1987a. "A Galaxy at a Redshift of 3.215." *Astronomical Journal 93:* 1318.

————. 1987b. "Discovery of a Probable Binary Quasar." *Astrophysical Journal 321:* L17.

Dressler, A., and D. Richstone. 1988. *Astrophysical Journal 324,* in press.

Eddington, A. 1933. *The Expanding Universe.* Cambridge: Cambridge University Press.

Graham, J., and others. 1990. "The Double Nucleus of Arp 220 Unveiled." *Astrophysical Journal Letters 354:* L5.

Gunn, J. E. 1979. "Feeding the Monster: Gas Discs in Elliptical Galaxies." In *Active Galactic Nuclei,* eds. C. Hazard and S. Mitton. London: Cambridge University Press.

Hayes, J., and A. Sadun. 1987. "CCD Observations of the Jet of the Quasar 3C 273." *Astronomical Journal 94:* 871.

Hubble, E. P. 1936. "A Relation Between Distance and Radial Velocity Among Extra-Galactic Nebulae." Reprinted in *Source Book in Astronomy, 1900–1950,* ed. H. Shapley. Cambridge, Mass.: Harvard University Press, 1960.

Kochanek, C., S. Shapiro, and S. Teukolsky. 1987. "How Big Are Supermassive Black Holes Formed from the Collapse of Dense Star Clusters?" *Astrophysical Journal 320:* 73.

Kühr, H., and others. 1983. "The Most Luminous Quasar: S5 0014+81." *Astrophysical Journal 275:* L33.

Osmer, P. 1982. "Quasars as Probes of the Distant and Early Universe." *Scientific American,* February, p. 126.

Peterson, B. 1982. "PKS 2000-300: A Quasi-Stellar Radio Source with a Red Shift of 3.78." *Astrophysical Journal 260:* L27.

Rees, M. J. 1990. " 'Dead Quasars' in Nearby Galaxies?" *Science 247:* 817.

Saslaw, W. C. 1974. "Theory of Galactic Nuclei." In *The Formation and Dynamics of Galaxies,* ed. J. Shakeshaft. Dordrecht, Netherlands: Reidel.

Shu, F. H. 1982. *The Physical Universe: An Introduction to Astronomy.* Mill Valley, Calif.: University Science Books.

Simkin, S., and others. 1987. "Markarian 348: A Tidally Distorted Seyfert Galaxy." *Science 235:* 1367.

Uson, Juan, S. Boughn, and J. Kuhn. 1990. "The Central Galaxy in Abell 2029: An Old Supergiant." *Science 250:* 539.

Chapter 26

*Anderson, P. 1987. "Ripples in the Universal Hubble Flow." *Physics Today,* October, p. 17.

Burns, J. O. 1986. "Very Large Structures in the Universe." *Scientific American,* June, p. 38.

*de Vaucouleurs, G. 1970. "The Case for a Hierarchical Cosmology." *Science 167:* 1203.

Einstein, A. 1952. "Cosmological Considerations on the General Theory of Relativity." In *The Principle of Relativity.* New York: Dover. (Originally published 1917.)

*Gregory, S., and L. Thompson. 1982. "Superclusters and Voids in the Distribution of Galaxies." *Scientific American,* March, p. 106.

*Harrison, E. R. 1974. "Why the Sky Is Dark at Night." *Physics Today 27:* 5.

Hawking, S. W. 1988. *A Brief History of Time.* New York: Bantam Books.

*North, J. D. 1965. *The Measure of the Universe.* Oxford: Oxford University Press.

Palca, J. 1991. "In Search of 'Dark Matter.' " *Science 251:* 22.

Raychaudhuri, A. K. 1979. *Theoretical Cosmology.* Oxford: Clarendon Press.

*Schatzman, E. S. 1965. *The Origin and Evolution of the Universe.* New York: Basic Books.

Shu, F. 1982. *The Physical Universe.* Mill Valley, Calif.: University Science Press.

Silk, J., 1980. *The Big Bang.* San Francisco: W. H. Freeman.

*Silk, J., A. Szalay, and Y. Zel'dovich. 1983. "The Large-Scale Structure of the Universe." *Scientific American,* October, p. 72.

Thomsen, D. E. 1987. "Neutrino Mass: A Tritium Disagreement." *Science News 131:* 342.

Tyson, J. A., F. Valdes, and R. Wenk. 1990. "Detection of Systematic Gravitational Lens Galaxy Image Alignments." *Astrophysical Journal 349:* L1–L4.

van den Bergh, Sidney. 1990. "Cosmology—In Search of a New Paradigm?" *Journal of the Royal Astronomical Society of Canada 84:* 275.

*Waldrop, M. M. 1986. "In Search of Dark Matter." *Science 234:* 152.

Wesson, P., K. Valle, and R. Stabell. 1987. "The Extragalactic Background Light and a Definitive Resolution of Olbers' Paradox." *Astrophysical Journal 317:* 601.

Chapter 27

*Barrow, J., and M. Turner. 1982. "The Inflationary Universe—Birth, Death, and Transfiguration." *Nature 298:* 801.

*Burns, J. O. 1986. "Very Large Structures in the Universe." *Scientific American,* June, p. 38.

Butcher, H. 1987. "Thorium in G-dwarf Stars as a Chronometer for the Galaxy." *Nature 328:* 127.

Gamow, G. 1952. *The Creation of the Universe.* New York: Viking Press.

Giovanelli, R., and Martha Haynes. 1989. "A Protogalaxy in the Local Supercluster." *Astrophysical Journal Letters 346:* L5.

Linde, A. 1987. "Particle Physics and Inflationary Cosmology." *Journal of Physics Today,* September, p. 61.

Ozernoy, L. M. 1974. "Dynamics of Superclusters as the Most Powerful Test for Theories of Galaxy Formation." In *The Formation and Dynamics of Galaxies,* ed. J. Shakeshaft. Dordrecht, Netherlands: Reidel.

Peebles, P. J. E. 1971. *Physical Cosmology.* Princeton, N.J.: Princeton University Press.

Penzias, A. A., and R. W. Wilson. 1965. "A Measurement of Excess Antenna Temperature at 4080 Mc/s." *Astrophysical Journal 142:* 419.

Peterson, Ivars. 1991. "State of the Universe." *Science News 139:* 232.

Reeves, H., and others. 1972. "On the Origin of Light Elements." Cal Tech Preprint OAP-296.

Schmidt, M. 1982. "Quasar Boundary." *Science 216:* 6.

*Silk, J. 1980. *The Big Bang.* San Francisco: W. H. Freeman.

Chapter 28

Ahrens, T. J., and J. D. O'Keefe. 1982. "Impact of an Asteroid or Comet in the Ocean and Extinction of Terrestrial Life." *Lunar Planet Science Abstracts 13:* 3.

*Alvarez, W., and others. 1984. "Impact Theory of Mass Extinctions and the Invertebrate Fossil Record." *Science 223:* 1135.

Brock, T. D. 1985. "Life at High Temperatures." *Science 230:* 132.

*Covey, C. 1984. "The Earth's Orbit and the Ice Ages." *Scientific American,* February, p. 58.

*Dole, S. H. 1964. *Habitable Planets for Man.* Waltham, Mass.: Blaisdell.

*Edmund, J. M., and K. Von Damm. 1983. "Hot Springs on the Ocean Floor." *Scientific American,* April, p. 78.

Goldsmith, D., and T. Owen. 1979. *The Search for Life in the Universe.* Menlo Park, Calif.: Benjamin/Cummings.

*Gould, S. J. 1978. "An Early Start." *Natural History,* February, p. 10.

Harrington, R. S. 1977. "Planetary Orbits in Binary Stars." *Astronomical Journal 82:* 753.

Horowitz, P. 1978. "A Search for Ultra-Narrowband Signals of Extraterrestrial Origin." *Science 201:* 733.

Imbrie, J., and J. Z. Imbrie. 1980. "Modeling the Climatic Response to Orbital Variations." *Science 207:* 943.

Kyte, F., L. Zhou, and J. Wasson. 1988. "New Evidence on the Size and Possible Effects of a Late Pliocene Oceanic Asteroid Impact." *Science 241:* 63.

Miller, S. L. 1955. "Production of Some Organic Compounds Under Possible Primitive Earth Conditions." *Journal of the American Chemical Society 77:* 2351.

O'Leary, B. 1980. "Searching for Other Planetary Systems." *Sky and Telescope 60:* 111.

Oparin, A. I. 1962. *Life: Its Nature, Origin, and Development.* New York: Academic Press.

Ponnamperuma, C., ed. 1983. *Cosmochemistry and the Origin of Life.* Dordrecht, Netherlands: D. Reidel.

Raup, D. M. 1979. "Size of the Permo-Triassic Bottleneck and Its Evolutionary Implications." *Science 206:* 217.

*Russell, D. A. 1982. "The Mass Extinctions of the Late Mesozoic." *Scientific American,* January, p. 82.

Sagan, C., ed. 1973. *Communication with Extraterrestrial Intelligence.* Cambridge, Mass.: M.I.T. Press.

*Schneider, S. H. 1989. "The Greenhouse Effect: Science and Policy." *Science 243:* 771.

Sekanina, Z., and D. Yeomans. 1984. "Close Encounters and Collisions of Comets with the Earth." *Astronomical Journal 89:* 154.

*Siegel, S. M. 1972. "Experimental Biology of Extreme Environments and Its Significance for Space Bioscience—2." *Spaceflight 12:* 256.

Simpson, G. G. 1964. "The Nonprevalence of Humanoids." *Science 143:* 769.

———. 1973. Added comments on "The Nonprevalence of Humanoids." In *Communication with Extraterrestrial Intelligence,* ed. C. Sagan. Cambridge, Mass.: M.I.T. Press.

Stern, D. K. 1975. "First Contact with Nonhuman Cultures." *Mercury,* September/ October, p. 14.

Turco, R., and others. 1981. "Tunguska Meteor Fall of 1908: Effects on Stratospheric Ozone." *Science 212:* 19.

Walker, J. C. G. 1988. "Origin of an Inhabited Planet." In *Conference on the Origin of the Earth* (Abstracts), Berkeley. Houston: Lunar and Planetary Institute.

Windley, B. F. 1977. *The Evolving Continents.* New York: Wiley and Sons.

Enrichment Essay A

*Abetti, G. 1951. *The History of Astronomy.* New York: Henry Schuman.

*Condon, E. U., and others. 1969. *Scientific Study of Unidentified Flying Objects.* New York: Dutton.

Emiliani, C., and others. 1975. "Paleoclimatological Analysis of Late Quaternary Cores from the Northeastern Gulf of Mexico." *Science 189:* 1083.

*Gould, S. J. 1975. "Velikovsky in Collision." *Natural History,* February, p. 20.

*Numbers, R. L. 1982. "Creationism in 20th-Century America." *Science 218:* 538.

Sagan, C., and J. Agel. 1975. *Other Worlds.* New York: Bantam Books.

*Shklovskii, I. S., and C. Sagan. 1966. *Intelligent Life in the Universe.* San Francisco: Holden-Day.

Velikovsky, I. 1950. *Worlds in Collision.* New York: Doubleday.

Index

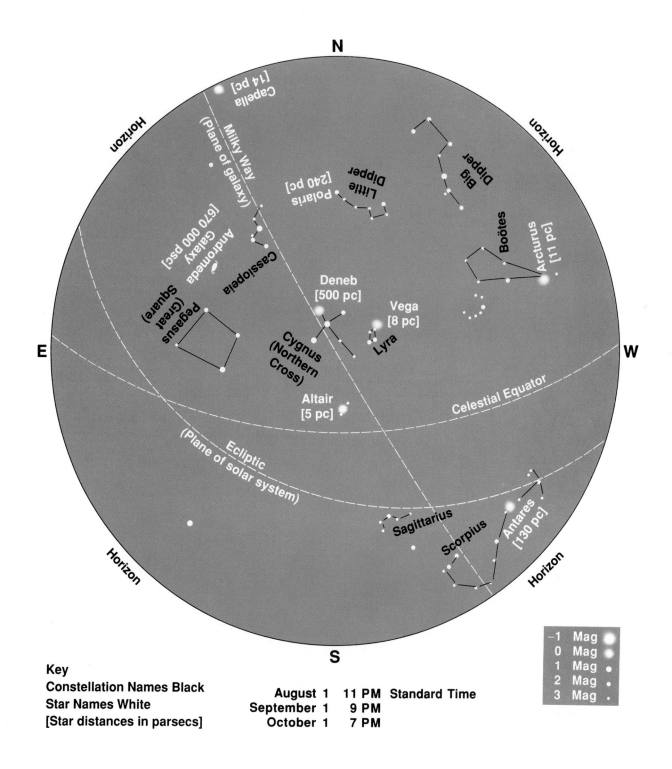

N

Capella [14 pc]

Horizon

Milky Way
(Plane of galaxy)

Horizon

Big Dipper

Little Dipper

Polaris [240 pc]

Boötes

Andromeda Galaxy [670 000 psc]

Arcturus [11 pc]

Cassiopeia

Deneb [500 pc]

Pegasus (Great Square)

Vega [8 pc]

E

Lyra

W

Cygnus (Northern Cross)

Altair [5 pc]

Celestial Equator

Ecliptic (Plane of solar system)

Sagittarius

Scorpius

Antares [130 pc]

Horizon

Horizon

S

Key
Constellation Names Black
Star Names White
[Star distances in parsecs]

August 1	11 PM Standard Time
September 1	9 PM
October 1	7 PM

−1	Mag
0	Mag
1	Mag
2	Mag
3	Mag

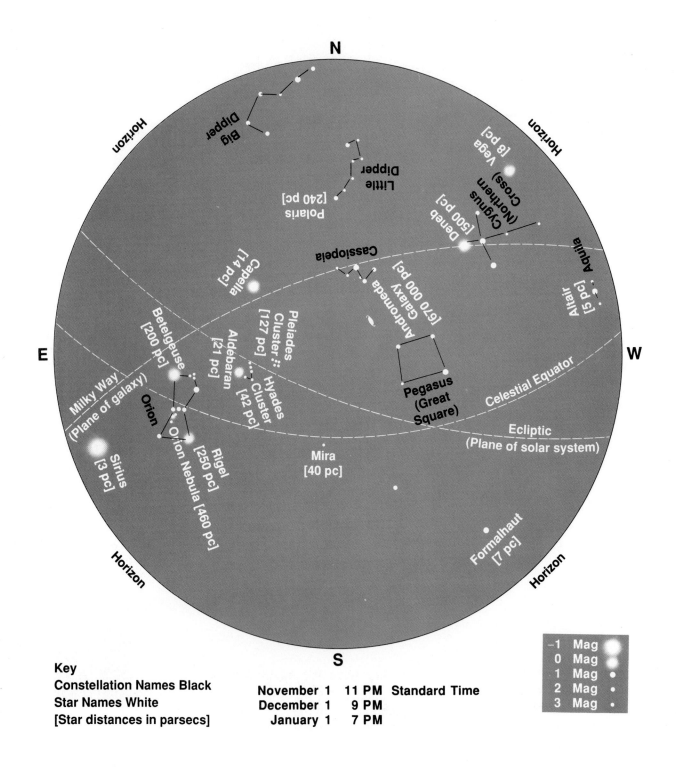

Key
Constellation Names Black
Star Names White
[Star distances in parsecs]

November 1   11 PM   Standard Time
December 1   9 PM
January 1   7 PM

−1	Mag
0	Mag
1	Mag
2	Mag
3	Mag

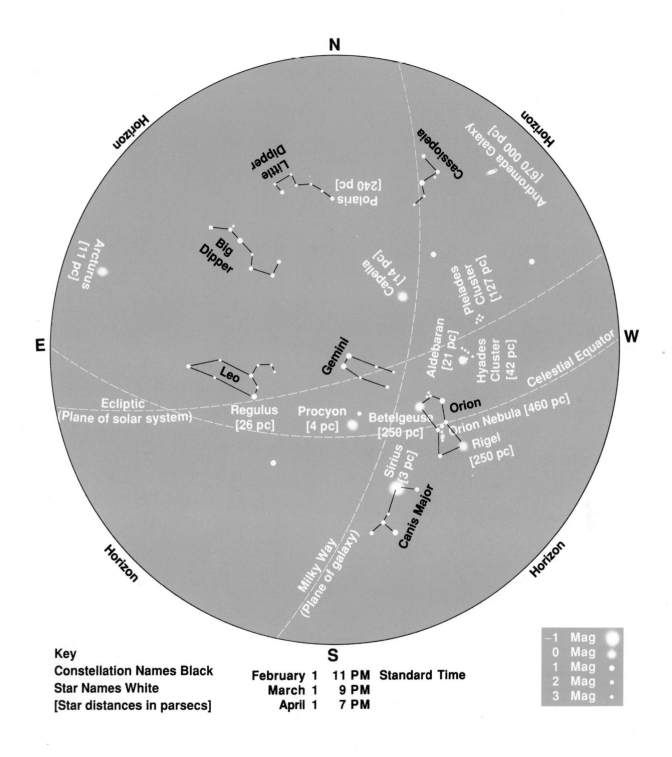

N

Horizon

Horizon

Little
Dipper

Cassiopeia

Andromeda Galaxy
[670 000 pc]

Polaris
[240 pc]

Big
Dipper

Pleiades
Cluster
[127 pc]

Arcturus
[11 pc]

Capella
[14 pc]

Aldebaran
[21 pc]

E

Hyades
Cluster
[42 pc]

W

Celestial Equator

Gemini

Leo

Orion

Regulus
[26 pc]

Procyon
[4 pc]

Betelgeuse
[250 pc]

Orion Nebula [460 pc]

Rigel
[250 pc]

Ecliptic
(Plane of solar system)

Sirius
[3 pc]

Canis Major

Milky Way
(Plane of galaxy)

Horizon

Horizon

S

Key
Constellation Names Black
Star Names White
[Star distances in parsecs]

February 1   11 PM   Standard Time
March 1   9 PM
April 1   7 PM

−1	Mag	●
0	Mag	●
1	Mag	·
2	Mag	·
3	Mag	·

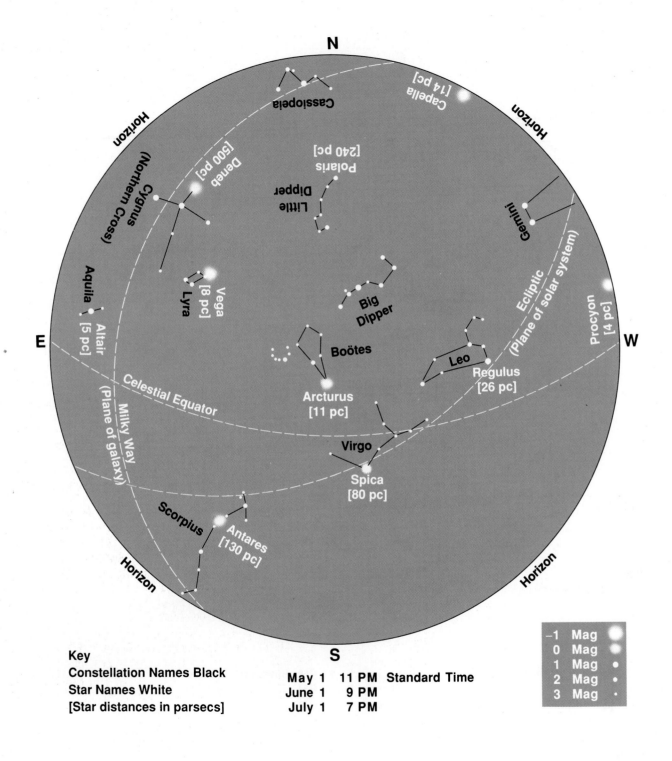

N

Horizon

Capella [14 pc]

Horizon

Cassiopeia

Deneb [500 pc]

Polaris [240 pc]

Cygnus (Northern Cross)

Little Dipper

Gemini

Aquila

Vega [8 pc]

Lyra

Big Dipper

Ecliptic (Plane of solar system)

Altair [5 pc]

Procyon [4 pc]

E

Boötes

Leo

W

Celestial Equator

Arcturus [11 pc]

Regulus [26 pc]

Milky Way (Plane of galaxy)

Virgo

Scorpius

Spica [80 pc]

Antares [130 pc]

Horizon

Horizon

S

−1 Mag
0 Mag
1 Mag
2 Mag
3 Mag

Key
Constellation Names Black
Star Names White
[Star distances in parsecs]

May 1   11 PM   Standard Time
June 1   9 PM
July 1   7 PM